T0348516

DESIGN, FABRICATION AND ECONOMY OF WELDED STRUCTURES

International Conference Proceedings 2008

Miskolc, Hungary, April 24-26, 2008

Horwood Publishing
Chichester, UK

About the Editors

Dr Károly Jármai is a professor at the Faculty of Mechanical Engineering at the University of Miskolc, where he graduated as a mechanical engineer and received his doctorate (dr.univ.) in 1979. He teaches design of steel structures, welded structures, composite structures and optimization in Hungarian and in the English language for foreign students. His research interests include structural optimization, mathematical programming techniques and expert systems. Dr. Jármai wrote his C.Sc. (Ph.D.) dissertation at the Hungarian Academy of Science in 1988, became a European Engineer (Eur. Ing. FEANI, Paris) in 1990 and did his habilitation (dr.habil.) at Miskolc in 1995. Having successfully defended his doctor of technical science thesis (D.Sc.) in 1995, he subsequently received awards from the Engineering for Peace Foundation in 1997 and a scholarship as Széchenyi professor between the years 1997-2000. He is the co-author (with Farkas) of two books in English *Analysis and Optimum Design of Metal Structures*, *Economic Design of Metal Structures* and one in Hungarian, and has published over 300 professional papers, lecture notes, textbook chapters and conference papers. He is a founding member of ISSMO (International Society for Structural and Multidisciplinary Optimization), a Hungarian delegate, vice chairman of commission XV and a sub-commission chairman XV-F of IIW (International Institute of Welding). He has held several leading positions in GTE (Hungarian Scientific Society of Mechanical Engineers) and has been the president of this society at the University of Miskolc since 1991. He was a visiting researcher at Chalmers University of Technology in Sweden in 1991, visiting professor at Osaka University in 1996-97, at the National University of Singapore in 1998 and at the University of Pretoria several times between 2000-2005.

Dr József Farkas is Professor Emeritus of metal structures at the University of Miskolc, Hungary. He graduated from the Faculty of Civil Engineering at the Technical University of Budapest and moved to the University of Miskolc where he became an assistant professor in 1950, an associate professor in 1966 and a university professor in 1975. He obtained degrees as a Candidate of Technical Science in 1966 and Doctor of Technical Science in 1978. Dr. Farkas's research field is the optimum design of metal structures, residual welding stresses and distortions, tubular structures, stiffened plates, vibration damping of sandwich structures. He has written expert opinions for many industrial problems, especially on storage tanks, cranes, welded press frames and other metal structures. He is the author of a Hungarian university textbook on metal structures, a book in English *Optimum Design of Metal Structures* (Ellis Horwood Ltd, Chichester 1984), the first author of two books in English *Analysis and Optimum Design of Metal Structures* (Balkema, Rotterdam-Brookfield 1997), *Economic Design of Metal Structures* (Millpress, Rotterdam 2003) and about 260 scientific articles in journals and conference proceedings. He is a Hungarian delegate of IIW, member of ISSMO and honorary member of GTE. The University of Miskolc has also honoured him as doctor honoris causa.

DESIGN, FABRICATION AND ECONOMY OF WELDED STRUCTURES

International Conference Proceedings 2008

Miskolc, Hungary, April 24-26, 2008

Edited by

Dr. Károly Jármai
Professor of Mechanical Engineering
University of Miskolc, Hungary

Dr. József Farkas
Professor Emeritus of Metal Structures
University of Miskolc, Hungary

Horwood Publishing
Chichester, UK
HORWOOD PUBLISHING LIMITED

International Publishers in Science and Technology
Coll House, Westergate, Chichester, West Sussex, PO20 3QL, England

First published in 2008.

COPYRIGHT NOTICE
All Rights Reserved. No part of this publication may be reproduced, stored in a retrieval system, or transmitted in any form or by any means, electronic, mechanical, photocopying, recording, or otherwise, without the permission of Horwood Publishing Limited, Coll House, Westergate, Chichester, West Sussex, PO20 3QL, England.

© Horwood Publishing Limited, 2008.

British Library Cataloguing in Publication Data
A catalogue record of this book is available from the British Library

ISBN: 978-1-904275-28-2

Cover design by Jim Wilkie.

Table of Contents

Section 4 Fatigue design

Section 5 Frames

Section 6 Hollow sections

Section 7 Plated structures

Section 8 Residual stresses and distortions

Section 9 Static stresses in welded connections

Section 10 Application

Section 11 Welding technology I

Section 12 Welding technology II

Section 13 Applied mechanics

Preface

A question arises, how to select the most suitable load-carrying engineering structures, since – due to the great technological development – a large selection of materials, profiles, structural types and joining technologies gives a lot of available structural versions.

To answer this question one should define the requirements for structures. In the case of welded structures the most important requirements are the safety, fitness for production and economy.

The safety against fracture, instability, large deformations, fatigue, earthquake, fire can be formulated as design constraints, the fabrication constraints express the limitations of sizes and residual welding distortions and economy is achieved by minimization of a cost function.

The title of our conference "Design, Fabrication and Economy" expresses these main aspects for the development of modern welded structures.

The structural optimization system can help this work, since – considering all he important engineering aspects - it can found the best structural version by mathematical methods of constrained function minimization, which fulfil the design and fabrication constraints and minimize the cost function.

The optimization is not a pure mathematical problem, since – in the case of welded structures – fabrication and economy need special analysis and formulations.

The aim of our conference is to collect the new developments in the field of design, fabrication and economy of welded structures.

The conference is dedicated to Prof. József Farkas in the occasion of his 80th birthday.

Károly Jármai, Editor-in-Chief

József Farkas, Assistant Editor

University of Miskolc, April, 2008

Acknowledgement

The editors would like to acknowledge the co-operation and help of the following organizations

International Institute of Welding (IIW),

International Society of Structural and Multidisciplinary Optimization (ISSMO),

Hungarian Academy of Science (MTA),

National Office for Research and Technology (NKTH),

Hungarian Association of Steel Producers and Builders (MAGESZ),

Scientific Society of Mechanical Engineers in Hungary (GTE),

Foundation for the Structural Optimization at the University of Miskolc,

and last but not least the

University of Miskolc, Hungary, which hosted the conference.

The editors would like to acknowledge the help of the following persons:

László Kota, research fellow,

György Kovács, assistant professor.

Károly Jármai, Editor-in-Chief

József Farkas, Assistant Editor

University of Miskolc, April, 2008

Section 1
Introductory papers

1.1 Memorable Moments in the Structural Optimization Research

József Farkas

University of Miskolc, Hungary, altfar@uni-miskolc.hu

Abstract

Experiences are described, which helped in development of the research in the field of optimum design of metal structures. The author participated in many conferences and visited a lot of universities in abroad as well as met persons and had useful discussions with them. Visits in Stanford, Berkeley, Osaka, Matsuyama, Kumamoto, Pretoria and Coimbra are mentioned.

Keywords: *structural optimization, steel bridges, cost engineering, seismic-resistant design, beam-to-column connections*

1 Introduction

Experiences during conferences and research visits play an important role in our scientific work, they have given us good ideas as well as helped to formulate new research concepts and to solve them. In this paper some such experiences are mentioned similar than in my review paper for our last conference in 2003. It is a very good feeling to remember these moments.

2 Stanford and Berkeley, USA, 1992

I took part with a paper in the ISOPE (International Society of Offshore and Polar Engineering) Conference in San Francisco, 1992 (Farkas 1992). During my stay I have visited three famous universities (UCLA University of California Los Angeles, Stanford and Berkeley). In Stanford I have visited the Hungarian born professor George Springer (Department of Aero- and Astronautics) and held a lecture on minimum cost design in a very famous place – in Timoshenko room.

In the laboratory of this department I have obtained a piece of a fiber-reinforced plastic (FRP) plate fabricated for aircraft industry (a two mm thick 16-layer plate). This new structural component was for me so interesting that I begun to research in this field. Later Dr. György Kovács my PhD student worked out his doctoral dissertation about the optimum design of FRP sandwich structures (Kovács et al. 2004).

It should be mentioned that I have enjoyed the beautiful and large Stanford campus with a chapel and a four-storey book shop with a lot of new books about metal structures.

Another experience was my visit to the Earthquake Research Center of the University of Berkeley, where professor Vitelmo Bertero has found for me a very useful literature about the hysteretic behaviour of braced steel frames. This research report helped me to work out my paper for the next Tubular Structures Symposium (Nottingham 1993, Farkas 1993, Farkas & Jármai 1997).

Another memorable moment was to meet professor P.E. Popov in his room in Berkeley. He was in this time over 80 years old but very active speaking about the seismic reinforcement of the Golden Gate bridge (he was a student of professor Timoshenko).

3 Japan, 1999 and 2002

In the frame of common research with the University of Osaka and Ehime University (Matsuyama), we have visited professors K. Horikawa, S. Ohkubo and K. Taniwaki in 1999 and with the University of Kumamoto professor Y. Kurobane in 2002.

3.1 *Ten new bridges*

A new road has been opened between Honshu and Shikoku islands and we have travelled by bus from Matsuyama to Fukuyama through ten new bridges. It was a beautiful trip, we had the opportunity to photo all these bridges. One of them – Tatara bridge – is the world longest cable-stayed bridge with span length of 850 m. It should be mentioned that we have seen these bridges also in construction in 1998.

3.2 *Shipyard Onishi and the formulae of Okerblom*

During our visit in this shipyard near Matsuyama I have described our problem to have new welding data for the formulae of Okerblom. Colleagues in this shipyard have carried out some measurements and later they have sent me their data, which have been very useful for this calculation.

3.3 *Aesthetical aspect in the optimum design*

Professors Ohkubo and Taniwaki in the University of Matsuyama have worked out the structural optimization of a cable-stayed bridge. In this work they include also the aesthetical aspect. Since this aspect cannot be formulated mathematically, they have computed more optimum solutions with different main geometry and asked a number of people for their opinion. The best solution was selected on the basis of these opinions (Taniwaki 1997).

3.4 *Research results realized in the industry*

After the catastrophic earthquake in Kobe (1995) a series of theoretical and experimental research is performed in seven Japanese universities to develop new seismic-resistant beam-to-column connections for welded steel frameworks of buildings. In the University of Kumamoto professor Kurobane and his research group have verified by experiments their advanced joint types and in 2002 we have seen in a private steel factory the production of frames with these new joint types (Fig. 1). This is an excellent example how to realize the research results in the industry (Jármai,K. et al. 2006).

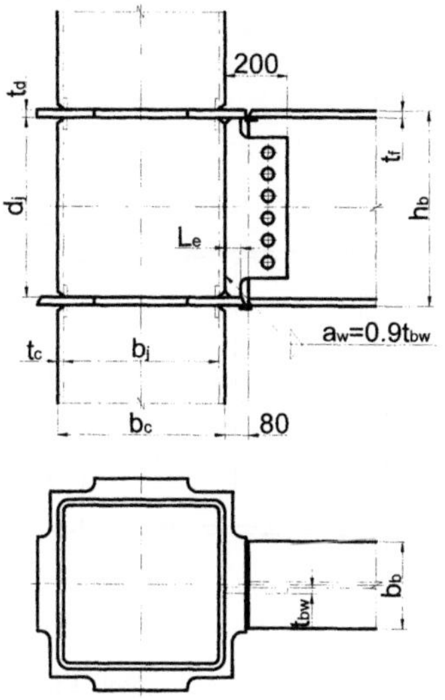

Figure 1. A beam-to-column connection improved for seismic resistance

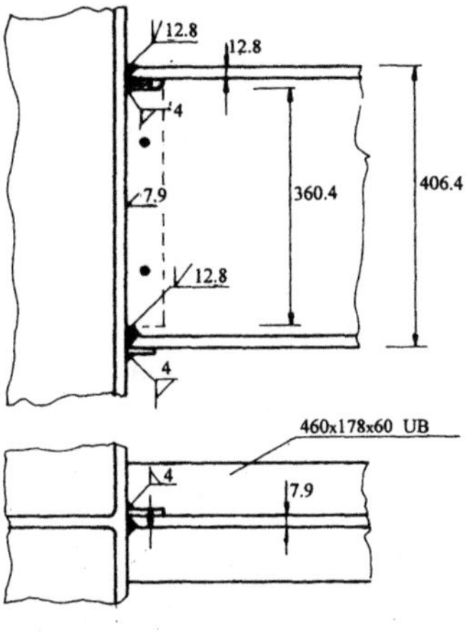

Figure 2. Fully welded connection

4 South Africa, 2004 and 2006

4.1 *International Cost Engineering Congress in Cape Town, 2004*

In this congress we have presented a paper about the cost comparison of welded and bolted beam-to-column steel frame connections (Figures 2 and 3). A manager had a question about the time comparison of these connections, since for managers the time of a frame construction is very important for the construction management (Farkas et al. 2003, Farkas et al. 2004).

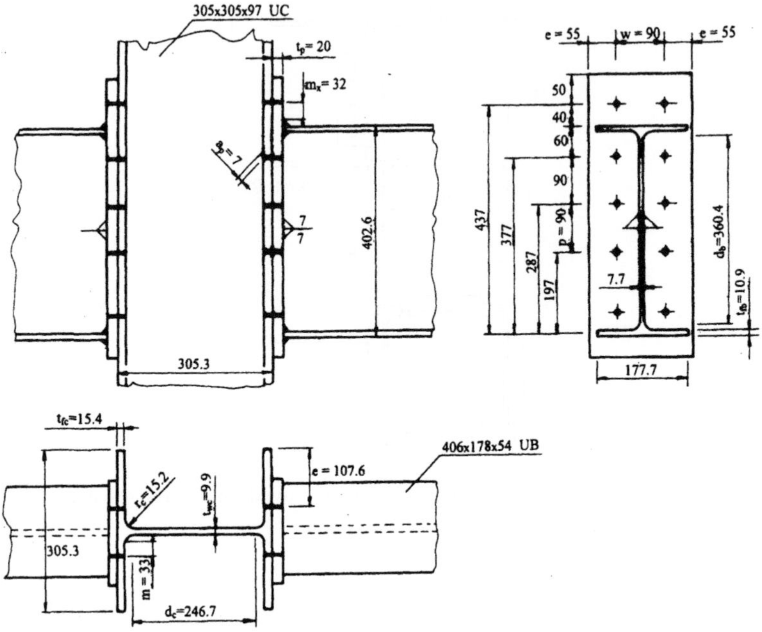

Figure 3. A double-sided bolted beam-to-column connection with extended end-plates

4.2 *International Welding Congress Stellenbosch, 2006*

IIW (International Institute of Welding) has organized this congress together with the South-African Welding Institute in a special "conference village Spier" near Stellenbosch (30 km from Cape Town). Since our students have had problems concerning the calculation of residual welding distortions, I have worked out a study, how to calculate more complicated cases of distortions (Farkas 2006b) (Fig. 4).

4.3 *Research discussion in the University of Pretoria, 2006*

After our very fruitful research cooperation in recent years, we have discussed with professor Jan Snyman our next optimization problem. My first proposition was not very suitable for professor Snyman, so I have proposed a more complicated problem of the optimum design of a stiffened plate supported at four corners with the application of gridwork analysis (Fig.5.).The gridworks of different number of stiffeners should be analyzed separately, so this problem needs special integer programming technique. This problem has been accepted for our further common research.

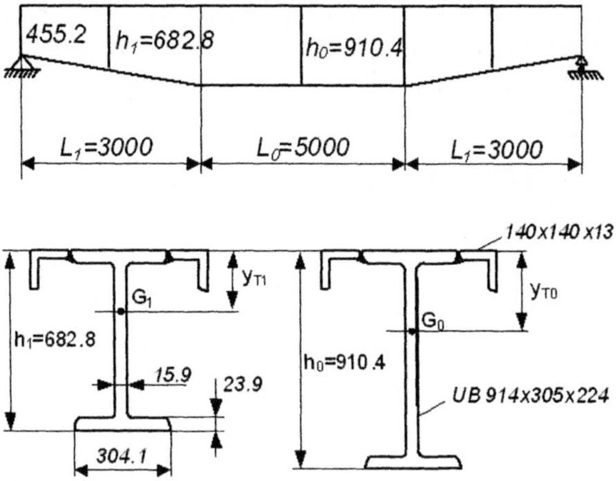

Figure 4. Numerical example of a welded beam of variable cross-section

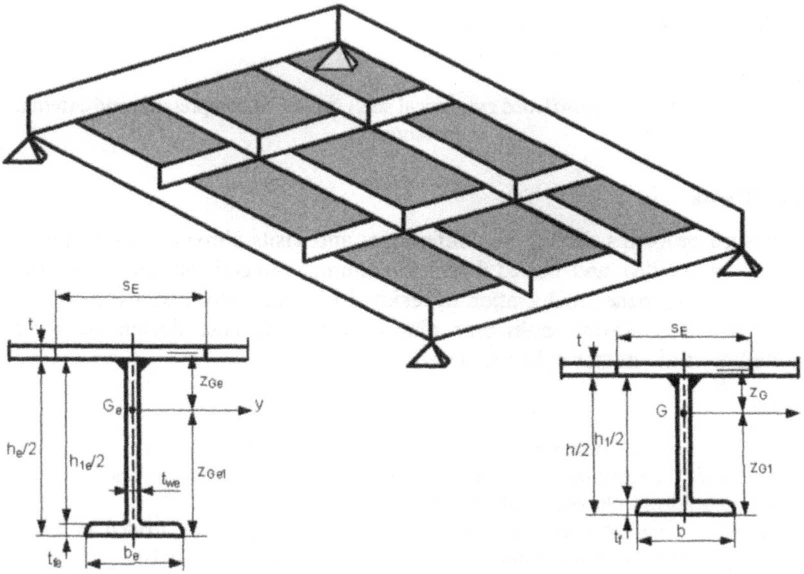

Figure 5. A schematic illustration of a stiffened square plate supported at four corners as well as the cross-sections of the edge and the internal stiffeners

5 Eurosteel Conference, Coimbra, Portugal, 2002

In this conference we have had a paper about the optimum design of a ring-stiffened cylindrical shell loaded by external pressure. Our calculation shows that the use of ring stiffeners in this case is very effective, the cost difference between ring-stiffened and unstiffened shell is about 50%. This difference was so significant that I decided to work out a series of studies about the economy of stiffened plates and shells (Farkas et al. 2002).

These studies have shown that this economy depends on some characteristics such as load (buckling behaviour in the case of compression, external pressure, bending), stiffening geometry (ring-, stringer-or orthogonal stiffening), constraints (buckling strength, deflection). I think that these cost comparisons are very useful for designers to select the most suitable structural versions (Farkas & Jármai 2003, 2005, 2006a, Farkas 2005). (Fig. 6.)

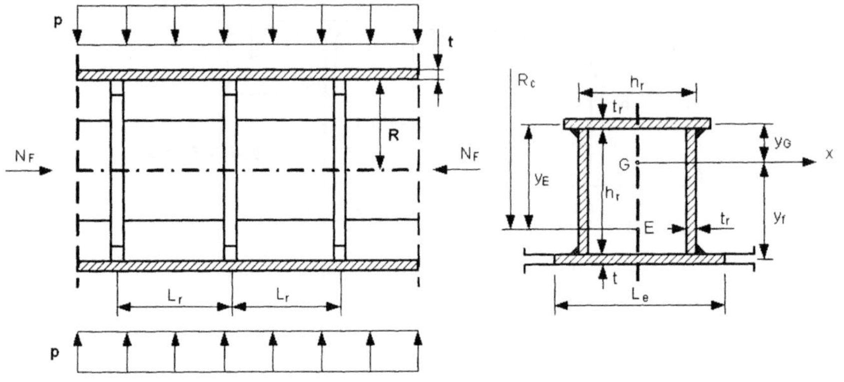

Figure 6. Stringer and ring stiffened cylindrical shell subject to compression and external pressure

6 Conclusions

The discussed problems during the conferences and visits abroad relate to the cost calculation of welded and bolted beam-to-column connections, seismic-resistant design of building frames, calculation of residual welding deformations, application of fiber-reinforced plastic multilayer plates, minimum cost design of welded stiffened plates and circular cylindrical shells.

References

Farkas,J. (1992) Optimum design of circular hollow section beam-columns. In *Proc. 2nd Int. Offshore and Polar Engineering Conference,* San Francisco Int.Soc. Offshore and Polar Engineers, Golden, Colorado, USA. Vol. IV. 494-499.

Farkas,J. (1993) Absorbed energy of CHS and SHS braces cyclically loaded in tension-compression. In *Tubular Structures V.* Coutie, M.G. and G.Davies. (eds) E & FN Spon, London etc. 607-614.

Farkas,J. & Jármai,K. (1997) *Analysis and optimum design of metal structures.* Balkema, Rotterdam-Brookfield.

Farkas,J., Jármai,K., Snyman,J.A. & Gondos,Gy. (2002) Minimum cost design of ring-stiffened welded steel cylindrical shells subject to external pressure. In *Proc. 3rd European Conf. Steel Structures,* Coimbra, Lamas,A. and Simoes da Silva, L.(eds) Universidade de Coimbra, 2002. 513-522.

Farkas,J. & Jármai,K. (2003) *Economic design of metal structures.* Rotterdam, Millpress.

Farkas,J., Jármai,K. & Visser-Uys,P. (2003) Cost comparison of bolted and welded frame joints. *Welding in the World* **47** No.1-2. 12-18.

Farkas,J., Uys,P. & Jármai,K. (2004) Cost analysis of bolted and welded frame connections. In *Proc. 4th International Cost Engineering Council (ICEC) World Congress*, Cape Town, 17-21. CD-ROM.

Farkas,J. & Jármai,K. (2005) Optimum design of a welded stringer-stiffened cylindrical steel shell loaded by bending. In *Proc. Eurosteel 2005. 4th Eurospean Conference on Steel and Composite Structures.* Maastricht, The Netherlands Proceedings. Hoffmeister,B. and Hechler,O. (eds) Aachen, Druck und Verlagshaus Mainz GmbH. Volume A, 1.3-15 – 1.3-22.

Farkas,J. & Jármai,K. (2005) Optimum design of a welded stringer-stiffened steel cylindrical shell subject to axial compression and bending. *Welding in the World* **49** No.5-6, 85-89.

Farkas,J. (2005) Economy of welded stiffened steel plates and cylindrical shells. *Journal of Computational and Applied Mechanics* (Univ. of Miskolc) **6** No.2. 183-205.

Farkas,J. & Jármai,K. (2006a) Optimum design and cost comparison of a welded plate stiffened on one side and a cellular plate both loaded by uniaxial compression. *Welding in the World* **50** No.3-4. 45-51.

Farkas,J. & Jármai,K. (2006b) Special cases of the calculation of residual welding distortions. In *IIW Congress Stellenbosch, South Africa.* Southern African Institute of Welding, . CD-Rom.

Jármai,K., Farkas,J. & Kurobane,Y. (2006) Optimum seismic design of a multistorey steel frame. *Engineering Structures* **28** No.7. 1038-1048.

Jármai,K., Snyman,J.A. & Farkas,J. (2006) Minimum cost design of a welded orthogonally stiffened cylindrical shell. *Computers and Structures* **84** No.12. 787-797.

Kovács,G., Groenwold,A.A., Jármai,K. & Farkas,J. (2004) Analysis and optimum design of fibre-reinforced composite structures. *Struct.Multidisc.Optim.* **28** No.2-3. 170-179.

Taniwaki,K. (1997) *Total optimal synthesis method for frame structures dealing with shape, sizing, material variables and prestressing.* Doctor dissertation, Matsuyama, 1997.

1.2 Our Research Relating to the Minimum Cost Design of Welded Structures

Károly Jármai, József Farkas

University of Miskolc, H-3515 Miskolc, Hungary, altjar@uni-miskolc.hu

Abstract

The main components of our structural optimization system for economic design of welded structures are the design and fabrication constraints, the cost function as well as the efficient mathematical methods for constrained function minimization. This structural optimization system has been applied for stiffened plates and shells as well as for tubular structures. The aim is to give designers useful aspects for cost savings, since welding is an expensive joining technology. Using realistic numerical models, cost comparisons have been worked out for optimized plates stiffened in one or two directions, for cellular plates as well as for circular and conical shells stiffened by rings, stringers or orthogonally. The economic design has been worked out for tubular structures as follows: a triangular truss beam, a strengthened pipeline, tubular frames and a wind turbine tower of circular shell or truss structure.

Keywords: *stiffened plates, stiffened cylindrical shells, tubular structures, minimum cost design, structural optimization*

1 Introduction

Optimization means a search for better solutions, which better fulfil the requirements. For load-carrying engineering structures the main requirements are the safety, fitness for fabrication and economy. In our optimization system the safety and producibility are guaranteed by design and fabrication constraints and economy is achieved by minimization of a cost function. For the constrained function minimization efficient mathematical methods should be used.

It can be concluded that this structural synthesis has four main components as follows: design constraints, fabrication constraints, cost function and mathematical methods.

Design constraints relate to the maximum stresses, stability, fatigue, displacements and can be formulated according to rules of Eurocode 3 for steel structures (Eurocode 3 2002) and Eurocode 9 for aluminium ones (Eurocode 9 2002). In come cases, e.g. for stiffened shells Eurocode 3 does not give suitable rules, thus the Det Norske Veritas (DNV 2002) formulae are used.

In the calculation of structural stability the effect of initial imperfections and residual welding distortions should be taken into account. We have shown that, in the case of overall flexural buckling of a compressed strut the use of Euler formula can cause an unsafe design of about 30% error. It has been shown that in the flexural buckling of compressed struts the cross-sectional type has also an important effect, e.g. a circular hollow section is much more economic than an open angular profile (Farkas & Jármai 1997).

Fabrication constraints relate to the size limitations as well as to the residual welding deformations. The available profile selection should also be taken into account. We have worked out a suitable relative simple calculation method for the residual welding stresses and distortions on the basis of Okerblom's method (Farkas & Jármai 1998) and formulae are developed for the calculation of deformations caused by shrinkage of circumferential welds in circular cylindrical shells (Farkas 2002).

For the calculation of welding times for different weld types and welding technologies formulae have been worked out on the basis of the Pahl-Beelich method as well as using a COSTCOMP software (Farkas & Jármai 2003). The cost function is formulated according to the fabrication sequence. Some efficient mathematical methods have been adapted, developed and used such as SUMT, backtrack, Rosenbrock hillclimb and particle swarm optimization (PSO) algorithm. (Farkas & Jármai 1997, Farkas & Jármai 2003).

Our aim is to give designers useful aspects for cost savings, since the welding is an expensive joining technology. The most suitable structural versions are selected by cost comparisons of optimized solutions. Since only optimized versions can be compared realistically, the structural optimization is the best basis for cost savings. In the following this structural synthesis is systematically applied for three main structural types i.e. for stiffened plates, stiffened shells and tubular structures. In the case of stiffened structures the main question is: a thicker unstiffened or a thinner stiffened version is more economic? To answer this question a systematic research is necessary, since the economy depends on loads and type of stiffening.

2 Stiffened plates

In the case of stiffened plates an unstiffened plate has a very small bending stiffness, thus, the stiffened version is in all cases more economic than the unstiffened one. For stiffened plates some useful comparisons have been worked out. Figure 1 illustrates these cases.

A plate stiffened on one side and a cellular one are compared to each other in the case of longitudinal stiffeners and uniaxial compressive force and it is concluded that the cellular plate is more economic because of its larger torsional stiffness Figs.1a and 1b) (Farkas & Jármai 2006).

It is also found that, in the case of uniaxial compression an orthogonally stiffened plate is more economic than a longitudinally stiffened one (Fig.1c) (Farkas & Jármai 2007). A calculation method is worked out for the optimum design of an orthogonally stiffened plate loaded by bending (Fig.1d) (Jármai et al 2006).

Special gridwork design has been applied to find the optimum solution of a stiffened plate supported at four corners (Fig.1e) (Farkas et al. 2007).

For comparison the same problem has been solved for a cellular plate supported also at four corners (Fig.1f). It has been found that the cellular plate is more economic than the stiffened one (Farkas & Jármai – to be published)..

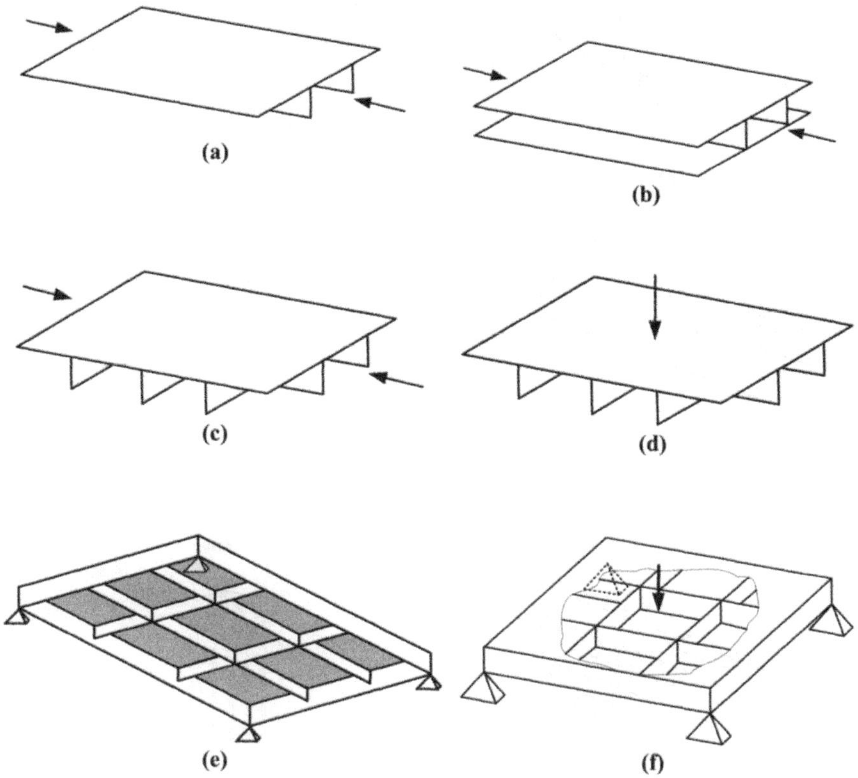

Figure 1. a) Longitudinally stiffened plate, b) longitudinally stiffened cellular plate, c) orthogonally stiffened plate, d) stiffened plate loaded by bending, e) stiffened plate supported at four corners, f) cellular plate supported at four corners

3 Stiffened circular cylindrical and conical shells

A ring-stiffened circular cylindrical shell is more economic than an unstiffened one in the case of external pressure, since these shells are very sensitive against buckling for this load (Fig.2a) (Farkas et al. 2002).

On the other hand, a cylindrical shell is very stiff against buckling for compressive load or bending, thus, the ring-stiffening cannot be economic (Fig.2b) (Farkas et al. 2004).

A stringer stiffening is economic only in that case, when a deflection constraint for the whole structure is active (Figs 2c and 2d) (Farkas & Jármai 2005a). For stiffening halved rolled I-section stringers should be used welded outside of the shell.

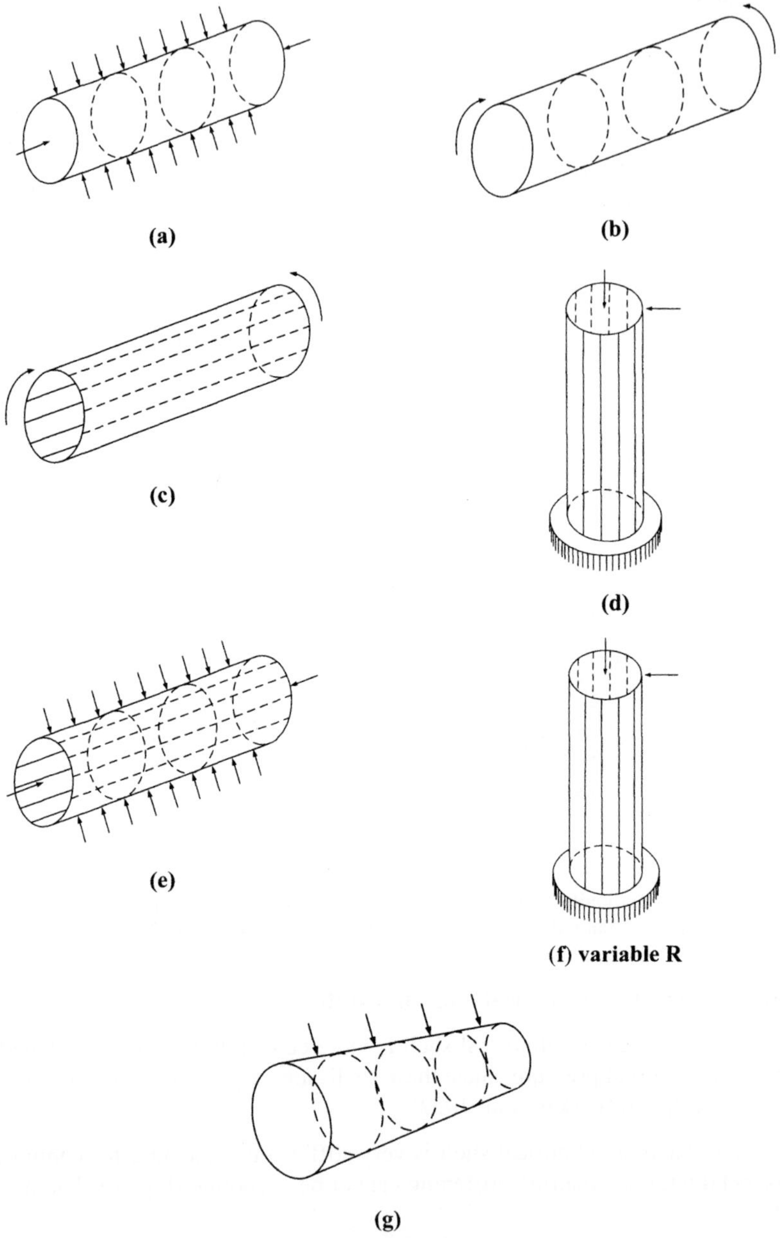

Figure 2. a) ring-stiffeners, external pressure, b) ring-stiffeners, bending, c) stringer stiffeners, bending, d) stringer stiffeners, bending, displacement constraint, e) orthogonal stiffening, external pressure and axial compression, f) radius also optimized, g) slightly conical shell, external pressure, ring-stiffeners

In the case of a column loaded by compression and bending, the stringer stiffening can be economic only when a displacement of the column top is restricted and the

shell radius is kept constant (Fig.2d) (Farkas & Jármai 2005b). When the radius is also optimized, the stiffening cannot be economic (Fig.2f) (Farkas et al.2007).

In order to generalize the problem of stiffened shells the case of orthogonally stiffened shell is worked out for axial compression and external pressure (Fig.2e) (Jármai et al. 2006).

Similarly to a cylindrical shell, a slightly conical one can be economic with ring-stiffeners for external pressure (Fig.2g) (Farkas et al.2007).

4 Tubular structures

The height of a triangular tubular truss beam can be optimized, since, increasing it the chord forces decrease but the length of braces increases (Fig.3a) (Farkas & Jármai 2001).

Optimum dimensions of a welded tubular truss are determined, which strengthen a column-supported oil` pipeline for a larger span length. The cost comparison shows that the cost of the strengthened pipe is much lower than that of the larger pipe without strengthening (Fig.3b) (Farkas & Jármai 2004).

A simple tubular frame supporting a pressure vessel is optimized for seismic loads (Fig.3c) (Farkas & Jármai 2006).

An earthquake-resistant design is worked out for a multi-storey frame with tubular columns, rolled I-beams and seismic resistant beam-to-column connections (Fig.3d) (Jármai et al. 2006).

A wind turbine tower is designed as a ring-stiffened cylindrical shell as well as a tubular truss structure (Figs 3e and 3f). The cost comparison of the two structural optimized versions show that the tubular truss version is much more cheap, than the shell structure (Farkas & Jármai 2006).

5 Conclusions

Our research is focused to the economy of welded structures. This problem needs a systematic research, since the economy depends on structural characteristics as follows: loads, type of structure, design and fabrication constraints, type of profiles, costs.

Our method is to work out numerical problems using realistic structural models and compare the costs of candidate versions. Since only optimized versions can be realistically compared to each other, the structural optimization system should be used. For this system we have developed calculation methods for residual welding distortions and for fabrication costs as well as adapted some efficient mathematical methods.

The investigated problems of stiffened plates and shells as well as tubular structures are briefly overviewed.

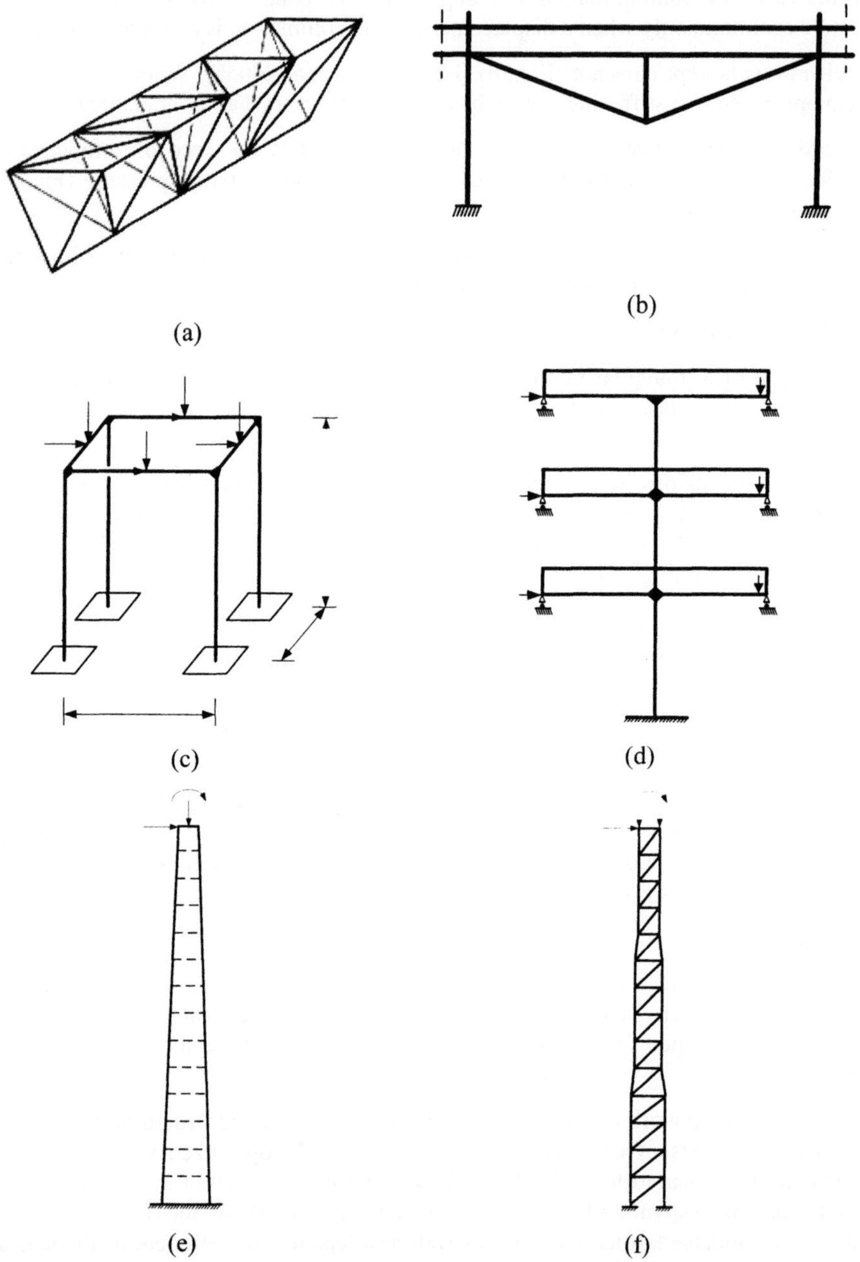

Figure 3. a) A tubular truss beam, b) a column-supported pipeline strengthened by a tubular truss, c) a simple tubular frame supporting a pressure vessel and loaded also by seismic forces, d) a multistory steel frame with a tubular column and rolled-I-section beams , e) a wind turbine tower of ring-stiffened shell structure, f) a wind turbine tower of tubular truss structure

References

Eurocode 3 (2002) Design of steel structures. Part 1-1: General structural rules. CEN Brussels.

Eurocode 9 (2002) Design of aluminium structures. CEN Brussels.

Det Norske Veritas (DNV) (2002) *Buckling strength of shells.* Recommended practice. DNV-RP-C202.

Farkas,J. & Jármai,K. (1997) *Analysis and optimum design of metal structures.* Balkema, Rotterdam-Brookfield

Farkas,J. & Jármai,K. (1998) Analysis of some methods for reducing residual beam curvatures due to weld shrinkage. *Welding in the World* **41** No.4, 385-398

Farkas,J. & Jármai,K. (2001) Height optimization of a triangular CHS truss using an improved cost function. In: *Tubular Structures IX.* Proceedings of the 9th International Symposium on Tubular Structures, Düsseldorf, 2001. Balkema, Lisse etc. pp. 429-435

Farkas,J. (2002) Thickness design of axially compressed unstiffened cylindrical shells with circumferential welds. *Welding in the World* **46** No.11/12, 26-29

Farkas,J., Jármai,K., Snyman,J.A. & Gondos,Gy. (2002) Minimum cost design of ring-stiffened welded steel cylindrical shells subject to external pressure. In: *Proc. 3rd European Conf. Steel Structures*, Coimbra, Lamas,A. & Simoes da Silva, L. (Eds), Universidade de Coimbra, 2002. pp.513-522

Farkas,J. & Jármai,K. (2003) *Economic design of metal structures.* Rotterdam, Millpress

Farkas,J., Jármai,K. & Virág,Z. (2004) Optimum design of a belt-conveyor bridge constructed as a welded ring-stiffened cylindrical shell. *Welding in the World* **48** No.1-2, 37-41

Farkas,J. & Jármai,K. (2005a). Optimum design of a welded stringer-stiffened cylindrical steel shell loaded by bending. In: *Eurosteel 2005. 4th Eurospean Conference on Steel and Composite Structures.* Maastricht, The Netherlands 2005. Proceedings. Hoffmeister,B. & Hechler,O. (Eds), Aachen, Druck und Verlagshaus Mainz GmbH. Volume A, pp. 1.3-15 – 1.3-22

Farkas,J. & Jármai,K. (2005b) Optimum design of a welded stringer-stiffened steel cylindrical shell subject to axial compression and bending. *Welding in the World* **49** No.5-6, 85-89

Farkas,J. & Jármai,K. (2006) Optimum design and cost comparison of a welded plate stiffened on one side and a cellular plate both loaded by uniaxial compression. *Welding in the World* **50** No.3-4, 45-51

Farkas,J. & Jármai,K. (2006) Optimum strengthening of a column-supported oil pipeline by a tubular truss. *Journal of Constructional Steel Research* **62** No.1-2, 116-120

Farkas,J. & Jármai,K. (2006) Seismic resistant optimum design of a welded steel frame supporting a pressure vessel. In: *ICMS 2006. 11st Internat. Conf. Metal Structures* Rzeszów 2006. Progress in Steel, Composite and Aluminium Structures. Gizejowski,M.A. et al. (Eds). Proceedings. Taylor & Francis, London, etc. 2006. 328-329. CD-Rom. IIW-doc. XV-1226-06, XV-F-78-06.

Farkas,J. & Jármai,K. (2006) Cost comparison of a tubular truss and a ring-stiffened shell structure for a wind turbine tower. In: *Tubular Structures XI.* Proc. 11th Int. Symposium and IIW Int. Conf. on Tubular Structures, Québec City, Canada, 2006. Packer,J.A. & Willibald,S. (Eds), Taylor & Francis, London etc. pp. 341-349.

Farkas,J. & Jármai,K. (2007) Economic orthogonally welded stiffening of a uniaxially compressed steel plate. *Welding in the World* **51** No.7-8. 74-78.

Farkas,J., Jármai,K. & Snyman,J.A. (2007) Global minimum cost design of a welded square stiffened plate supported at four corners. In: *7th World Congress on Structural and Multidisciplinary Optimization,* Seoul, Korea, 2007. Paper AO381, Abstract p.84. CD-ROM.

Farkas,J., Jármai,K. & Rzeszut,K. (2007) Optimum design of a welded stringer-stiffened steel cylindrical shell of variable diameter subject to axial compression and bending. In: *17th Internat. Conf. Computer Methods in Mechnaics CMM-2007*, Lódz-Spala. Short papers pp.143-144. CD-ROM.

Farkas,J., Jármai,K. & Orbán,F. (2007) Cost minimization of a ring-stiffened conical shell loaded by external pressure. IIW-Doc.XV-1248-07. XV.F80-07. Dubrovnik, 2007.

Farkas,J. & Jármai,K. Minimum cost design of a welded steel square cellular plate supported at four corners (to be published)

Jármai,K., Farkas,J. & Groenwold,A. (2006). Economic welded stiffening of a steel plate loaded by bending. In: *IIW Congress* Stellenbosch, South Africa. Southern African Institute of Welding, 2006. CD-Rom.

Jármai,K., Snyman,J.A. & Farkas,J. (2006) Minimum cost design of a welded orthogonally stiffened cylindrical shell. *Computers and Structures* **84** No.12, 787-797

Jármai,K., Farkas,J. & Kurobane,Y. (2006) Optimum seismic design of a multistorey steel frame. *Engineering Structures* **28** No.7. June, 1038-1048

1.3 Test and Numerical Verification of Ice-Hockey Hall

Stanislav Kmeť [1], Ján Kanócz [1]

[1] *Technical University of Košice, Vysokoškolská 4, 040 01Košice,Slovakia, stanislav.kmet@tuke.sk, jan.kanocz@tuke.sk*

Abstract

The structural analysis and tests of the steel bearing structure with double curved roof of the new ice-hockey hall in Kosice named Steel Arena are described. Special features of the large-span bearing structure include the main middle three-side spatial truss tubular arch with 120 m span. Deformation of the main bearing system of hall was continuously monitored during the construction using geodetic measurements. Experimental results are compared with those obtained by nonlinear FEM analysis. To ensure structural reliability of the main arch the behaviour of tubular joints were investigated. Numerical results obtained by finite element technique were verified by experiment.

Keywords: *steel structure, large-span tubular arch, computational models, non-linear FEM analysis, stability analysis.*

1 Introduction

Before the final building approval of reconstructed ice hockey hall in Košice, named Steel Arena, tests of the main steel bearing structure were carried out. The goal of the monitoring was to determine the increase of stress level in most stressed members of steel bearing structure from the several load states.

Figure 1. The hall under reconstruction

Special features of the large-span bearing structure include the main middle three-side spatial truss tubular arch with 120 m span and the spatial frame systems of the side stands. These two systems support the roof truss girders which are creating the spatial curvature of the roof. Typical roof construction includes double cold-formed profiled sheet as a roof cladding with an intermediate thermal insulation on steel truss purlins on truss girders.

The shape of the structure was strongly influenced by the environmental constraints in the form of defined space in the built-up area. The structural system of the main hall consists of two following subsystems: the middle truss arch which is located in the direction of longitudinal axis of the hall and the systems of spatial supports - columns supporting the stands. The truss girders of the roof lay on one side on the supporting middle arch and on the other side the spatial columns of stands support them. This solution gives the dynamic looks to the hall, because the roof follows the shape of an arch in the middle part and the roof edges are horizontal in the both sides (Figure 1). All foundations are supported by drilled reinforced concrete piles (Kanocz, J. & Kmet, S. 1996).

2 Numerical calculation of bearing structure

In order to accurately understand the structural behaviour and the response of the structure 3D spatial model simulating all structural members was analyzed (Figure 2). The SCIA ESA PT software was used for the analyses.

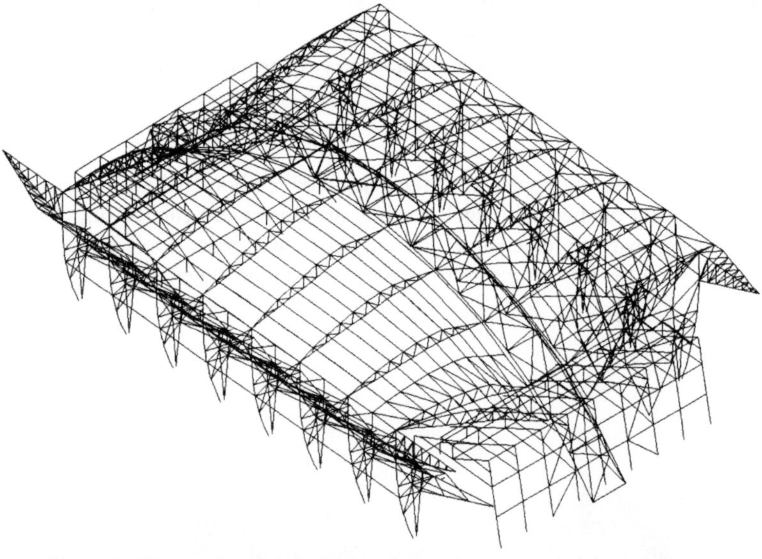

Figure 2. 3D computational model of a bearing system of the structure

Verification of the stability of the structure and its parts was carried out considering imperfections and second order effects. Second order effects and imperfections were accounted for both totally by the global analysis. Geometrical non-linear analysis with incremental and iterative procedures was used for elastic region. Because second order effects in individual members and relevant member imperfections were totally accounted for in the geometrical non-linear global analysis of the structure, no individual stability check for the members was necessary (verification of a compression member against buckling).

In order to verify results of structural numerical analysis the tensiometric measurement of stress and force behaviour of the selected tubular joints and members of main arch in situ under variable load were performed.

3 Test and monitoring of bearing structure

The experimental investigation of bearing structure to the main longitudinal arch (Figure 3) and to the central cross frame was focused. On the arch four cross section was selected in which the top and bottom chord and also the diagonal rod were monitored. In the cross frames the stresses in the tribune elements and columns were measured. To determine increased level of stresses in selected members the strain gauges - measuring accuracy elements (one-line and two-line Hottinger tensiometers) were used (Figure 4) (Kmet, S. & Kanocz, J. 2006).

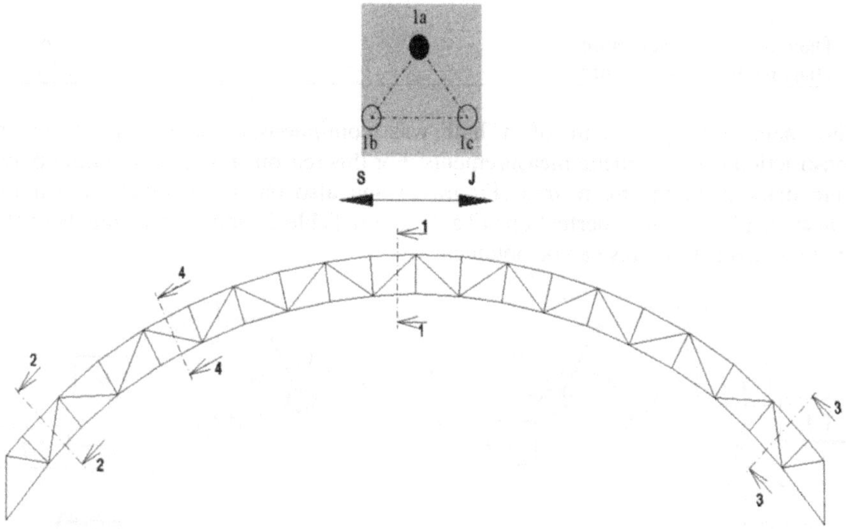

Figure 3. Locations of the monitoring members on the main arch

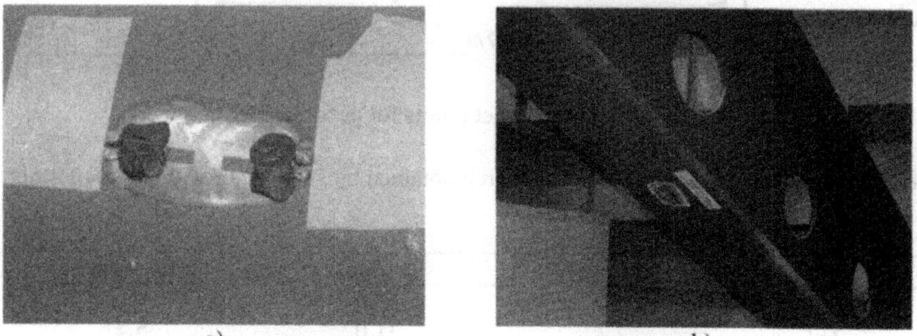

a) b)

Figure 4. Strain gauges on the arch (a) and tribune member (b)

The tensiometers were introduced into selected members of the main arch after hall completion as is shown in Figure 4. At the time of an initial measuring the hall was loaded by self weight of the structure and by permanent action. The next measurement was realized after completion of information system (screens and sound box) and in the time when maximum snow load $s=0{,}72kNm^{-1}$ (in year 2006) was acted. During this measurement the seats of the tribunes were empty. Values of the received stresses in the monitored members of the arch (for location see Figure 3) under increment of the snow and info action are presented in Table 1.

Table 1. Stresses in the monitored members of arch

Member	Action	Stresses [MPa]			
		Section 1-1	Section 2-2	Section 3-3	Section 4-4
Top chord a)	Snow+info	-6,3	1,26	1,19	
Bottom chord b)	Snow+info	-0,42	-6,72	-7,46	
Bottom chord c)	Snow+info	-3,15	-9,56	-9,03	
Diagonal a)	Snow+info				3,99
Diagonal b)	Snow+info				7,04

The main bearing system of a hall was continuously monitored during the construction using geodetic measurements. For this reason, a number of target points were marked on the main arch (Figure 5) and also on the members of tribune. Consequently, initial imperfections (Table 2 and Table 3) and displacements (Table 4) of the main structures can be obtained.

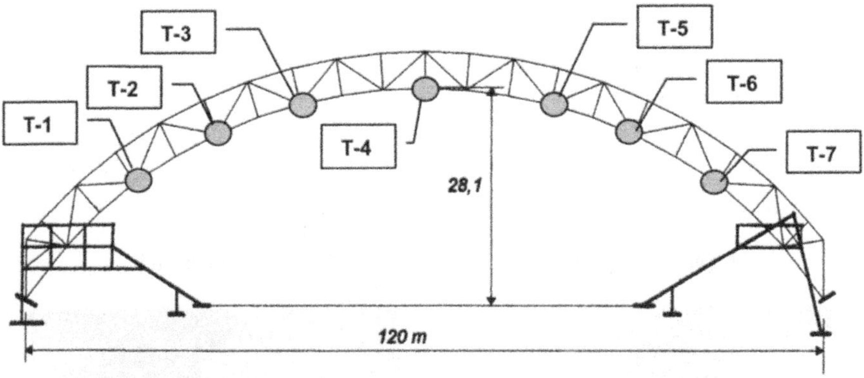

Figure 5. Scheme of the target points for geodetic measurement

Table 2. Global imperfections of the main arch obtained by geodetic measurement (bottom chord - west part)

Designation	e_x (mm)	e_y (mm)	e_z (mm)
T-1	-3,9	50,0	-3,41
T-2	7,2	31,0	-5,4
T-3	6,4	18,2	-8,1
T-4	5,6	9,0	-12,4
T-5	1,7	-26,0	-53,5
T-6	8,3	-17,2	-67,0
T-7	6,7	-54,2	-33,1

x - transversal axe, y - longitudinal axe (in plane of the arch), z - vertical axe

Displacements in marked points into three main directions were measured. The vertical displacement in the middle span of the arch (in z-axe direction) under increment of the snow and info action is relatively small compared to the displacement limit of the 120 m span arch as shown in Table 4.

Table 3. Global imperfections of the main arch obtained by geodetic measurement (bottom chord - east part)

Designation	e_x (mm)	e_y (mm)	e_z (mm)
T-1	-4,5	39,2	-21,9
T-2	7,2	33,3	-5,6
T-3	4,0	19,9	-2,5
T-4	5,4	4,4	-13,1
T-5	0,4	-31,6	-16,5
T-6	0,9	-36,4	-2,7
T-7	9,2	-43,6	-44,6

Table 4. Characteristic nodal displacements of the main bearing arch

Location of measurements	Displacements [mm]		
	Δu	Δv	Δw
T-1	2,9	0,4	-5,1
T-4	-1,2	0,9	-6,3
T-7	-2,2	0,6	-4,2

4 Comparison of theoretical and experimental results

Comparison of theoretically and experimentally obtained results (stresses in the monitored members of the arch) is shown in Tables 5 - 8. Computational bar model of the structure with realistic geometry when imperfections obtained from geodetic measurement was used for theoretical geometrical non-linear analysis.

Table 5. Stresses in the monitored members of the arch (section 1-1) - comparison of theoretical and experimental results

Member	Action	Stresses (MPa) Section 1-1	
		Theory	Test
Top chord a)	Permanent	-107,3	
	Info+Snow	-11,3	-6,3
	Total	-118,6	
Bottom chord b)	Permanent	-29,4	
	Info+Snow	-0,9	-0,42
	Total	-30,3	
Bottom chord c)	Permanent	-28,5	
	Info+Snow	-3,0	-3,15
	Total	-31,5	

Table 6. Stresses in the monitored members of the arch (section 2-2) - comparison of theoretical and experimental results

Member	Action	Stresses (MPa) Section 2-2 Theory	Test
	Permanent	9,8	
Top chord	Info+Snow	1,7	1,26
a)	Total	11,5	
Bottom chord	Permanent	-119,2	
b)	Info+Snow	-10,5	-6,72
	Total	-129,7	
Bottom chord	Permanent	-119,4	
c)	Info+Snow	-10,6	-9,56
	Total	-130,0	

Table 7. Stresses in the monitored members of the arch (section 3-3) - comparison theoretical and experimental results

Member	Action	Stresses (MPa) Section 3-3 Theory	Test
	Permanent	9,6	
Top chord	Info+Snow	1,5	1,19
a)	Total	11,1	
Bottom chord	Permanent	-117,6	
b)	Info+Snow	-10,5	-7,46
	Total	-128,1	
Bottom chord	Permanent	-118,0	
c)	Info+Snow	-10,4	-9,03
	Total	-128,4	

Table 8. Stresses in the monitored members of the arch (section 4-4) - comparison of theoretical and experimental results

Member	Action	Stresses (MPa) Section 4-4 Theory	Test
Diagonal	Permanent	34,8	
a)	Info+Snow	4,7	3,99
	Total	39,5	
Diagonal	Permanent	30,8	
b)	Info+Snow	8,1	7,04
	Total	38,9	

Tables show that experimentally obtained stresses from a load-increment represented by the investigated snow load and by the load from the information system never exceed theoretically obtained stress levels when FEM was applied at the same load condition (no partial reliability factors were used).

5 Investigation of load-bearing and/or ultimate capacity of tubular joints

To ensure structural reliability of the main arch the behaviour of tubular joints without use of joint sheets and with rather different diameters needed to be investigated and well understood. The finite element method was a useful and powerful tool for studying this problem. However, this numerical technique had to be verified by experimental results. To establish a reference point simple tubular T-joints and doubler plate-reinforced tubular T-joints were fabricated, instrumented and tested to failure under axial compression brace loading (Figure 6 and Figure 7). The structural behaviours were observed and analysed. It was found that the ultimate capacity was significantly improved by the inclusion of double plate Fung et al. (1999), (Figure 7).

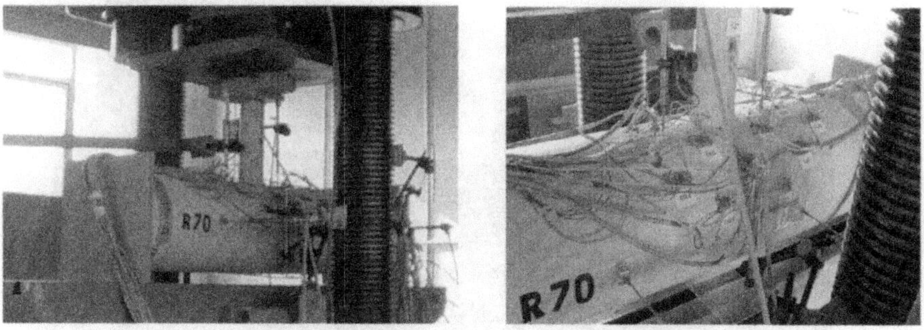

Figure 6. Specimen of the tubular T-joint during the test

The experimental load-displacement curves were compared with the finite-element prediction and good agreement was obtained. The finite element techniques (Figure 7 a) were then verified and could be used to study the load-bearing capacity and behaviour of tubular joints with various types of brace loading and different geometric parameters.

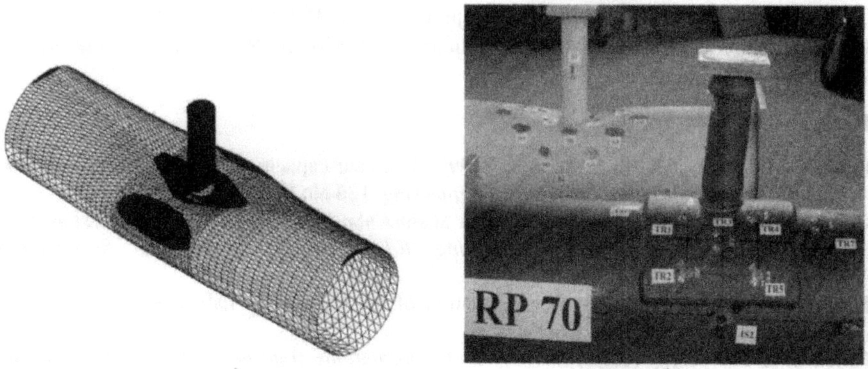

a) b)

Figure 7. Finite-Element model (a) and failure modes of specimens (b)

6 Conclusion

The Steel Arena is a multi-purpose sporting, social and entertainment hall. It becomes the dominant building of the city (Figure 8). The building was opened on May 2006 and has received international attention. The Steel Arena has demonstrated its reliability and safety during the one year since the stadium opening, when the stadium was repeatedly filled to capacity with an event taking place.

Figure 8. Steel Arena

The aesthetics of the hall is strongly enhanced by the structural solution, which was created by means of combination of inspiration of architects and logic of structural designers.

Acknowledgements

This work is a part of the Research project No. 1/3345/06, founded by the Scientific Grant Agency of the Ministry of Education of Slovak Republic and the Slovak Academy of Sciences.

References

Fung, T.C. Chang, T.K. & Soth, C.K. (1999) Ultimate capacity of double plate-reinforced tubular joints, *Journal of Structural Engineering,* **125** No 8, 891-899.

Grierson, D.E. (2003) Designing buildings against abnormal loading, In: *Progress in Civil and Structural Engineering Computing, Edited by B.H.V. Topping, Saxe-Coburg Publications,* pp. 37-62.

Kanocz, J. & Kmet, S. (1996) Reconstruction of the ice hockey hall in Košice, Structural analysis, CZC Consulting, Košice.

Kmet, S. & Kanocz, J. (2006) *Stresses and deformations state monitoring of load-bearing structure of Steel Arena in Košice,* Faculty of Civil Engineering, Technical University in Košice, Košice.

Kmet, S., Kanocz, & J. Tomko, M. (2006) *Reliability assessment of load-bearing structure of Steel Arena in Košice against accidental action*, Faculty of Civil Engineering, Technical University in Košice, Košice.

1.4 Optimization of an Orthogonally Stiffened Plate Considering Fatigue Constraints

Luis M.C. Simões[1], József Farkas[2] and Károly Jármai[2]
[1]*University of Coimbra, Portugal, lcsimoes@dec.uc.pt*
[2]*University of Miskolc, Hungary,altjar@uni-miskolc.hu*

Abstract

The aim of this work is the optimization of a uniaxially compressed stiffened plate subjected to static and fatigue loading. The design variables are the thickness of the base plate, the number and stiffeners of the orthogonally stiffened plate. The constraints deal with the static overall plate buckling, the stiffener failure and the fatigue strength of the welded connections between the stiffeners and the interaction of the two types of failure. The cost function includes the cost of material, assembly, welding and painting. Randomness is considered both in loading and material properties. A level II reliability method (FORM) is employed. The overall structural reliability is obtained by using Ditlevsen method of conditional bounding. The costs of the plate designed to ensure a stipulated probability of failure will be compared with the solutions obtained for a code based method, which employs partial safety factors.

Keywords: *reliability-based optimization, stiffened plates, fatigue*

1 Introduction

Stiffened plates are often the main structural components of load-carrying structures such as bridges, columns, towers, platforms, vehicles etc. The aim of this work is the optimization of a uniaxially compressed stiffened plate. The thickness of the base plate as well as the numbers and dimensions of the longitudinal and transverse stiffeners are sought, which fulfil the design and fabrication constraints and minimize the cost function. The constraints relate to the static overall plate buckling, to the stiffener induced failure and to the fatigue strength of welded connections between the stiffeners. Interaction of the two types of failure, buckling and fatigue can be more dangerous than each individually: the fatigue crack propagation might affect the development of buckling.

The buckling constraints are formulated according to the Det Norske Veritas design rules, the fatigue strength constraint is expressed using the data of Eurocode 3. The fabrication constraints limit the maximal number of stiffeners in one direction to ensure the welding of welds connecting the stiffeners to the base plate. The cost function includes the cost of material, assembly, welding and painting and is formulated according to Farkas & Jármai (2003).

Stresses and displacements can be computed given the deterministic parameters of loads, geometry and material behaviour. Some structural codes specify a maximum probability of failure within a given reference period (lifetime of the structure). This probability of failure is ideally translated into partial safety factors and combination factors by which variables like strength and load have to be divided or multiplied to find the so called design values. The structure is supposed to have met the reliability requirements when the limit states are not exceeded. The advantage of code type level I method (using partial safety factors out of codes) is that the limit states are to be checked for only a small number of combinations of variables. The safety factors are often derived for components of the structure disregarding the system behaviour. The disadvantage is lack of accuracy. This problem can be overcome by using more sophisticated reliability methods such as level II (first order second order reliability method, FOSM and level III (Monte Carlo) reliability methods. In this work FOSM was used and the sensitivity information was obtained analytically. Besides stipulating maximum probabilities of failure for the individual modes, the overall probability of failure, which account for the interaction by correlating the modes of failure is considered.

A branch and bound strategy coupled with a entropy-based algorithm is used to solve the reliability-based optimization. The entropy-based procedure is employed to find optimum continuous design variables giving lower bounds on the decision tree and the discrete solutions are found by implicit enumeration. Results are given comparing deterministic and reliability-based solutions and show how the optimum solution changes with the axial force and loading amplitude used to describe fatigue.

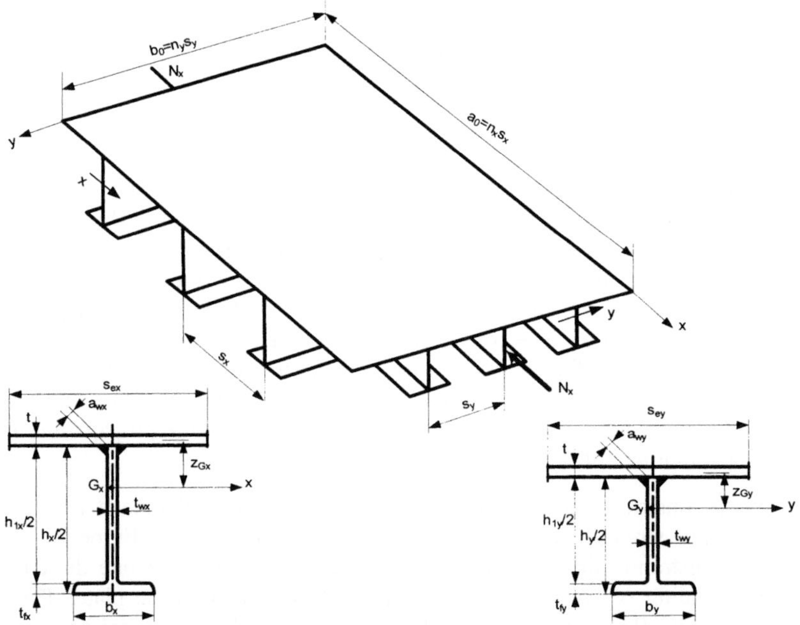

Figure 1. Orthogonally stiffened plate loaded by uniaxial compression

2 Design variables

The design variables are the base plate thickness t, sizes and number of stiffeners in both directions: h_y, h_x, n_y, n_x.
Ranges of unknowns: $4 < t < 20$ mm, $152 < h < 1016$ mm, $4<n<n_{max}$. The maximum values of n_i is given by the fabrication constraints Eq. (1).

$$\frac{b_0}{n_y} - b_y \geq 300 \text{ mm}, \quad \frac{a_0}{n_x} - b_x \geq 300 \text{ mm}. \tag{1}$$

3 Constraints

3.1 *Overall buckling constraint* according to DNV

$$\sigma = \frac{N_x}{n_y A_{ey}} \leq \sigma_{cr} = \frac{f_{y1}}{\sqrt{1+\lambda^4}}, f_{y1} = \frac{f_y}{1.1} \tag{2}$$

$$\lambda = \sqrt{\frac{f_{y1}}{\sigma_E}}, \sigma_E = \frac{N_E s_y}{A_{ey}}, N_E = \frac{\pi^2}{b_0^2}\left(B_x \frac{b_0^2}{a_0^2} + B_y \frac{a_0^2}{b_0^2} \right) \tag{3}$$

It can be seen from the load-carrying capacity formula N_E that, when $a_0 > b_0$, to have a larger N_E, B_x (h_x) should be larger than B_y (h_y). From the theoretical buckling strength σ_E the critical strength σ_{cr} is calculated by using a slenderness λ to take into account the effect of initial imperfections. The factored compressive force is calculated as

$$N_x = \gamma_{stat} N_{xstat} + \gamma_F \Delta N/2, \tag{4}$$

where $\gamma_{stat} = 1.1$ and $\gamma_F = 1.35$ are safety factors, N_{xstat} is the static component and a variable component has an amplitude of $\Delta N/2$.

3.2 *Constraint on stiffener torsional buckling* according to DNV

The constraint is formulated as

$$\sigma_1 = \frac{N_x}{n_y A_{ey1}} \leq \sigma_{acr} = \frac{\sigma_k}{\phi + \sqrt{\phi^2 - \lambda_S^2}} \tag{5}$$

3.3 *Constraint on fatigue strength of welded connections of stiffeners*

The constraint on fatigue strength is defined by

$$\alpha_0 \frac{\Delta N}{n_y A_{ey}} \leq \frac{\Delta\sigma_N}{\gamma_{Mf}} \tag{6}$$

where α_0 is the interaction factor to avoid the danger of interaction of the buckling and fatigue phenomena, ΔN is the variable load range, $\Delta\sigma_N$ is the fatigue stress range corresponding to the number of cycles N_C, γ_{Mf} is the safety factor for fatigue.

4 Cost function

The cost function includes the cost of material, assembly, welding as well as painting and is formulated according to the fabrication sequence.
The cost of material

$$K_M = k_M \rho V_2; k_M = 1.0 \text{ \$/kg}. \tag{7}$$

Welding of the base plate from butt welds (3 in direction of a_0 and 3 in direction of b_0) (SAW - submerged arc welding) Farkas et al. (2007):
The fabrication cost factor is taken as $k_F = 1.0$ \$/min, the factor of complexity of the assembly $\Theta_W = 2$:

$$K_0 = k_F \left[\Theta_W \sqrt{16 \rho V_0} + 1.3 C_W t^n (3a_0 + 3b_0) \right], \tag{8}$$

Welding (n_x-1) stiffeners to the base plate in y direction with double fillet welds (GMAW-C - gas metal arc welding with CO_2):

$$K_{W1} = k_F\left[\Theta_W\sqrt{n_x\rho V_1} + 1.3x0.3394x10^{-3}a_{wx}^2 2b_0(n_x - 1)\right] \qquad (9)$$

Welding of (n_y – 1) stiffeners to the base plate in x direction with double fillet welds. These stiffeners should be interrupted and welded with fillet welds to the stiffeners in the y direction.

$$K_{W2} = k_F\left[\Theta_W\sqrt{(n_y n_x - n_x + 1)\rho V_2} + 1.3x0.3394x10^{-3}a_{wy}^2 2a_0(n_y - 1) + T_1\right] \qquad (10)$$

which is rounded to $0.4t_{wy}$.

Painting $K_P = k_P\Theta_P S_P$ $k_P = 14.4x10^{-6}$ \$/mm^2, $\Theta_P = 2$, (11)

Surface to be painted

$$S_P = 2a_0b_0 + a_0(n_y - 1)(h_{1y} + 2b_y) + b_0(n_x - 1)(h_{1x} + 2b_x) \qquad (12)$$

The total cost $K = K_M + K_0 + K_{W1} + K_{W2} + K_P$ (13)

5 Reliability-based optimization

A failure event may be described by a functional relation, the limit state function, in the following way

$$F=\{g(x)\leq 0\} \qquad (14)$$

In the case the limit state function g (x) is a linear function of the normally distributed basic random variables x the probability of failure can be written in terms of the linear safety margin M as:

$$P_F = P\{g(x) \leq 0\} = P(M \leq 0) \qquad (15)$$

which reduces to the evaluation of the standard normal distribution function

$$P_F = \Phi(-\beta) \qquad (16)$$

where β is the reliability index given as

$$\beta = \mu_M / \sigma_M \qquad (17)$$

The reliability index has the geometrical interpretation as the smallest distance from the line (or the hyperplane) forming the boundary between the safe domain and the failure domain. The evaluation of the probability of failure reduces to simple evaluations in terms of mean values and standard deviations of the basic random variables.

When the limit state function is not linear in the random variables x, the linearization of the limit state function in the design point of the failure surface represented in normalised space u. was proposed in Hasofer & Lind (1974),

$$u_i = (x_i - \mu_{x_i}) / \sigma_{x_i} \qquad (18)$$

As one does not know the design point in advance, this has to be found iteratively in a number of different ways. Provided that the limit state function is differentiable, the following simple iteration scheme may be followed:

$$\alpha_i = -\partial g(\beta\alpha)/\partial u_i \left[\sum_{j=1}^{n} \partial g(\beta\alpha)^2 / \partial u_i\right] \qquad (19)$$

$$G(\beta\alpha_1, \beta\alpha_2, ...\beta\alpha_v) \qquad (20)$$

which will provide the design point u* as well as the reliability index β.

The reliability assessment requires an enumeration of the reliability indices associated with limit state functions to evaluate the structural system probability of failure. Collapse modes are usually correlated through loading and resistances. For this reason, several investigators considered this problem by finding bounds for p_F. By taking into account the probabilities of joint failure events such as $P(F_i \cap F_j)$ which means the probability that both events F_i and F_j will simultaneously occur. The resulting closed-form solutions for the lower and upper bounds are as follows:

$$p_F \geq (F_1) + \sum_{i=2}^{m} Max\left\{\left[P(F_i) - \sum_{j=1}^{i-1} P(F_i \cap F_j)\right];0\right\} \tag{21}$$

$$p_F \leq \sum_{i=1}^{m} P(F_i) - \sum_{i=2}^{m} \underset{j<i}{Max} P(F_i \cap F_j) \tag{22}$$

The above bounds can be further approximated using Ditlevsen (1979) method of conditional bounding to find the probabilities of the joint events. This is accomplished by using a Gaussian distribution space in which it is always possible to determine three numbers β_1, β_2 and the correlation coefficient ρ_{ij} for each pair of collapse modes F_i and F_j such that if ρ_{ij} >0 (F_i and F_j positively correlated):

$$P(F_i \cap F_j) \geq Max\ \{\Phi(-\beta_j)\ \Phi(-\frac{\beta_i - \beta_j\rho_{ij}}{\sqrt{1-\rho_{ij}^2}})\ ;\Phi(-\beta_i)\ \Phi(-\frac{\beta_j - \beta_i\rho_{ij}}{\sqrt{1-\rho_{ij}^2}})\} \tag{23}$$

$$P(F_i \cap F_j) \leq \Phi(-\beta_j)\ \Phi(-\frac{\beta_i - \beta_j\rho_{ij}}{\sqrt{1-\rho_{ij}^2}})\ +\Phi(-\beta_i)\ \Phi(-\frac{\beta_j - \beta_i\rho_{ij}}{\sqrt{1-\rho_{ij}^2}}) \tag{24}$$

In which β_i and β_j are the safety indices of the ith and the jth failure mode and Φ() is the standardized normal probability distribution function.

The probabilities of the joint events $P(F_i \cap F_j)$ in (8) and 89) are then approximated by the appropriate sides of (23) and (24). For example, if F_i and F_j are positively dependent for the lower (21) and upper (22) bounds it is necessary to use the approximations given by the upper (24) and lower (23) bounds, respectively.

6 Optimization strategy

6.1 *Branch and bound*

The problem is non-linear and the design variables are discrete. Given the small number of discrete design variables an implicit branch and bound strategy was adopted to find the least cost solution. The two main ingredients are a combinatorial tree with appropriately defined nodes and some upper and lower bounds to the optimum solution associated the nodes of the tree. It is then possible to eliminate a large number of potential solutions without evaluating them.

Three levels were considered in the combinatorial tree. The plate thickness is fixed at the top of the tree, the remaining levels corresponding to n_x (and the appropriate UB profile h_x) and n_y associated with h_y. A strong branching rule was employed. Each node can be branched into n_s new nodes, each of these being associated with the number of stiffeners needed in the x direction. This requires using continuous values close to the geometric characteristics of an UB section, (A_s, b, t_f, t_w), which are approximated by curve-fitting functions written as a function of h. The stiffener height is also obtained from a curve fitting of the heights h. Care has to be taken to find geometrical properties leading to convex underestimates of the actual UB section, so that the solution obtained by using the real UB geometric characteristics is more costly than the solution given by using continuous approximations. In the second level of the tree the branches correspond to different stiffener UB profiles. At the third level the resulting minimum discrete solution becomes the incumbent solution (upper bound). Any leaf of the tree whose bound is strictly less than the incumbent is active. Otherwise it is designated as terminated and need not to be considered further. The B&B tree is developed until every leaf is terminated. The branching strategy adopted was breadth first, consisting of choosing the node with the lower bound.

6.2 *Optimum design with continuous design variables*

For solving each relaxed problem with continuous design variables the simultaneous minimization of the cost and constraints is sought. All these goals are cast in a normalized form. For the sake of simplicity, the goals and variables described in the following deal with stiffened shells. If a reference cost K_0 is specified, this goal can be written in the form,

$$g_1(t,n,h) = K(t,n,h)/K_0 - 1 \le 0 \tag{25}$$

Another two goals arise from the constraint on overall buckling and single panel buckling:

$$g_3(t,n,h) = \sigma_c/\sigma_{cr} - 1 \le 0 \tag{26}$$

$$g_2(t,n,h) = \sigma_l/\sigma_{acr} - 1 \le 0 \tag{27}$$

The remaining goal deals with the fatigue strength of the stiffeners connections:

$$g_4(t,h) = \Delta\sigma/\Delta\sigma_n - 1 \le 0 \tag{28}$$

The objective of this Pareto optimization is to obtain an unbiased improvement of the current design, which can be found by the unconstrained minimization of the convex scalar function Simões & Templeman (1989):

$$F(t,h) = \frac{1}{\rho}.\ln\left[\sum_{j=l}^{3}\exp\rho(g(t,h))\right] \tag{29}$$

This form leads to a convex conservative approximation of the objective and constraint boundaries. Accuracy increases with ρ.

The strategy adopted was an iterative sequence of explicit approximation models, formulated by taking Taylor series approximations of all the goals truncated after the linear term. This gives:

$$\text{Min } F(t,h) = \frac{1}{\rho}.\ln\left[\sum_{j=l}^{3}\exp\rho\left(g_0(t,h) + \frac{\partial g_{0j}(t,h)}{\partial t}dt + \frac{\partial g_{0j}(t,h)}{\partial h}dh\right)\right] \tag{30}$$

This problem has an analytic solution giving the design variables changes dt and dh. Solving for a particular numerical value of g_{oj} forms an iteration of the solution to problem (30). Move limits must be imposed on the design variable changes to guarantee the accuracy of the approximations. Given the small number of design variables an analytic solution is available. During the iterations the control parameter ρ, which should not be decreased to produce an improved solution, is increased.

7 Numerical examples

Numerical data (Figure 1): a_0 = 24000, b_0 = 8000 mm, steel yield stress f_y = 355 MPa, elastic modulus E = 2.1x10^5 MPa, shear modulus G = 0.8x10^5, density ρ = 7.85x10^{-6} kg/mm^3, selected rolled I-sections UB profiles. Consistent with the traditional limit state design (level 1 approach), the following solutions consider a deterministic behaviour of all the variables for several load combinations of the original compressive force N_{xstat} and the load range for fatigue ΔN.

	N_{xstat}	ΔN	h_x	h_y	t	n_x	n_y	*Cost*
1	2x10^7	6x10^6	403.2	257.2	13	6	19	54198
2	3x10^7	6x10^6	533.1	403.2	16	5	11	57097
3	4x10^7	6x10^6	533.1	403.2	18	6	13	64831
4	2x10^7	9x10^6	454.6	353.4	19	4	15	64499
5	2x10^7	12x10^6	403.2	353.4	27	4	12	76865

By adopting coefficients of variation of 0.15 for the compressive force, 0.25 for the load amplitude and 0.10 for the design stress, solution 1 was used to tune mean values for the static force, load amplitude and design stresses. The Gaussian distribution was adopted for all the random variables. Although the randomness of Young modulus also plays an important role in the structural reliability, this was not considered here for the sake of simplicity. In this example the probability of failure will be describe with the overall buckling stresses, the stiffener torsional buckling and the fatigue strength of the welded connections of stiffeners induced by the loadings. A maximum individual probability of failure $p_f \leq$ 1.0E-4 (beta larger than 3.72) was established. The following reliability based optimum solutions were obtained:

	μ_{Nxstat}	$\mu_{\Delta N}$	h_x	h_y	t	n_x	n_y	*Cost*
1	1.145x10^7	2.95x10^6	403.2	257.2	13	6	19	54418
2	1.172x10^7	2.95x10^6	454.6	203.2	13	8	19	56336
3	2.29x10^7	2.95x10^6	607.6	403.2	15	6	14	61045
4	1.145x10^7	4.43x10^6	403.2	308.7	18	5	19	64389
5	1.145x10^7	5.9x10^6	403.2	403.2	24	4	16	75479

The reliability-based solutions are generally least costly than the deterministic and ensure the safety level adopted. If the all the modes are considered and the overall probability of failure $p_f \leq$ 1.0E-4 is imposed by using Ditlevsen improved second order bounds, the reliability based solutions for problems 4 and 5 are changed as in the first both the local buckling and fatigue have small β values and in the later the overall buckling and fatigue are critical.

	μ_{Nxstat}	$\mu_{\Delta N}$	h_x	h_y	t	n_x	n_y	*Cost*
4	1.145x10^7	4.43x10^6	403.2	257.2	19	6	18	64977
5	1.145x10^7	5.9x10^6	403.2	353.4	25	4	15	76336

The solutions are 1% more costly, being thicker but reducing the number and sizes of the stiffeners in the y direction. The influence of the coefficient of variation of the static compressive force on solution 1 was also studied. If the coefficient of variation is increased by 10% to 0.165 the reliability-based solution becomes:

	μ_{Nxstat}	$\mu_{\Delta N}$	h_x	h_y	t	n_x	n_y	*Cost*
1	$1.145\text{x}10^7$	$2.95\text{x}10^6$	403.2	308.7	14	6	16	54419

The fatigue constraint is now more important, replacing the local bucking constraint obtained in the previous reliability-based solution. There is almost no change in the cost, the solution now being thicker and with larger stiffeners in the y direction.
Fatigue is usually associated with a Weibull-type II probability distribution and it is usually more demanding in terms of design than the normal distribution. If the probability distribution functions of the random variables are not Gaussian, the Rosenblatt transformation may be used. It consists of finding for each random value an "equivalent" Gaussian distribution function. An increased coefficient of variation for load variation of of 0.275 was specified in problem 1. The following result was obtained:

	μ_{Nxstat}	$\mu_{\Delta N}$	h_x	h_y	t	n_x	n_y	*Cost*
1	$1.145\text{x}10^7$	$2.95\text{x}10^6$	403.2	257.2	13	6	20	55072

This reliability-based solution is now 1% more costly. However, this is obtained by increasing the number of stiffeners in the y direction.

References

Farkas,J. & Jármai,K.(2007): Economic orthogonally welded stiffening of a uniaxially compressed steel late. *Welding in the World* **51** No.3-4, 80-84.

Det Norske Veritas (1995): *Buckling strength analysis*. Classification Notes No.30.1. Høvik, Norway.

Eurocode 3 (2002), Part 1-9. *Fatigue strength of steel structures*. Brussels.

Farkas,J. & Jármai,K.(2003): *Economic design of metal structures*. Rotterdam, Millpress.

Farkas,J. & Jármai,K. (1997): *Analysis and optimum design of metal structures*. Rotterdam, Balkema. Section 9.1.4.4.

Farkas,J. & Jármai,K. (1996) Fatigue constraints in the optimum design of welded structures. *Int.Conf. Fatigue of Welded Components and Structures*. Senlis (France). Eds. Lieurade,H.P. and Rabbe,P. Les Editions de Physique, Les Ulis, France, 49-56.

Hasofer, A.M. & Lind, N.C. (1974): Exact and invariant second moment code format, *J.Eng. Mech Div.* **100** No.1, 111-121.

Profil Arbed Sales program (2001), *Structural shapes*. Arcelor Long Commercial.

Farkas,J.(2202) Thickness design of axially compressed unstiffened cylindrical shells with circumferential welds. *Welding in the World*, **46** No.11/12. 26-29.

European Convention of Constructional Steelwork (ECCS) (1988), *Recommendations for Construction. Buckling of steel shells*. No.56. Brussels.

Ditlevsen, O.(1979): Narrow reliability bounds for structural systems, *J.Struct. Mech*, **7** No. 4, 453-472.

Simões, L.M.C. & Templeman, A.B.(1989): Entropy-based synthesis of pretensioned cable net structures, *Engineering Optimization*, **15**, 121-140

Rosenblatt, M (1952) Remarks on a Multivariate Transformation, *Annals of Math. Stat.*, **23** 470-472.

Simões,L.M.C., Farkas, J & Jarmai, K. (2006), Reliability-based optimum design of a welded stringer-stiffened steel cylindrical shell subject to axial compression and bending, *Struct. Multidisciplinary Optim*, Springer Verlag **31** No. 2, 147-155.

Farkas,J., Jármai,K. & Kozuh, Z. (2007) Cost minimization of an orthogonally stiffened welded steel plate subject to static and fatigue load, *Welding in the World*, **51**, special issue, pp. 357-366.

1.5 Shape and Composite Layup Design of Multilayer Composite Structures Using the Snyman-Fatti Algorithm

Daniel N. Wilke[†], **Jan A. Snyman**[†] **and Schalk Kok**[†]

[†] Department of Mechanical and Aeronautical Engineering,
University of Pretoria, Pretoria, 0002, South Africa
nico.wilke@up.ac.za, jan.snyman@up.ac.za and schalk.kok@up.ac.za

Abstract

We consider the simultaneous optimal shape and composite layup design of two structures. The cost function is multimodal since we consider the composite layup design problem. In addition the cost function is step discontinuous resulting from the remeshing strategy used to discretize the various geometrical domains. The global search gradient based Snyman-Fatti algorithm is used to optimize the resulting multimodal and step discontinuous cost functions.

Keywords: *shape optimization, composite layup, step discontinuity, global optimization.*

1 Introduction

In this study we consider the optimal shape design of multilayer composite structures using the Snyman-Fatti (SF) gradient based optimization algorithm Snyman & Fatti (1987). Instead of conducting multilevel design optimization where the shape design is conducted independent from the laminate design Wu & Burgueno (2006), we consider the simultaneous shape and laminate design problem. The advantage of the simultaneous approach is that the coupling between the shape design and laminate design comes directly into play during the optimization process.

It is well known that the optimization of multilayer laminate design renders optimization problems multimodal. The tendency is to employ computationally inefficient stochastic or evolutionary procedures Conceicao Antonio (2006), as few gradient based optimization algorithms have the ability to escape local regions of attractions. Here we use the global search gradient based SF algorithm to optimize the shape and laminate design problem. The advantages of the SF algorithm are its inherent ability to seek low local minima and its computational efficiency, which is augmented through the use of analytical sensitivities.

In addition to the local minima resulting from the laminate design, there may also be multiple local minima present in the shape design problem itself. The finite element method (FEM) is used to solve for structural equilibrium for the various geometrical domains. To do so spatial discretizations (meshes) are required. Instead of using conventional mesh adaption (moving) strategies, we use a remeshing strategy Wilke et al. (2006). The remeshing introduces step discontinuities in the displacement-based cost function due to changes in the nodal connectivities, and as a result of varying the number of elements between the various discretized geometrical domains. These step discontinuities may also result in local minima Wilke et al. (2006).

The advantages of remeshing strategies are that mesh distortion is avoided, especially for starting points far from the optimum. In addition, there is no need for complicated mesh adaption strategies which in itself may complicate sensitivity computations.

The two structural problems we consider are firstly, the design of a composite Michell-like structure and secondly, the design of a composite spanner. Both structures are considered under fixed loading conditions. We consider laminates of 1, 3 and 5 layers for each of the structures. We only consider layers of constant and equal thickness. In addition all structures considered have a fixed laminate thickness, therefore the more layers considered in a laminate the thinner the individual layers become.

2 Formal optimization problem

The unconstrained minimization problem is of the following form: find $\boldsymbol{x}^*$ with associated function value $f(\boldsymbol{x}^*)$ of an everywhere differentiable real-valued function $f : \mathcal{X} \subseteq \mathbb{R}^n \rightarrow \mathbb{R}$, such that

$$f^* = f(\boldsymbol{x}^*) \ \leq \ f(\boldsymbol{x}), \ \forall \, \boldsymbol{x} \in \mathcal{X}, \tag{1}$$

with $\mathcal{X}$ the convex set of all possible solutions.

3 Snyman-Fatti algorithm

The Snyman-Fatti algorithm is basically a multi-start technique in which several starting points are sampled in the search domain $\mathcal{X}$. A local search procedure is then applied at each starting point, after which the lowest minimum function value found after a finite number of searches is taken as an estimate of f^*.

3.1 *Local convergence of search trajectories*

The SF algorithm explores the variable space via local searches using trajectories derived from the differential equation:

$$\ddot{\boldsymbol{x}} = -\nabla f(\boldsymbol{x}(t)) \tag{2}$$

with $\nabla f(\boldsymbol{x}(t))$ the gradient vector of $f(\boldsymbol{x}(t))$, and the position vector $\boldsymbol{x}$ depending on time t. The differential equation given in (2) describes the motion of a *unit* mass particle in a conservative force field, where $f(\boldsymbol{x}(t))$ represents the potential energy of the particle at time t.

Integrating (2) from time 0 to t, for initial conditions: position $\boldsymbol{x}^0 = \boldsymbol{x}(0)$ and velocity $\boldsymbol{v}^0 = \boldsymbol{v}(0) = \dot{\boldsymbol{x}}(0) = \boldsymbol{0}$, implies the energy conservation relationship:

$$\frac{1}{2}\|\boldsymbol{v}(t)\|^2 + f(\boldsymbol{x}(t)) = \frac{1}{2}\|\boldsymbol{v}^0\|^2 + f(\boldsymbol{x}^0) = f(\boldsymbol{x}^0). \tag{3}$$

The first term containing the velocity $\boldsymbol{v}$ in (3) represents the kinetic energy of the particle, whereas the second term containing the position $\boldsymbol{x}$ represents its potential energy at any instant t. Clearly the particle will start moving in the direction of steepest descent and its kinetic energy will increase, consequently decreasing f as long as

$$-\nabla f \cdot \boldsymbol{v} > 0 \tag{4}$$

with $\cdot$ denoting the scalar product. If descent is violated along the generated path the magnitude of the velocity $\boldsymbol{v}$ will decrease until descent is re-established, implying a change in direction towards a local minimizer. In cases of multiple local minimizers the aim is to find the global minimizer. In this case a realistic global strategy is to monitor the trajectory and record the point $\boldsymbol{x}^m$, with its corresponding velocity $\boldsymbol{v}^m = \dot{\boldsymbol{x}}^m$ and function value f^m, at which the minimum along the path occurs, letting the particle continue uninterrupted along its path with conserved energy. This is done in the hope that it may surmount a ridge of height $f^r, f^m < f^r < f^0 = f(\boldsymbol{x}^0)$, continuing further along a path that may lead to an even lower value of f beyond the ridge. On the other hand it is necessary to terminate the trajectory before it retraces itself in indefinite periodic or ergodic (space-filling) motion.

The termination condition employed in the SF algorithm: a trajectory is stopped once it reaches a point with a function value f^r close to its starting value f^s while still moving uphill, i.e. $-\nabla f(\boldsymbol{x}(t))\cdot\boldsymbol{v} < 0$. The starting point for the first inner trajectory is set at $\boldsymbol{x}^s := \boldsymbol{x}^0$ (random sample point). Thereafter, for the subsequent inner trajectory, the best point $\boldsymbol{x}^b := \{\boldsymbol{x}^m : \min f(\boldsymbol{x}^m)$ over all previous inner trajectories$\}$, and a new inner trajectory is started from a new inner starting point $\boldsymbol{x}^s := \frac{1}{2}(\boldsymbol{x}^s + \boldsymbol{x}^b)$ with initial velocity $\frac{1}{2}\boldsymbol{v}^m$ and associated starting function value $f^s = f(\boldsymbol{x}^s)$. This generation of successive inner trajectories is continued until $\boldsymbol{x}^b$ converges, or the gradient vector $\nabla f(\boldsymbol{x}^b)$ is effectively zero.

3.2 *Numerical considerations*

Of course, the above strategy assumes that the trajectory obtained from the solution of (2) is exactly known at all time instances. In practice this is not possible, and the generation of the trajectories is done numerically by means of the leap-frog scheme Snyman (1982): Given initial position $\boldsymbol{x}^{\{0\}}$ and initial velocity $\boldsymbol{v}^{\{0\}} = \dot{\boldsymbol{x}}^{\{0\}}$ and a suitably chosen time step Δt, compute for $k = 0, 1, 2, \ldots$

$$\begin{aligned} \boldsymbol{x}^{\{k+1\}} &:= \boldsymbol{x}^{\{k\}} + \boldsymbol{v}^{\{k\}}\Delta t \\ \boldsymbol{v}^{\{k+1\}} &:= \boldsymbol{v}^{\{k\}} - \nabla f(\boldsymbol{x}^{\{k+1\}})\Delta t. \end{aligned} \tag{5}$$

The initial velocity over the first ($k = 0$) step (or on a restart) is taken as $\boldsymbol{v}^{\{k\}} := -\frac{1}{2}\nabla f(\boldsymbol{x}^{\{k\}})\Delta t$.

The numerical leap-frog integration scheme computes only approximately energy conserving trajectories. Additional tests are therefore introduced in the actual implementation to ensure termination of the inner trajectories, and convergence of $\boldsymbol{x}^b$ to a local minimum for each sample starting point $\boldsymbol{x}^0$. In particular an inner trajectory is terminated if the particle moves uphill and $f(\boldsymbol{x}^{\{k\}}) \approx f(\boldsymbol{x}^s)$. Further the sequence of inner trajectories is terminated if $\|\nabla f(\boldsymbol{x}^{\{k+1\}})\| < \epsilon_g$ or $\|\boldsymbol{x}^{\{k+1\}} - \boldsymbol{x}^{\{k\}}\| < \epsilon_x$. At this termination point the current minimizer $\boldsymbol{x}^{\{k+1\}}$ and associated function value $f^{\{k+1\}}$ are passed to a global evaluation procedure, which determines the probability of $f^{\{k+1\}}$ being the global minimum (see Snyman & Fatti (1987) for details). In the actual implementation of the SF algorithm a target probability q^* must be specified, typically $q^* := 0.99$. If the target probability is not met, a new random sample point is selected and a new sequence of inner trajectories is initiated.

4 Problem formulation

The problem under consideration is the general inequality constrained minimisation problem: Given a cost function $\mathcal{F}(\boldsymbol{x})$, find the minimum $\mathcal{F}^*$ such that

$$\mathcal{F}^* = \mathcal{F}(\boldsymbol{x}^*) = \min_{\boldsymbol{x} \in \mathbb{R}^n} \{\mathcal{F}(\boldsymbol{x}) : \boldsymbol{g}(\boldsymbol{x}) \leq \mathbf{0}\}, \tag{6}$$

where $\boldsymbol{x} \in \mathbb{R}^n$ is a real vector with n components. The vector of inequality constraints $\{\boldsymbol{g}(\boldsymbol{x}) \leq \mathbf{0}, \boldsymbol{g} \in \mathbb{R}^n\}$ may be incorporated into the cost function using the standard quadratic penalty formulation.

In this study the design variables $\boldsymbol{x}$ are partitioned into geometric variables $\boldsymbol{x}_g$ and laminate layer angles $\boldsymbol{x}_\alpha$. Thus the cost function $\mathcal{F}(\boldsymbol{x})$ is a scalar function of the geometric domain $\Omega(\boldsymbol{x}_g)$ (which is defined by the geometric variables $\boldsymbol{x}_g$), as well as the fibre layer angles $\boldsymbol{x}_\alpha$. The constraints are only functions of the geometric variables, i.e. $g_j(\boldsymbol{x}_g) \leq 0,\ j = 1, 2, \cdots, m$. We choose to represent the geometrical domain boundary $\partial\Omega$ by a simple piecewise linear interpolation between the geometric variables $\boldsymbol{x}_g$. Here the geometric variables $\boldsymbol{x}_g$ only have vertical degrees of freedom.

In our case, the cost function $\mathcal{F}(\boldsymbol{x}) = \mathcal{F}\big(\boldsymbol{u}(\boldsymbol{x}_g, \boldsymbol{x}_\alpha)\big)$ is an explicit function of the nodal displacements $\boldsymbol{u}(\boldsymbol{x}_g, \boldsymbol{x}_\alpha)$, which are obtained by solving the approximate finite element equilibrium equations for linear elasticity, formulated as

$$\boldsymbol{K}\boldsymbol{u} = \boldsymbol{F}, \tag{7}$$

where $\boldsymbol{K}$ represents the assembled structural stiffness matrix, and $\boldsymbol{F}$ is the consistent structural load vector.

The required discretization of Ω for the finite element analyses are obtained using an unstructured remeshing strategy Wilke et al. (2006), which is based on a truss structure analogy mesh generator proposed by Persson & Strang (2004). The resulting three noded triangular (constant strain triangular) meshes are transformed to six noded triangular (linear strain triangular) meshes, by introducing mid-side nodes such that the elemental stiffness matrices can be integrated analytically Moser & Swoboda (1978).

The resulting cost function is step discontinuous due to changes in element connectivity and the variation in number of nodes, resulting from our remeshing strategy when different geometrical domains are discretized Wilke et al. (2006). These step discontinuities in turn may manifest as local minima in the cost function. The use of the global gradient based SF algorithm is therefore not only to accommodate the multimodality resulting from the composite laminate design, but also to overcome the numerically induced step discontinuities resulting from the shape design.

4.1 *Orthotropic constitutive relation*

The stiffness matrix $\boldsymbol{K}$ depends explicitly on $\boldsymbol{x}_\alpha$ through an orthotropic constitutive relation. The elemental stiffness matrices $\boldsymbol{K}_e(\boldsymbol{x}_\alpha)$ are obtained by stacking unidirectional composite layers on top of each other to construct an elemental laminate. Only symmetric fibre layups are considered to avoid out of plane effects. Each unidirectional composite layer pair i has its own fibre orientation described by x^i_α which is measured counterclockwise from X_1 in the global coordinate system $\{X_1, X_2\}$.

We consider plane stress constitutive relation $\boldsymbol{C}(E_1, E_2, \nu_{12}, \nu_{21}, G_{12})$ described in the global coordinate system $\{X_1, X_2\}$. Here, E_i, $i = 1, 2$ denotes the Young's modulus in the longitudinal X_1 and transverse X_2 directions respectively, G_{12} denotes the shear modulus which relates the change in angle between the longitudinal and transverse directions. The Poisson's ratio ν_{21} describes the resulting strain in the longitudinal direction for applied tension in the transverse direction, etc.

The constitutive relation $\boldsymbol{C}(x^i_\alpha)$ for the ith fibre pair with orientation x^i_α, is obtained via the following transformation Chandrupatla & Belegundu (1997):

$$\boldsymbol{C}(x^i_\alpha) = \boldsymbol{T}^{\mathrm{T}}(x^i_\alpha)\boldsymbol{C}\boldsymbol{T}(x^i_\alpha), \tag{8}$$

where $\boldsymbol{T}(x^i_\alpha)$ is a 3×3 orthogonal matrix that depends on x^i_α. The elemental stiffness matrices $\boldsymbol{K}_e$ are obtained by adding the contributions of all the unidirectional composite layer pairs $\boldsymbol{x}_\alpha$, where each fibre layer has a constant and equal thickness.

The orthotropic material considered is Boron-Epoxy in a tape outlay, i.e. all the fibers are aligned in a single direction for a plane stress analysis. The material properties for Boron-Epoxy used in this study are $\mathrm{E}_1 = 228$ GPa, $\mathrm{E}_2 = 145$ GPa, $G_{12} = 48$ GPa, and $\nu_{12} = 0.23$, with ν_{21} following from the symmetry relation $\mathrm{E}_1\nu_{21} = \mathrm{E}_2\nu_{12}$.

4.2 *Sensitivities*

We only consider analytical gradients in this study. The analytical gradients are obtained by direct differentiation of u_F w.r.t. the geometric variables $\boldsymbol{x}_g$ and the fibre layup angles $\boldsymbol{x}_\alpha$. A detailed analysis of the analytical sensitivities w.r.t. to the geometric variables $\boldsymbol{x}_g$ for our unstructured remeshing strategy is given in Wilke et al. (2006). The sensitivities obtained for the fibre layup angles $\boldsymbol{x}_\alpha$ are simply obtained by direct differentiation of (8).

5 Example problem 1: Composite Michell structure

We now consider the Michell structure with the geometry, load F, appropriate boundary conditions and displacement of interest u_F, as depicted in Figure 1(1). The structure has a predefined length of 15 units. Our objective is to minimise the sum of the vertical displacement (magnified by 10^8) at the point of load application u_{F}, and the volume fraction $\frac{V}{V_0}$ with $V_0 = 150$ for a unit thickness structure. The magnitude of F is 10 N and the ideal element length of the mesh is $h_0 = 1.75$, where the ideal element length indicates the average element edge length. We use 16 geometric variables $\boldsymbol{x}_g$ to describe the linearly interpolated geometry. To avoid non-physical designs we include quadratic penalties for upper and lower geometric variables that are within 1 unit distance from each other, as well as for the two upper geometric variables on the edges when they are within 1 unit from the base. A penalty parameter of 10 is used for the quadratic penalty.

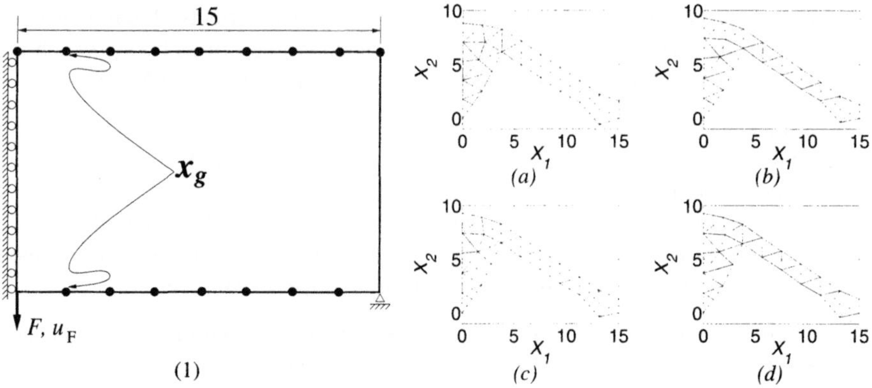

Figure 1: (1) Initial structure and definition of half the Michell-like structure. Optimal design obtained for the Michell structure with the SF algorithm for (a) $N_\alpha = 0$, (b) $N_\alpha = 1$, (c) $N_\alpha = 2$ and (d) $N_\alpha = 3$.

We consider 1, 3 and 5 layer composite laminates for the Michell structure design which, due to the laminate symmetry, adds respectively only $N_\alpha = 1, 2, 3$ fibre angle variables to the existing 16 geometric variables. The results obtained with the SF algorithm for the 1, 3, and 5 layer laminate Michell structures are presented in Table 1. Also included are the results obtained with a fibre layer angle fixed at 0° $(N_\alpha = 0)$. Listed for each case in Table 1 are the optimal function value $\mathcal{F}^*$, the norm of the gradient $\|\nabla\mathcal{F}^*\|$ at the optimum, the number of function evaluations N_f, the number of random starting points N_R, and the number of global minima N_S found. Independent of the number of layers considered, all optimal angles were found to be $\approx 29°$.

Table 1: Results obtained for the 1, 3, and 5 composite layer Michell structure using the Snyman-Fatti algorithm.

N_α	$\mathcal{F}^*$	$\|\nabla\mathcal{F}^*\|$	N_f	N_R	N_S	x_α^i (degrees)		
						x_α^1	x_α^2	x_α^3
0	5.727E-01	7.45E-04	768	6	5	-	-	-
1	5.183E-01	6.43E-04	1237	6	5	-29.20	-	-
2	5.183E-01	8.88E-04	1967	12	5	-29.26	-29.31	-
3	5.183E-01	9.98E-04	858	7	5	-29.18	-29.29	-29.29

The optimal shape designs obtained are depicted in Figures 1 (a)-(d) for respectively $N_\alpha = 0, 1, 2, 3$. Significant differences in final designs are obtained for $N_\alpha = 1, 2, 3$ in comparison to $N_\alpha = 0$. The optimal shape for a fibre angle of $\approx 29°$ is significantly different from the optimal shape for a prescribed constant angle of 0°. Since the optimal angle was found to be $\approx 29°$ independent of the number of layers, the final designs obtained for $N_\alpha = 1, 2, 3$ are similar. Note that the slope of the diagonal leg of the optimal structure is approximately equal to the optimum layup angle of 29°. This explains why the optimum angles were the same for all three layup cases, with the fibre aligning along the diagonal leg of the optimal geometric structure to minimize

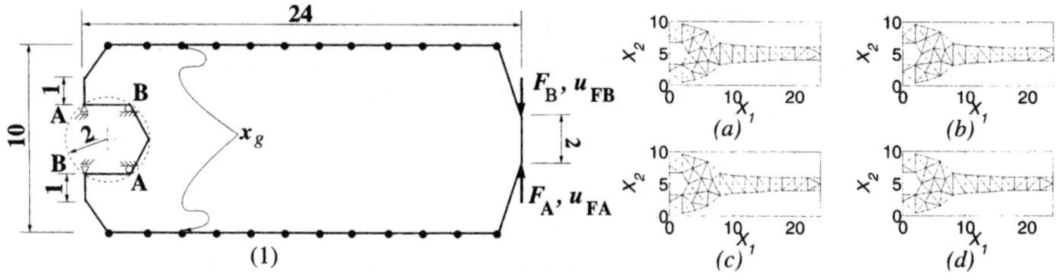

Figure 2: (1) Initial structure and loads for the full spanner problem. Optimal design obtained for the spanner design with the SF algorithm for (a) $N_\alpha = 0$, (b) $N_\alpha = 1$, (c) $N_\alpha = 2$ and (d) $N_\alpha = 3$.

the compliance of the structure.

6 Example problem 2: Composite Spanner design

We now consider the spanner design with the geometry, loads F_A and F_B, corresponding boundary conditions A and B, and displacements of interest u_{FA} and u_{FB}, as depicted in Figure 2(1). The structure has a predefined length of 24 units. Every cost function evaluation involves two finite element analyses, the first finite element analysis with load F_A and boundary condition A to obtain u_{FA}, and the second finite element analysis with load F_B and boundary condition B to obtain u_{FB}. Here, with the magnitudes of F_A and F_B equal, the solution to the problem should result in a symmetric geometry. However, symmetry is *not* enforced; deviations from symmetry are used to qualitatively evaluate the obtained designs.

Table 2: Results obtained for the 1, 3, and 5 composite layer spanner design using the Snyman-Fatti algorithm.

N_α	$\mathcal{F}^*$	$\|\nabla\mathcal{F}^*\|$	N_f	N_R	N_S	x^i_α (degrees)		
						x^1_α	x^2_α	x^3_α
0	1.372E+00	3.16E-03	6709	9	5	-	-	-
1	1.372E+00	3.99E-03	5702	8	5	-2.02	-	-
2	1.365E+00	1.63E-03	3965	8	5	-31.83	-171.66	-
3	1.365E+00	3.45E-03	3573	6	5	-96.34	-20.00	-162.16

Our objective is to minimize the sum of the displacements u_{FA} and u_{FB} (magnified by 10^6), and the volume fraction $\frac{V}{V_0}$ with $V_0 = 100$ for a unit thickness structure. The magnitude of F_A and F_B is 10 N and the ideal element length of the mesh is $h_0 = 2$. We use 22 geometric variables $\boldsymbol{x}_g$ to describe the linearly interpolated geometry. To avoid poor ergonomic designs we include quadratic penalties for the upper and lower geometric variables. The upper geometric variables have to be greater than 6 units and the lower geometric variables have to be less than 4 units. A penalty parameter of 10 is used for the quadratic penalty.

We consider 1, 3 and 5 layer composite laminates for the spanner design which, due to the laminate symmetry, adds respectively only $N_\alpha = 1, 2, 3$ fibre angles to the existing

22 geometric variables. The results obtained with the SF algorithm for the 1, 3 and 5 layer laminate spanner designs are presented in Table 2. Also included are the results obtained with a fibre layer angle fixed at 0° $N_\alpha = 0$. The results are listed in Table 2 with the symbols having the same meaning as given before.

The optimal shape designs obtained are depicted in Figures 2 (a)-(d) for respectively $N_\alpha = 0, 1, 2, 3$. Differences in final optimum designs obtained for $N_\alpha = 0, 1, 2, 3$ are evident from inspection of Figures 2 (a)-(d). The optimum designs for $N_\alpha = 0$ and $N_\alpha = 1$ are effectively identical (with $x^1_\alpha = -2° \approx 0°$), with the fibres of the layup aligning along the longitudinal axis of the spanner for optimal strength. The fibre angles for $N_\alpha = 3$ supply stiffness to resist the opening of the spanner head around the bolt with $x^1_\alpha \approx 90°$. The remaining two fibre angles supply stiffness to resist both the opening of the spanner head as well as bending of the structure for the two load cases with $x^2_\alpha \approx -20°$ and $x^3_\alpha \approx 20°$.

7 Conclusion

We showed that the Snyman-Fatti global optimization algorithm can be successfully employed in the optimal multimodal shape design of multiple composite layers, even when step discontinuities are present in the cost function. Significantly the optimal geometrical shapes obtained when the fibre angles were optimized differ from the geometrical shapes obtained for a constant fibre angle of 0°. This highlights the importance of conducting simultaneous shape and laminate design for efficient optimization. Conventionally, multilevel optimization strategies decouples the shape design problem from the laminate design problem, which may prolong convergence to the overall optimum design.

References

Chandrupatla, T. R. & Belegundu, A. D. (1997). *Introduction to finite elements in engineering*. Prentice-Hall, Upper Saddle River, NJ, USA, second edition

Conceicao Antonio, C. (2006). A hierarchical genetic algorithm with age structure for multimodal optimal design of hybrid composites. *Structural and Multidisciplinary Optimization*, **31** No. 4, 280–294

Moser, K. & Swoboda, G. (1978). Explicit stiffness matrix of the linearly varying strain triangular element. *Computers & Structures*, **8** No. 2, 311–314

Persson, P.-O. & Strang, G. (2004). A simple mesh generator in matlab. *SIAM Review*, **46** No. 2, 329–345

Snyman, J. A. (1982). A new and dynamic method for unconstrained minimization. *Applied Mathematical Modelling*, **6** No. 6, 449–462

Snyman, J. A. & Fatti, L. P. (1987). A multi-start global minimization algorithm with dynamic search trajectories. *J Optim Theory Appl*, **54** No. 1, 121–141

Wilke, D. N., Kok, S. & Groenwold, A. A. (2006). A quadratically convergent unstructured remeshing strategy for shape optimization. *International Journal for Numerical Methods in Engineering*, **65** No. 1, 1–17

Wu, J. & Burgueno, R. (2006). An integrated approach to shape and laminate stacking sequence optimization of free-form FRP shells. *Computer Methods in Applied Mechanics and Engineering*, **195** No. 33-36, 4106–4123

1.6 Stress Analysis in the Steel Tank Shell at its Connection with the Product Pipeline

Jerzy Ziółko [1], Ewa Supernak [2], Tomasz Mikulski [3]
[1, 2, 3] Gdańsk University of Technology, Poland,
e-mail: [1] jziolko@pg.gda,pl, [2] esuper@pg.gda.pl, [3] tomi@pg.gda.pl

Abstract

There are three analyzed methods referring to the strengthening the shell of the cylindrical vertical steel tank at the opening zone, where the product pipeline is to be installed. It is advisable to connect the pipeline as close to the bottom of the tank as possible since in that case, the "dead capacity", i.e. the zone from which it is not possible to use the stored fuel, is the smallest. However, the pipeline axis cannot be freely moved to the bottom of the tank as due to process aspects, a proper distance between the welds in the shell and bottom connection zone shall be kept.

The stress in the shell and bottom of the tank was analyzed taking into account three options of the shell strengthening. The solution providing the smallest stress concentration was advised.

Keywords: *steel tanks, pipeline-tank connection, tank shell stresses*

1 Introduction

The paper analyzes a structural element of a vertical cylindrical steel tank with flat bottom. The tanks of this type have significant representation among industrially-applied tanks, which results from the fact that single-curvature shells are the cheapest both in terms of their manufacture and assembly (Ziółko 1986). Product pipelines are connected to the tank in the first sheet ring of the shell, i.e. in the zone of boundary disturbances produced by the limited strain capacity of the cylindrical shell at its connection with the flat bottom. High stresses in the first ring of shell sheets require compensation of the loss of cross-sectional area caused by the opening cut-out for the product pipeline. A traditional solution for this structural element has been to strengthen the shell by welding a ring around the pipeline, the cross-section of which was equivalent to the section of the opening made (Fig. 1), (API Standard 650) and (API Standard 620).

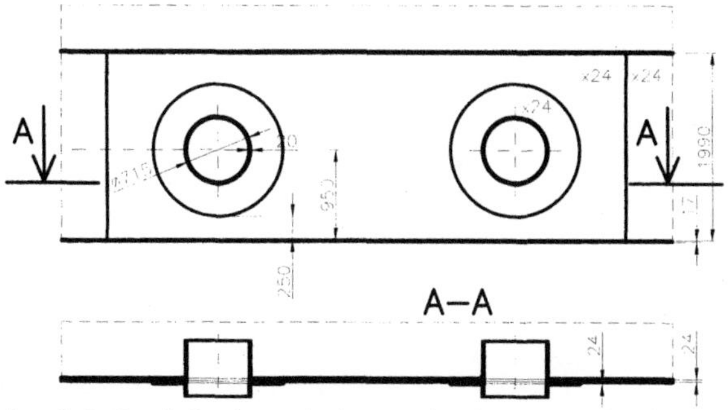

Figure 1. Traditional shell-reinforcing solutions – circular cover plate is welded around the pipeline.

High capacity tanks (50 000 m^3 and more) are most frequently connected to 700 mm pipelines. This enables rapid filling and emptying of the tanks. The investors more than ever require the designer to minimize the "dead capacity" of the tank – the space from which stored liquid fuel cannot be recovered during operation. This space extends from the tank bottom to the upper generating line of the pipeline since for safety reasons, the pipeline is not allowed to draw in the air. The pipeline introduced into the shell at small distance from the bottom would result in overlapping of post-welding effects in the zone between the bottom and the shell-reinforcing cover plate (Fig. 1). This paper analyzes two other design solutions enabling the reduction of the "dead capacity" of the tank (Fig. 2a and 2b), (*Правила* ПБ 03-381-00, 2001).

Figures 2a and 2b. Two alternative solutions to reinforce the shell in the opening area

2 Analyzed design solutions

The structural element mentioned in the introduction has two practical solutions:

Option I – in the bottom ring of the shell one double-thick sheet is used in the connection zone between the pipeline and the tank (Fig. 3).

Option II - cover plate reinforcing the shell is applied around the pipeline. The bottom edges of the plates are welded to the tank bottom perimeter sheets (Fig. 4).

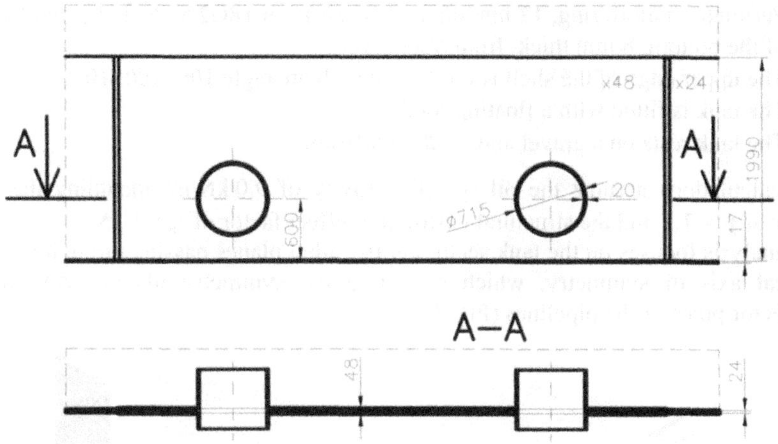

Figure 3. Option I – in the bottom shell ring one double-thick sheet is applied

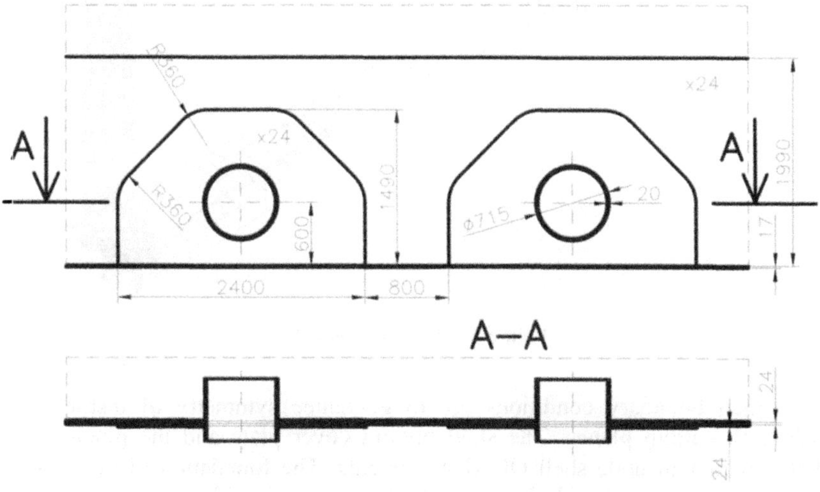

Figure 4. Option II – shell-reinforcing cover plates are applied to the pipeline

Both options enable keeping the technologically necessary distance between the welds joining the shell with the bottom and the pipeline with the shell. The distance

should be at least 250 mm. The stresses in the tank shell and bottom were identified by means of the finite element method – using the MSC NASTRAN software.
The numerical analysis presented in the paper concerns a steel tank with nominal capacity of V=50000 m^3, designed for storage of crude oil. Below, basic design data of the tank has been quoted:

- Internal shell diameter d_w = 64 840 mm, height h = 18 000 mm,
- Starting from the bottom, the tank shell is built using rings of the following thickness: 24; 22; 19; 17; 16; 16; 13; 10; 10 mm and height: 1990 mm each; three bottom rings are made from 18G2A steel, the remaining ones – St3V,
- Perimeter bottom ring, 17 mm thick, is made from 18G2A steel, the middle part of the bottom, 8 mm thick, from St3V steel,
- The upper edge of the shell is reinforced with an angle 100×100×10 mm,
- The tank is fitted with a floating roof,
- The tank rests on a gravel and sand foundation.

The calculations assume the oil specific gravity of 9.0 kN/m^3 including the load factor of $\gamma_f = 1.1$ and the structural destruction effect factor of $\gamma_n = 1.15$.
The analysis focuses on the tank sector cut by radial planes passing through the tank vertical axis of symmetry, which constitute the symmetry planes between the connector pipes of the pipelines (Fig. 5).

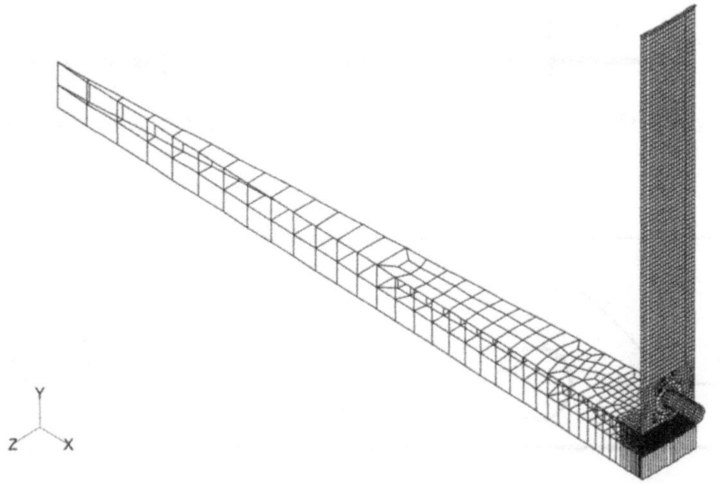

Figure 5. Numerical model

The assumed boundary conditions are to guarantee symmetry of distortions as regards both cutting planes. The shell bottom cover plate and the pipeline were modelled with four-node shell QUAD4 elements. The foundation of the tank was assumed to be elastic, Winkler's type, with bottom detachable from the foundation (unidirectional foundation). The foundation was assumed to be gravel and sand with constant elasticity of k =150 MN/m^3.

Below there are the results of the analysis:

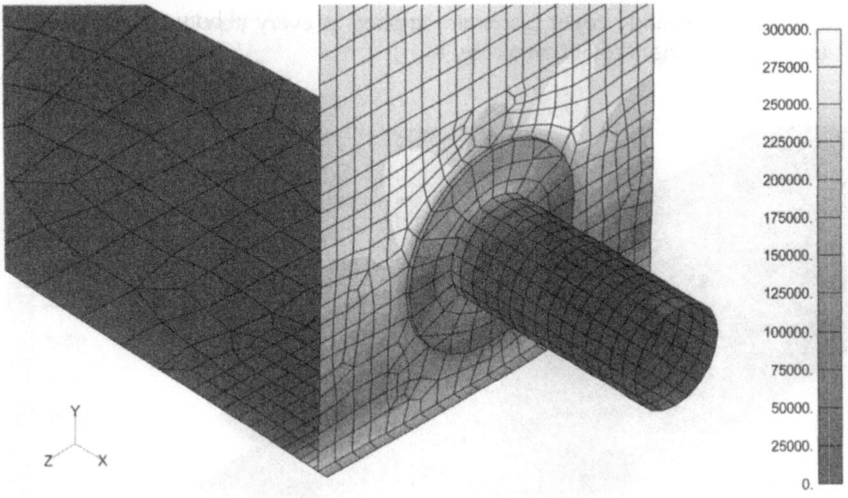

External side of the tank shell

Internal side of the tank shell

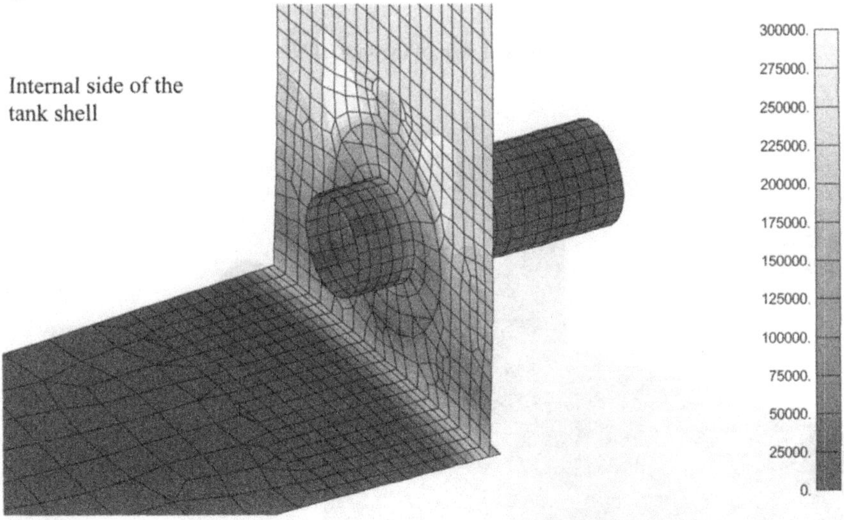

Figure 6. Traditional solution – reduced stress as according to the Huber-Mises hypothesis σ_R [kPa]

- Traditional case of a pipeline connected to the tank's shell, where the latter is reinforced with a circular cover plate – the pipeline's axis at 950 mm (Fig. 6)
- option I – double-thick sheet is applied in the bottom ring of the shell at the connection with the pipeline – the pipeline's axis is at the height of 600 mm (Fig. 7)

- option II – separate cover plates are applied at every pipeline – the pipeline's axis is at the height of 600 mm (Fig. 8)

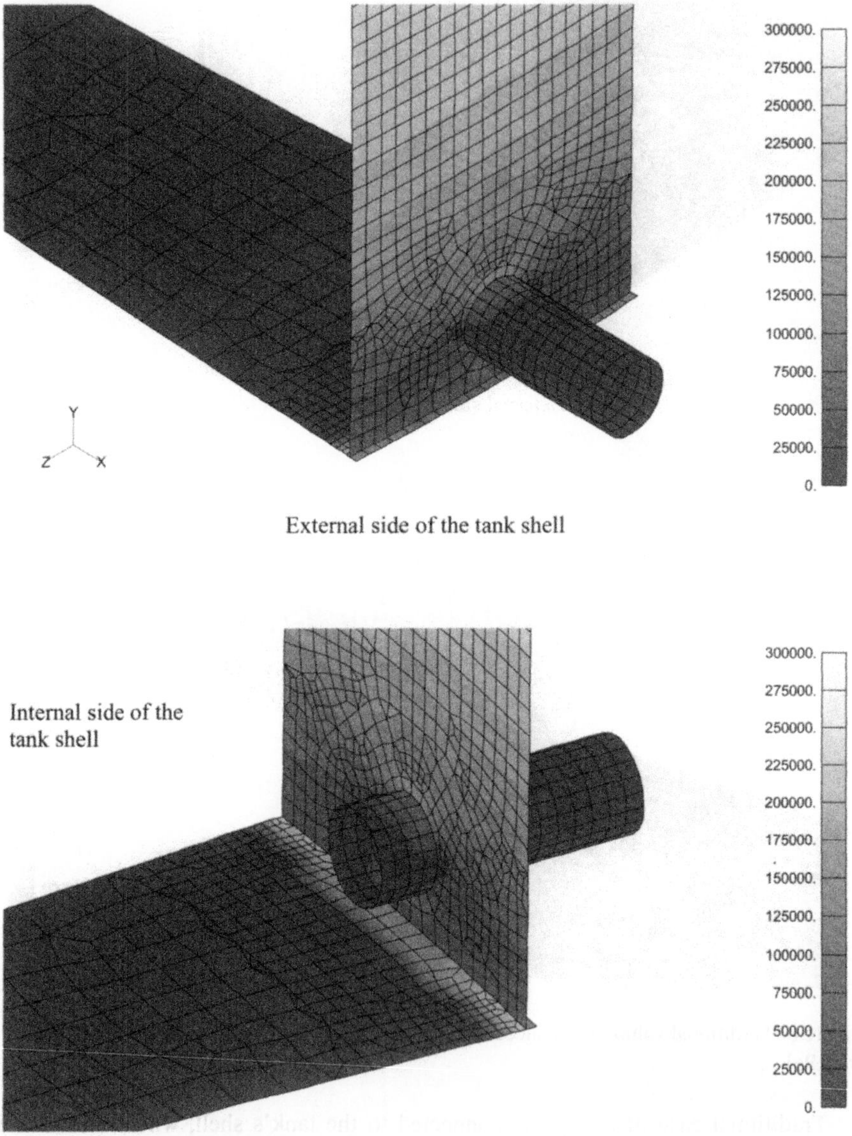

Figure 7. Option I – reduced stress as according to the Huber-Mises hypothesis σ_R [kPa]

3 Conclusions

3.1 Reduction of the "dead capacity" of liquid fuel tank is of significant importance in operational terms. In the case of the analyzed tank, of nominal capacity V = 50 000 m^3, the traditional solution of the shell-pipeline connection (pipeline axis at the height of 950 mm) results in "dead capacity" of ca. V_m = 4 300 m^3, which is ≈ 8.6 %.

3.2 "Dead capacity" of the tank can be reduced by lowering the axis of the product pipelines.

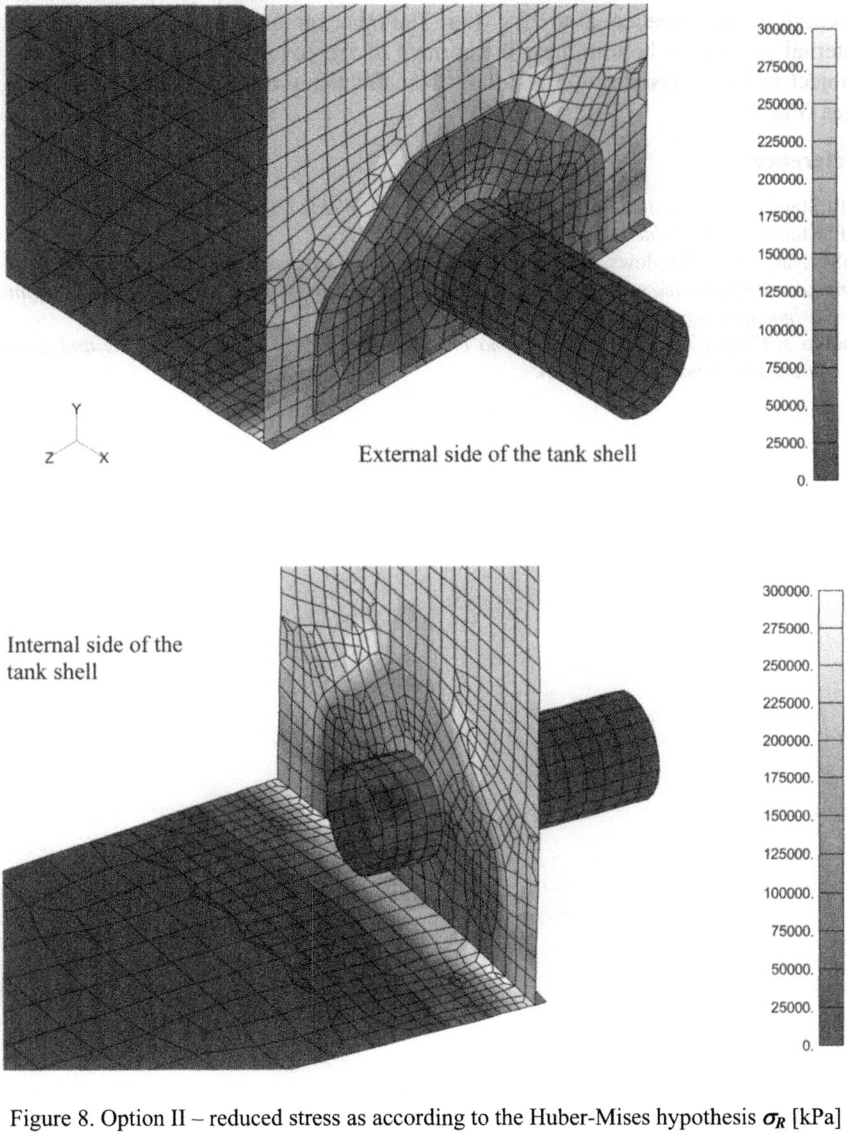

Figure 8. Option II – reduced stress as according to the Huber-Mises hypothesis σ_R [kPa]

Lowering the axis of the product pipelines by 350 mm enables the reduction of "dead capacity" of the analysed tank down to ca. $V_m = 3\ 100\ m^3$, i.e. by $1200\ m^3$, which is ca. 2.3 % of the nominal capacity. However, when lowering the pipeline axis, the technologically necessary distance between the welds in the shell-bottom connection zone must be maintained.

3.3 The static analysis of the two options of design solutions enabling the lowering of the axis of the pipeline introduced into the tank shell, has shown that option I (Fig. 3) is better in that since it leaves no point in the shell between the pipeline and the bottom of the tank where stress would exceed 200 MPa (Fig. 7) with analytical strength of used steel equal to 295 MPa. The other option (Fig. 4) is worse as in the internal part of the bottom, near the corners of the cover plate, the bottom sheet is subject to local plasticization resulting from fluctuation in the shell rigidity in this area (Fig. 8).

References

API Standard 620 *Design and Construction of Large, Welded, Low-Pressure Storage Tanks.*

API Standard 650 *Welded Steel Tanks for Oil Storage.*

MSC Nastran for Windows, *Version 2001, MSC Software Corporation,* Los Angeles US.

Правила устройства вертикальных цилиндрических стальных резервуаров для нефти и нефтопродуктов ПБ 03-381-00, 2001.

Ziółko J. (1986) *Zbiorniki metalowe na ciecze i gazy (Metal tanks for liquids and gases),* Warszawa, Arkady, (in Polish).

Section 2

Structural optimization I

2.1 Combined Shape and Size Optimization of Steel Bridges

Anikó Csébfalvi
University of Pécs, H-7624 Pécs, Hungary,
csebfalv@witch.pmmf.hu

Abstract

In this paper, a continuous hybrid meta-heuristic method (ANGEL) is presented for the minimal weight optimization problem of steel bridges. A combined shaping-sizing optimization problem is considered where the cross-sectional areas and the positions of the nodal points are simultaneously optimized. ANGEL combines ant colony optimization (ACO), genetic algorithm (GA), and local search strategy (LS). ACO and GA search alternately and cooperatively in the solution space. The discrete solutions are computed by an optimal rounding algorithm according to the given catalogue values. A simple but efficient local search procedure is proposed as well, which is able to improve the quality of the discrete solutions. Experimental results are presented for shaping-sizing optimization of steel bridges.

Keywords: *simultaneous optimization, steel bridges, ANGEL meta-heuristic method*

1 Introduction

The shape and sizing optimization of steel bridges where the design variables are the cross-sections and nodal coordinates is very complex problem. The different nature of the design variables usually causes unacceptable numerical instability for a single optimization algorithm. To solve this problem it is necessary to adequately combine different optimization algorithms, depending on the different types of design variables of the problem. The separate treatment of the two kinds of design variables is not an obligation but a simplification.

Several studies are discussed on the field of the simultaneous optimization from a general point of view. Achtziger (2007) presented an accurate selection as a brief review of the problem formulations and its solution methods. We have to mention some of the early works on simultaneous optimization of geometry and topology. Dobbs and Felton (1969) developed first time an alternating process. At the same time Pedersen (1969) presented a 2D truss optimization problem with stress and local buckling constraints under single load condition. The extended problem for multiple loading cases with stress and buckling constraints can be seen in Pedersen (1972) and Pedersen (1973). Vanderplaats and Moses (1972) proposed an automated design for optimum geometry of 3D truss problems. In the book of Kirsch (1981) a simultaneous optimization problem is considered based on the displacement method of finite element analysis (FEA) where Sequential Linear Programming (SLP) has been applied as solution method. Haftka et al (1990) discussed the simultaneous optimization problem as a multilevel optimization. The two-level form of the traditional nested approach is evident in problems where the structural analysis can be formulated as a minimal weight design subject to constraints on the collapse loads. More recently, Wang et al (2002) presented an evolutionary optimization method for weight minimum problem of a 3-dimensional truss structure in terms of nodal coordinates and element cross-sectional areas subject to stress, local buckling and displacement constraints in one load case. Two types of design variables with different natures are optimized separately: (1) a fully stressed design (FSD) and scaling techniques are applied to sizing variables and (2) the evolutionary node shift method is applied to shape variables. Alternating procedure is utilized to couple the two types of variables and to combine the results. The optimum solution is achieved gradually from the initial configuration design. The weakness of this method is its high computational cost in the shape optimization. Since in each loop, only a small number of nodes are shifted and the interval of the node shift is considerably small, which means that the FEA has to be applied repetitively in the evolutionary optimization process. In the work of Gil and Andreu (2001) a nodal co-ordinates optimization is driven by a conjugate- gradients strategy and the cross-section optimization is achieved inside the evaluation of the objective function with a FSD strategy.

In this paper, we present a hybrid meta-heuristic (ANGEL) method for combined sizing shaping optimization of truss bridges. ANGEL combines ant colony optimization (ACO), genetic algorithm (GA) and local search (LS) strategy. First ACO searches the solution space and generates designs to provide the initial population for GA. Next GA is executed and the pheromone set in ACO is updated when GA obtains a better solution. When GA terminates, ACO searches again by using a new pheromone set. ACO and GA search alternately and cooperatively in the solution space. Different heuristic methods are applied previously for nonlinear discrete optimization problems in Csébfalvi (1999), Csébfalvi & Csébfalvi (2006). The present method ANGEL has been used for 3D sizing problems with local and global stability constraints in Csébfalvi & Csébfalvi (2007a) and Csébfalvi & Csébfalvi (2007b). In this study, the sensitivity analysis is based on the evaluation of the eigenvalue structure of the stability matrix in terms of load parameter factor and sizing variables. Structural stability, displacements, and member buckling are monitoring continuously. Experimental results are presented for truss bridges.

2 The combined shape and sizing optimization procedure

In this chapter we present a new hybrid metaheuristic Angel algorithm for the continuous truss design problem with sizing and shaping variables. The presented algorithm is a slightly modified and simplified continuous version of the original metaheuristic algorithm developed by Tseng & Chen (2006) for the resource-constrained project scheduling problem. The ANGEL algorithm combines ant colony optimization (ACO), genetic algorithm (GA), and local search (LS) strategy. The main procedure of ANGEL follows the repetition of these two steps: (1) ACO with LS and (2) GA with LS. The initial population is a totally random design set. In other words, firstly ANGEL generates an initial population, after that, in an iterative process ACO and GE search alternately and cooperatively on the current design set. According to the systematic simplification, the algorithm consists of only three basic operators: random selection, random perturbation (ACO), and random combination (GA). In Figure 1 the pseudo-code of the continuous ANGEL algorithm is presented for continuous truss design with sizing and shaping variables. In this figure, the sets, functions, and subroutines are represented by "bold" symbols.

```
W* ← W̄
For Generation = 0 To Generations
If Generation = 0 Then
RandomPhase ← True : AntColonyPhase ← False
Else
RandomPhase ← False : AntColonyPhase ← Not AntColonyPhase
End If

For Member = 1 To PopulationSize
If RandomPhase Then
For G = 1 To MemberGroups
X(G) ← U(X^min, X^max)
Next G
For G = 1 To CoordinateGroups
Y(G) ← U(Y(G)^min, Y(G)^max)
Next Group
Else
If AntColonyPhase Then
{X, Y} ← RandomPerturbation(Generation)
Else
{X, Y}← RandomCombination(Generation)
End If
End If
{W, λ, X, Y, φ} ← LocalSearch(λ, X, Y)
If RandomPhase Then
W(Member) ← W : λ(Member) ← λ
X(Member) ← X : Y(Member) ← Y :  φ(Member) ← φ
Else
{Design^worst, φ^worst}← WorstDesignSelection
If φ > φ^worst Then
W(Design^worst) ← W : λ(Design^worst) ← λ
X(Design^worst) ← X : Y(Design^worst) ← Y : φ(Design^worst) ← φ
End If
End If
If λ = 1 And W* < W  Then W* ← W : X* ← X : Y* ← Y
Next Member
Next Generation

Exit Width { W*, X*, Y* }
```

Figure 1. The pseudo-code of the continuous ANGEL algorithm

The algorithm has two global parameters: PopulationSize and Generation, and two "tunable" parameter pairs $\{I^{min}, I^{max}\}$ and $\{R^{min}, R^{max}\}$. The progress of the iterative searching process, in the function of the generation index (Generation), is controlled by two variables $I(\text{Generation})$ and $R(\text{Generation})$. Variable

$I(\text{Generation})$, where $I^{\min} \le I(\text{Generation}) \le I^{\max}$ is the maximal number of the LS iterations. Variable $R(\text{Generation})$, where $R^{\min} \le R(\text{Generation}) \le R^{\max}$ has an effect similar to that of the pheromone evaporation rate in ACO. Our algorithm, according to its "robust" nature, is not so sensitive to the "fine tuning" of these parameters. In other words, $\{I^{\min}, I^{\max}\}$ and $\{R^{\min}, R^{\max}\}$ can be kept "frozen" in the algorithm, which results in a practically tuning-free algorithm. According to progress of the searching process, the "freedom of diversification" is decreasing $(R^{\max} \to R(\text{Generation}) \to R^{\min})$ but the "freedom of intensification" is increasing $(I^{\min} \to I(\text{Generation}) \to I^{\max})$ step by step.

In the algorithm, a design is represented by the set of $\{W, \lambda, \mathbf{X}, \mathbf{Y}, \varphi\}$, where W is the weight of the structure, λ is the maximal load intensity factor, $\boldsymbol{X}$ is the current set of the cross-sectional areas for member groups, and $\mathbf{Y}$ is the set of the shifted coordinates for coordinate groups, and φ is the current fitness function value. In the algorithm, the currently best (not necessarily optimal) feasible design is represented by the set $\{W^*, \mathbf{X}^*, \mathbf{Y}^*, \varphi^*\}$.

In the presented very simple pseudo-code, the first function (in top-down order), namely $C \leftarrow \mathbf{U}(C^{\min}, C^{\max})$, is a uniform random number generator, which generates a $C^{\min} \le C \le C^{\max}$ real random number.

The **RandomPerturbation**(Generation) and **RandomCombination(**Generation) procedures call the $\{\text{Design}, \mathbf{X}^{\text{Design}}, \mathbf{Y}^{\text{Design}}\} \leftarrow$ **RandomSelection** function, to select a "more or less good" design from the current population using the well-known discrete "inverse" method. The higher the fitness value φ, the higher the chance is that the design will be selected by the function. The essence of the discrete inverse method is shown in Figure 2. The selected design is identified by its index: $1 \le \text{Design} \le \text{PopulationSize}$ .The functions call $\mathbf{U}(C^{\min}, C^{\max})$ in the algorithm of the inverse method and use a $Y \leftarrow \text{Interpolation}(X_1, Y_1, X_2, Y_2, X)$ function for linear interpolation.

The main idea of the selection operator is the following: Any feasible solution is preferred to any infeasible solution. Between two feasible solutions, the one having a smaller weight is preferred. Between two infeasible solutions, the one having a larger load intensity factor is preferred.

Based on these criteria, fitness function φ $(0 \le \varphi \le 2)$ is defined as

$$\varphi = \begin{cases} 2 - \dfrac{W - \underline{W}}{\overline{\overline{W}} - \underline{W}} & \text{if } \lambda = 1 \\[2ex] \lambda & \text{if } \lambda < 1 \end{cases} \tag{1}$$

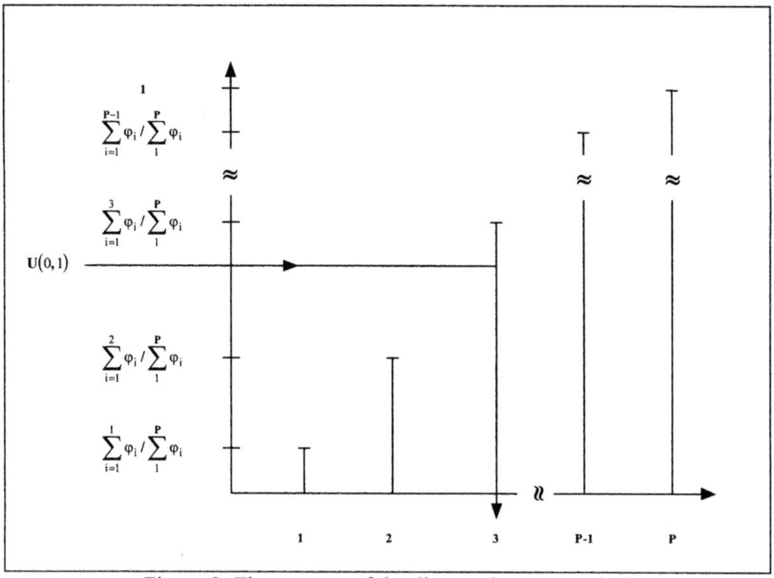

Figure 2. The essence of the discrete inverse method

The subroutine $\{\lambda, \mathbf{X}, \mathbf{Y}\} \leftarrow$ **LocalSearch(** $\lambda, \mathbf{X}, \mathbf{Y}$, Generation **)** is the central element of our algorithm. It is based on the local linearization of the feasibility constraints. The algorithm, in an iterative process, minimizes the weight increment needed to get a better (a lighter feasible or less unfeasible) discrete solution. The local search procedure calls a fast and efficient "state-of-the-art" interior point solver (BPMPD) to solve the linear programming problems.

The subroutine $\{\text{Design}^{\text{worst}}, \varphi^{\text{worst}}\} \leftarrow$ **WorstDesignSelection** selects the worst design from the current population. If the current design is better than the worst than the worst one will be replaced by the better one. The algorithm maintains the dynamically changing $\{W^*, \mathbf{X}^*, \mathbf{Y}^*\}$ set.

The $\{\mathbf{X}, \mathbf{Y}\} \leftarrow$ **RandomPerturbation** subroutine uses the continuous inverse method to generate the perturbated mean from the selected distribution. The essence of the continuous inverse method for random perturbation is shown in Figure 3.

The $\{\mathbf{X}, \mathbf{Y}\} \leftarrow$ **RandomCombination** subroutine uses the continuous inverse method to generate the child's mean from the combined distribution of the selected mother and father distributions. The combined distribution is the weighted sum of the parent's distributions. The essence of the continuous inverse method for combination is shown in Figure 4. In order to establish the value of the standard deviation in generation Generation, we calculate the average absolute distance from the selected design to other designs in the current population, and we multiply it by the parameter R. The higher the value of the parameter, the higher the variability of the searching process is.

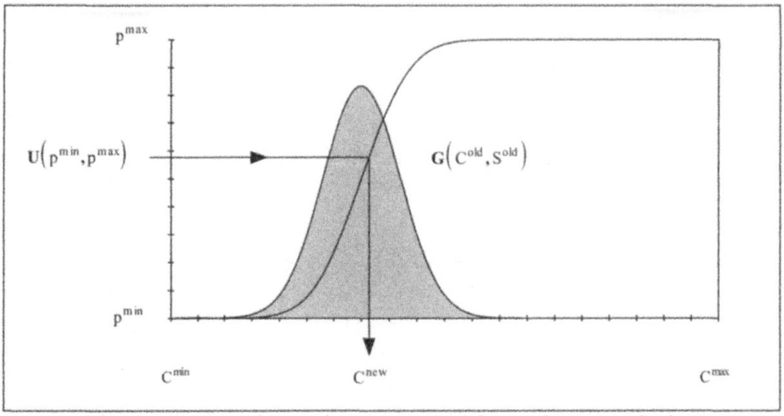

Figure 3. The continuous inverse method for random perturbation

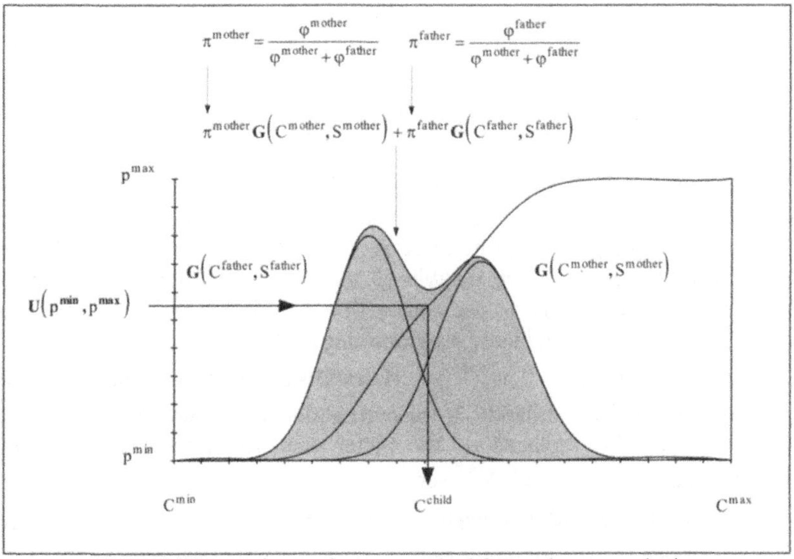

Figure 4. The continuous inverse method for random perturbation

3 Numerical examples

For combined shaping-sizing problem the example of Figure 5 is presented. The applied material is (S 235 – EC2 (10025-2)). The member stresses, the nodal displacements, the local buckling and the structural instability are considered as structural constraints. The stress constraints for tension are $\sigma_{max} = +235\ Mpa$. The stress constraints for compression are $\sigma_{min} = min\left\{-223; - E\pi^2 i_m^2 / l_m^2\right\}$ The modulus of elasticity is 210000 MPa. The bridge is subjected to five alternative load conditions P_q $(q = 1,2...,5)$ where the load intensity is $P_q = 300\,kN$. The displacement constraints are vertically and horizontally $5\,cm$ at each free node. The symmetry of the structure is kept during the optimization process. However, the displacements are computed as single variables. The design variables are grouped into eleven sizing and five shaping group variables where node 7 is shifted vertically and nodes 5, 6 and nodes 11, 12 are shifted vertically and horizontally as well. The applied cross-sections are tubular and its value is changing in between $4.29\,\text{cm}^2$ and $134.41\,\text{cm}^2$. First, only displacement constraints are considered, Secondly, both, the member buckling and nodal displacements are imposed. The results, namely the best weight and the shifting variables are presented in Figure 6 and Figure 7.

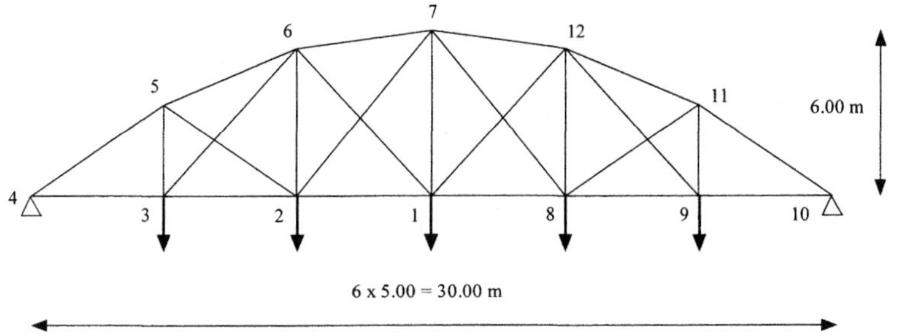

Figure 5. Initial layout of the bridge

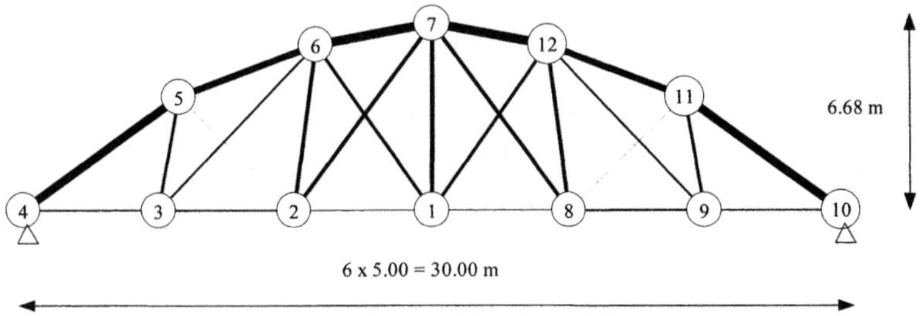

Figure 6. Optimal layout under consideration of displacement constraints
($W = 3264.599\,\text{kg}; Y = \{68.022; 59.185; 73.439; 78.852; 73.863\}$)

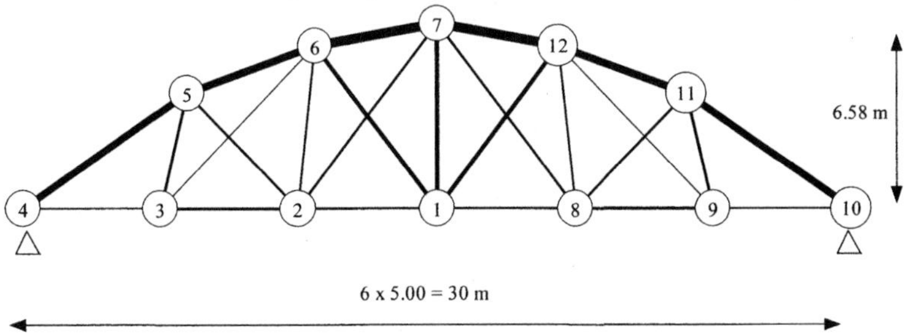

Figure 7. Optimal layout under consideration of buckling and displacement constraints
($W = 3285.149\,\text{kg}; Y = \{57.902; 46.172; 77.335; 61.714; 96.910\}$)

4 Conclusion

In this study, a new hybrid metaheuristic method named ANGEL has been proposed for combined sizing shaping optimal design of truss bridges. The computational results of a plane truss example after Pedersen (1972) reveal the fact that the proposed ANGEL method produces high quality solutions.

Acknowledgement

This work was supported by the Hungarian National Science Foundation No.T046822.

References

Achtziger, W. (2007) On simultaneous optimization of truss geometry and topology, *Struc Multidisc Optim* **33** 285-304.

Csébfalvi, A. (1998) Nonlinear path-following method for computing equilibrium curve of structures, *Annals of Operation Research* **81** 15-23.

Csébfalvi, A. (1999) Discrete optimal weight design of geometrically nonlinear truss-structures, *Computer Assisted Mechanics and Engineering Sciences*, **6** 313-320.

Csébfalvi, A. (2005) Evolution Methods for Discrete Minimal Weight Design of Space Trusses with Stability Constraints, *J. Computational and Applied Mechanics*, **6** No. 2, 159-173.

Csébfalvi, A. & Csébfalvi, G. (2006) A new hybrid meta-heuristic method for optimal design of space trusses with elastic-plastic collapse constraints, in *Proceedings of The Eighth International Conference on Computational Structures* Technology, Edited by B.H.V. Topping, G. Montero and R. Montenegro, Civil-Comp Press, Stirling, UK, paper 199.

Csébfalvi, A. & Csébfalvi, G. (2007a) Optimum Design and Sensitivity Analysis of Shallow Space Structures using an Improved Meta Heuristic Method, in *Proceedings of the 15th UK Conference of the Association of Computational Mechanics in Engineering*, B.H.V. Topping (Editor), Civil-Comp Press, Stirlingshire, Scotland, paper 59.

Csébfalvi, A. & Csébfalvi, G. (2007b) An ANGEL meta-heuristic method for combined shape and sizing truss optimization, in *Proceedings of the 7th World Congress on Structural and Multidisciplinary Optimization*, Seoul, Korea, ISBN 978-959384-2-3 98550

Dobbs, M.W. & Felton, L.P (1969) Optimization of truss geometry, *ASCE J Struct Div* **95** 2105–2118.

Gil,L. & Andreu, A. (2001) Shape and cross-section optimisation of a truss structure, *Computers and Structures* **79** 681-689.

Haftka, R.T, Gürdal, Z. & Kamat, M.P (1990) *Elements of Structural Optimization*, Second Revised Edition, Kluwer Academic Publisher

Kirsch, U (1981) *Optimium structural design*, McGraw-Hill Book Company

Pedersen, P. (1970) On the minimum mass layout of trusses, In: *AGARD Conference Proceedings No. 36 (AGARDCP-36-70), NATO Research and Technology Organization*, pp 11.1–11.17

Pedersen ,P. (1972) On the optimal layout of multi-purpose trusses. *Comput Struct* **2** 695–712

Pedersen, P. (1973) Optimal joint positions for space trusses. *ASCE J Struct Div* **99** 2459–2476

Vanderplaats, G. & Moses, F. (1972) Automated design of trusses for optimum geometry. *ASCE J Struct Div* **98** 671–690.

Tseng, L.Y. & Chen,S.C. (2006) A hybrid metaheuristic for the resource-constrained project scheduling problem, *European J. of Operation Research* **175** 707-721.

Wang, D., Zhang, W.H. & Jiang, J.S. (2002) Combined shape and sizing optimization of truss structures, *Comput Mech* **29** 307-312.

2.2 Minimum Cost Design of a Square Box Column Composed from Orthogonally Stiffened Welded Steel Plates

József Farkas, Károly Jármai
University of Miskolc, Hungary, altfar@uni-miskolc.hu

Abstract

In the previous research the minimum cost design of the uniaxially compressed orthogonally stiffened welded steel plate has been worked out. Based on this result the cost minimization of a cantilever stub column of square box cross section with orthogonally stiffened side plates is formulated and solved. The constraints relate to the overall and stiffener torsional buckling of side plates and to the horizontal displacement of the column top due to a horizontal force. Halved rolled I-section stiffeners are used in both directions. The cost function includes the cost of material, assembly, welding and painting. The variables are the thickness and width of the side plates as well as the dimensions and numbers of stiffeners in both directions. The optimization is performed using the particle swarm algorithm.

Keywords: *buckling of stiffened plates, economy of welded structures, minimum cost design, fabrication costs, stiffened box columns*

1 Introduction

Box beams and columns of large load-carrying capacity are widely applied in bridges, buildings, highway piers, pylons etc. Since the thickness required for an unstiffened box column can be too large, stiffened plate elements should be used.

Steinhardt (1975) has proposed a design method for box beams with stiffened flange plates using formulae for effective plate width. Nakai et al. (1985) have worked out empirical formulae for stiffened box stub-columns subject to combined actions of compression and bending.

Ge et al. (2000) and Usami et al. (2000) have studied the cyclic behaviour and ductility of stiffened steel box columns used as bridge piers. Longitudinal flat plate stiffeners and diaphragms as well as constant compressive axial force and cyclic lateral loading have been considered. Empirical formulae have been proposed for ultimate strength and ductility capacity. Other papers about bridge piers can be found in a conference proceedings as follows: Yamao et al. (2004), Ohga et al.(2004) and Hirota et al. (2004).

In our previous studies it has been shown that, in the case of uniaxial compression, an orthogonal stiffening is more economic than a longitudinal one (Farkas & Jármai 2006). In a study we have worked out a minimum cost design of an orthogonally stiffened plate subject to uniaxial compression (Farkas & Jármai 2007).This method is used in present paper for a square box column constructed from four equal orthogonally stiffened plates.

A cantilever column is loaded by a compression force and a horizontal load, thus, it is subject to compression and bending. From this loading a compression force is calculated for two opposite plate elements, while the remaining plate elements are subject to compression and bending. Since this loading is not so dangerous for the buckling of remaining side plate elements, it is sufficient to design only the two main plate elements. Halved rolled I-section stiffeners are used in both directions.

To show the necessity of stiffening let us design an unstiffened square box column using the following data (Fig.1): L = 15 m, N_F = 34000 kN, $H_F = 0.1N_F$, the limit of the horizontal displacement on the column top $w_0 = L/1000$ = 15 mm, the steel yield stress f_y = 355 MPa, elastic modulus $E = 2.1\text{x}10^5$ MPa.

The limiting plate slenderness is expressed according to Eurocode 3 (2002)

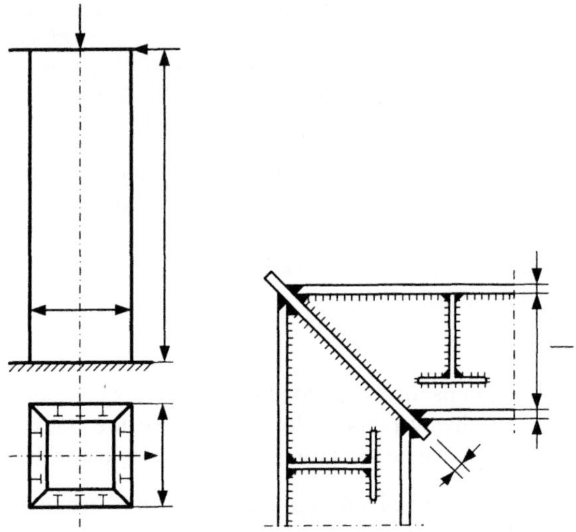

Figure 1. A cantilever stub-column of square box section with orthogonally stiffened side plates and the welded corner

$$b/t \le 1/\delta = 42\varepsilon, \varepsilon = \sqrt{235/f_y}, 1/\delta = 34, t \ge \delta b \qquad (1)$$

Taking the last inequality as equality, the cross section area, moment of inertia and section modulus are defined as

$$A = 4bt = 4\delta b^2, I_x = 2\delta b^4/3, W_x = 4\delta b^3/3 \qquad (2)$$

The stress and displacement constraints are written as

$$\frac{N_F}{A} + \frac{H_F L}{W_x} \le \frac{f_y}{1.1}, \frac{H_F L^3}{3EI_x} \le w_0 \qquad (3)$$

Since the displacement constraint is governing, the required box section width can be calculated as

$$b \ge \sqrt[4]{\frac{H_F L^3}{2\delta E w_0}} = 2805, t = 2805/34 = 82.5 \text{ mm} \qquad (4)$$

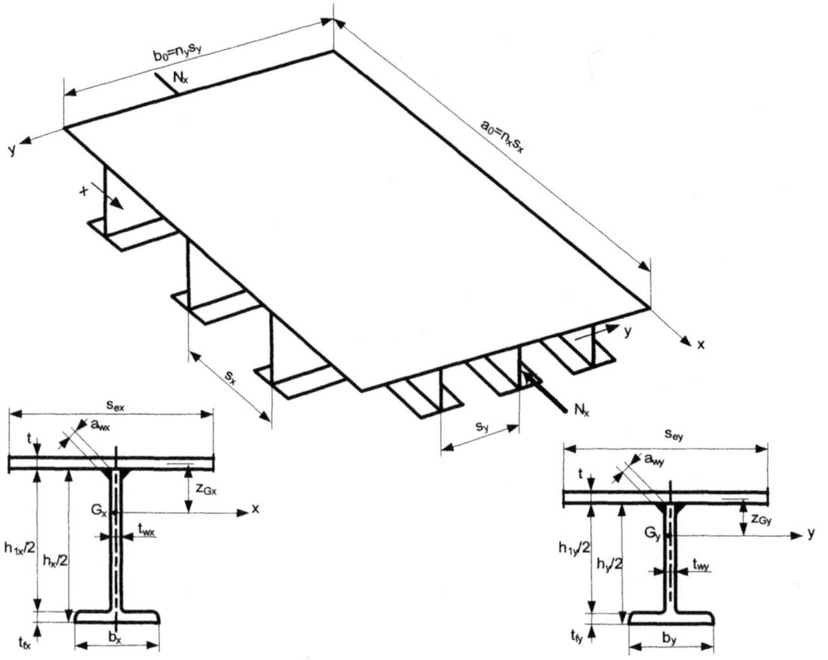

Figure 2. Orthogonally stiffened plate

This thickness is unrealistically large, thus, stiffening is needed.

In the optimum design the following variables should be optimized: the column width b_0, the base plate thickness t, dimensions and number of stiffeners in both directions h_x, h_y, n_x and n_y. It is sufficient to determine the heights h_x and h_y, since the other profile dimensions (b, t_w and t_f) can be calculated using approximate functions determined for a selected series of UB sections according to the Arcelor Mittal catalogue (Sales program 2007).

The buckling constraints are formulated according to the Det Norske Veritas rules (1995).

2 Constraint on overall buckling of a plate element (Fig.1)

$$\sigma = \frac{N_F}{4A_{ey}(n_y - 1)} + \frac{0.1 N_F a_0}{W_\xi} \le \sigma_{cr} = \frac{f_{y1}}{\sqrt{1+\lambda_e^4}} \tag{5}$$

Effective cross-sectional areas ($i = x,y$)

$$A_{ei} = \frac{h_{1i} t_{wi}}{2} + b_i t_{fi} + s_{ei} t, s_y = \frac{b_0}{n_y}, s_x = \frac{a_0}{n_x}. \tag{6}$$

Effective plate widths in two directions

$$s_{ey} = s_y C_y C_\tau, s_{ex} = s_x C_x C_\tau, \ s_y = b_0 / n_y, s_x = a_0 / n_x \tag{7}$$

$$C_y = \frac{1.8}{\beta_y} - \frac{0.8}{\beta_y^2}, C_x = \frac{1.8}{\beta_x} - \frac{0.8}{\beta_x^2} \tag{8}$$

$$C_\tau = \sqrt{1 - 3\left(\frac{\tau}{f_{y1}}\right)^2}, \ \tau = \frac{0.1 N_F}{2 b_0 t} \tag{9}$$

$$\beta_y = \frac{s_y}{t}\sqrt{\frac{f_y}{E}} \quad \text{if} \quad \beta_y \geq 1 \tag{10a}$$

$$\beta_y = 1 \qquad \text{if} \quad \beta_y < 1$$

$$\beta_x = \frac{s_x}{t}\sqrt{\frac{f_y}{E}} \quad \text{if} \quad \beta_x \geq 1 \tag{10b}$$

$$\beta_x = 1 \qquad \text{if} \quad \beta_x < 1$$

The distances of the gravity centres G_i

$$z_{Gi} = \frac{1}{A_{ei}}\left[\frac{h_{1i} t_{wi}}{2}\left(\frac{h_{1i}}{4} + \frac{t}{2}\right) + b_i t_{fi}\left(\frac{h_i + t - t_{fi}}{2}\right)\right], \tag{11}$$

The moments of inertia

$$I_i = s_{ei} t z_{Gi}^2 + \frac{h_{1i}^3 t_{wi}}{96} + \frac{h_{1i} t_{wi}}{2}\left(\frac{h_{1i}}{4} + \frac{t}{2} - z_{Gi}\right)^2 + b_i t_{fi}\left(\frac{h_i + t - t_{fi}}{2} - z_{Gi}\right)^2 \tag{12}$$

The bending stiffnesses

$$B_x = \frac{E I_y}{s_y}; B_y = \frac{E I_x}{s_x} \tag{13}$$

$$\sigma_E = \frac{N_E s_y}{A_{ey}}, N_E = \frac{\pi^2}{b_0^2}\left(B_x \frac{b_0^2}{a_0^2} + B_y \frac{a_0^2}{b_0^2}\right) \tag{14}$$

$$\lambda_e^2 = \frac{f_{y1}}{\sigma_e}\left[\left(\frac{\sigma}{\sigma_E}\right)^c + \left(\frac{\tau}{\tau_E}\right)^c\right]^{1/c}, \ c = 2 - \frac{b_0}{s_x} \tag{15}$$

$$\sigma_e = \sqrt{\sigma^2 + 3\tau^2}, \tau_E = \frac{C_1 \pi^2 E}{12(1 - \nu^2)}\left(\frac{t}{b_0}\right)^2 \tag{16}$$

$$C_1 = 5.34 + 4\left(\frac{b_0}{s_x}\right)^2 \tag{17}$$

$$W_\xi = \frac{I_\xi}{\frac{b_0}{2} - z_{Gy}} \tag{18}$$

$$I_{\xi} = 2\left\langle \left[I_y + A_{ey}\left(\frac{b_0}{2} - z_{Gy}\right)^2 \right](n_y - 1) + I_{\xi 0} \right\rangle \tag{19}$$

If n_y is even

$$I_{\xi 0} = 2\sum_{i=1}^{\frac{n_y}{2}-1}\left(I_0 + A_{ey}s_y^2 i^2\right) \tag{20a}$$

if n_y is odd

$$I_{\xi 0} = 2\sum_{i=1,3,5}^{n_y-2}\left[I_0 + A_{ey}\left(\frac{s_y}{2}\right)^2 i^2 \right] \tag{20b}$$

$$I_0 = \frac{b_y^3 t_{fy}}{12} + \frac{s_y^3 t}{12} \tag{21}$$

3 Constraint on stiffener induced failure according to DNV [6]

$$s_{ey1} = (1.1 - 0.1\beta_y)s_y \tag{22}$$

but $s_{ey1.\max} = 1$

$$A_{ey1} = \frac{h_{1y}t_{wy}}{2} + b_y t_{fy} + s_{ey1}t \tag{23}$$

$$z_{Gy1} = \frac{h_{1y}t_{wy}}{2A_{ey1}}\left(\frac{h_{1y}}{4} + \frac{t}{2}\right) + \frac{b_y t_{fy}}{2A_{ey1}}\left(h_{1y} + t - t_{fy}\right) \tag{24}$$

$$I_{y1} = s_{ey1}tz_{Gy1}^2 + \frac{h_{1y}^3 t_{wy}}{96} + \frac{h_{1y}t_{wy}}{2}\left(\frac{h_{1y}}{4} + \frac{t}{2} - z_{Gy1}\right)^2 + I_{y11} \tag{25}$$

$$I_{y11} = b_y t_{fy}\left(\frac{h_y + t - t_{fy}}{2} - z_{Gy1}\right)^2 \tag{26}$$

$$\sigma_{Ex} = \frac{\pi^2 EI_{y1}}{A_{ey1}s_x^2} \tag{26a}$$

Torsional buckling strength is expressed by

$$\sigma_{ET} = \frac{A_w + A_f\left(\frac{t_f}{t_w}\right)^2}{A_{wf}} G\left(\frac{2t_w}{h_1}\right)^2 + \frac{3x2.6\pi^2 EI_z}{A_{wf}s_x^2} \tag{27}$$

where $A_w = \frac{h_1 t_w}{2}, A_f = b_y t_{fy}, A_{wf} = A_w + 3A_f, I_z = \frac{b_y^3 t_{fy}}{12}$ (28)

$$\lambda_T = \sqrt{\frac{f_y}{\sigma_{ET}}} \tag{29}$$

$$\sigma_T = \frac{f_{y1}}{\phi_T + \sqrt{\phi_T^2 - \lambda_T^2}}, \phi_T = 0.5\left(1 + \mu_T + \lambda_T^2\right) \tag{30}$$

$$\mu_T = 0.007(\lambda_T - 0.6) \tag{31}$$

$$\lambda_S = \sqrt{\frac{\sigma_k}{\sigma_{Ex}}} \tag{32}$$

where $\sigma_k = f_y \quad \text{if} \quad \lambda_T < 0.6$ (33)

$$\sigma_k = \sigma_T \quad \text{if} \quad \lambda_T \geq 0.6$$

The constraint is formulated as

$$\sigma_1 = \frac{N_x}{n_y A_{ey1}} \leq \sigma_{acr} = \frac{\sigma_k}{\phi + \sqrt{\phi^2 - \lambda_S^2}} \tag{34}$$

where $\phi = 0.5\left(1 + \mu + \lambda_S^2\right)$ (35)

$$\mu = \frac{\delta z_t A_{ey1}}{I_{y1}} \quad \text{if} \quad \lambda_T < 0.6 \tag{36}$$

$$\mu = \frac{2.3\delta z_t A_{ey1}}{I_{y1}} \quad \text{if} \quad \lambda_T \geq 0.6$$

$$\delta = 0.0015 s_x \tag{37}$$

$$z_t = z_{Gy1} + \frac{t_{fy}}{2} \tag{38}$$

4 Constraint on horizontal displacement of the column top

$$w_{\max} = \frac{H_F}{\gamma_M} \frac{L^3}{3EI_\xi} \leq \frac{L}{\phi}, \gamma_M = 1.5, \phi = 1000 \tag{39}$$

5 Constraint on local buckling of face plates connecting the transverse stiffeners

$$t_C \geq \frac{\sqrt{2}h_x}{2x14\varepsilon}, \varepsilon = \sqrt{\frac{235}{\sigma}} \tag{40}$$

6 Numerical data (Fig. 1)

$a_0 = 24000$, $b_0 = 8000$ mm, $N_x = 3x10^7$ [N], steel yield stress $f_y = 355$ MPa, elastic modulus $E = 2.1x10^5$ MPa, shear modulus $G = 0.81x10^5$, density $\rho = 7.85x10^{-6}$ kg/mm^3, Poisson ratio $\nu = 0.3$, selected rolled I-sections UB profiles.

Ranges of unknowns: $4 < t < 20$ mm, $152 < h < 1016$ mm, $4<n<n_{max}$, n_{max} are determined by the following fabrication constraints:

$$\frac{b_0}{n_y} - b_y \geq 300 \text{ mm}, \quad \frac{a_0}{n_x} - b_x \geq 300 \text{ mm}. \tag{41}$$

The other dimensions of a halved rolled I-section are given by approximate functions of h in Appendix.

$$h_1 = h - 2t_f .$$

The discrete values of h are as follows: 152.4, 177.8, 203.2, 257.2, 308.7, 353.4, 403.2, 454.6, 533.1, 607.6, 683.5, 762.2, 840.7, 910.4, 1016 mm.

7 Cost function

The cost function includes the cost of material, assembly, welding as well as painting and is formulated according to the fabrication sequence.

The cost of material

$$K_M = k_M \rho V_2 ; k_M = 1.0 \text{ \$/kg}. \tag{42}$$

Welding of the base plate with butt welds (SAW - submerged arc welding) (Farkas & Jármai 2003). A fabricated plate element has sizes of 6000x1500 mm or less, thus the number of butt welds in x direction is the rounded up value of $n_{px} = a_0/6000$ and in y direction $n_{py} = b_0/1500$.

The fabrication cost factor is taken as $k_F = 1.0$ \$/min, the factor of complexity of the assembly $\Theta_W = 2$:

$$K_{F0} = k_F \left[\Theta_W \sqrt{(n_{px}+1)(n_{py}+1)\rho V_0} + 1.3 C_W t^n \left(n_{px} a_0 + n_{py} b_0 \right) \right], \tag{43}$$

$$V_0 = a_0 b_0 t, \tag{44}$$

$$\text{for } t < 11 \qquad C_W = 0.1346x10^{-3}; n = 2, \tag{45a}$$

$$\text{for } t \geq 11 \qquad C_W = 0.1033x10^{-3}; n = 1.904 . \tag{45b}$$

Welding $(n_x$-1) stiffeners to the base plate in y direction with double fillet welds (GMAW-C - gas metal arc welding with CO_2):

$$K_{W1} = k_F \left[\Theta_W \sqrt{n_x \rho V_1} + 1.3x0.3394x10^{-3} a_{wx}^2 2 b_0 \left(n_x - 1 \right) \right], \tag{46}$$

$a_{Wx} = 0.4\, t_{wx}$ but $a_{wx.min} = 3$ mm,

$$V_1 = a_0 b_0 t + \left(\frac{h_{1x} t_{wx}}{2} + b_x t_{fx} \right) b_0 \left(n_x - 1 \right) \tag{47}$$

Welding of $(n_y - 1)$ stiffeners to the base plate in x direction with double fillet welds. These stiffeners should be interrupted and welded with fillet welds to the stiffeners in the y direction.

$$K_{W2} = k_F \left[\Theta_W \sqrt{(n_y n_x - n_x + 1)\rho V_2} + 1.3x0.3394x10^{-3} a_{wy}^2 2 a_0 \left(n_y - 1 \right) + T_1 \right], \tag{48}$$

$$T_1 = 1.3x0.3394x10^{-3} a_{wy}^2 4 \left(n_y - 1 \right)\left(n_x - 1 \right)\left(\frac{h_{1y}}{2} + b_y \right), \tag{49}$$

$a_{Wy} = 0.4\, t_{wy}$ but $a_{Wy.min} = 3$ mm,

$$V_2 = V_1 + \left(\frac{h_{1y} t_{wy}}{2} + b_y t_{fy} \right) a_0 (n_y - 1) \tag{50}$$

Painting cost is calculated as

$$K_P = k_P \Theta_P S_P \tag{51}$$

$k_P = 14.4x10^{-6}$ \$/mm^2 , $\Theta_P = 2$,

Surface to be painted

$$S_P = 2a_0 b_0 + a_0 (n_y - 1)(h_{1y} + 2b_y) + b_0 (n_x - 1)(h_{1x} + 2b_x) \tag{52}$$

The total cost of a side stiffened plate element

$$K_S = K_M + K_0 + K_{W1} + K_{W2} + K_P \tag{53}$$

In the corners the horizontal stiffeners should be welded together by means of face plates (Fig.1). The cost of these corner welds is expressed as

$$K_C = k_w \left\langle \Theta_w \sqrt{8\rho V} + 1.3x0.3394x10^{-3} x8 \left[0.7a_0 t + (n_x - 1)\left(\frac{h_{1x}}{2} \sqrt{2} a_{wc} + 1.4 b_x t_{fx} \right) \right] \right\rangle \tag{54}$$

$$V = 4V_2 + 4a_0 t_C \left(\frac{h_x}{2} \sqrt{2} + 3t_C \right) \tag{55}$$

$$a_{wc} = 0.4 t_{wx} \text{ but } a_{wc.min} = 3 \text{ mm.} \tag{56}$$

The cost of the whole column is given by

$$K = 4K_S + K_C \tag{57}$$

8 Optimization and results

For the optimization the Particle Swarm Optimization (PSO) algorithm is used (Farkas & Jármai 2003). Table 1 shows the results of the optimization. It can be seen that the optimum column width is b_0 = 4500 mm and all structural versions fulfil the design constraints.

Table 1. Results of the optimization. The optimum is marked by bold letters. Dimensions in mm, stresses in MPa. For each version h_y = 152.4 and h_x = 177.8 mm. The allowable deflection is w_{allow} = 15 mm

b_0	t	n_y	n_x	σ_e	σ_{ecr}	σ_{acr}	w_{max}	K \$
4000	17	9	8	232	323	250	14.0	81020
4300	13	11	9	261	316	287	13.9	77780
4500	**12**	**11**	**9**	**276**	**310**	**287**	**13.7**	**76990**
4700	11	11	10	297	298	297	13.9	78200
5000	11	11	9	284	284	286	12.0	80340

9 Conclusions

A realistic numerical model of a cantilever stub column of square box section is optimized. The column is subject to compression and bending and is constructed from four equal orthogonally stiffened side plates. The thickness and width of side plates as well as the dimensions and numbers of stiffeners in both directions are calculated to fulfil the constraints and minimize the cost function.

The constraints on overall and stiffener torsional buckling are formulated according to the Det Norske Veritas design rules. The horizontal displacement of the column top is limited. The minimum distance between stiffeners is prescribed to ease the welding of stiffeners to the base plates.

Halved rolled I-profile stiffeners are used in both directions. Their height characterizes the whole profile, since the other dimensions can be expressed by height using approximate functions derived from the data of a profile series selected from available sections.

The cost function is formulated according to the fabrication sequence. The particle swarm mathematical method is suitable for this constrained function minimization problem.

It should be mentioned that the same structure has been optimized as a unstiffened and stringer-stiffened circular cylindrical shell (Farkas et al. 2007) and the result is as follows: the cost of the unstiffened shell with a diameter of 5400 and thickness of 26 mm is 82177 $, the cost of the stringer-stiffened shell is 83309 $. It means that the square box column constructed from stiffened plates is more economic than the shell columns.

References

Det Norske Veritas (DNV) (1995) *Buckling strength analysis.* Classification Notes No.30.1. Høvik, Norway.

Eurocode 3. *Steel structures.* Part 1-1. (2002).

Farkas J. & Jármai K.(2003) *Economic design of metal structures*, Rotterdam: Millpress.

Farkas J. & Jármai K. (2006) Optimum design and cost comparison of a welded plate stiffened on one side and a cellular plate both loaded by uniaxial compression, *Welding in the World* **50** No. 3-4. 45-51.

Farkas,J. & Jármai,K. (2007) Economic orthogonally welded stiffening of a uniaxially compressed steel plate. *Welding in the World* **51** No. 7-8. 74-78.

Farkas,J., Jármai,K., Rzeszut,K. (2007) Optimum design of a welded stringer-stiffened steel cylindrical shell of variable diameter subject to axial compression and bending. 17th Internat. Conf. Computer Methods in Mechanics CMM-2007, Lódz-Spala. Short papers pp.143-144. CD-ROM.

Ge,H., Gao,Sh. & Usami,Ts. (2000) Stiffened steel box columns. Part 1. Cyclic behaviour. *Earthquake Engineering and Structural Dynamics* **29** 1691-1706.

Hirota,T., Sakimoto,T.,Yamao,T. & Watanabe,H. (2004): Experimental study on hysteretic behaviour of inverted L-shaped steel bridge piers filled with concrete. In *Thin-walled Structures. Proc. 4th Int. Conf. on Thin-walled Structures, Loughborough,UK.* 2004. Ed. J. Loughlan. Institute of Physics Publ., Bristol & Philadelphia, pp. 373-380.

Nakai,H., Kitada,T. & Miki,T. (1985) An experimental study on ultimate strength of thin-walled box stub-columns with stiffeners subjected to compression and bending. *Proc. JSCE Structural Eng./Earthquake Eng.* **2** No. 2. 87-97.

Ohga,M.,Takemura,Sh. & Imamura,S. (2004): Nonlinear behaviours of round corner steel box-section piers. In *Thin-walled Structures. Proc. 4th Int. Conf. on Thin-walled Structures, Loughborough,UK.* 2004. Ed. J. Loughlan. Institute of Physics Publ., Bristol & Philadelphia, pp. 365-372.

Sales program (2007) *Commercial sections.* Arcelor Mittal. Long Carbon Europe, http://www.arcelor.com/sections/upload/diglib/PDFs/190_en.pdf

Steinhardt,O. (1975) Berechnungsmodelle für ausgesteifte Kastenträger. In Beiträge zum Beulproblem bei Kastenträgerbrücken. Deutscher Ausschuss für Stahlbau. Berichtsheft 3. 27-35.

Usami,Ts., Gao,Sh. & Ge,H. (2000) Stiffened steel box columns. Part 2. Ductility evaluation. *Earthquake Eng. and Structural Dynamics* **29** 1707-1722.

Yamao,T.,Matsumara,S.,Hirayae,M. & Iwatsubo,K. (2004): Steel tubular bridge piers stiffened with inner cruciform plates under cyclic loading. In *Thin-walled Structures. Proc. 4th Int. Conf. on Thin-walled Structures, Loughborough,UK.* 2004. Ed. J. Loughlan. Institute of Physics Publ., Bristol & Philadelphia, pp. 357-364.

Appendix

Table 2. Approximate formulae for the calculation of UB rolled I-profile dimensions b, t_w and t_f in function of x = h

t_w	b	t_f
y=a+bx+cx^2+dx^3+ex^4+ fx^5+gx^6+hx^7+ix^8	y=a+bx+c/x+dx^2+e/x^2+ fx^3+g/x^3+hx^4+ i/x^4+jx^5+k/x^5	y=a+bx+cx^2+dx^3+ex^4+ fx^5+gx^6+hx^7+ix^8
a= 4.598131496764401D0	a= -1108926.658794802D0	a= -26.93816005910891D0
b= -0.1667245062310966D0	b= 2054.96457373585D0	b= 0.7030053260773679D0
c= 0.002662252625070477D0	c= 394347552.4221416D0	c= -0.005693338027675875D0
d= -1.662919418563092D-05	d= -2.475920494568994D0	d= 2.383106288900282D-05
e= 5.425706060478163D-08	e= -91315532919.66857D0	e= -5.605511692214832D-08
f= -1.003562929221022D-10	f= 0.001858445891156483D0	f= 7.662794440441443D-11
g= 1.063362615303672D-13	g= 13189053888762.85D0	g= -5.902409222905948D-14
h= -6.028516555302632D-17	h= -7.856977790442618D-07	h= 2.267417977644635D-17
i= 1.419727611913505D-20	i= -1073670362507492D0	i= -2.999371468428559D-21
	j= 1.422535840934241D-10	
	k= 3.744384150518803D+16	

2.3 Multicriteria Tubular Truss Optimization

Jussi Jalkanen
Tampere University of Technology, P.O.Box 589, 33101 Tampere, Finland, jussi.jalkanen@tut.fi

Abstract

The topic of study is multicriteria optimization of tubular trusses that are made of standard RHS or SHS steel sections. Structural hollow sections have several good mechanical properties and optimization makes it possible to choose profiles optimal way from the large commercial available selection of profiles. Particle swarm optimization (PSO) is chosen as a solution algorithm which is a population based heuristic optimization method capable to solve various different problems. The example problem deals with the simultaneous cost and deflection minimization.

Keywords: *tubular truss, multicriteria, particle swarm optimization, discrete*

1 Introduction

Welded tubular steel trusses have become common in the applications of structural and mechanical engineering due to their high efficiency. The selection of commercially available profiles is large and in addition a lot of research has been done to ensure the safety of the design codes of tubular members and joints. In the design of tubular trusses the next natural step is to move from the analysis to the optimization. Previously for example Farkas & Jármai (1997) and (2006), Jármai et. al. (2004), Iqbal & Hansen (2006) and Saka (2007) have optimized tubular trusses.

In the optimization of welded steel structures the target is usually the minimization of the cost of the structure and the role of the design constraints is to take care that the structure remains useful and it is possible to manufacture. Other possibilities as the objective function are e.g. displacements and the maximum stress. In literature the cost optimization of welded steel structures has been considered in the articles of Jármai & Farkas (1999) and Pavlovčič et. al. (2004).

The aim of this paper is to consider the multicriteria optimization of tubular truss in which cost and deflection are considered as the criteria. The solution for a multicriteria optimization problem with conflicting criteria is a set of Pareto-optima. Since the solution is not unique the decision-maker can choose the final design from the set of mathematically equally optimal solutions based on some additional information which was not possible to include to the original optimization problem. This enables the trade-off between competing criteria in the final decision-making and thus brings more potential to the optimization. For example Koski (1994) and Osyczka (1984) give an introduction to multicriteria structural optimization.

In the current problem the task is to select a suitable profile for each member in the truss from given selection of RHS and SHS steel sections so that the demands of design code (Eurocode 3) and other relevant recommendations are fulfilled. Also the topology and the shape of the truss can be changed. Heuristic multipurpose optimization algorithms like genetic algorithm (GA) and simulated annealing (SA) are suitable methods for solving such difficult optimization problems. In this paper a rather new population based algorithm particle swarm optimization (PSO) is used. Previously e.g. Fourie & Groenwold (2002) and Venter & Sobieszczanski-Sobieski (2003) have used PSO in structural optimization.

2 Tubular trusses

In the closed thin walled cross-section of structural hollow section the material is distributed far away from centroid and the bending rigidities according to both principal axes and the torsional rigidity are high. Thus structural hollow sections are especially suitable for compressing and torsion members.

In the tubular truss design the amount of joints should be kept as small as possible to reduce the needed labour in machine workshop. On the other hand it is advantageous if all the loads act in joints so that the bending moment stays small in flanges. The joints should also design so that there is no need to reinforce or stiffen joints afterwards. The safe and usable tubular truss has to fulfil the demands of appropriate steel design code (Eurocode 3) and other recommendations given by e.g. CIDECT and manufacturer (SFS-ENV 1993-1-1 (1993), Rondal et. al. (1992), Packer et.al (1992) and Rautaruukki Metform (1997)).

The cost of tubular truss can be divided into material, manufacturing, transportation, erection and maintenance costs. In this paper only the material and the manufacturing costs are considered. The material cost K_m can be calculated simply multiplying the mass of the structure ρV by the current material cost factor k_m [\$/kg]. Jármai & Farkas (1999) have suggested that the manufacturing cost K_f is the sum of production times multiplied by a single manufacturing cost factor k_f [\$/min]. Thus the cost of welded steel structure can be expressed as

$$K = K_{\mathrm{m}} + K_{\mathrm{f}} = k_{\mathrm{m}} \rho V + k_{\mathrm{f}} \sum_i T_i \quad . \tag{1}$$

T_1 is the time for preparation, assembly and tack welding

$$T_1 = C_1 \Theta_{\mathrm{d}} \sqrt{\kappa \rho V} \quad . \tag{2}$$

C_1 is a constant, Θ_d is a difficulty factor ($\Theta_{\mathrm{d}} = 1, \ldots, 4$) and κ is the number of structural elements to be assembled.

T_2 and T_3 are the times for welding and additional fabrication costs like chancing the electrode, deslagging and chipping. The formula for the sum of T_2 and T_3 is

$$T_2 + T_3 = 1{,}3 \cdot \sum_i C_{2_i} a_{w_i}^n L_{w_i} \tag{3}$$

where a_{w_i} is the size [mm] and L_{w_i} the length [m] of weld i. Constant C_{2_i} and exponent n depends on the welding technology, the type of the weld and position.

T_4 is the time for surface preparation which means the surface cleaning, sand-spraying, etc. The formula for T_4 is

$$T_4 = \Theta_{\mathrm{ds}} a_{\mathrm{sp}} A_{\mathrm{s}} \quad . \tag{4}$$

Θ_{ds} is a difficulty factor, a_{sp} is a constant [min/mm^2] and A_s means the area of surface to be cleaned [mm^2].

T_5 means the painting time of ground and topcoat. It can be calculated using formula

$$T_5 = \Theta_{dp}\left(a_{gc} + a_{tc}\right)A_s \quad (5)$$

where Θ_{dp} is the difficulty factor for painting, a_{gc} and a_{tc} are constants [min/mm^2] and A_s is the area of surface [mm^2].

The time for cutting and edge grinding T_6 can be calculated using formula

$$T_6 = \Theta_{dc}\sum_i L_{c_i}\left(4{,}5 + 0{,}4\,t_i^2\right). \quad (6)$$

It is a modified version from the formula presented Farkas & Jármai (2006). Θ_{dc} is the difficulty factor, L_{c_i} is the length of cut i [m] and t_i is the wall thickness [mm].

3 Particle swarm optimization, PSO

The basic idea of stochastic PSO is to model the social behaviour of a swarm (e.g. birds or fishes) in nature. A swarm of particles tries to adapt to its environment by using previous knowledge based on the experience of individual particles and the collective experience of the swarm. It is useful for a single member, and at the same time for the whole swarm, to share information among other members to gain some advantage.

In PSO the new position $\mathbf{x}_{k+1}^i$ for particle i depends on the current position $\mathbf{x}_k^i$ and velocity $\mathbf{v}_{k+1}^i$

$$\mathbf{x}_{k+1}^i = \mathbf{x}_k^i + \mathbf{v}_{k+1}^i \quad (7)$$

where the velocity is calculated as follows

$$\mathbf{v}_{k+1}^i = w\mathbf{v}_k^i + c_1 r_1\left(\mathbf{p}_k^i - \mathbf{x}_k^i\right) + c_2 r_2\left(\mathbf{p}_k^g - \mathbf{x}_k^i\right). \quad (8)$$

$\mathbf{p}_k^i$ is the best ever position for particle i and $\mathbf{p}_k^g$ is the best ever position for the whole swarm. w is so called inertia, r_1 and r_2 are uniform random numbers $r_1, r_2 \in [0,1]$ and c_1 and c_2 are the scaling parameters. The value of w controls how widely the search process is done in the search space. The idea of the last two terms connected to c_1 and c_2 in the Eq. 8 is to direct the optimization process towards good potential areas in the search space. Usually $0{,}8 \le w \le 1{,}4$ and $c_1 = c_2 = 2$ are selected. The value of w can be changed dynamically so that it is bigger during early iteration rounds and becomes smaller later.

Basically, PSO is an algorithm for continuous unconstrained optimization problems. Discrete design variables can be taken into account simply by rounding each design variable to closest allowed value in Eq. 7 and constraints can be handled by penalizing unfeasible solutions according to the unfeasibility. The suitable size of the swarm depends on the amount of design variables so that the more variables the bigger the swarm. Usually the initial swarm is chosen randomly. As a terminating criteria the fixed iteration round amount can be used.

4 Cost and deflection minimization

The example problem concerns the simultaneous cost and deflection minimization of the tubular plane truss presented in Figure 1. These criteria are clearly conflicting because the most economic and at the same time very light truss can not be also the stiffest. The topology design variable $x^{t} \in \{1, 2, 3, 4\}$ indicates directly the ordinal number of chosen topology. The height of truss $x^{sh} \in \{1{,}0, 1{,}1, \ldots, 5{,}0\}$ m is the only shape design variable and the size design variables $x_i^{s} \in \{1, 2, \ldots, 112\}$ ($i = 1,2,3,4$) represent the ordinal numbers of profiles which are chosen for the group of members (x_1^{s} upper chord, x_2^{s} lower chord, x_3^{s} tension and x_4^{s} compression bracing members) in the given set of SHS and RHS sections (Table 2).

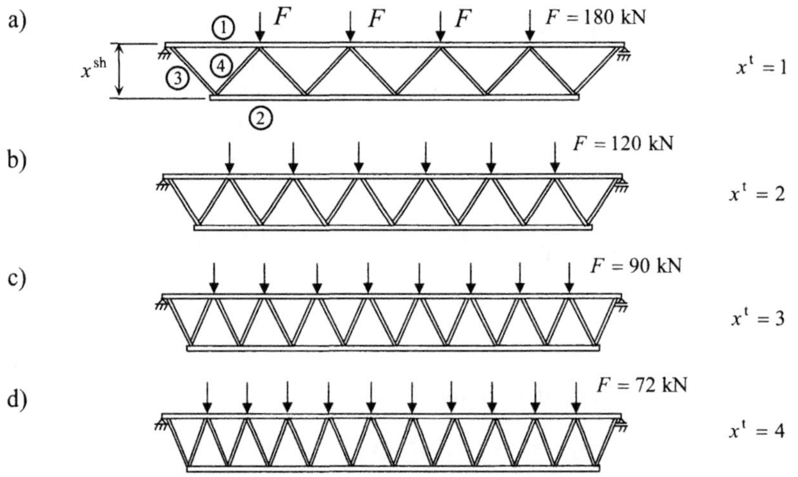

Figure 1. The four available topologies a), b), c) and d), the shape design variable x^{sh} and four different cross-sections. The span of truss is 20 m and it is made of steel S355.

The structural analysis of truss is done by using finite element method. It is assumed that chords are continuous beams and they both consist of three pieces welded together. Joints between web bars and chords are hinges and the eccentricity is zero. The compressed upper chord is supported in out of plane direction in all joints.

The mathematical form of the discussed discrete topology, shape and sizing optimization problem is

$$
\begin{aligned}
& min \begin{bmatrix} K(\mathrm{x}) \\ v(\mathrm{x}) \end{bmatrix} \\
& g^{\mathrm{s}}(\mathrm{x}) \leq 0 \\
& g^{\mathrm{b}}(\mathrm{x}) \leq 0 \\
& g^{\mathrm{j}}(\mathrm{x}) \leq 0 \\
& \mathrm{x} = \left[x^{\mathrm{t}} \; x^{\mathrm{sh}} \; x_1^{\mathrm{s}} \; x_2^{\mathrm{s}} \; x_3^{\mathrm{s}} \; x_4^{\mathrm{s}} \right]^{\mathrm{T}} \in \left\{ \mathrm{x}_1, \mathrm{x}_2, \ldots, \mathrm{x}_{n_{\mathrm{cs}}} \right\}
\end{aligned}
\quad . \qquad (9)
$$

$K(\mathbf{x})$ is the cost of truss and $v(\mathbf{x})$ is the deflection in the middle. The strength constraint $g^{\mathrm{s}}(\mathrm{x}) \leq 0$ takes care that all truss members fulfil the strength requirements of Eurocode 3. The strength is checked in the both ends of each member and the most critical cross-section determines the value of $g^{\mathrm{s}}(\mathrm{x})$. The buckling constraint $g^{\mathrm{b}}(\mathrm{x}) \leq 0$ prevents members from loosing their stability according to Eurocode 3. The buckling strength is checked for each truss member and the most critical member determines the value of $g^{\mathrm{b}}(\mathrm{x})$ (tension members can not be critical). $g^{\mathrm{j}}(\mathrm{x}) \leq 0$ is the strength constraint for welded joints. Connections can be either gap or overlap joints. Again each joint is checked and the most critical determines the value of $g^{\mathrm{j}}(\mathrm{x})$. The constraints have been explained more detailed in Jalkanen (2007).

$n_{\mathrm{cs}} \approx 2,58 \cdot 10^{10}$ is the number of possible candidate solutions in the current pure discrete optimization problem. It is, as usually, a huge number and it is not to check all the candidate solutions one by one and to choose the best feasible solution.

The material cost factor is assumed to be $k_{\mathrm{m}} = 0,6$ \$/kg and value $k_{\mathrm{f}} = 0,9$ \$/min is used for the fabrication cost factor. In the manufacturing cost calculation the values of parameters (Table 1) have been chosen based on Jármai & Farkas (1999).

Table 1. The values of parameters in the manufacturing cost calculation.

Preparation, assembly and tacking	$C_1 = 1$, $\Theta_{\mathrm{d}} = 3$ and $\kappa = 16$
Welding and additional fabrication costs	$C_2 = 0,4$ and $n = 2$ (GMAW-C)
Surface preparation	$\Theta_{\mathrm{ds}} = 2$ and $a_{\mathrm{sp}} = 3 \cdot 10^{-6}$
Painting	$\Theta_{\mathrm{dp}} = 2$, $a_{\mathrm{gc}} = 3 \cdot 10^{-6}$ and $a_{\mathrm{tc}} = 4,15 \cdot 10^{-6}$
Cutting	$\Theta_{\mathrm{dc}} = 2$ or $\Theta_{\mathrm{dc}} = 3$ (two cuts for overlapping bracing members)

Table 2. The set of 112 possible RHS- and SHS-profiles in the ascending order according to the cross-section area, Rautaruukki Metform (1997).

No	*h* [mm]	*b* [mm]	*t* [mm]	No	*h* [mm]	*b* [mm]	*t* [mm]	No	*h* [mm]	*b* [mm]	*t* [mm]
1	40	40	2,5	39	100	80	6	77	200	100	8
2	60	40	2,5	40	90	90	6	78	140	140	8,8
3	50	50	2,5	41	140	70	5	79	200	200	6
4	50	50	3	42	140	80	5	80	200	120	8
5	80	40	2,5	43	110	110	5	81	160	160	8
6	60	60	2,5	44	120	80	6	82	140	140	10
7	70	50	2,5	45	160	80	5	83	220	120	8
8	80	60	2,5	46	120	120	5	84	150	150	10
9	70	70	2,5	47	150	100	5	85	180	180	8
10	60	60	3	48	110	110	6	86	200	120	10
11	70	50	3	49	140	80	6	87	160	160	10
12	80	80	2,5	50	140	80	6,3	88	250	150	8
13	80	60	3	51	120	120	5,6	89	260	140	8
14	90	50	3	52	140	140	5	90	200	200	8
15	70	70	3	53	160	80	6	91	220	120	10
16	100	50	3	54	120	120	6	92	180	180	10
17	60	60	4	55	100	100	8	93	260	180	8
18	100	60	3	56	150	100	6	94	220	220	8
19	80	80	3	57	150	150	5	95	250	150	10
20	80	60	4	58	150	100	6,3	96	260	140	10
21	70	70	4	59	140	140	5,6	97	200	200	10
22	100	80	3	60	160	160	5	98	250	250	8
23	90	90	3	61	180	100	6	99	180	180	12,5
24	100	100	3	62	200	80	6	100	260	180	10
25	100	60	4	63	140	140	6	101	220	220	10
26	80	80	4	64	160	90	7,1	102	260	260	8,8
27	100	80	4	65	150	150	6	103	250	150	12,5
28	120	60	4	66	200	100	6	104	200	200	12,5
29	90	90	4	67	120	120	8	105	300	200	10
30	80	80	5	68	150	100	8	106	250	250	10
31	120	80	4	69	180	100	7,1	107	260	260	11
32	100	100	4	70	140	140	7,1	108	300	200	12,5
33	100	80	5	71	200	120	6	109	250	250	12,5
34	90	90	5	72	160	160	6	110	300	300	10
35	110	110	4	73	180	100	8	111	300	300	12,5
36	120	120	4	74	140	140	8	112	400	200	12,5
37	120	80	5	75	180	180	6				
38	100	100	5	76	150	150	8				

The solution of a multicriteria optimization problem is the set of Pareto-optima. Design variable vector **x*** is Pareto-optimal if there exists no feasible vector **x** which would improve some criterion without causing a simultaneous worsening in at least one criterion. Correspondingly, vector **x*** is weakly Pareto-optimal if there exists no feasible vector **x** which would improve all criteria simultaneously.

The current optimization problem has been solved using the constraint method. One criterion is chosen as the minimized objective function and the other is converted into constraint. By varying the bound of constraint weakly Pareto-optimal solutions can be generated but not necessarily Pareto-optima. Some of solutions have been calculated by minimizing the cost K with varying the deflection upper limit v^{max} and the rest by minimizing deflection v with varying the cost upper limit K^{max}.

In PSO the size of the swarm is 50 particles but it will be increased by one randomly chosen particle every time the best known feasible objective function value has not improved during 5 previous iteration rounds. However, if the best known objective function value stays the same 20 iteration rounds the increase of swarm stops. The initial inertia is $w_0 = 1,4$ and the new inertia for the next 5 iteration rounds is calculated from previous one by multiplying with coefficient 0,85. The values of scaling parameters are $c_1 = c_2 = 2$. The number of iteration rounds is 300 which means approximately 6000 FEM-analysis per each optimization run.

Figure 2 represents the best found solutions in criterion space. Each single criterion problem has been solved 20 times using PSO and the best solution has been picked. Table 3 represents the minimum cost truss and the minimum deflection truss.

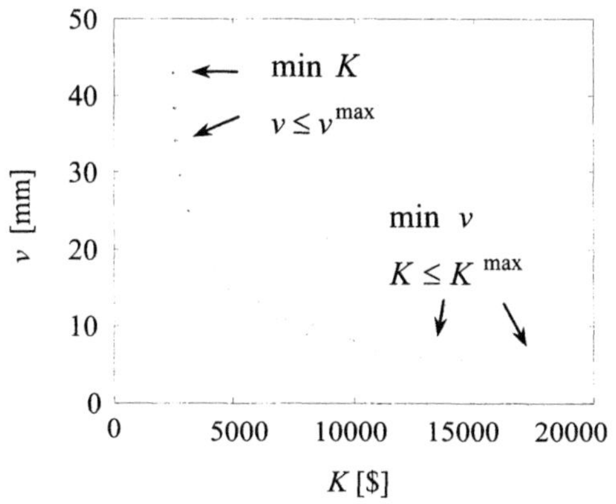

Figure 2. The best found solutions in criterion space. The stars are solutions from the cost minimization and circles are solutions from the deflection minimization.

Table 3. The best found truss in a) the cost minimization and b) the deflection minimization.

	x^t	x^{sh} [m]	x_1^s	x_2^s	x_3^s	x_4^s	K [$]	v [mm]
a)	1	3,8	57	38	29	56	2500,9	42,9
b)	1	5,0	112	112	112	108	17294,7	5,1

All calculated solutions are equally optimal in Figure 2 if the single criterion problems of the constraint method have been able to solve exactly using PSO. In the decision making phase the final design, i.e. the best compromise solution, can be chosen from this set.

5 Conclusions

In the tubular truss design it is better to exploit the advantage of optimization than to abide by the analysis of few candidate structures selected based on designer's experience and intuition. Due to excellent versatility particle swarm optimization is a suitable algorithm for the discrete tubular truss optimization. Unfortunately PSO cannot guarantee the global optimum but it can be used to find good solutions which are relatively close to the global optimum.

The example problem illustrates the potential of multicriteria optimization in tubular truss design: Several conflicting criteria can be taken into account in one problem, a constraint which allowable limit is fuzzy can be treated as a criterion, trade-off considerations can be made between conflicting criteria and the final decision making is not limited only to one solution. The price of potential is the increased computational burden compared to the traditional single criterion optimization.

References

Farkas, J. & Jármai, K. (1997) *Analysis and Optimum Design of Metal Structures*, Rotterdam: A. A. Balkama.

Farkas, J. & Jármai, K. (2006) Optimum strengthening of a column-supported oil pipeline by a tubular truss. *Journal of Constructional Steel Research* **62** 116-120.

Fourie, P. C. & Groenwold, A. A. (2002) The particle swarm optimization algorithm in size and shape optimization, *Structural and Multidisciplinary Optimization* **23** 259-267.

Iqbal, A. & Hansen, J. (2006) Cost-based, integrated design optimization. *Structural and Multidisciplinary Optimization* **32** 447-461.

Jalkanen, J. (2007) *Tubular Truss Optimization Using Heuristic Algorithms*, Dissertation, Tampere University of Technology

Jármai, K. & Farkas, J. (1999) Cost calculation and optimization of welded steel structures. *Journal of Constructional Steel Research* **50** 115-135.

Jármai, K., Snyman, J. A. & Farkas, J. (2004) Application of nov.el constrained optimization algorithms to the minimum volume design of planar CHS trusses with parallel chords. *Engineering Optimization* **36** 457-471.

Koski, J. (1994) *Multicriterion structural optimization.* In: Adeli H. (ed.) *Advances in Design Optimization*

Osyczka, A. (1984) *Multicriterion Optimization in Engineering with FORTRAN Program*, Chichester, Ellis Horwood

Packer, J. A., Wardenier, J., Kurobane, Y., Dutta, D. & Yeomans N. (1992) *Design guide for rectangular hollow section (RHS) joints under predominantly static loading.* Verlag TÜV Rheinland

Pavlovčič, L., Kranjc A. & Beg, D. (2004) Cost function analysis in the structural optimization of steel frames. *Structural and Multidisciplinary Optimization* **28** 286-295

Rautaruukki Metform (1997) *Rautaruukki's structural hollow section manual.* (in finnish), Hämeenlinna

Rondal, J., Würker, K. G., Dutta, D., Wardenier, J. & Yeomans, N. (1992) *Structural stability of hollow sections.* Verlag TÜV Rheinland

Saka, M. P. (2007) Optimum topological design of geometrically nonlinear single layer latticed domes using coupled genetic algorithm. *Computers&Structures* Article in press.

SFS-ENV 1993-1-1 (1993) *Eurocode 3: Design of steel structures. Part 1-1: General rules for buildings.*

Venter, G. & Sobieszczanski-Sobieski, J. (2003) Particle swarm optimization. *AIAA Journal* **41** 1583-1589.

2.4 Optimization of a Steel Frame for Fire Resistance with and without Protection

Károly Jármai
University of Miskolc, H-3515 Miskolc Egyetemváros, Hungary
e-mail: altjar@uni-miskolc.hu

Abstract

The main aim of the paper is to show the calculation and optimization of a steel frame according to Eurocode 1 and 3 with and without fire resistance. A comparison is made using square hollow section (SHS) columns and SHS or rectangular hollow section (RHS) for beams at a pressure vessel supporting frame (Figure 1). Optimizing for fire resistance for a given time it shows the prize of safety, the relation between mass and safety. To increase fire resistance we have to put more steel into the structure. A relatively new and promising optimization technique is introduced, the particle swarm optimization (PSO). In this evolutionary technique the social behaviour of birds is mimicked. The technique is modified in order to be efficient in technical applications. It calculates discrete optima, uses dynamic inertia reduction and craziness at some particles.

Keywords: *steel frames, fire resistance, structural optimization, evolutionary technique, particle swarm optimization*

1 Introduction

Fire research has tended to lag behind other fields of scientific and technological endeavour. This is due, no doubt partly to its extreme complexity but also due to the relatively low perceived importance of the topic in man's progress towards industrial development. Safety in general and fire safety in particular, after several major disasters, has become a subject of increasing importance in recent years. A general definition for the fire resistance of construction elements can be the following: the time after which an element, when submitted to the action of a fire, ceases to fulfil the functions for which it has been designed (Kay et al. 1996, Cox 1999, Rodrigues et al. 2000).

Steel structures have been used in industrial and residential buildings because they offer a wide range of advantages. However, these structures, when unprotected, behave poorly in fire situation. The high thermal conductivity of steel, together with the deterioration of its mechanical properties as a function of temperature, can lead to large deformations of structural elements and the premature failure of the buildings. The calculation of these steel frames can be according to Eurocode 1 and 3 (2005). The steel can be protected by materials such as mineral fibres, gypsum boards, concrete, intumescent paints and water-filled structures. In this study the optimal fire design of a steel frame structure is investigated. Using a relatively simple frame model it is shown how to apply the optimum design system for the case of fire resistance of a welded steel structure. Hollow sectional columns and beams are designed for minimum volume and weight. Overall and local buckling constraints are considered.

In the first design phase the structural mass is used as an objective function. A refined objective function is the material cost. A final objective function is the total cost including the steel mass, fabrication, fire protection technique costs.

2 Calculation of the frame members

Beams are made of RHS or SHS, unknowns are h_2, b_2, t_{f2}, columns are made of SHS, unknowns are h_1, t_{f1}.

The cross-section area of a RHS beam profile with a height h, width b and thickness t, considering rounded corners of corner radius of $R = 2t$ and supposing that $b_2 = h_2/2$, using the formulae given by Eurocode 3 Part 1.3 (2005), can be calculated as

$$A_2 = 2t_2(1.5h_2 - 2t_2)\left(1 - 0.43\frac{4t_2}{1.5h_2 - 2t_2}\right), \tag{1}$$

For SHS column it is

$$A_1 = 4t_1(h_1 - t_1)\left(1 - 0.43\frac{2t_1}{h_1 - t_1}\right), \tag{2}$$

For RHS beams the second moments of area are as follows (Figure 3).

$$I_{x2} = \left[\frac{(h_2 - t_2)^3 t_2}{6} + \frac{t_2}{2}\left(\frac{h_2}{2} - t_2\right)(h_2 - t_2)^2\right]\left(1 - 0.86\frac{4t_2}{1.5h_2 - 2t_2}\right), \tag{3}$$

$$I_{y2} = \left[\frac{(0.5h_2 - t_2)^3 t_2}{6} + \frac{t_2}{2}\left(\frac{h_2}{2} - t_2\right)^2(h_2 - t_2)\right]\left(1 - 0.86\frac{4t_2}{1.5h_2 - 2t_2}\right). \tag{4}$$

For SHS columns

$$I_{x1} = I_{y1} = \left[\frac{2(b_1 - t_1)^3 t_1}{3}\right]\left(1 - 0.86\frac{2t_1}{b_1 - t_1}\right). \tag{5}$$

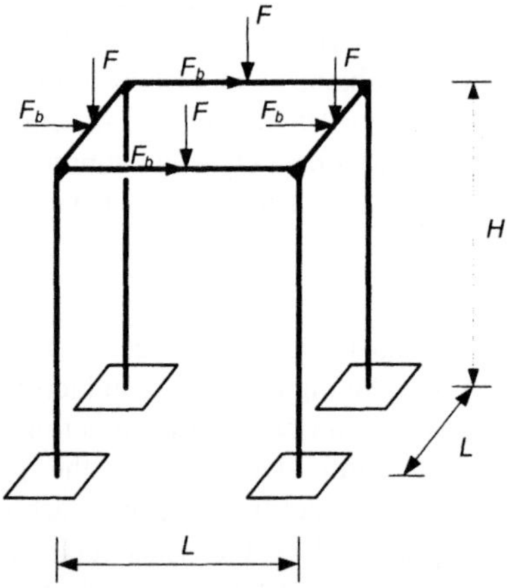

Figure 1. Supporting frame structure with vertical and horizontal forces

2.1 *Bending moments and forces from the vertical lo*ads F can be seen on Figure 2 and their calculations according to Glushkov et al. (1975) are as follows (Farkas & Jármai 1997, 2003)

$$H_A = \frac{3M_A}{H};\ M_A = \frac{M_B}{2};\ M_B = \frac{FL}{4(k+2)};\ k = \frac{I_{y2}H}{I_{y1}L} \tag{6}$$

$$M_E = \frac{FL}{4} - M_B, \qquad M_1 = \frac{F_b H 3k}{2(6k+1)} \tag{7}$$

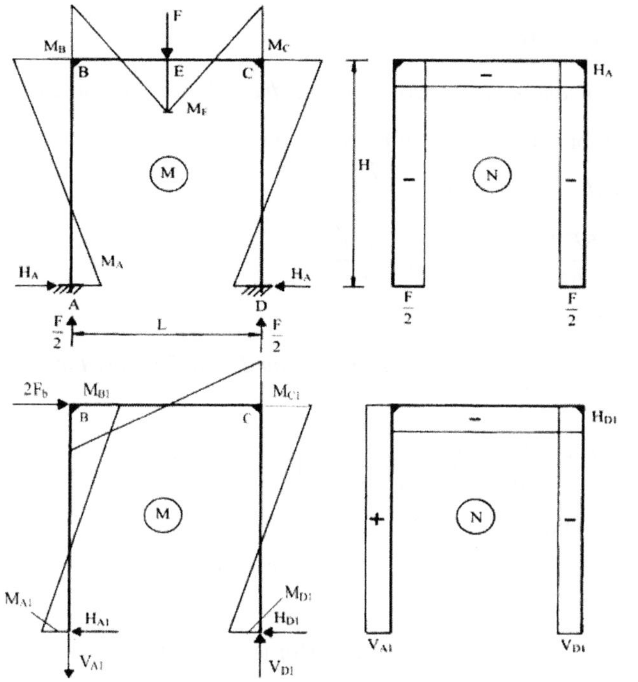

Figure 2. Diagrams for the bending moments and normal forces of a frame

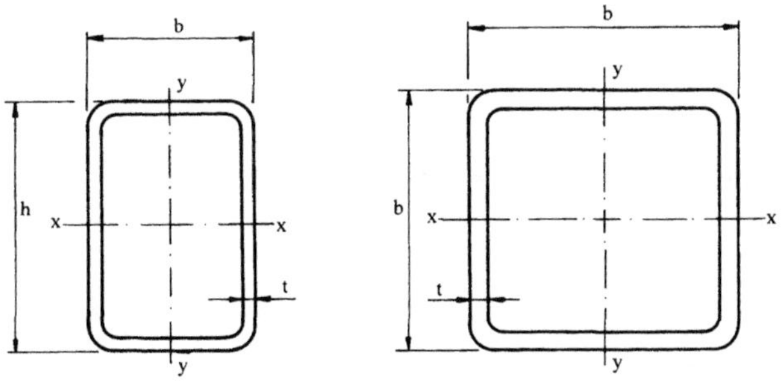

Figure 3. Dimensions of RHS and SHS profiles

Both members are made of hollow sections.

$$V_{D1} = \frac{2M_1}{L}, \qquad N_1 = F + V_{D1}, \; H_{D1} = \frac{k+1}{k+2}\frac{F_b}{2}, \tag{8}$$

$$M_{A1} = \frac{3k+1}{6k+1}\frac{F_b}{2}H, \quad M_{B1} = \frac{3k}{6k+1}\frac{F_b}{2}H, \tag{9}$$

$$H_2 = \frac{3k}{6k+1}H, \quad M_{Bt} = M_B + M_{B1} \tag{10}$$

$$M_{At} = M_A + M_{A1} \tag{11}$$

2.2 *Bending moment in the horizontal frame due to horizontal force* F_b

The horizontal force is the tenth of the vertical one.

$$F_b = 0.1F, \quad M_{Bz1} = \frac{F_b L}{4}, \tag{12}$$

$$M_{Bz2} = \frac{5F_b L}{64}, \quad M_{Bz3} = \frac{F_b L}{64}, \tag{13}$$

$$M_{Bz} = M_{Bz1} - \left(M_{Bz2} + M_{Bz3}\right) \tag{14}$$

2.3 *The stress constraint for the* ***beam*** *(point E, no fire resistance) according to Eurocode 3, Part 1-2 (2005)*

$$\frac{H_A + H_{D1}}{\chi_{2.min} A_2 f_{y1}} + \frac{k_{yy2} M_E}{W_{y2} f_{y1}} + \frac{k_{yz2} M_{Bz}}{W_{z2} f_{y1}} \le 1, \quad f_{y1} = \frac{f_y}{\gamma_{M1}} \tag{15}$$

The flexural buckling factor is

$$\chi_i = \frac{1}{\phi_i + \left(\phi_i^2 - \bar{\lambda}_i^2\right)^{0.5}}; \quad \phi_i = 0.5\left[1 + 0.34\left(\bar{\lambda}_i - 0.2\right) + \bar{\lambda}_i^2\right] \tag{16}$$

$$\bar{\lambda}_{y2} = \frac{K_{y2} L}{r_{y2}\lambda_E}; \text{ the effective length factor is } K_{y2} = 0.5 \tag{17}$$

$$r_{y2} = \left(\frac{I_{y2}}{A_2}\right)^{0.5}; \lambda_E = \pi\left(\frac{E}{f_y}\right)^{0.5}; \ E \text{ is the elastic modulus} \tag{18}$$

$$\bar{\lambda}_{z2} = \frac{K_{z2} L}{r_{z2}\lambda_E}; \text{ the effective length factor is } K_{z2} = 0.5 \tag{19}$$

$$r_{z2} = \left(\frac{I_{z2}}{A_2}\right)^{0.5} \tag{20}$$

$\chi_{2.min}$ is calculated from $\bar{\lambda}_{2.max} = \max\left(\bar{\lambda}_{y2}, \bar{\lambda}_{z2}\right)$.

$$C_{my2} = 0.4 \tag{21}$$

$$k_{yy2} = \min\left(C_{my2}\left(1 + \frac{0.6\lambda_{y2}\left(H_A + H_{D1}\right)}{\chi_{y2} A_2 f_{y1}}\right), C_{my2}\left(1 + \frac{0.6\left(H_A + H_{D1}\right)}{\chi_{y2} A_2 f_{y1}}\right)\right) \tag{22}$$

$$C_{mz2} = 0.4 \tag{23}$$

$$k_{zz2} = \min\left(C_{mz2}\left(1 + \frac{0.6\lambda_{z2}\left(H_A + H_{D1}\right)}{\chi_{z2} A_2 f_{y1}}\right), C_{mz2}\left(1 + \frac{0.6\left(H_A + H_{D1}\right)}{\chi_{z2} A_2 f_{y1}}\right)\right) \tag{24}$$

$$k_{yz2} = 0.8 k_{yy2} \tag{25}$$

2.4 *The stress constraint for the* ***beam*** *(point E, with fire resistance) according to Eurocode 1, Part 1-2 (2005)*

Member with Class 3 cross-sections, subject to combined bending and axial compression

$$\frac{H_A + H_{D1}}{\chi_{2.min} k_{y,\Theta} A_2 f_{y1}} + \frac{k_{yy2} M_E}{W_{y2} k_{y,\Theta} f_{y1}} + \frac{k_{yz2} M_{Bz}}{W_{z2} k_{y,\Theta} f_{y1}} \le 1, \tag{26}$$

The value of $\chi_{i,min\,fi}$ (i = 1,2) should be taken as the lesser of the values of $\chi_{y,fi}$ and $\chi_{z,fi}$ determined according to:

$$\chi_{fi} = \frac{1}{\varphi_\Theta + \sqrt{\varphi_\Theta^2 - \bar{\lambda}_\Theta^2}} \tag{27}$$

with $\varphi_\Theta = \frac{1}{2}\left(1 + \alpha\bar{\lambda}_\Theta + \bar{\lambda}_\Theta^{\,2}\right)$, and $\alpha = 0.65\sqrt{\frac{235}{f_y}}$ (28)

The non-dimensional slenderness for the temperature Θ_a, is given by:

$$\bar{\lambda}_\Theta = \bar{\lambda}\left(\frac{k_{y,\Theta}}{k_{E,\Theta}}\right)^{0.5} \tag{29}$$

Due to the application of hollow section we need not consider the lateral torsional buckling.

$$k_y = 1 - \frac{\mu_y N_{fi,Ed}}{\chi_{y,fi} A k_{y,\Theta} \frac{f_y}{\gamma_{M,fi}}} \le 3 \tag{30}$$

with $\mu_y = \left(1.2\beta_{M,y} - 3\right)\bar{\lambda}_{y,\Theta} + 0.44\beta_{M,y} - 0.29 \le 0.8$ (31)

for beam $\beta_{M,y} = 1.4$, $k_z = 1 - \frac{\mu_z N_{fi,Ed}}{\chi_{z,fi} A k_{y,\Theta} \frac{f_y}{\gamma_{M,fi}}} \le 3$ (32)

with $\mu_z = \left(1.2\beta_{M,z} - 5\right)\bar{\lambda}_{z,\Theta} + 0.44\beta_{M,z} - 0.29 \le 0.8$, $\bar{\lambda}_{z,\Theta} \le 1.1$, $\beta_{M,z} = 1.4$ (33)

2.5 *Stress constraint for* ***columns*** *(point C, no fire resistance) according to Eurocode 3, Action on structures, Part 1-2*

$$\frac{N_1}{\chi_{1.min} A_1 f_{y1}} + \frac{k_{yy1}(M_C + M_{B1})}{W_{y1} f_{y1}} + \frac{k_{zz1}(M_C)}{W_{z1} f_{y1}} \le 1, \qquad C_{my1} = 0.4, \tag{34}$$

$$k_{yy1} = min\left(C_{my1}\left(1 + \frac{0.6\lambda_{y1}(H_A + H_{D1})}{\chi_{y1} A_1 f_{y1}}\right), C_{my1}\left(1 + \frac{0.6(H_A + H_{D1})}{\chi_{y1} A_1 f_{y1}}\right)\right), \qquad C_{mz1} = 0.4, \tag{35}$$

$$k_{zz1} = min\left(C_{mz1}\left(1 + \frac{0.6\lambda_{z1}(H_A + H_{D1})}{\chi_{z1} A_1 f_{y1}}\right), C_{mz1}\left(1 + \frac{0.6(H_A + H_{D1})}{\chi_{z1} A_1 f_{y1}}\right)\right), \qquad k_{yz1} = 0.8 k_{yy1}, \tag{36}$$

$$r_{y1} = \left(\frac{I_{y1}}{A_1}\right)^{0.5}; r_{z1} = \left(\frac{I_{z1}}{A_1}\right)^{0.5}; \bar{\lambda}_{y1} = \frac{K_{y1} H}{r_{y1}\lambda_E}; K_{y1} = 2.19; \bar{\lambda}_{z1} = \frac{K_{z1} H}{r_{z1}\lambda_E}; K_{z1} = 0.5 \tag{37}$$

$$\bar{\lambda}_{1.max} = max(\bar{\lambda}_{y1}, \bar{\lambda}_{z1}), \qquad \chi_{i.min} = \frac{1}{\phi_i + \left(\phi_i^2 - \bar{\lambda}_{i.max}^{\ 2}\right)^{0.5}}; \tag{38}$$

$$\phi_i = 0.5\left[1 + 0.34\left(\bar{\lambda}_{i.max} - 0.2\right) + \bar{\lambda}_{i.max}^2\right] \tag{39}$$

2.6 *Stress constraint for* ***columns*** *(point C, with fire resistance) according to Eurocode 1, Action on structures, Part 1-2*

Member with Class 3 cross-sections, subject to combined bending and axial compression

$$\frac{N_1}{\chi_{1.min.fi} A_1 k_{y,\Theta} f_{y1}} + \frac{k_y (M_C + M_{B1})}{W_{y1} k_{y,\Theta} f_{y1}} + \frac{k_z M_C}{W_{z1} k_{y,\Theta} f_{y1}} \le 1 \tag{40}$$

Calculation of the parameters is according to Eqs. (27-33).

for column $\beta_{M.\psi} = 1.8 - 0.7\psi$, $\psi = -1$ (41)

Due to the application of hollow section we need not to consider the lateral torsional buckling.

3 Local buckling of plates

For the local buckling calculation we use limit slendernesses, given by *Eurocode 3, Action on structures, Part 1-2* (2005).

For the beam flange $\frac{b_2}{t_{f2}} - 3 \le 42\varepsilon$ (42)

For the beam web $\frac{h_2}{t_{w2}} - 3 \le 69\varepsilon$ (43)

For the column flange $\frac{b_1}{t_{f1}} \le 42\varepsilon$ (44)

For the column web $\frac{h_1}{t_{w1}} - 3 \le 42\varepsilon$ (45)

where for fire resistance design $\varepsilon = 0.85\sqrt{\frac{235}{f_y}}$ (46)

4 Calculation of temperature

For unprotected structure the calculation of temperature is as follows:

The time at the beginning of the fire is $t_i = 0$, and every time period: $\Delta t_i = 5$ we calculate it $t_i = t_i + \Delta t_i$ [sec], (47)

Chancing the time from $0 \le t_i \le t_{max}$ [sec], (48)

where t_{max} can be ½, 1, 1 ½, 2 , 4 hours, means 1800, 3600, 5400, 7200, 14400 [sec].

The temperature of the steel can be between 20 [°C] $\le \Theta_a \le$ 1200 [°C] (49)

Starting values for temperature and density are as follows:

$\Theta_a = 20$ [°C], $\Delta\Theta_a = 0$ [°C], $\rho_m = 7850$, $\rho = 7.85x10^{-6}$ (50)

The specific heat of steel can be calculated as a function of different temperature according to Eurocode.

The gas temperature in the vicinity of the fire exposed member (standard temperature-time curve)

$$\Theta_g = 20 + 345 log\left(8\frac{t_i}{60}+1\right) \text{ [°C]}, \tag{51}$$

The net *convection* heat flux

$$\dot{h}_{netc} = \alpha_c(\Theta_g - \Theta_a), \tag{52}$$

Where the coefficient of heat transfer by convection $\alpha_c = 25$ [W/m^{2}K]

The net *radiative* heat flux

$$\dot{h}_{netr} = \Phi\varepsilon_m\varepsilon_f\sigma\left[(\Theta_g + 273)^4 - (\Theta_a + 273)^4\right] \text{ [W/m}^2\text{]} \tag{53}$$

where the configuration factor $\Phi = 1$, the surface emissivity of the member $\varepsilon_m = 0.8$, the emissivity of the fire $\varepsilon_f = 1.0$, the Stephan Boltzmann constant $\sigma = 5.67x10^{-8}$ [W/m^{2}K^4],

The total net heat flux can be calculated as the sum of convection and radiative heat fluxes

$$\dot{h}_{netd} = \dot{h}_{netc} + \dot{h}_{netr} \tag{54}$$

$$A_m V_m = \frac{1}{10^{-3} t_2} \tag{55}$$

The temperature changing

$$\Delta\Theta_a = k_{sh}\frac{A_m V_m h_{netd}\Delta t_i}{c_a \rho_m}, \quad \text{where } k_{sh} = 1 \tag{56}$$

The surface temperature of the steel member

$$\Theta_a = \Theta_a + \Delta\Theta_a \tag{57}$$

7. Calculation of material properties

The calculation of the yield stress and Young modulus on higher temperature is according to Eurocode 1 (2005). Figure 4 shows the reduction factors in the function of temperature between 20 and 1200 C°.

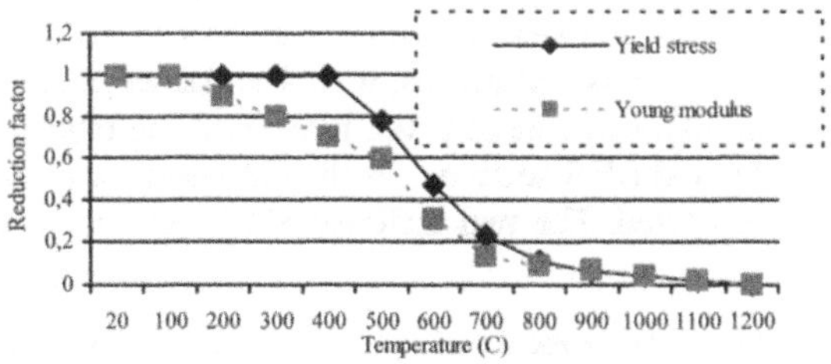

Figure 4. The yield stress and the Young modulus reduction factors in the function of temperature

7.1 *Calculation of yield strength*

The yield strength at a given temperature can be calculated by $k_{y,\Theta}$ reduction factor

$$f_{y,\Theta} = k_{y,\Theta} f_y \tag{58}$$

7.2 *Calculation of Young modulus*

The yield strength at a given temperature can be calculated by $k_{E,\Theta}$ reduction factor

$$E_{a,\Theta} = k_{E,\Theta} E_a \tag{59}$$

values of $k_{y,\Theta}$ and $k_{E,\Theta}$ can be calculated according to Table 1 and Figure 4.

The objective function is the mass of the frame to be minimized.

$$M = \rho(4HA_1 + 4LA_2) \tag{60}$$

Particle Swarm Optimization (PSO) techniques is used (Kenedy & Eberhardt 1995).

9 Numerical optimization results

9.1 *Numerical data*

The sizes of the frame are $H = 4000$, $L = 4000$ mm. The vertical and horizontal loads are $F = 75$ kN, $F_b = 0.1F$ for the normal design and $F = 0.74\text{x}75$ kN, $F_b = 0.1\ F$ for the fire resistant design. The Young modulus and the shear modulus and the yield stress $E = 2.1\text{x}10^5$ MPa, $G = 0.8\text{x}10^5$ MPa, $f_y = 355$ MPa respectively. The frame is a sway one with class 3 section.

The objective function is the structural mass M according to Eq. (60). *The unknowns* are the dimensions of SHS columns (b_1, t_1) and those of RHS beams (h_2, t_2). If SHS beams are taken into account, the formulae for SHS columns should be used with subscript 2 and their unknowns are b_2 and t_2.

Fabrication limitation

$$b_2 = \frac{h_2}{2} \le b_1 \tag{61}$$

To ease the fabrication, the solution of $b_2 = b_1$ is recommended. In this case the number of unknowns is 3.

9.2 *Optimization results*

Table 1 shows the optimum sizes of the frame, when we consider the same SHS section for the column and beam members, means 3 variables (SHS 3v), or different SHS sections for columns and beams, with 4 variables (SHS 4v), or different SHS and RHS sections for columns and beams, with 4 variables, assuming that the width of RHS section is the half of its height. We have used the tables of Dutta (1999) to get the available SHS and RHS sections. Both continuous (unrounded) and discrete optima have been calculated. The two different SHS sections version gives the best solution.

Table 1. Optimization results for the frame (no fire resistance has been taken into account)

Section		h_1 **(mm)**	t_1 **(mm)**	h_2 **(mm)**	t_2 **(mm)**	K **(kg)**
SHS 3v	discrete	180	5	-	4	775.57
SHS 4v	**discrete**	**200**	**5**	**150**	**4**	**765.53**
SHS-RHS 4v	discrete	180	5	200	5	782.24

For the frame with the same SHS section at columns and beams we have calculated the optima considering fire resistance. The fire resistance time vary from 225 sec up to 4500 sec. Both continuous and discrete optima have been calculated. Optima show that increasing the time of fire resistance, considerable increment of mass can be detected. If increase the time from 450 sec to 4500 sec (10 times more) we get an increment of mass from 1561 up to 4703 kg (3 times more). One more hour safety means three times more steel in the structure (Table 2).

Table 2. Optimization results for the frame (with fire resistance considerations)

Fire resistance time (sec)	h_1 (mm)	t_1 (mm)	t_2 (mm)	K (kg)
225	250	8	6.3	1699.19
450	250	8	6.3	1699.19
900	250	8	6.3	1699.19
1800	250	12	8	2317.63
2700	220	20	12	3028.55
3600	220	25	18	3865.90
4500	220	35	22	4703.10

Using intumescent painting, we can compare the efficiency of the painting. If we calculate the cost of the structure and consider only the material and the painting costs, than we can calculate on the following way:

$$K = K_m + K_p \tag{62}$$

$$K_m = k_m M, \tag{63}$$

where $k_m = 1$ \$/kg.

$$K_p = k_p A_p, \tag{64}$$

where $k_p = 14$ \$/m^2 which is the normal painting cost in two layers. The additional intumescent painting cost is either 20 \$/m^2, or 60 \$/m^2 depending on the fire safety time, which is half an hour, or one hour. A_p is the total painted surface

$$A_p = 16\, h_1 H + 16\, h_2 L \tag{65}$$

Table 3 shows the cost optimization results for the frame with and without intumescent painting. It shows, than in case of half an hour fire resistance the cost saving is about 5 %. In case of one hour resistance the cost saving can be about 27 %.

Table 3. Cost optimization results for the frame with and without intumescent painting

	Fire resistance time (sec)	h_1 (mm)	t_1 (mm)	t_2 (mm)	K (\$)
no protection	1800	250	12	8	2317.6
no protection	3600	220	25	18	3865.9
intumescent painting	1800	250	7	7	2210.8
intumescent painting	3600	230	10	6	2811.0

9 Conclusion

Optimization of steel frames for fire safety is a relatively new area. We have calculated the members of a high pressure vessel supporting frame without fire resistance. Using different cross sections (SHS, RHS) the mass of the frame is also different. The best solution occurred, when both columns and beams were made of SHS sections, with four variable sizes. When we consider fire resistance, the time after which its elements still work, needs more material (steel) to be built into the structure. The present example shows, that about 1 hour increment in fire safety needs 3 times more material in the structure. For a designer it is important to know the relation between mass and fire safety. Using intumescent painting and considering only the material and painting cost, the cost savings can be between 5 – 27 % depending on the fire resistance time. The applied optimization technique was very robust, the modified particle swarm optimization. It calculated both the continuous and discrete optima.

References

Cox G. (1999) Fire research in the 21st century, *Fire Safety Journal*, **32** 203-219.

Correia Rodrigues, J.P., Cabrita Neves,I. & Valente,J.C. (2000) Experimental research on the critical temperature of compressed steel elements with restrained thermal elongation, *Fire Safety Journal*, **35** 77-98.

Dutta,D. (1999) *Hohlprofil-Konstruktionen*. Ernst & Sohn, 532 p. ISBN 3-433-01310-1

Eurocode 1, Action on structures, Part 1-2 (2002) General actions – Actions on structures exposed to fire, Final Draft, CEN prEN 1991-1-2. 60 p. Bruxelles

Eurocode 3, Design of steel structures, Part 1-2 (2002) General rules –Structural fire design, Final Draft, CEN prEN 1993-1-2, 2002. 74 p. Bruxelles

Farkas,J.,Jármai,K. (1997) *Analysis and optimum design of metal structures*, Balkema Publishers, Rotterdam, Brookfield, 347 p. ISBN 90 5410 669 7.

Farkas,J.& Jármai,K. (2003) *Economic design of metal structures*. Rotterdam, Millpress, 340 p. ISBN 90 77017 99 2

Glushkov,G., Yegorov,I., Yermolov ,V. (1975) *Formulas for designing frames,* MIR Publishers, Moscow.

Kay,T.R.,Kirby,B.R.& Preston,R.R. (1996) Calculation of the heating rate of an unprotected steel member in a standard fire resistance test, *Fire Safety Journal*, **26** 327-350.

Kennedy J. & Eberhardt R. (1995) *Particle swarm optimization*. Proc. Int. Conf. on Neural Networks, Piscataway, NJ, USA. 1942-1948.

International Standards Organisation: *ISO 834* (1975) "*Fire Resistance Test. Elements of Building Construction*".

2.5 The MINLP Approach to Cost Optimization of Structures

Stojan Kravanja, Tomaž Žula, Uroš Klanšek
University of Maribor, Faculty of Civil Engineering, Smetanova 17, SI-2000 Maribor, Slovenia, e-mail: stojan.kravanja@uni-mb.si

Abstract

The paper presents the Mixed-Integer Non-Linear Programming (MINLP) approach to cost optimization of structures. The MINLP is a combined discrete/continuous optimization technique, where the discrete binary 0-1 variables are defined for the optimization of the discrete topology, material and standard sizes alternatives and the continuous variables for the optimization of continuous parameters. An economic objective function of the manufacturing material and labour costs is subjected to structural analysis and dimensioning constraints. The Modified Outer-Approximation/Equality-Relaxation (OA/ER) algorithm is used for the optimization. The accompanied Linked Multilevel Hierarchical Strategy (LMHS) accelerates the convergence of the algorithm. Three examples of the cost optimization of steel structures are presented at the end of the paper.

Keywords: *Cost optimization, Structural optimization, MINLP, Mixed-Integer Non-Linear Programming, Steel structures*

1 Introduction

The paper presents cost optimization of structures using the Mixed-Integer Non-Linear Programming (MINLP) approach. The MINLP handles continuous and discrete binary 0-1 variables simultaneously. While the continuous variables are defined for the continuous optimization of parameters (stresses, deflections, weights, costs, etc.), the discrete variables are used to express discrete decisions, i.e. usually the existence or non-existence of structural elements inside the defined structure. Different discrete materials, standard sizes and rounded dimensions may also be defined as discrete alternatives. Since the continuous and discrete optimizations are carried out simultaneously, the MINLP approach also finds optimal continuous parameters, structural topology, material, standard and rounded dimensions simultaneously.

The MINLP discrete/continuous optimization problems are in most cases comprehensive, non-convex and highly non-linear. The MINLP optimization approach is thus proposed to be performed through three steps: i.e. the generation of a mechanical superstructure, the modelling of an MINLP model formulation and the solution of the defined MINLP problem. Many different methods for solving MINLP problems have been developed in the near past. This paper reports the experience in solving MINLP problems by using the Modified Outer-Approximation/Equality-Relaxation (Modified OA/ER) algorithm by Kravanja & Grossman (1994), see also Kravanja et al. (1998, Part I). The Linked Multilevel Hierarchical strategy (LMHS) has been developed to accelerate the convergence of the mentioned algorithm. Since the number of discrete alternatives and defined binary 0-1 variables are usually too high for normal solution of the MINLP, a special reduction procedure is developed to reduce automatically the number of binary variables on a reasonable level.

An economic objective function of the manufacturing material and labour costs is defined for the optimization. The objective function is subjected to structural analysis and dimensioning constraints. The design constraints are defined according to Eurocodes. Three examples at the end of the paper show the efficiency of the proposed MINLP approach to the cost optimization of structures.

2 Mechanical superstructure

The MINLP optimization approach to structural synthesis requires the generation of an MINLP mechanical superstructure composed of various topology and design alternatives that are all candidates for a feasible and optimal solution. While the topology alternatives represent different selections and interconnections of corresponding structural elements, the design alternatives include different materials, standard and rounded dimensions.

The superstructure is typically described by means of unit representation: i.e. structural elements and their interconnection nodes. Each potential topology alternative is represented by a special number and a configuration of selected structural elements and their interconnections; and each structural element may in addition have different material, standard and rounded dimension alternatives. The main goal is thus to find within the given superstructure a feasible structure that is optimal with respect to manufacturing costs, topology, material, standard and rounded dimensions.

3 MINLP model formulation

A general non-linear and non-convex discrete/continuous optimization problem can be formulated as an MINLP problem in the following form:

$$
\begin{aligned}
\min \quad & z = \boldsymbol{c}^{\mathrm{T}} \boldsymbol{y} + f(\boldsymbol{x}) \\
\text{s.t.} \quad & \boldsymbol{h}(\boldsymbol{x}) = \boldsymbol{0} \\
& \boldsymbol{g}(\boldsymbol{x}) \leq \boldsymbol{0} \\
& \boldsymbol{B}\boldsymbol{y} + \boldsymbol{C}\boldsymbol{x} \leq \boldsymbol{b} \\
& \boldsymbol{x} \in X = \{\boldsymbol{x} \in R^{\mathrm{n}}: \boldsymbol{x}^{\mathrm{lo}} \leq \boldsymbol{x} \leq \boldsymbol{x}^{\mathrm{up}}\} \\
& \boldsymbol{y} \in Y = \{0,1\}^{m}
\end{aligned}
\qquad \textbf{(MINLP)}
$$

where $\boldsymbol{x}$ is a vector of continuous variables specified in the compact set X and $\boldsymbol{y}$ is a vector of discrete, mostly binary 0-1 variables. Functions $f(\boldsymbol{x})$, $\boldsymbol{h}(\boldsymbol{x})$ and $\boldsymbol{g}(\boldsymbol{x})$ are non-linear functions involved in the objective function z, equality and inequality constraints, respectively. Finally, $\boldsymbol{By}+\boldsymbol{Cx}\leq\boldsymbol{b}$ represents a subset of mixed linear equality/inequality constraints.

The above general MINLP model formulation has been adapted for the optimization of structures. In the context of structural optimization, continuous variables $\boldsymbol{x}$ define structural parameters (dimensions, strains, stresses, costs, weight...) and binary variables $\boldsymbol{y}$ represent the potential existence of structural elements as well as the choice of materials, standard and rounded dimensions.

The economic objective function z involves fixed costs charges in the term $\boldsymbol{c}^T\boldsymbol{y}$ for manufacturing, while the dimension dependant costs are included in the function $f(\boldsymbol{x})$. Non-linear equality and inequality constraints $\boldsymbol{h}(\boldsymbol{x})=\boldsymbol{0}$, $\boldsymbol{g}(\boldsymbol{x}) \leq \boldsymbol{0}$ and the bounds of the continuous variables represent the rigorous system of the design, loading, stress, deflection and dimensioning constraints known from the structural analysis. Mixed logical constraints $\boldsymbol{By}+\boldsymbol{Cx}\leq \boldsymbol{b}$ describe relations between binary variables and define the structure's topology, materials, standard and rounded dimensions. It should be noted, that the comprehensive MINLP model formulation for mechanical structures may be found elsewhere, e.g. Kravanja et al. (1998, Part II).

4 Solving the MINLP problem

Since the discrete/continuous optimization problem is non-convex and highly non-linear, the Outer-Approximation/Equality-Relaxation (OA/ER) algorithm by Kocis and Grossmann (1987) has been used for the optimization. The OA/ER algorithm consists of solving an alternative sequence of Non-linear Programming (NLP) optimization subproblems and Mixed-Integer Linear Programming (MILP) master problems. The former corresponds to continuous optimization of parameters for a mechanical structure with fixed topology, standard and rounded dimensions and yields an upper bound to the objective to be minimized. The latter involves a global approximation to the superstructure of alternatives in which a new topology, discrete materials, standard and rounded dimensions are identified so that its lower bound does not exceed the current best upper bound. The search is terminated when the predicted lower bound exceeds the upper bound. The OA/ER algorithm guarantees the global optimality of solutions for convex and quasi-convex optimization problems.

The OA/ER algorithm as well as all other mentioned MINLP algorithms do not generally guarantee that the solution found is the global optimum. This is due to the presence of nonconvex functions in the models that may cut off the global optimum. In order to reduce undesirable effects of nonconvexities the Modified OA/ER algorithm was proposed by Kravanja and Grossmann (1994) by which the following modifications are applied for the master problem: deactivation of linearizations, decomposition and deactivation of the objective function linearization, use of the penalty function, use of the upper bound on the objective function to be minimized as well as a global convexity test and a validation of the outer approximations.

The optimal solution of comprehensive non-convex and non-linear MINLP problem with a high number of discrete decisions is in general very difficult to be obtained. For this purpose, the Linked Multilevel Hierarchical strategy (LMHS) strategy has been developed to accelerate the convergence of the OA/ER algorithm, see Kravanja et al. (2005). Using the LMHS strategy, we decompose the original integer space and original MINLP problem in a hierarchical manner into several subspaces and corresponding MINLP levels. Each time the next MINLP optimization level is performed, the current integer subspace is extended by the next integer subspace and prescreened, while the discrete decisions belonging to all of the remaining subspaces are approximated by the relaxed 0-1 variables. The levels are linked by accumulating outer-approximations and yield lower bounds to their next level objective functions to be minimized, which considerably improve the efficiency of the search. Decision levels are hierarchically classified into four levels: from the topology level to the material, standard sizes and rounded dimension levels.

Higher levels give lower bounds to the original objective function to be minimized while lower levels give upper bounds. The MINLP subproblems are iterated about each level until there are no improvements in the NLP solution. Thus, we start with the discrete topology optimization at the relaxed materials and standard dimensions. When the optimal topology is reached, the process proceeds with the discrete topology and material optimization at the second level. After obtaining the optimal result, the calculation continues with the simultaneous discrete topology, material and standard dimension optimization at the third level. Finally, after the optimal topology, materials and standard dimensions are obtained, the MINLP is carried out once more for the complete discrete decisions at the fourth level (all the continuous dimensions are additionally rounded to the discrete values in mm or cm).

The optimization model may contain up to some thousands or ten thousands of binary 0-1 variables of alternatives. Most of them are subjected to rounded dimensions. Since this number of 0-1 variables is too high for a normal solution of the MINLP, a special reduction procedure has been developed, which automatically reduces the binary variables for rounded dimension alternatives into a reasonable number. In the optimization at the fourth level are included only those 0-1 variables which determine rounded dimension alternatives close to the continuous dimensions, obtained at the previous third MINLP optimization level. This procedure can be similarly applied also for the reduction of standard dimension alternatives or others.

5 Numerical examples

The MINLP optimization approach is illustrated by three examples. The first example presents the cost, material and standard dimension optimization of a composite floor system, the second one introduces the cost and standard section optimization of a three-storey steel frame and the third example shows the cost, topology and standard section optimization of a 60 m long industrial building.

The MINLP optimization models for the mentioned different structures were developed. As an interface for mathematical modeling and data inputs/outputs GAMS (General Algebraic Modeling System) by Brooke et al. (1988), a high level language, was used. The optimizations were carried out by a user-friendly version of the MINLP computer package MIPSYN, the successor of programs PROSYN by Kravanja & Grossmann (1994) and TOP by Kravanja et al. (1992). The Modified OA/ER algorithm and the LMHS strategy were applied, where GAMS/CONOPT2 (Generalized reduced-gradient method), see Drud (1994), was used to solve NLP subproblems and GAMS/Cplex 7.0 (Branch and Bound) was used to solve MILP master problems.

5.1 *Cost, material and standard dimension optimization of a composite I beam floor*

The first example presents the simultaneous cost, material and standard dimension optimization of a composite I beam floor system with the span of 28 m, subjected to the self-weight and to the uniformly distributed imposed load of 10 kN/m^2.

The material and labour costs for the composite beams were accounted for in the economical type of the objective function, subjected to the given design, material, resistance and deflection constraints, defined in accordance to Eurocodes 4 (1992). The material and labour costs for the composite beams considered are shown in Table 1. The superstructure comprised 6 different concrete strengths (C25, C30, C35, C40, C45, C50), 3 different structural steel grades (S 235, S 275, S 355), 48 various standard reinforcing steel sections as well as 9 different standard thickness of sheet-iron plates (from 8 mm to 40 mm) for webs and flanges separately.

Table 1. Material and labour costs

Material costs for structural steel S 235-S 355	1.0-1.2	EUR/kg
Material costs for reinforcing steel S 400	1.2	EUR/kg
Material costs for concrete C 25/30-C 50/60	90.0-120.0	EUR/m^3
Sheet-iron cutting costs	5.0	EUR/m^1
Welding costs	7.5	EUR/m^1
Anti-corrosion resistant painting costs (R30)	20.0	EUR/m^2
Panelling costs	15.0	EUR/m^2

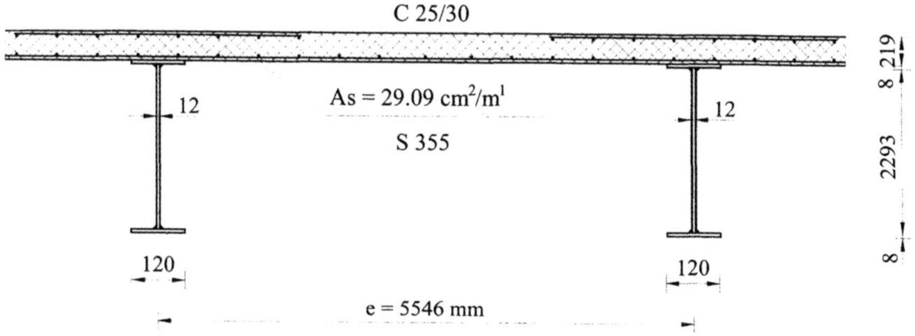

Figure 1. Optimal cross-section of the composite I-beam.

The optimal result of 113.42 EUR/m^2 was obtained in the 3rd MINLP iteration, see Figure 1. Beside the optimal self-manufacturing costs, the optimal concrete strength C25/30, steel grade S 355, standard reinforcing wire mesh and the optimal standard thickness of webs and flanges were obtained.

5.2 *Cost and standard section optimization of a three-storey/three bay steel frame*

The second example presents the cost and standard section optimization of an unbraced three-storey, three-bay steel frame. The frame was subjected to the self-weight, to the horizontal concentrated variable loads of 5 kN and to the uniformly distributed variable load of 50 kN/m^1, see Figure 2.

The objective function represented the material costs (mass) of the structure. The finite element equations were defined in the set of constraints for the calculation of the internal forces and the deflections, while the constraints for the dimensioning were determined in accordance to Eurocode 3 (1995). The ultimate and serviceability limit states were checked. The second-order elastic structural optimization was performed by considering a geometric nonlinearity due to P-δ and P-Δ effects. The frame superstructure was generated in which all possible structures were embedded by different standard section variation. The superstructure comprised 17 different standard hot rolled HEA sections (from HEA 100 to 500) for each beam and column separately. The material used was steel S 355 (1.0 EUR/kg).

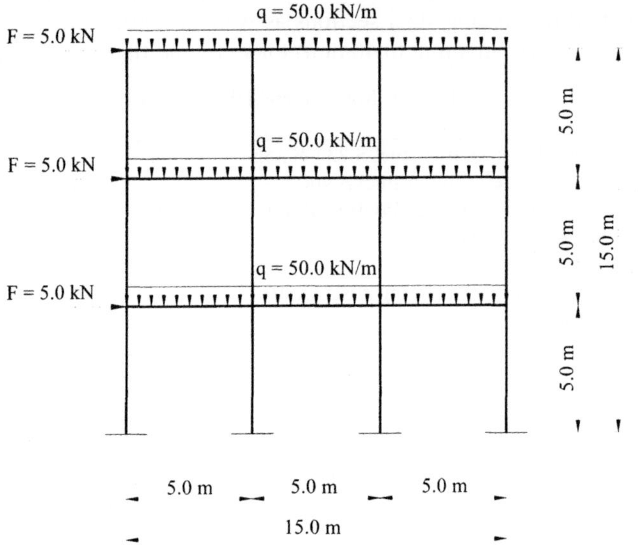

Figure 2. Three-storey, three-bay steel frame

The final optimal solution of 5047 EUR (5047 kg) was obtained in the 61st main MINLP iteration. The obtained structure and standard sizes are shown in Figure 3.

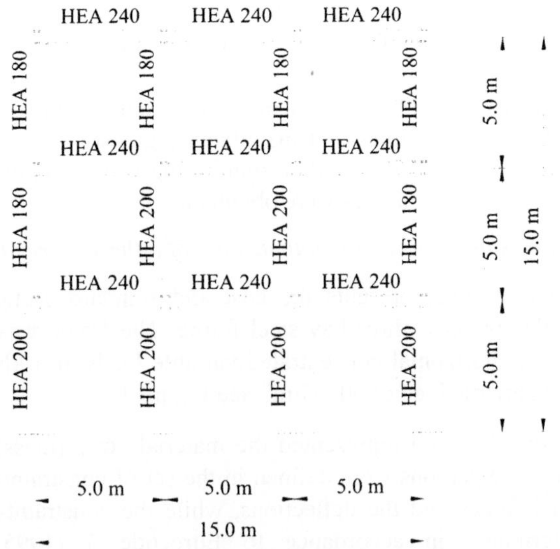

Figure 3. Optimal structure of the steel frame

5.3 *Cost, topology and standard section optimization of an industrial building*

The last example presents the cost, topology and standard section optimization of a single-storey industrial building. The structure was consisted from equal non-sway steel portal frames, which are mutually connected with purlins. The building was 22 meters wide, 60 meters long and 6 meters high. The structure was subjected to the self-weight and to the variable load of snow and wind. The mass of the roof was 0.20 kg/m^2. The variable imposed loads: 2.50 kN/m^2 (snow), 0.13 kN/m^2 (vertical wind) and 0.55 kN/m^2 (horizontal wind) were defined in the model input data.

The economic objective function was defined for the optimization. The material and labour costs considered are shown in Table 2. The fabrication costs of steel elements were calculated to be equal to 40 % of the obtained material costs. The internal forces and deflections were calculated by the elastic first-order analysis. The design/dimensioning constraints were defined in accordance with Eurocode 3. The structure was checked for both the ultimate and serviceability limit states. The superstructure was generated in which all possible building's structures were embedded by 30 portal frame alternatives, 20 purlin alternatives and 24 different alternatives of standard hot rolled HEA sections (from HEA 100 to 1000) for each column, beam and purlin separately. The material used was steel S 355.

Table 2. Material and labour costs

Material costs for structural steel S 355	1.0	EUR/kg
Anti-corrosion resistant painting costs (R30)	20.0	EUR/m^2
Erection costs per portal frame	400.0	EUR/frame
Erection costs per purlin	200.0	EUR/purlin

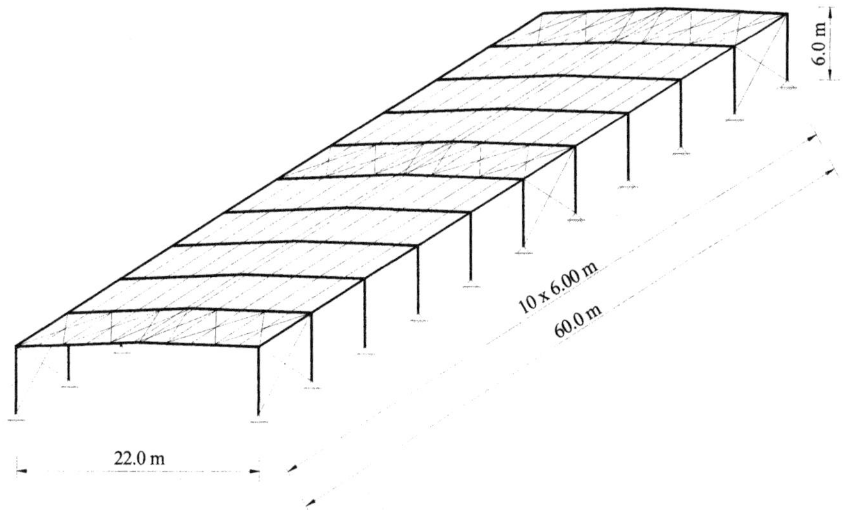

Figure 4. Optimal design of the single-storey industrial building.

The final optimal solution of 159131 EUR was obtained in the 6th main MINLP iteration. The optimal solution represents the obtained »minimal« self material and labour costs of the considered steel industrial building structure. The selling price may be at least twice higher.

The solution also comprises the calculated structure mass of 86.75 tons as well as the building topology of 11 portal frames and 12 purlins, see Figure 4. Columns are designed from HEA 800, the frame beams from HEA 500 and purlins from HEA 160 standard sections.

6 Conclusions

The paper presents the MINLP approach to cost optimization of structures. The MINLP was found to be successful optimization technique for solving large-scale non-linear, discrete, continuous and non-convex cost optimization problems.

References

Brooke, A., Kendrick, D. & Meeraus, A. (1988) *GAMS - A User's Guide*, Scientific Press, Redwood City, CA.

CPLEX User Notes, ILOG inc.

Drudd, A.S. (1994) CONOPT – A Large-Scale GRG Code, *ORSA Journal on Computing* **6** No. 2, 207-216.

Eurocode 3 (1995) *Design of steel structures*, Brussels: European Committee for Standardization.

Eurocode 4 (1992) *Design of composite structures*, Brussels: European Committee for Standardization.

Kocis, G.R. & Grossmann, I.E. (1987) Relaxation Strategy for the Structural Optimization of Process Flowsheets. *Ind. Engng. Chem. Res.* **26** 1869-80.

Kravanja, S., Kravanja, Z., Bedenik, B.S. & Faith, S. (1992) Simultaneous Topology and Parameter Optimization of Mechanical Structures, In: *Proceedings of the First European Conference on Numerical Methods in Engineering (ed Ch. Hirsch et al.), Brussels, Belgium, Elsevier, Amsterdam,* pp. 487-495.

Kravanja, S., Kravanja, Z. & Bedenik, B.S. (1998) The MINLP optimization approach to structural synthesis. Part I: A general view on simultaneous topology and parameter optimization. *International Journal for Numerical Methods in Engineering* **43** 263-292.

Kravanja, S., Kravanja, Z. & Bedenik, B.S. (1998) The MINLP optimization approach to structural synthesis. Part II: Simultaneous topology, parameter and standard dimension optimization by the use of the Linked two-phase MINLP strategy. International Journal for Numerical Methods in Engineering **43** 293-328.

Kravanja, S., Šilih S. & Kravanja, Z. (2005) The multilevel MINLP optimization approach to structural synthesis: the simultaneous topology, material, standard and rounded dimension optimization. *Internatinal Journal on Advances in Engineering Software* **36** 568-583.

Kravanja, Z. & Grossmann, I.E. (1994) New Developments and Capabilities in PROSYN - An Automated Topology and Parameter Process Synthesizer. *Computers chem. Engng.* 18 No. 11/12, 1097-1114.

2.6 Finite Element Analysis and Optimization of a Car Seat under Impact Loading

Ferenc János Szabó
University of Miskolc, H-3515 Miskolc, Hungary,
e-mail: machszf@uni-miskolc.hu

Abstract

Programming possibilities of finite element program systems make possible to enlarge the applicability of the finite element models: all the possibilities of the program system remain usable and the new user defined programs and macros can give new possibilities for the analysis. A macro has been developed for the analysis of structural behaviour for impact loads and has been integrated into an optimization program written under the COSMOS/M finite element system. By this way one can solve special multidisciplinary optimization problem types which could be very difficult or impossible to solve in any other program system. A numerical example is shown for the optimization of the metal structure in a car seat back. For the optimization the author developed his new algorithm, the Random Virus Algorithm (RVA Algorithm), based on the simulation of reproduction process of biological or computer viruses

Keywords: *Optimization, car seat, impact loading, programming FEM*

1 Introduction

By using the built-in programming language of the COSMOS/M finite element program system a macro has been developed for the modelling, analysis and optimization of structures loaded by impact load. The impact load is supposed as the dynamic load during the impact of another body having its mass (m) and dropped from a given height (h). All the necessary data are calculated by the program and the results of displacements and stresses caused by the impact load can be post-processed as conventional finite element results. Main steps of the calculation process:

- Build-up the finite element model of the structure;
- Calculation of static displacement caused by the weight of the dropped body at the contact area of the target body;
- Determine the mass reduction factor by using numerical integration of the displacement field determined in the previous step;
- Calculation of the equivalent mass and the mass ratio;
- Determine the dynamic factor.

Multiplying the static stress and displacement results by the dynamic factor, one can get the results for impact load. The macro contains optimum searching algorithm developed for multidisciplinary optimization. The algorithm is the Random Virus Algorithm (RVA) which has been developed specially for finite element programming usage. The theoretical basis of the algorithm is the simulation and modelling of the behaviour and quick reproduction of the biological (and computer) viruses. The computer code of the algorithm is very simple, easy to program on any programming language, the working and thinking of the algorithm is very efficient regarding the number of objective function evaluations and the number of constraint checking until reaching the optimum result. By the combination of the method for the analysis of the impact load and the RVA optimization algorithm, it is possible to solve several special multidisciplinary optimization problems because the macro can use all the facilities and possibilities of the original finite element program. For example, it will be possible to combine the multiphysics possibilities with optimization and impact load, which could lead to the solution of special complex problems, e.g. thermal effects (fire) with impact load and optimization, as well as magnetic structure or fluids combined with impact load and optimization. The solution of these kinds of problems could be very difficult or impossible in the conventional finite element program packages.

The results of the method and of the macro are presented through the numerical example of the metal structure of a car seat back loaded by the person who sits in there in case of an impact of 30 kph. (It is supposed that a person sits in the car and another car hits the car from the backside and the seat will deform because of the inertial force exerted by the person sitting in the car). Before the optimization, the results for the structural behaviour are compared by experimental ones in order to verify the accuracy of the numerical calculations.

2 Analysis of structures subjected to impact load, by programming FEM

Let us suppose that the investigated structure is subjected to a load caused by an object of mass m_o dropped from height h. After the collision of the object into the structure, the structure starts a vibrating motion. Supposing uniform mass distribution of the structure, this motion can take place through many possible eigenfrequencies (infinite number of possible eigenfrequencies). According to Carnot- rule, if we investigate only the first eigenfrequency for this motion, we loose an amount of energy comparing to the situation if we suppose the structure to move with the impact speed in every point, Ponomarjov (1965). According to this rule, the error made in stress results is four times higher than in case of the results for displacements. For this reason, it is always recommended to verify the computational results by experiments. These experiments will show the limits of applicability of the method for higher speeds and nonlinear deformations, too. Taking into account the higher eigenfrequencies will increase the accuracy of the method.

Although the method is able to handle general three dimensional structures, for simplicity reasons let us investigate a plate-like structure. These kinds of structures are commonly used in many fields of engineering science (cars, sidewalls of buildings, roof structures, covers of machines, etc.). Many of more complicated problems can be modelized by or can be originated in plate structures. The following equations are derived for non-elastic impact (when the structure and the falling object can move together after the impact). In case of elastic impact, when the falling object will rebound from the structure, before the rebounce they move together for a short time and during this period the equations will be usables. In these equations, the structure is substituted by a one-mass vibrating system, with a spring constant calculated by the static displacement of the structure caused by the object's mass:

$$c = \frac{w_{s\max}}{m_o g} \quad . \tag{1}$$

Where $w_{s\max}$ is the static displacement of the structure in the point of impact, g is the gravity. For the calculation of the kinetic energy of the structure it is supposed that the velocity of the different points of the structure is proportional to the displacement:

$$\frac{v_p}{w_p} = \frac{v_{\max}}{w_{\max}} , \tag{2}$$

$v_p(x,y)$ is the velocity of a point of the structure and $w_p(x,y)$ is the displacement of the same point, $v_{\max}$ is the velocity of the impact point on the structure. The kinetic energy can be calculated:

$$E_k = \int_{x=0}^{a} \int_{y=0}^{b} \frac{q v_p^2}{2} dxdy = \frac{v_{\max}^2}{2w_{\max}^2} \int_{x=0}^{a} \int_{y=0}^{b} q w_p^2 dxdy \quad , \tag{3}$$

q is the specific mass of the structure:

$$\int_{x=0}^{a} \int_{y=0}^{b} q dxdy = m \tag{4}$$

The kinetic energy of the substituting one-mass vibrating system should be equal to the energy calculated in Eq. 3. Supposing uniform mass- distribution:

$$m_e \frac{v_{\max}^2}{2} = \frac{v_{\max}^2}{2w_{\max}^2} m \int_{x=0}^{a} \int_{y=0}^{b} w_p^2 dxdy \quad , \tag{5}$$

and

$$\frac{m_e}{m} = k_m = \int_{x=0}^{a} \int_{y=0}^{b} \left(\frac{w_p}{w_{\max}} \right)^2 dxdy \quad ; \quad m_e = m\, k_m \tag{6}$$

where k_m is the coefficient of mass reduction. The equivalent mass is denoted by m_e. Supposing non- elastic impact, the point of impact on the structure and the falling object are moving together after the impact by $v_{\max}$ velocity. This means that the velocity of the structure will change from the original 0 [m/s] to $v_{\max}$, and the velocity of the falling object will change from the maximum v_1 speed reached during the falling down from the height h [m] to $v_{\max}$. The equilibrium equation of momentum leads to the following:

$$m_o v_1 = (m_0 + m_e) v_{\max} \; ; \; v_{\max} = \frac{m_o}{m_0 + m_e} v_1 \; ; \; v_1^2 = 2gh \; . \tag{7}$$

The total kinetic energy of the structure:

$$E_{kr} = \frac{m_0 + m_e}{2} v_{\max}^2 = \frac{m_0 v_1^2}{2} \frac{m_0}{m_0 + m_e} \; . \tag{8}$$

The following equation states that the total kinetic energy of the structure plus the work of the gravity must be equal to the work of deformation:

$$\frac{m_0 v_1^2}{2}\frac{m_0}{m_0+m_e}+(m_0+m_e)gw_{\max}=\frac{m_0 g}{2}\frac{w_{\max}^2}{w_{s\max}} \quad . \tag{9}$$

Introducing Ψ dynamic factor and μ mass ratio:

$$\Psi=\frac{w_{\max}}{w_{s\max}} \quad , \qquad \mu=\frac{m_e}{m_0} \quad . \tag{10}$$

Using these factors in Eq. 9, we get:

$$A\Psi^2-B\Psi-C=0 \quad , \tag{11}$$

where:

$$A=\frac{m_0 g}{2}w_{s\max} \;\; ; \;\; B=(m_0+m_e)gw_{s\max} \;\; ; \;\; C=\frac{m_0^2 v_1^2}{2(m_0+m_e)} \quad .$$

The solution of Eq. 11:

$$\Psi=(1+\mu)\pm\sqrt{(1+\mu)^2+\frac{2h}{(1+\mu)w_{s\max}}} \quad . \tag{12}$$

By using this method, one can determine the dynamic deformation caused by the falling object and it is possible to calculate the stresses, too.

Summarizing the steps of the method:

1. Calculation of the mass of the structure, Eq. 4.
2. Finite element calculation of the static deformations caused by the weight of the falling object, w_{smax} .
3. Calculation of the mass reduction factor, Eq. 6.
4. Determining of the equivalent mass: $m_e = m\, k_m$.
5. Calculation of the mass ratio, Eq. 10.
6. Solving the dynamic factor, Eq. 12.
7. Result for the dynamic deformation: $w_{max} = \Psi\, w_{smax}$.

These steps could be steps of a program code, because the method is easy to program in any programming language. Writing this code in a built-in programming language of a finite element program system (eg. COSMOS/M, or ANSYS APDL), it gives the possibility to perform the necessary static calculation of static deformations inside of the same finite element program system. By using numerical integration, this program can be developed into a complete program for the analysis of any three dimensional structure for impact load. The code of this program was written in the built-in programing language of COSMOS/M finite elements program system of SRAC (1997) by applying a numerical integration technique, performing the steps from 1 to 7. An important advantage of this method is that we always can use all the possibilities of the original finite element program system and one can combine these new features of impact with all the multidisciplinary and multiphysics features of the original program system. By this way it will be possible to solve complex and special problems, which could be very difficult or impossible to solve in any other program system. For example, one can combine the impact load and optimization with nonlinear structural behaviour, or with thermal effects and temperature dependent material characteristics (e.g. in case of fire), magnetic effects, fluids, etc. The solution of these kinds of special and complex problems could be very useful during the design and optimization of structures subjected to extreme loads in case of disasters (fire, tornado, etc.), accidents or explosions due to natural catastrophes or terrorist attacs. The results and solutions of these problems could lead to the design, optimization and fabrication of new and safe products, building elements, car parts, human protection clothes or helmets, etc.

3 The RVA algorithm for multidisciplinary optimization

Evolutionary algorithms are very efficient and robust algorithms for the optimum searching and they are capable to handle large number of design variables and/or computer capacity consuming calculations of the objective functions and design constraints. These characteristics make it possible to apply these algorithms in multidisciplinary optimization problems. The computation time necessary to reach the final optimum depends on the thinking and strategy of the algorithm and also on the computation time necessary to evaluate the objective function and to check the fulfilment of the design constraints. Therefore, it is very important that the algorithm should be very efficient in point of view of the necessary objective function evaluations and design constraints checks until reaching the final optimum. This fact gives the possibility and the necessity of developing newer algorithms having higher and higher efficiency in this context.

Genetic Algorithm, Particle Swarm Algorithm and Ant Algorithm are based on investigations of biological systems and the thinking of these algorithms simulates the behaviour of these systems. The efficiency of these algorithms is because of the efficiency of the given biological system behind the theory of the algorithm. The given biological system is a result of several thousands of years of evolution, the existence and development of these systems is the prove of their success. This success is transferable into the optimization process and the given algorithm will be successive too. Let us investigate a very efficient biological construction: a virus. The efficiency of this biological system is in his very fast reproduction capacity. If the circumstances and conditions (temperature, light, oxigen, food) are good, viruses can reproduce themselves in a very high speed and they will cover almost all the possible places in the area of investigation. If the life conditions are not good, or the conditions show a non-uniform distribution, higher number of viruses can be found in an area of better conditions than in some areas giving poor conditions of life for viruses. Therefore, these biological structures are very efficient in finding the best conditions of life, the highest number of virus entities will be found in the area having the best life conditions. Another very important thing is that a virus is always a very simple construction, contains only the most important information necessary for life and reproduction. This simplicity gives a very high flexibility to a virus in changing and mutation, therefore they can accomodate to several conditions very easily. Therefore, the efficiency of a virus is very high in point of view of behaviour, construction, life reproduction and changing. This efficiency is applied during the development of computer viruses, which show several similarities to real biological viruses (simple structures, very fast reproduction, easy changing). Mainly these characteristics give that a computer virus is also a very efficient system. Applying this multi-form efficiency in development of an optimization algorithm could result in high efficiency in the optimum searching process, too.

During the build-up of the algorithm, the first step is to find the starting points fulfilling all the explicit and implicit constraints. It is possible to generate coordinates using the explicit constraints and check the generated points against the implicit constraints. In order to keep the simplicity of the virus algorithm, the number of the starting points proposed is very low. The coordinates of the starting points can be denoted by x_i , i = 1,2,…,n where n is the number of design variables. In this case the points can be denoted as P_j , j= 1,2,…,m , where m is the number of the starting points. The explicit constraints of the design variables:

$$l_i \leq x_i \leq h_i \ , \qquad i=1,2,\ldots,n\ , \tag{13}$$

l_i are the lower limits, h_i are upper limits of the constraints. The implicit constraints can be written in the following form:

$$u_k \leq f_k(x_i) \leq v_k\ , \qquad i=1,2,\ldots,n \quad , \quad k=1,2,\ldots,p\ , \tag{14}$$

where p is the number of implicit constraints. The starting points are vectors in the design space:

$$P_j = \{x_i\}_j \qquad . \tag{15}$$

They can be found in the feasible region of explicit constraints by using random numbers and after they are checked against the implicit constraints. If a point is unfeasible, a new one should be generated. The goal of the optimization process is to find the extremum value (maximum or minimum) of the objective function:

$$\Omega = extr[F(x_i)] \qquad , \tag{16}$$

where F is an arbitrary non-linear function of the design variables.
Once the starting points generated, the reproduction procedure is starting:

$$y_i^{\alpha} = x_i + R_i q(h_i - l_i) \qquad . \tag{17}$$

Here y_i are the coordinates of the new point generated, R_i are random numbers between the values of 0 and 1 and α is the number of new entities generated in the reproduction procedure, q is the spreading parameter. Proposed value of spreading parameter is between 0.5 and 0.8 in case of the first three generations and between 0.2 and 0.4 afterwards. The reproduction step is executed for each starting point. The new generation created in this step can be denoted as generation α. The next generation (we can call it generation *β*, after *γ* and so on) can be created using the reproduction formula of Eq. 17. for each point of the previous generation. In order to prevent the overwhelming number of points controlled at the same time, it is necessary to select the points having the best objective function value and destroy the points having the worst objective function value. This procedure can be continued until a given number of generations is reached or the procedure can be ended if the maximum difference in objective function values regarding a generation will be under a given small value.

4 Numerical example – optimization of the metal structure of a car set back for impact loading

Supposing a car accident, where a car hits another car from the back, the person sitting in the “target car” will be pressed into his seat due to his inertial forces. This load is an impact-like, sudden load exerted to the back of the seat. Before the optimization, numerical and experimental investigations were made in order to observe the behaviour of the car seat structure subjected to impact loading and to verify the calculation method introduced in this paper, in chapter 2. The three dimensional model of the seat was made in SolidEdge program and the finite element model was built in the COSMOS/M program system. Figure 1 and 2 show the 3D and FEM model used for the investigations.

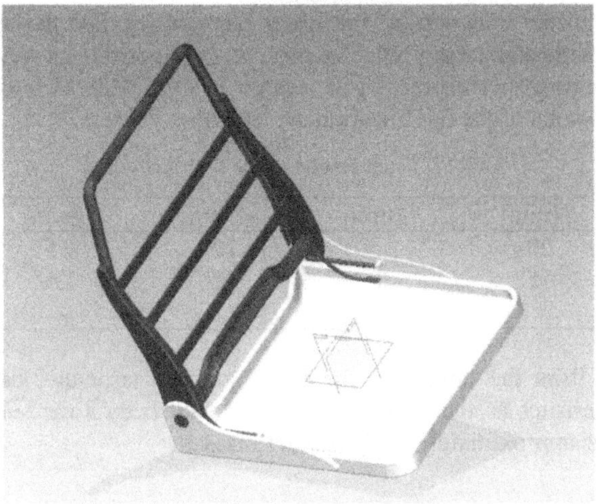

Figure 1. Three dimensional model of the car seat.

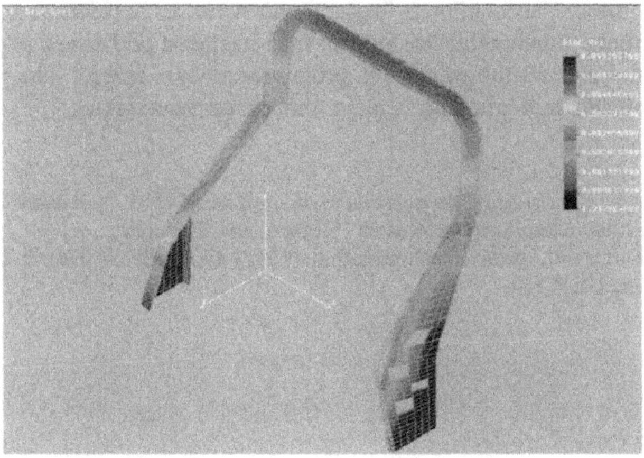

Figure 2. Finite element results of the optimized structure (displacements).

During the collision it is supposed that a 50 kg part of the person's body is hitting the car seat back by 30 km/h velocity. This is the impact loading to the seat back. As experimental investigation, a mass of 50 kg was dropped from 3,5 m height to the seat back, while the seat was positioned so that the back was horizontal. The experimental result for the maximum deformation of the seat back due to this loading was 250 mm, the finite element result using the method described in the chapter 2 was 220 mm. An important part of the difference was caused by the deformation of axial element placed at the bottom of the seat back which was not so rigid as supposed in the finite element model. It can be said that the experimental and finite element results are in good agreement, therefore the proposed method can be used for the optimization. During the optimization the objective function was the mass of the structure because it is placed in a car. The constraints of maximum deformation and maximum stress were applied. The design constraints were the diameter of the (upper) pipe element (*d*) of the seat, thickness of the pipe element (*t*), width (*b*) and the thickness of the lower element (*c*). For the optimization the Random Virus Algorithm was used. The code of the algorithm as well as the code of the method described in chapter 2 was written in COSMOS/M built-in parametric language. The results of the optimization are shown in Table 1.

Table 1. Final results of the optimization.

	d [mm]	*t* [mm]	*b* [mm]	*c* [mm]	mass [kg]
original	20	1,5	88	1,5	0,85
optimal	18	1	85	1,5	0,69
diff. [%]	-	-	-	-	18

It can be seen from the table that the final optimum structure, having the same rigidity characteristics as the original structure taken from a car selected from the marketplace, has approximately 18 % smaller mass.

5 Conclusions

The metal structure of a car seat back has been optimized by using RVA algorithm, for minimum mass. The loading is impact load when an accident is supposed. The analysis of structural behaviour for impact load has been performed in COSMOS/M program system and all the process is programmed as a macro. The results of the numerical analysis are in good agreement with experimental ones.

References

Sz. D. Ponomarjov (1965) *Szilárdsági számítások a gépészetben 6. , (Strength calculations in machine design, in Hungarian),* Műszaki Könyvkiadó, Budapest.

Structural Research and Analysis Corporation (1997) *COSMOS/M User's Guide*, SRAC, Santa Barbara CA, USA.

2.7 Experimental Testing of Space Framework Jib of a Crane

Imre Timár, Pál Horváth, István Lisztes
University of Pannonia, H-8100 Veszprém, Hungary,
e-mail: timari@almos.vein.hu

Abstract

We composed the objective function and the restrictions of the steel structure with spatially changing geometry. Making use of genetics algorithm we optimized a jib of crane with variable height. Based on the similitude theory we calculated a small model dimension and we fabricated it. On the model we measured the stresses, the deflections of jib and the eigenfrequency. We examined the strength of the correlation between the calculated results and the measured ones.

Keywords: *optimization, genetic algorithm, spatially variable geometry, model.*

1 Introduction

Based on our Faculty research results we composed an objective function suitable for optimizing a steel structure with spatially variable geometry, which takes into account the following cost components: material costs, cutting, chopping expenses and the cost of preparation, welding, cleaning and painting. The restrictions refer to the geometrical sizes, local buckling, maximum stresses, bending out, welding stresses, plasticity of the flange member, shear tearing out of the nodes, maximum deflection, eigenfrequency and dimensions due to production technology. The calculation are with 10 variables (diameters and thickness of walls of flange, column bar in horizontal plane, cross-bar in horizontal plane and cross-bar in the sloping plane, minimum and maximum heights of jib).

2 The optimization process

We carried out the calculations with the help of genetic algorithm *(GEATbx)* and the *COMSOL* finite element program, both of which run under *MATLAB* surroundings. Optimization is carried out by the *GEATbx* program part, whilst the *COMSOL* program checks the stresses. The forces were taken with the value of $F_x = 4256$ [N], $F_y = 49050$ [N], $F_z = 9750$ [N] (see Figure 1). We also completed the calculation with fixed height, which showed that a saving of 5.7 % can be achieved by variable height.

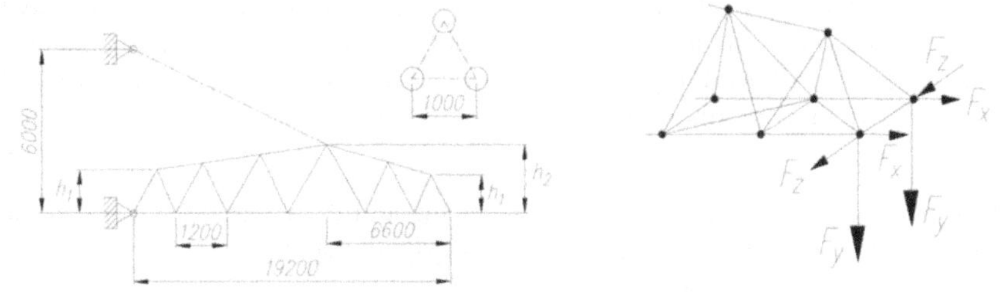

Figure 1. Dimensions and load of the jib

The material was S235 steel. Table 1 shows the results of optimization. We got similar results when using another material, too [Timár et al. (2003a), (2003b)].

Table 1. Optimization results calculating of the steel S235

h_1 [mm]	h_2 [mm]	$d_1 \times v_1$ [mm]	$d_2 \times v_2$ [mm]	$d_3 \times v_3$ [mm]	$d_4 \times v_4$ [mm]	Cost [$]
725	1 775	Ø 108×3.6	Ø 44.5×2.6	Ø 28×2.3	Ø 25×2.3	2 267

Here d_1 and v_1 are dimensions of flange, d_2 and v_2 are dimensions of cross-bar in the sloping plane, d_3 and v_3 are dimensions of cross-bar in horizontal plane, d_4 and v_4 are dimensions of column bar in horizontal plane.

3 Development of model

In order to check the results calculated on the basis of the mechanical model we prepared the physical model of the jib. We defined the main dimensions of the jib model structure on the basis the theory of similitude (l, 1_1, l_2, d_1 v_1, and h_1), where l is the length of the jib, l_1 is the length of the jib during suspension, l_2 the length of the console part of the jib and the other letters are as earlier. During the calculation (Eq.1.) we chose the maximal stress of outside fiber of the jib as the basis for the similitude, since it is not significantly influenced by the diameter and wall thickness of the jib member.

$$\frac{Fl}{K_x E} = \frac{F'l'}{K_x' E} = \frac{F c_F l c_l}{K_x c_K E}, \qquad (1)$$

where $F{=}F_y$, K_x is section module of jib, E modulus of elasticity of material, the letters with commas are the similar dimension of jib and c_F, c_l, c_K are scale factors. We had taken down F'=1075 [N] and l'=1500 [mm], so would be c_F=0.0219 and c_l=0.0765. After simplification we got:

$$\frac{c_F c_l}{c_K} = 1, \qquad (2)$$

from Eq.2 we got c_K=0.0016. We had chosen d_1' x v_1' dimensions Ø 14x1.5 mm and after we calculated h_1' value. We got h_1'=56 mm and chose h_1'=60 mm.

We made our calculation with the force of 1075 N as shows Figure 2. We calculated the other dimensions of the model construction, the rod strengths and the displacements using COMSOL program. Figure 3 shows the dimension of optimized model. At the model the dimensions of $d_2 \times v_2$, $d_3 \times v_3$, $d_4 \times v_4$ are the same for the fabrication.

We have calculated the stresses, the displacements and the eigenfrequencies in the rods of jib with AxisVM 7 program. Figure 4 shows the shape of loaded jib and its deflections in the vertical plane.

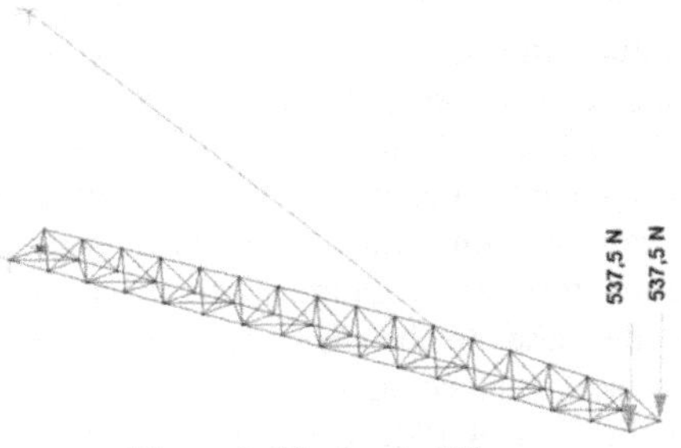

Figure 2. The load of jib

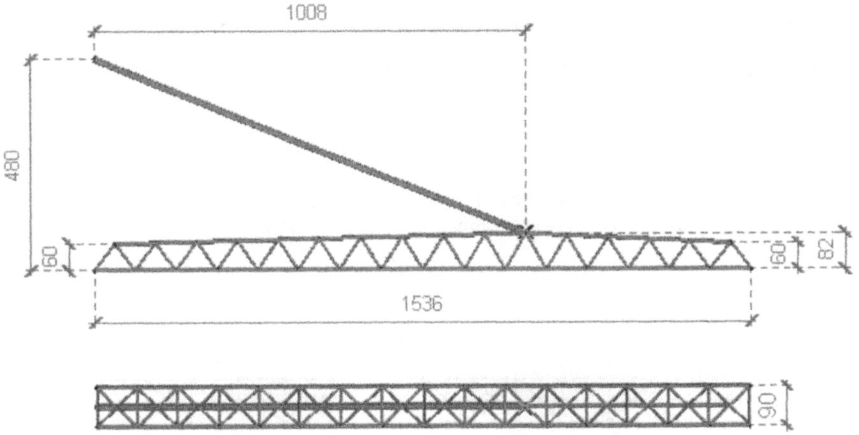

Figure 3. The measurement of optimized model

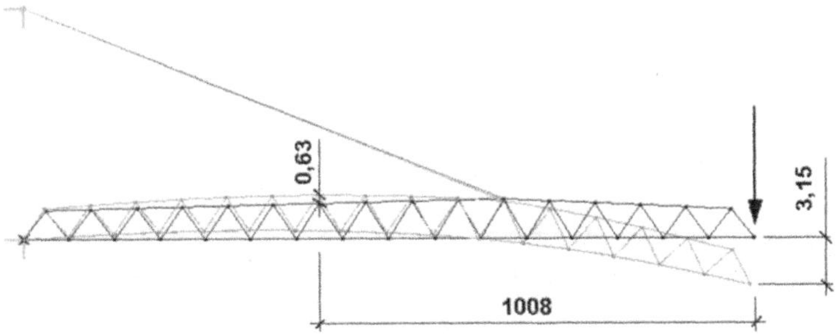

Figure 4. The calculated deflection of jib

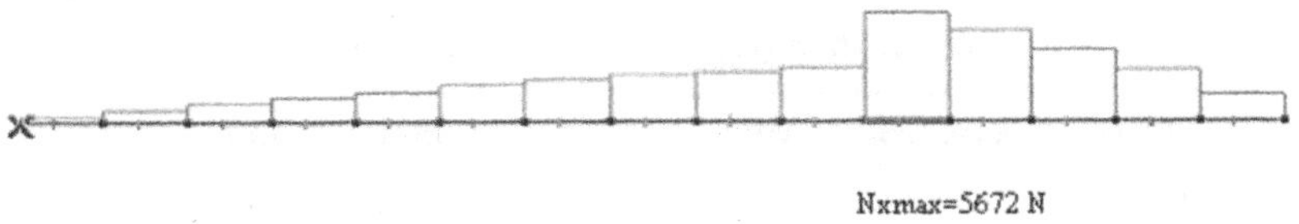

Figure 5. The rod forces in the upper flange
On the rod the size of rectangle is proportional to the force.

Figures 5 and 6 show the calculated rod forces in the upper and lower flanges and Figure 7 and 8 show the form of the eigenfrequencies. In the end we have made a comparison of the calculated values and the measured values. For the measurement we used Hottinger HBM Spider 8 device and its software that is CATMAN. The device could measure by using 8 channels at the same time. We have used only three channels in the electronic connection with half Weathstone bridge. For the measuring of the strain two strain gauges have been used, one of them on the unstressed place and the other to the measured point. The device of displacement was the WI/10 instrument that could measure 10 mm-s. We have measured the eigenfrequency with the sign of B12 acceleration signal device and its measuring frequency was 4800 Hz. The measured value is showed in the table 2.

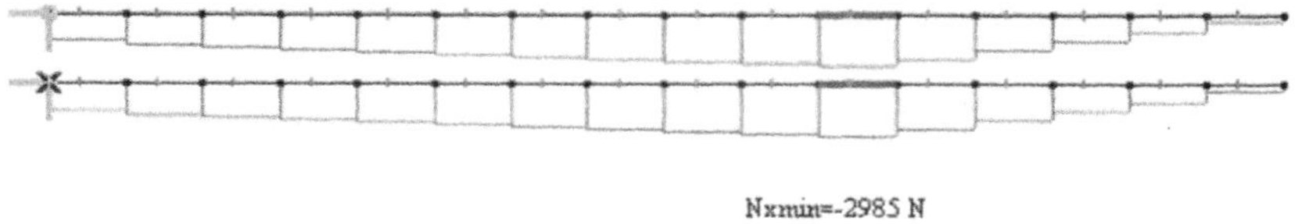

Figure 6. The rod forces in the lower flanges
On the rod, size of rectangle is proportional to the force.

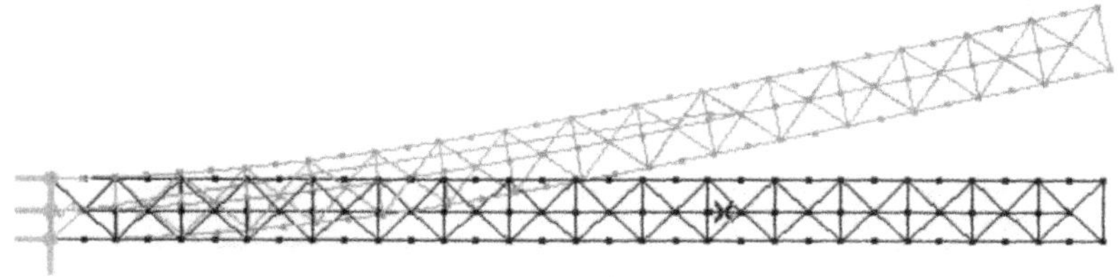

Figure 7. The shape of first eigenfrequency in the horizontal plane

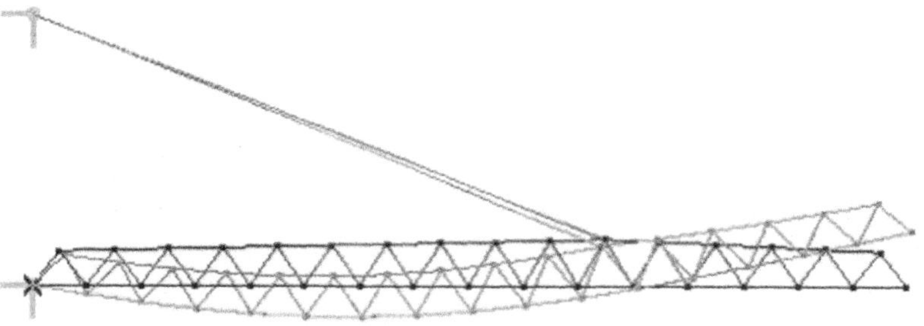

Figure 8. The shape of first eigenfrequency in the vertical plane

Table 2. The measured stresses and deflections

unit of measurement	MPa	mm	mm	mm
average	108.9	-0.21	-3.09	-4.67
dissipation	0.67718	0.02538	0.17165	0.08651

The first deflection value is at the highest point of jib, the second value is at the clamping and the third is at the end of jib.

Figure 9. The measurement

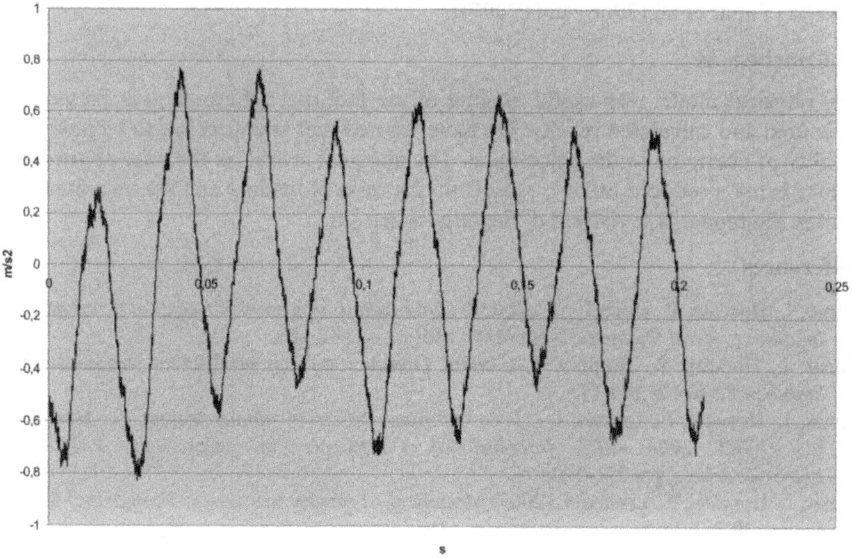

Figure 10. The measured first eigenfrequency in the horizontal plane

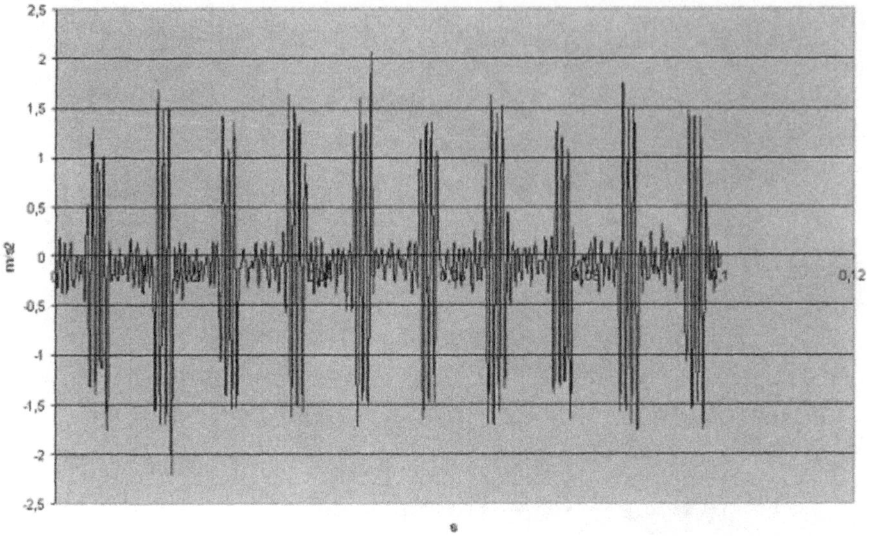

Figure 11. The measured first eigenfrequency in the vertical plane

The measuring time in the vertical plane was 0.1 [s] and 0.2 [s] in the horizontal one. The results are shown in Figure 10 and Figure 11. So we got that the measured eigenfrequency in the horizontal plane is 100 Hz and the calculated one is 105.5 Hz. The values got for the horizontal plane are 39 Hz and 30.1 Hz. The reason of deviation could be the rigidity of clamping. The accuracy of the measured stresses is 0.45 % [Timár et al. (2006) and (2007)].

4 Conclusions

The physical model was useful in spite of the fact that difference was between the measured and calculated results. We have learned that attention has to be paid to the rigidity of clamping in the calculation. The physical model on the base of similitude theory is not a reduced jib. It is seen from the ratio of loading and the own mass. The load in the model is sevenfold of the large-sized jib.

References

Timár, I., Horváth, P., Borbély, T. (2003a) *Optimization of a welded I-section frame with size limitation. Metal Stuctures,* Rotterdam: Millpress, 183-188.

Timár, I., Horváth, P., Borbély, T. (2003b) Optimierung von profilierten Sandwichbalken. *Stahlbau* **72** No. 2, 109-113.

Timár, I., Horváth, P., Lisztes, I. (2006) Optimum design of tubular trusses. *(in Hungarian)* In: *OGÉT 2006, XIV. International Conference in Mechanical Engineering, Marosvásárhely*, pp. 336-340.

Timár, I., Horváth, P., Lisztes, I. (2007) Modelling of tubular trusses. *(in Hungarian) Műszaki Szemle* **38** 383-389.

Section 3

Structural optimization II

3.1 Optimization of Steel Beams and Columns for Variable Rib Configuration

Marcin Chybiński[2], József Farkas[1], Andrzej Garstecki[2], Károly Jármai[1] and Katarzyna Rzeszut[2]

[1]*University of Miskolc, H-3515 Miskolc, Hungary,* altfar@uni-miskolc.hu,

[2]*Poznan University of Technology, Poznan, Poland*

Abstract

Optimal configuration of stiffening ribs in the steel welded I – beams and columns is studied. A wide class of designs with traditional orthogonal configuration implementing transverse and/or longitudinal ribs is considered and compared with a new configuration with diagonal ribs. The critical load and the total manufacturing cost, composed of costs of steel, cutting and welding are computed for all designs. The configuration with diagonal ribs proved to be more efficient than the traditional ones.

Keywords: *stability analysis, local buckling, steel welded girders, optimization.*

1 Introduction

Welded steel girders have found wide application in modern steel structures. Steel beams produced from very thin steel sheets are advantageous from economical reasons. However, these structures demonstrate great tendency towards local instability and are sensitive to initial geometric imperfections. Moreover, when local buckling load coincides with, or is closed to global one, we face an interactive buckling. It can result in high sensitivity to imperfections and often to unstable post-buckling behaviour.

The disadvantageous influence of local instability can be reduced by application of transverse stiffeners. The optimal detailing of transverse stiffeners is not a trivial task. In optimal designing of girders in 4-th class of cross section, the aim is to find configuration of ribs, which minimizes the influence of local instability and simultaneously it must provide a design which is less expensive than a girder in 3-rd class of cross section.

Actual design codes provide practical design recommendations with respect to local instability of web only when the transverse stiffeners are orthogonal to girder's web and flange plates. In this case the problem of local instability can be limited to the square or rectangular plate representing the part of the web between ribs. Influence of the local instability is usually accounted for by the coefficient χ as a function of plate slenderness $\bar{\lambda}$ (ENV 1993-1-1 1992). This approach cannot be applied to diagonal ribs. Moreover, it does not take into consideration the interactive buckling, when local and global instability appears at the same load level.

The influence of rib configuration on the load capacity of the beam was discussed in Chybiński et al. (2007), Rzeszut et al. (2004) and Shahabian & Haji-Kazemi (2005). The dependency of stiffness of the head plates on the critical bending moment in the lateral buckling was studied experimentally in Lindner & Gietzelt (1984) and numerically in Kurzawa et al. (2005). In Szymczak et al. (2001) a supper element suitable for numerical modelling of the thin-walled members with different rib configuration was proposed. The influence of initial imperfections on buckling and post-buckling behaviour was studied in Rzeszut et al. (2004) using non-linear stability analysis with Riks method.

In this paper stability analysis of steel girders with transverse and longitudinal stiffeners for a wide range of beam lengths and rib dimensions and configurations is presented. One can expect simultaneous local and global buckling. Therefore, FEM is used employing shell elements implemented in general proposed program ABAQUS (2001). Hence, the global and local buckling modes can be captured for all rib configurations.

2 Formulation of the problem

The linear stability analysis is performed using FEM with rectangular four-node shell elements S4R with reduced integration and 6 DOF in each node. Vlasov beam theory in continuous formulation was used for comparison. The linear eigenvalue problem had the classical matrix form:

$$(\mathbf{K}^{O} + \lambda \mathbf{K}^{G*})\mathbf{U} = \mathbf{0}, \tag{1}$$

where: $\mathbf{U}$, λ and $\mathbf{K}^{G*}$ denote eigenvector, load multiplier and geometric matrix, respectively. The efficiency of various configurations and dimensions of ribs is studied for beams and columns. However, the major part of the study is devoted to beams, since the state of stress in webs of beams and local buckling phenomena are more complex due to high shear effects.

In numerical analyses concerning beams the I-section 1250x10x260x24mm was assumed with varying span lengths L = 5.0; 7.5 and 10.0m. The calculations were carried out for different rib dimensions and configuration including diagonal ones (Fig. 2). Theoretical “fork” supports with unconstrained warping were modelled (Fig. 1).

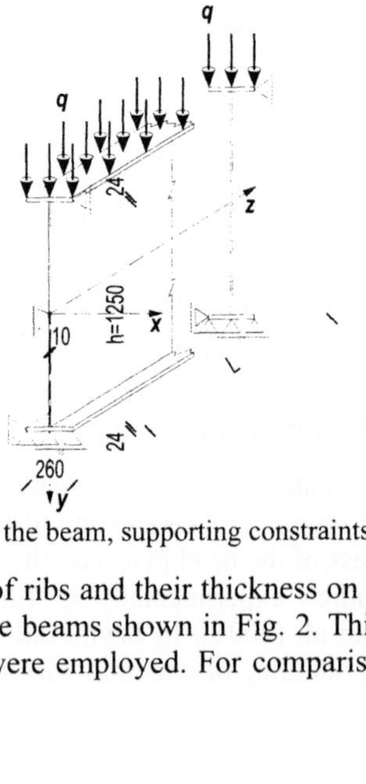

Figure 1. Dimensions of the beam, supporting constraints and the loading q.

The influence of the number of ribs and their thickness on the critical buckling load was carried out on a set of five beams shown in Fig. 2. Thickness of ribs was 4 and 12mm. FEM shell elements were employed. For comparison Vlasov buckling load was computed, too.

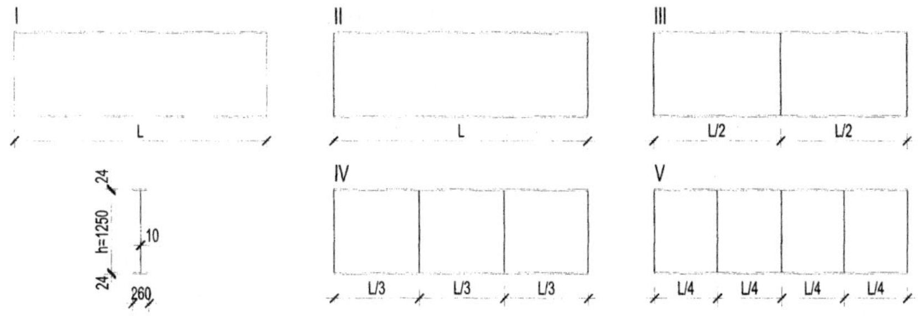

Figure 2. The set of 5 beams. Thickness of ribs 4 and 12mm. (I) Beam without ribs. (II) Ribs (head-plates) at supports. (III)-(V) Ribs at supports and in span.

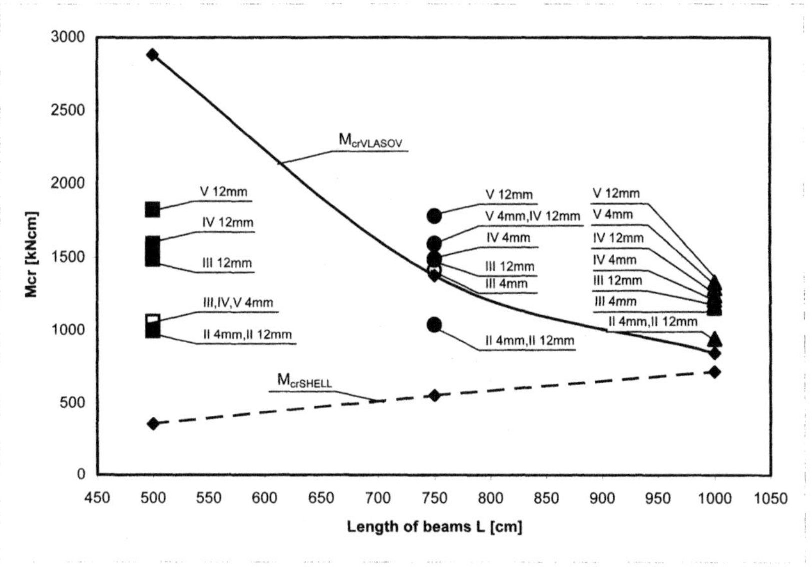

Figure 3. Critical bending moments M_{cr}. Solid line: Vlasov theory. Dashed line and points: shell elements. Greek numbers II –V refer to beams shown in Fig. 2. Open circles, squares and triangles refer to webs 4 mm.

The results of the stability analysis are shown in Fig 3. Note that the vertical axis in Fig. 3 represents the bending moment of simply supported beam M= 0.125·qsL^2, where s=26cm is the width of flange. Convergence of solid line (Vlasov theory) to the dashed line (shell elements) for increasing length of beams confirms the well known assumption of Vlasov theory, that the length of beam is sufficiently long, namely $L\geq$10H. In our case H=1298+48=1298mm and $L\leq$13.0m, therefore the Vlasov theory cannot be applied. Analyzing the results represented by points one observes radical increase of critical moment with increase of number of ribs. This result was expected, since in case of short beams the essential role is played by the local buckling. Rather unexpected were small differences in M_{cr} due to variable thickness of head-plates and ribs (4mm and 12mm).

The main part of the study concerns the efficiency of ribs with variable configuration. More than 20 configurations were analyzed. Eight of them will be discussed in the paper. Fig. 4 illustrates four beams with diagonal ribs, which will be compared with four traditional orthogonal rib configurations. The length of all beams is 10.0m. The cross-sectional dimensions, supporting constraints and the loading q is shown in Fig. 2. The thickness of all ribs is 12mm. The longitudinal stiffeners in models 7 and 8 have the same thickness and width as transverse ribs in all models. The longitudinal stiffeners are located in the zone of compressive stress, namely at $2h/3$ from the bottom flange. Therefore, in model 7 they are concentrated in middle part of the span, where maximum bending moment appears.

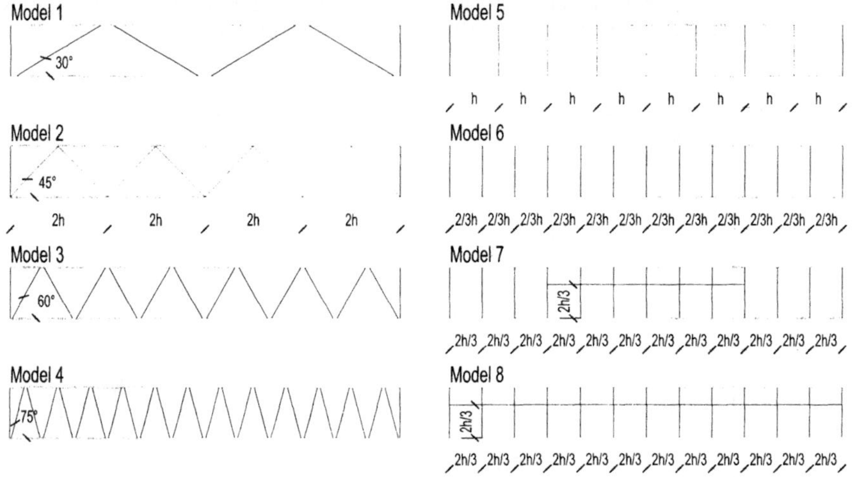

Figure 4. Various configurations of ribs. Models 1-4: diagonal ribs with inclination angles α= 30°, 45°, 60°, 75°. Models 5-8 orthogonal ribs.

3 Influence of ribs' configuration on critical buckling moment and on manufacturing cost

The models considered in the analysis and the results are presented in Table 1. The first three columns specify the model, namely its number referring to Fig. 4, the angle of inclination of ribs α and the total number of stiffening ribs n_s as the sum of numbers of vertical ribs n_v, horizontal n_h and diagonal n_d .

The buckling load multipliers λ_{cr} and hence the buckling bending moments M_{cr} were computed for all beams using FEM with shell elements. The critical buckling moment is presented in 4-th column of the table 1. The 5-th column presents the total length of fillet welds connecting flanges to the web and ribs to the I-section. The next column gives the total mass of the beam. The round bracketed numbers in columns 5 and 6 specify the total length of welds connecting ribs and the total mass of ribs, respectively. The last column gives the total manufacturing cost of beams.

The total manufacturing cost was assessed using the specific unit costs of material (steel), cutting and welding. These unit costs were obtained from a medium-sized company in Poland, specialized in production of steel structures. The prizes were actual in October 2007. Cutting cost of steel sheets was estimated for gas cutting.

The cost of fillet welding refers to metal active gas welding (MAG welding) in process (process number is 135) EN ISO 4063 (2000), which provides continuous welds, both fillet and butt welds. It was assumed that for the connection of flanges to the web the double-bevel butt welds 12mm, are used, whereas stiffening ribs are connected to I-section using fillet welds 5mm. Specific unit cost of the butt weld 12 mm was 1.2€/m. For the fillet weld 5mm it was 0.51€/m. Cost of cutting was assessed basing on specific unit costs of cutting steel sheets: thickness 24mm – 0.84€/kg, thickness 12mm and 10mm – 0.11€/kg. Unit cost of material also depended on the thickness of steel sheets, namely: thickness 24mm – 0.82€/kg, thickness 12mm and 10mm – 0.64€/kg.

Table 1. Efficiency of various configurations of stiffening ribs

model	α [°]	$n_s = [n_v, n_h, n_d]$	M_{cr} [*kNcm*]	l_w (l_{fw}) [*cm*]	m_t (m_r) [*kg*]	*Total cost* [€]
Model 1	30	12 = [4,0,8]	195471	9600 (7600)	2255 (294)	1949
Model 2	45	20 = [4,0,16]	282107	11664 (9664)	2353 (392)	2033
Model 3	60	28 = [4,0,24]	312508	13328 (11328)	2427 (467)	2097
Model 4	75	52 = [4,0,48]	323370	20023 (18023)	2751 (790)	2371
Model 5	90	18 = [18,0,0]	159276	9400 (7400)	2225 (265)	1927
Model 6	90	26 = [26,0,0]	171351	11800 (9800)	2343 (383)	2026
Model 7	0, 90	38 = [26,12,0]	207021	14370 (12370)	2459 (499)	2126
Model 8	0, 90	50 = [26,24,0]	218119	16940 (14940)	2575 (615)	2225

Table 1 demonstrates that the configuration of diagonal ribs is surprisingly efficient in comparison with traditional orthogonal ribs. Let us compare the values of buckling moment M_{cr} of beams with diagonal ribs with beams with orthogonal configuration. Compare diagonal model 1 (cost 1949EU, M_{cr} = 1955kNm) with model 5 (cost 1927EU, M_{cr} = 1593kNm). We observe 23% increase of M_{cr} by approximately same price. Compare now diagonal model 2 (cost 2033EU, M_{cr} = 2821kNm) with model 6 (cost 2026EU, M_{cr} = 1714kNm). We observe 65% increase of M_{cr} by approximately same price. Finally compare diagonal model 3 (cost 2097EU, M_{cr} = 3125kNm) with model 7 (cost 2126EU, M_{cr} = 2070kNm). The increase of M_{cr} is 51%. The first author has also developed a study of the post buckling response of the beams with diagonal ribs. The results confirmed the advantage of this configuration. The problem will be further studied.
The results of the analysis of columns will be presented at the conference.

4 Concluding remarks

A study of the efficiency of various configurations and thickness of stiffening ribs in welded steel beams of I-section is presented in the paper. The simply supported beams subjected to uniformly distributed load were considered. The dimensions of beams (4-th slenderness class) showed that the local buckling played an essential role. The bearing capacity was computed from linear stability analysis using FEM with shell elements.

The configuration of diagonal ribs was studied as an alternative to traditional orthogonal transverse and/or longitudinal stiffening ribs. The bifurcation critical bending moments M_{cr} and the total manufacturing costs were computed for a set of different configurations of ribs. In the assessment of the total cost, special attention was paid to implement actual specific unit costs of steel, cutting and welding.

The analyses demonstrated that the configuration with diagonal ribs provides surprisingly efficient solutions. The beams with diagonal ribs demonstrate 20% - 65% increase of M_{cr} compared to the beams of the same cost but with orthogonal ribs. This was already shown in Chybiński et al. (2007). The originality of this paper is that it demonstrates that the increase of M_{cr} is gained without increase of the total cost, which was assessed basing on actual unit prices.

Acknowledgement

Financial support by Poznań University of Technology grant – DS 11-957/2007 is kindly acknowledged. The project was also supported by Hungarian – Polish Intergovernmental Scientific and Technological Cooperation Program for 2006/2007 under No. PL 4/2005. The Hungarian partner is the Research and Technological Innovation Fund NKTH and KPI, the Polish partner is the Polish Ministry of Science and Informatics.

References

ABAQUS/Standard (2001) Hibbitt, Karlsson & Sorensen, Inc.

Chybiński, M., Garstecki, A. & Rzeszut, K., (2007) Influence of stiffening ribs' topology on local stability of steel welded girders (in Polish). In: *The First Congress of Polish Mechanics, Warsaw, Poland.*

EN ISO 4063 (2000) *Welding and allied processes – Nomenclature of processes and reference numbers*, CEN, Brussels, 2000.

ENV 1993-1-1 (1992) Eurocode 3: *Design of Steel Structures, Part 1.1, General Rules and Rules for Buildings*, CEN, Brussels.

Kurzawa, Z., Szumigała, M., Rzeszut, K. & Chybiński, M. (2005) Behaviour of I-sections with head plates in stability problems considering initial geometrical imperfection. In: *CMM-2005 – Computer Methods in Mechanics, Częstochowa, Poland*, pp. 197-198

Lindner, J. & Gietzelt, R. (1984) Stabilising of I-sections under bending with welded head plates. *Stahlbau* **53** No. 3. 69-74. (in German)

Rzeszut, K., Garstecki, A. & Kąkol, W. (2004) Local-sectional imperfections in coupled instability problems of steel thin-walled cold formed Σ members. In: *Proc. of 4th International Conference on Coupled Instabilities in Metal Structures, Rome, Italy*, pp. 21-30

Shahabian, F. & Haji-Kazemi, H. (2005) A modified formula to evaluate shear resistance of plate girders to combined shear and patch loading. In: *Proc. of Fourth International Conference on Instabilities Problems in Metal Structures, Rome, Italy*, pp. 297-306

Szymczak, Cz., Kreja, I. & Mikulski, T. (2001) Numerical modelling of thin-walled I-beam with battens (in Polish). In: *XLVII, PAN, Krynica, Poland*, pp. 111-118

Vlasov, V.Z. (1963) *Thin-walled elastic rods*, Moscow: Izd. Akad. Nauk SSSR.

3.2 Metamodels for Optimum Design of Laser Welded Sandwich Structures

Kaspars Kalnins, Edgars Eglitis, Gints Jekabsons, Rolands Rikards
Riga Technical University, 1 Kalku St. LV-1658, Riga, Latvia,
kasisk@latnet.lv, gintsj@svnets.lv, edgars.eglitis@bf.rtu.lv, rikards@latnet.lv

Abstract

All-metal sandwich panels, made by a process of laser welding faceplates to core-stiffeners, show advanced cost/weight properties compared with the conventional structural applications of stiffened plates. However, optimal design of these advanced structures requires a fast simulation procedure that should have the same level of reliability compared to finite element calculations and natural tests, while being more time effective and less complex. It was shown that different polynomial functions together with design of computer experiments can contribute to such an aim by providing simple however reliable metamodels. The validation procedure indicated an average of 10% relative root mean square error prediction accuracy, and due to this precision the procedure is capable to be used for further (cost/weight) design optimisation together with structural sizing studies and parametric sensitivity analysis.

Keywords: *meta modelling, different core type sandwich panels.*

1 Introduction

The development of new materials and new manufacturing techniques has accelerated during the last several years, and this has made an impact on innovative structural solutions introduced in industrial production. One of these new ideas is the laser welding technique, which has started to find increasing application among different methods of joining components of ship structures Roland (2006). Laser welding is one of the newest welding techniques, and has been available since the 60's. The main advantages of laser welding are low welding distortions, high productivity and easy automation, and these have opened new opportunities in the design of steel structures. The latest advances in sandwich structures compiled by researchers from the aerospace, wind turbine, marine, rail/road transport industries have been summarised recently by Shenoi et al. (2005).

All-metal sandwich panels, made by a process of laser welding of faceplates to core-stiffeners, show advanced cost/weight properties compared with conventional structural applications of stiffened plates. The main benefits of a sandwich structure are caused by the high stiffness and bending strength properties due to the location of the material as far as possible from the neutral axis of the panels Zenkert (1997).

Progress in sandwich structures has been enabled by the development of a straightforward and inexpensive manufacturing technology for different core types Wadley et al. (2003). Cores of interest include honeycombs Cote et al. (2004), pyramidal and tetrahedral trusses Chiras et al. (2002), as well as diamond ducts and corrugated prismatic cores Pokharel et al. (2005). Full-scale application requires that the structural performance be characterised using a combination of analytical and numerical results, validated by experiments. The structural analysis of sandwich panels with thin flat faces was undertaken as early as the 1940's, particularly for aeronautical applications. The theoretical foundation and governing differential equations for the analysis of sandwich panels were presented in detail by Allen

(1969) and Plantema (1966). Design formulations for different core type all-metal sandwich panels filled with core material or empty, and with symmetric or asymmetric faceplates were recently summarised Romanoff & Vasta (2006) where relations necessary to calculate the stiffness and stress of sandwich panels were presented for application in an equivalent 2D for full 3D finite element (FE) analysis. This procedure of different core type sandwich design was implemented into the commercially available software code ESAComp. However, a significant disadvantage compared to a full 3D FE analysis is the estimation accuracy of the total stresses. In real structures the total stress and strain would be the sum of local and global stresses, so neglecting these local stresses leads to underestimation in the presented analysis procedure with respect to the real structure under experimental testing.

Currently sandwich panels composed of I-core and V-core stiffeners are among the most extensively used in manufacturing, however other core-stiffener of Z-core, C-core, Osquare-core, Ocircle-core as seen in Figure 1, continue to retain interest for further investigation. The different core type panels represent different manufacturing and material supply strategies, which have a number of benefits including added value from innovative manufacturing or seamless welding joints if the stiffeners are joined through the core structure to the top plate.

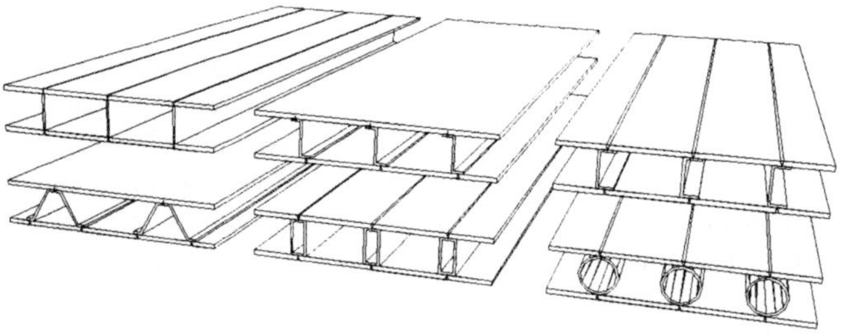

Figure 1. I, Z, C, V, Osquare and Ocircle core type sandwich panels.

2 Meta modelling

2.1 *Design of computer experiments*

The main issue related to meta modelling of structural responses is how to achieve good accuracy of approximated models with a reasonable number of sample experiments. When FE analyses are used to determine stress/strain responses the use of classical design of experiments (DOE), which needs repeated runs, is not effective. Instead deterministic computer experiments sampled according to the space-filling criteria, for example the Latin Hypercube (LH) design McKay et al. (1979), should be used as a basis for evaluation of parametric/non-parametric approximation functions. Typically LH design sample points tend to spread out to the corners of the unit cube, which can be avoided by introducing optimality criteria such as Audze & Eglajs (1977), Minimax and Maximin designs Johnson et al.(1990), Mean Square Error (MSE) and uniform designs Fang & Wang, (1994), Morris & Mitchell (1995), which is a generalisation of Eglajs' criterion. All these designs require pre-knowledge regarding the actual amount of experiments needed for fixed-size design, thus the sampled design space cannot be extended or narrowed without affecting the optimality criteria. Considering this, a more efficient strategy is proposed Auzins (2004), by arranging and adding new experimental points to an already existing design of experiments according to a space-filling criterion, thus achieving a good balance between the space filling quality in the whole design space and quantitative improvement by adding sample points. Moreover sequential designs can be obtained by adding new points to the already existing design space or by arranging the points in optimised large sample quantity design spaces. An advantage of the proposed approach is the fine sampling quality even before all experiment runs are performed, which once elaborated could be made publicly available (www.rtu.lv/mmd/).

2.2 *Polynomials as approximation functions*

Originally meta modelling was associated with low-order polynomial regression models which have global nature in describing numerical responses. They have been well accepted in engineering practice, as requiring low number of sample points, and are computationally very efficient. On other hand they are loosing efficiency when highly nonlinear behaviour should be approximated. Instead the higher-order polynomials can be employed however, if no special care is taken, they tend to overfit the data and produce high errors in regions where the sample points are relatively sparse. As a possible remedy for overfitting particular problems, a subset selection (or model building) techniques (e.g., see Mayers & Montgomery (2002)) may be used. They are aimed to identify the best subset of polynomial terms (or basis functions) to include in the model and to remove the unnecessary ones, in this manner increasing model's predictive performance. However the approach of subset selection assumes that the chosen *fixed* full set of *predefined* (usually just by fixing the maximal order of the polynomials) basis functions contains a subset that is sufficient to describe the target relation sufficiently well. Hence the effectiveness of subset selection largely depends on whether or not the predefined set of basis functions contains such a subset. A short outline of proposed approach of adaptive construction of basis functions is described in the next subsection. It should be noted that proposed approach does not require for user to choose a set of basis functions (or to set the maximal order of the polynomials) – instead the required basis functions can be evaluated automatically.

2.3 *Adaptive basis function construction of polynomial metamodels*

Generally a polynomial model can be defined by a linear summation of basis functions:

$$\hat{y} = \sum_{i=1}^{k} a_i f_i(x) \tag{1}$$

where k is the number of the basis functions included in the model (equal to the number of model's parameters); and $f(x)$ are the basis functions which generally may be defined as a product of the input variables each raised to some order:

$$f_i(x) = \prod_{j=1}^{d} x_j^{r_{ij}} \tag{2}$$

where r_{ij} is the order of the j-th variable in the i-th basis function (a non-negative integer). It should be noted that when all r_j's of a basis function are equal to 0, we have the intercept term.

The Adaptive Basis Function Construction (ABFC) approach Jekabsons et al. (2007) allows generating polynomials of arbitrary complexity without the requirement to predefine any basis functions. In ABFC the standard model refinement operators of subset selection, namely addition and deletion of basis functions, are replaced with other operators, which not only allow adding or deleting basis functions but also allow changing the basis functions themselves (increasing and decreasing orders). Thus in ABFC the search operates directly with the matrix r in the Eq.2.

Still the refinement operators of ABFC allow using the same search algorithms as in subset selection – in Jekabsons et al. (2007) or using of ABFC together with Sequential Floating Forward Selection proposed by Pudil et al. (1994). In order to achieve the trade-off between simplicity and predictive performance of models the Corrected Akaike's Information Criterion was used Hurvich & Tsai (1989).

2.4 *Meta model validation*

Presented research focuses on validation of the selected approximating functions in metamodel building of sandwich structure stress/deformation responses. A total of five hundred sequential design sampling points has been elaborated for training the different core sandwich panel metamodels. The Cross-Validation (CV) technique has been used, where validation procedure has been applied to 400 training points and 100 validation points, we named it 5-fold CV. In order to assess the alteration in prediction performance a half of the sample points where selected for training and half for validation purpose, we named it 2-fold CV. The test sample accuracy measure used is the Relative Root Mean Square Error:

$$RMSE\% = \sqrt{\frac{\sum_{i=1}^{n}(y_i - \hat{y}_i)^2}{STD}} \tag{3}$$

where y_i is i-th test point, $\hat{y}_i$ is predicted value of i-th test point, n is the number of test sample points, and STD is the standard deviation in test sample:

$$STD = \sqrt{\frac{\sum_{i=1}^{n}(y_i - \bar{y})^2}{n}} \tag{4}$$

It should be noted that RMSE% and STD are calculated using strictly only the test sample and averaged over the Cross-Validation runs.

3 Case Study

The present paper deals with derivation of metamodels for a fast simulation tool that should have the same level of reliability compared to FE calculations and natural tests, however, required to be more time effective and less complex. Moreover the developed simulation procedure should be applicable for derivation of optimal design guidelines. A six different core type sandwich panels under bending loading were studied for application as deck panels in a modularized ship concept. Initial studies where metamodels for I-core and V-core type panels Kalnins et al. (2004) and Barkanov (2006) were used in design optimization revealed explicit cost/weight efficiency for certain panel applications. The choice of design variables depended on the core type of all-metal sandwich panels and industrial demands. The geometrical design variables of all considered sandwich core types are shown in Figure 2. All core type stiffeners were similarly positioned in plates at a distance measured to the plate or core profile neutral vertical axis. Also, the V-core stiffener had a constant 60^0 opening angle, thus besides the spacing factor as used for the other core analysis a constant was added in order to avoid stiffener crossing.

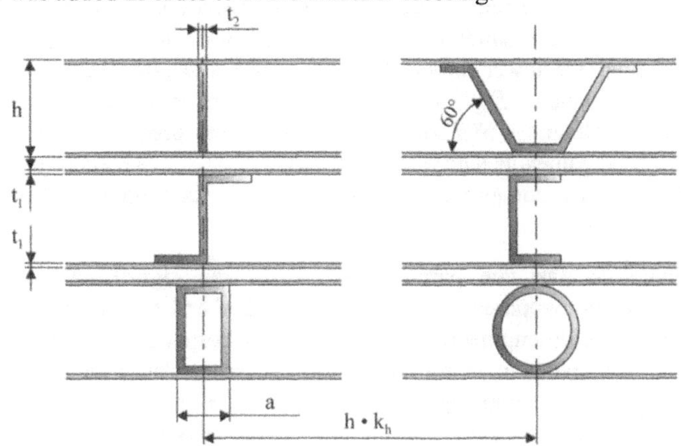

Figure 2. Geometrical parameters for different core type panels

A design process was conducted linking the width of the panel B with the symmetrical number of stiffeners n and the stiffener spacing parameter. Thus, multiplying the panel height h and the core stiffener spacing factor k_h a stiffener spacing parameter can be established. Furthermore the panel length L parameter and two corresponding plate thicknesses are taken as design variables: t_1 – cover plate thickness and core stiffener thickness t_2. The full domain of interest representing lower and upper bounds of the design parameters is outlined in Table 1.

Table 1. Geometrical variables of different core type panels

Name	Notation	Design boundaries		Dimension
		Lower	Upper	
Panel length	L	3	7	M
Panel height	h	4	16	Mm
Top and bottom plate thickness	t_1	2	4	Mm
Core stiffener thickness	t_2	1.5	4	Mm
Core stiffener spacing factor	k_h	1.5	4	
Symmetrical number of core stiffeners	n	2	6	

All-steel sandwich panel numerical experiments were conducted using FEM commercial software ANSYS employing SHELL 181 - 4-node shell element. Initial model verification was performed comparing deflection and stress results obtained in physical tests Kozak (2004). Simply supported boundary conditions were applied to the transverse edge bottom nodes corresponding to the boundaries conditions used in the testing rig. A combined loading has been applied in particular uniformly distributed pressure load of 3 kPa on the top plate and a concentrated load of 1 kN was applied in the centre of the sandwich panel. This corresponded to the load levels required for certification of deck designs corresponding to the DNV (2003) design guidelines.

4 Results

A cross-validation procedure has been carried out comparing different order of full polynomials and polynomials of adaptive basis functions. The prediction errors of the essential structural response metamodels have been compared. In particular the global deflection of the sandwich panel – *DEF_BOT*, the local deflection ratio between the upper and lower sandwich plates – *DEF_DIF*, the equivalent stresses at the upper cover plate – *EQV_TOP*, and the maximum shear stresses from the sandwich core stiffeners – *SHEAR*. Comparison of the prediction accuracy by six different core type panels is summarized in Tables 2 and 3, where 2nd, 3rd, and 4th order polynomials are compared with partial polynomials elaborated by means of ABFC approach.

One can conclude that the partial polynomials can significantly improve the prediction precision compared to the conventional 2nd order polynomials, which are mostly associated with engineering problems of the response surface methodology. For example, the precision of the deflection responses could be improved by an order of magnitude compared to the 2nd order polynomials. In contrary improvement in the equivalent stresses and share stresses characteristics is less efficient. By analysing 5-fold and 2-fold CV results, it could be outlined that, by decreasing amount of the training points, the most decrease of the approximation performance is the property of 4th order full polynomials. In contrast the performance of lower order and partial polynomials reduced in average by only 1%.

Table 2. Metamodel validation accuracy with 5-fold CV

Polynomial	2nd order	3rd order	4th order	Adpt.	Core – type design		2nd order	3rd order	4th order	Adpt.
Response	RMSE%						RMSE%			
DEF_BOT	35.51	20.51	13.27	1.73	I-core	C-core	33.94	18.25	11.04	1.45
DEF_DIF	29.75	13.06	5.73	1.21			31.07	14.22	5.94	1.56
EQV_TOP	17.43	9.57	9.99	7.54			18.11	10.63	13.08	8.71
SHEAR	12.53	8.10	10.52	6.69			11.96	6.60	7.94	5.21
DEF_BOT	33.15	17.22	9.50	2.72	Z-core	V-core	33.09	17.48	10.89	4.17
DEF_DIF	31.32	13.38	5.90	1.57			37.49	18.29	16.12	3.34
EQV_TOP	20.57	13.51	17.03	12.08			38.98	38.75	60.27	35.42
SHEAR	12.06	7.28	8.65	6.45			18.28	13.09	13.35	11.63
DEF_BOT	34.62	18.67	11.79	1.72	Os-core	Oc-core	37.36	21.79	16.18	3.41
DEF_DIF	34.02	15.18	7.10	1.62			30.87	14.34	6.45	1.40
EQV_TOP	18.08	10.29	12.40	8.82			20.04	11.40	11.22	7.83
SHEAR	15.31	9.24	11.40	8.06			22.14	15.77	22.31	15.79

Table 3. Metamodel validation accuracy with 2-fold CV

Polynomial	2nd order	3rd order	4th order	Adpt.	Core – type design		2nd order	3rd order	4th order	Adpt.
Response	RMSE%						RMSE%			
DEF_BOT	37.73	22.28	22.05	4.01	I-core	C-core	35.01	19.27	19.20	2.48
DEF_DIF	30.33	13.74	9.88	1.28			31.10	14.18	10.12	1.57
EQV_TOP	17.76	10.62	22.39	8.24			18.34	11.42	26.29	10.12
SHEAR	12.19	8.38	20.85	7.02			11.82	7.19	17.26	6.66
DEF_BOT	34.94	18.83	19.01	3.39	Z-core	V-core	34.09	17.98	21.03	7.25
DEF_DIF	31.19	14.26	10.56	1.80			38.01	18.58	13.21	3.92
EQV_TOP	20.26	13.88	30.42	11.81			44.97	49.12	91.77	46.29
SHEAR	12.15	7.35	18.69	6.62			19.13	15.39	25.88	13.47
DEF_BOT	35.05	19.13	18.93	2.47	Os-core	Oc-core	39.87	24.36	27.32	6.11
DEF_DIF	34.36	16.19	11.57	1.48			30.88	14.25	10.47	1.44
EQV_TOP	19.16	11.37	26.02	9.18			18.76	11.47	23.72	8.28
SHEAR	15.20	10.52	24.18	10.24			22.05	17.78	38.22	17.38

Conclusion

It was concluded that the elaborated metamodels of adaptive basis function construction as different parametrical polynomials are efficient in surrogating FE analysis of different core type sandwich structures. The approximations obtained, by their precision, are capable of serving in the development process for design guidelines of new sandwich or different composite structures. Moreover evaluated metamodels will be used for further (cost/weight) design optimization together with structural sizing studies and parametric sensitivity analysis.

Acknowledgments

This work was partly supported by the European Social Fund with the National Programme "Support for the development of doctoral studies at Riga Technical University".

References

Allen, H.G. (1969) *Analysis and design of structural sandwich panels*, Pergamon Press.

Auzins J. (2004) Direct optimization of experimental designs. In: *Proc. 10th AIAA/ISSMO Conf., Albany*, NY, AIAA 2004-4578.

Barkanov, E. (2006) Optimal design of laser-welded sandwich modules. *Schiffbauforschung*, **45** No.1, 21-33

Chiras, S., Mumm, D.R., Evans, A.G., Wicks, N., Hutchinson, J.W., Dharmasen, K., Wadley, H.G., Fichter, S. (2002) The structural performance of near-optimized truss core panels. *Solids and Structures* **39** No.15 4093-4115

Cote, F., Deshpande, V., Fleck, N.A., Evans, A.G. (2004) The out-of-plane compressive behavior of metallic honeycombs. *Materials Science and Engineering* **380** No.1–2, 272–280

Det Norske Veritas (2003) In: *Technical report, Project Guidelines for Metal-Composite Laser- Welded Sandwich Panels*, Nr: 2003-0751.

ESACOMP (2006) *User Manual Version 3.6,* Helsinki: www.componeering.com

Fang, K.T. & Wang, Y. (1994) *Number-Theoretic Methods in Statistics*, London: Chapman & Hall

Hurvich, C.M. & Tsai C-L. (1989) Regression and time series model selection in small samples. *Biometrika* **76** 297-307

Jekabsons, G., J. Lavendels, J., Sitikov, V. (2007) Model evaluation and selection in multiple nonlinear regression analysis. *The Baltic Journal on Mathematical Applications* **12** No.1, 81-90

Johnson, M.E., Moore, L.M., Ylvisaker, D. (1990) Minimax and Maximin Distance Designs. *Statistical Planning and Inference* **26** 131-148

Kalnins K., Skukis E., Auzins A. (2005) Metamodels for I-core and V-core sandwich panel optimisation. In: *Proc. 8th Int. Conf. (SSTA-05) Jurata*, Poland, London:Taylor & Francis, pp. 569-572

Kozak J. (2004) Strength tests of steel sandwich panel. *In Proc. 9th Int. Symp. on Practical Design of Ship and Other Floating Structures. Luebech-Travemuende,* Germany, available on web: www.prads2004.de

Lok, T.S., Cheng, Q. (1999) Elastic Deflection of Thin-Walled Sandwich Panel. *Sandwich Structures and Materials*, **1** 279-298

Myers, R.H. & Montgomery, D.C. (2002) *Response Surface Methodology: Process and Product Optimization Using Designed Experiments, 2nd ed.*, New York: John Wiley & Sons

McKay, M.D., Conover, W.J., Beckman, R.J. (1979) A comparison of three methods for selecting values of input variables in the analysis of output from a computer code. *Technometrics* **21** No.2, 239-245

Morris, M.D. & Mitchell, T.J. (1995) Exploratory Designs for Computational Experiments. *Statistical Planning and Inference* **43** No.3, 381-402

Plantema, F.J. (1966) *Sandwich Construction*, New York: John Wiley & Sons.

Pokharel, N. & Mahendran, M. (2005) An investigation of lightly profiled sandwich panels subject to local buckling and flexural wrinkling effects. *Constructional Steel Research* **61** 984-1006

Pudil, P., Novovicova, J., Kittler, J. (1994) Floating search methods in feature selection. *Pattern Recognition Letters* **15** 1119-1125

Roland, F. (2006) Lightweight structures in the maritime industries – an overview of European research projects. *Schiffbauforschung* **45** No.1, 5-20

Romanoff, J. & Varsta, P. (2006) Bending response of web-core sandwich beams. *Composite Structures* **73** No.4, 478-487

Shenoi, R., Groves, A., Rajapakse, Y. (2005) *Theory and Applications of Sandwich Structures*, Southampton: Dorst Press.

Wadley, H.G., Fleck, N.A., Evans, A.G. (2003) Fabrication and structural performance of periodic cellular metal sandwich structures. *Composites Science and Technology* **63** No.16, 2331–2343

Zenkert, D. (1997) *Handbook of Sandwich Construction*, London: EMAS.

3.3 Optimal Design of a Composite Cellular Plate Structure

György Kovács, Károly Jármai, József Farkas
University of Miskolc, H-3515 Miskolc, Hungary, altkovac@uni-miskolc.hu

Abstract

This study shows single and multi-objective optimization of a new complex structural model [laminated carbon fiber reinforced plastic (CFRP) deck plates with aluminium (Al) stiffeners] which is depicted in Figure 1. The structure was designed for both minimal cost and minimal weight. Design constraints on maximum deflection of the total structure, buckling of the composite plates, buckling of the Al webs, stress in the composite plates, stress in the Al stiffeners and eigenfrequency of the structure are considered in the calculation. The flexible tolerance method was used in the single objective optimization and particle swarm algorithm in the multiobjective optimization process.

Keywords: *optimal design, composite cellular plate, cost calculation*

1 Introduction

Sandwich structures utilize the advantages of different structural components. These components can have different structural configurations (e.g. plates or beams) or different material properties (e.g. density or damping coefficients). In the design of layered beams, plates and shells, one can exploit the different beneficial characteristics of these components. Prime examples are orthotropic sandwich structures, which have a high ratio of bending stiffness to density. Hence they are often used in light-weight structures.

Recent literature reviews (Noor & Burton & Bert 1996, Vinson 2001) highlight the significant effort directed at the design, analysis, and applications of sandwich structures. Examples include a bending theory for sandwich beams with thick faces in Stam & Witte (1974). Notable work is reflected by the book of Zenkert (1995). The optimum design of specialized welded sandwich panels for ship floors was treated in Jármai et al (1999), while a five layer beam was analysed and optimized in Farkas & Jármai (1998, 2003). This beam consists of a rubber layer, two aluminium profile beams and two CFRP deck layers.

In the present study a new structural model is investigated. Sandwich plates have deck layers made of metal or FRP (fiber reinforced plastic) plates, and their inner layer is usually made of foam or honeycomb. On the other hand, cellular plates consist of metal deck plates and metal stiffeners welded into the deck plates. Our new structural model combines the sandwich and cellular plates, since it has FRP deck plates and two or more aluminium square hollow section stiffeners riveted into the deck plates. So it is a new combination of materials, stiffeners and fabrication technology.

The multi-cellular sandwich plate is constructed from number of longitudinal Al (aluminium) square hollow section beams and two laminated CFRP deck plates (Fig.1). The connection between the beams and deck plates is effected through riveting. This type of sandwich plate can be applied in many engineering load carrying structures such as ship floors, bridges, airplanes, building floors, etc.

The main aim of the present study is to work out an optimum design procedure for such a structural model. In doing so, design constraints are formulated on the buckling strength of the compressed deck plate, the local buckling of the aluminium square hollow section plate elements, stress in the composite plates and in the Al stiffeners, deflection of the simply supported beams as well as the eigenfrequency of the structure subjected to distributed pressure acting at the total surface.

In order to achieve cost savings in the design stage, a cost function is formulated on the basis of material and fabrication cost analysis. The mass function used in the optimization process includes the sum of the mass of CFRP plates and beams. Mathematical programming methods for constrained function minimization are an integral part of the procedure. The flexible tolerance method (Farkas & Jármai 1997) is used for the determination of the optimal dimensions of the structural model.

2 A new cellular sandwich plate model

The sandwich plate model under consideration is depicted in Figure 1. The *CFRP* plates are constructed from laminated layers. The fiber volume fraction is 61% and the matrix volume fraction is 39%. All of the fibers of a layer and laminate are arranged in the longitudinal direction. Plates are riveted to the upper and lower flanges of the aluminium square hollow section (*SHS*) profiles.

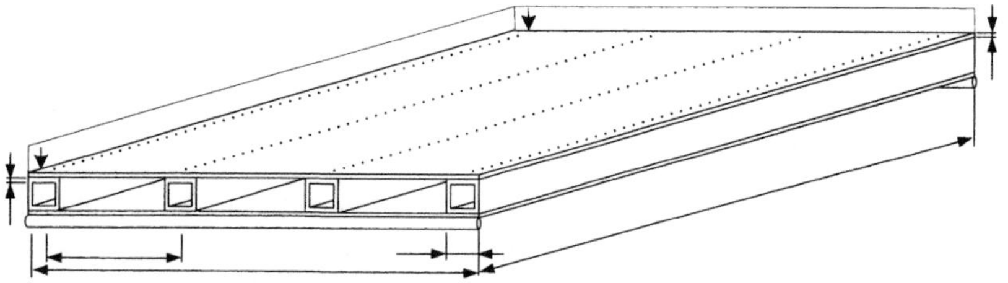

Figure 1. Cellular sandwich plate structure.

The structure is simply supported, and a uniformly distributed loading of $3{,}5\cdot10^{-3}$ N/mm^2. (p = 7 N/mm line pressure) acts on the total surface of the structure. The dimensions of the structure are: $L = 2250$ mm, $B = 2000$ mm.

The material parameters of a pre-impregnated *CFRP* layer are given as follows: the thickness of a layer $t^* = 0{,}2$ mm, the longitudinal Young's modulus $E_x = E_c = 120$ GPa and the transverse modulus $E_y = 9$ GPa. The specific mass of the *CFRP* plate ρ_c = 180 g/m^2, and Poisson's ratios $\nu_{xy} = 0{,}25$ and $\nu_{yx} = 0{,}019$.

3 Objective functions and constraints

3.1 *Cost function*

The structure is optimized with respect to minimum cost K, which can be formulated as the sum of the material and manufacturing costs (Farkas & Jármai 1997), i.e.

$$f(x) = K = K_{CFRP} + K_{Al} + K_{heat\ treatment} + K_{manufacturing}$$

$$K\,(€)= 2\cdot(n\cdot 31{,}047)+ k_{Al}\,[n_s\,(\rho_{Al}\,4\,h_{Al}\,t_w\,L)]+ 2\cdot n\frac{525}{528} +k_f\,[n\cdot 14_{\min} +$$

$$+ n_s\cdot 26_{\min}+ 110_{\min}] \tag{1}$$

where n represents the number of *CFRP* layers, n_s the number of stiffeners, ρ_{Al} the density of the *Al* profile, h the height and t_w thickness of the SHS *Al* profiles.

The main contribution to the material cost arises from the raw material for the composite plates. In our case this cost reached 31,047 €/layer. The cost of the *Al* profile is 4,94 €/kg. The specific fabrication cost k_f=0,6 €/min. The cost of heat treatment depends on the volume of deck plates to be heat treated and type of resin matrix. In our case these cost components can be calculated as a function of layer number and plate dimension. Heat treatment cost of a manufactured 220x1200x2mm *CFRP* plate is known, so compared to it the cost of the examined plates based on volume can be calculated. The resulted ratio can be seen in Eq. (1).

The total fabrication cost (as the function of time [min]) is the sum of the cost required for the manufacturing of the *CFRP* plates ($n\cdot 14_{\min}+110_{\min}$), the cutting cost of the *Al* profiles ($n_s\cdot 6_{\min}$) and the total assembly costs ($n_s\cdot 20_{\min}$). The time associated with manufacturing of the *CFRP* plates consists of the time lost in press form preparation, layer cutting, layer sequencing and final working. Final assembly consists of drilling of the *CFRP* plates and the *Al* profiles, and also riveting. Drilling of the holes is an implicit function of the number of layers. The design variables are the height h and thickness t_w of the SHS *Al* profiles, the number of layers n of the *CFRP* plates and the number of stiffeners n_s. The fiber orientation is fixed for all layers (0°) as described earlier.

3.2 *Mass function*

The total cost of the structure is the sum of the *CFRP* and *Al* components:

$$m= 2\,\rho_c\,[B\,L(\,n\,t^*)] + n_s\,\rho_{Al}\,[L\,(4\,h_{Al}\,t_w - 4\,t_w^{\,2})] \tag{2}$$

where t^* is the thickness of a laminate.

3.3 *Constraints*

3.3.1 *Deflection of the total structure*

$$w_{\max} = \frac{5p\,L^4}{384(E_c I_c + E_{AL} n_s I_{AL})} + \frac{5\Delta M\,L^2}{48(E_c I_c + E_{AL} n_s I_{AL})} \le \frac{L}{200} \tag{3}$$

where: I_c, I_{Al}: moment of inertia of the *CFRP* plate and *Al* profile,

E_c, E_{Al}: reduced modulus of elasticity of the *CFRP* lamina and Young's modulus of *Al* profile.

There is the effect of the relative movement between the components, and is expressed as a function of the differences in predicted stresses in the middle of *Al* profile and *CFRP* plate. Due to difference in stress *($\Delta\sigma$)* there is a corresponding difference in the equivalent applied moment *(ΔM)*. So the second term of the equation is the additional deflection due to the sliding.

3.3.2 *Composite plate buckling* (Barbero 1999)

$$\left(\frac{b_c}{nt^*}\right) \le \sqrt{\frac{\pi^2}{6\sigma_{\max}\left(1-\nu_{xy}\nu_{yx}\right)}\left[\sqrt{E_x E_y}+E_x\nu_{xy}+2G_{xy}\left(1-\nu_{xy}\nu_{yx}\right)\right]} \tag{4}$$

where b_c: plate width between stiffeners, σ_{max}: maximal stress in the *CFRP* lamina E_x, E_y, G_{xy}: laminate moduli, ν_{xy}, ν_{yx}: Poisson's ratios.

3.3.3 *Web buckling in the Al profiles* (Farkas & Jármai 1997)

$$\frac{h_{Al}}{t_w} \le 42\sqrt{\frac{235E_{Al}}{240E_{Steel}}} \tag{5}$$

where: E_{Al}, E_{Steel}: Young's modulus of elasticity of *Al* and *Steel.*

3.3.4 *Stress in the composite plates*

The moment acting on the total structure is distributed on the components of the structure. X_cM is the part of total moment which is acting on composite plate, $X_{Al}M$ is the part of total moment which is acting on stiffeners.

$$\frac{X_c M}{I_c}\cdot\frac{h_{Al}+nt}{2} \le \sigma_{Call} \tag{6}$$

where: $X_c = \dfrac{E_c I_c}{E_{Al} n_s I_{Al} + E_c I_c}$; $M = \dfrac{pL^2}{8}$; $\sigma_{Call} = \dfrac{\sigma_T}{\gamma_c}$: allowable stress, X_cM: moment acting on composite plate, σ_T: tensile strength of composite lamina, γ_c: safety factor (=2)

Because of the high number of stiffeners in the case of optimum design, the stress due to the transversal bending moment can be neglected.

3.3.5 *Stress in the Al stiffeners*

$$\frac{X_{Al} M}{n_s I_{Al}}\cdot\frac{h_{Al}}{2} \le \sigma_{Alall} \tag{7}$$

where: $X_{Al} = \dfrac{E_{Al} n_s I_{Al}}{E_{Al} n_s I_{Al} + E_c I_c}$; $\sigma_{Alall} = \dfrac{f_y}{\gamma_{Al}}$: allowable stress, X_AM: moment acting on *Al* tube, f_y: yield stress of *Al,* γ_{Al}: safety factor (=2)

3.3.6 *Eigenfrequency of the total structure*

$$f_1 = \frac{\pi}{2L^2}\sqrt{\frac{10^3(E_{Al}I_{Al}+E_kI_k)}{m}} \ge f_0 \tag{8}$$

m: weight/unit length of the structure [kg/m], f_0: limitation for eigenfrequency (50 Hz)

Eigenfrequency constraint was not taking into consideration during the optimization, but the optimal structure parameters obtained after the optimization were checked and in all cases satisfied the inequality constraints.

3.3.7 *Size constraints for design variables*

$$10 \le h_{Al} \le 100$$
$$2 \le t_w \le 6 \qquad (9)$$
$$16 \le n \le 32$$
$$7 \le n_s \le 20$$

These represent physical limitations on the design variables [mm], taking economical and manufacturing aspects into consideration.

3.4 *Flexible tolerance optimization method*

Flexible tolerance optimization method was used during the optimization process.

This method is a constrained random search technique. The Flexible Tolerance algorithm (Himmelblau 1982) improves the value of the objective function by using information provided by feasible points, as well as certain nonfeasible points termed near-feasible points. The near-feasibility limits are gradually made more restrictive as the search proceeds toward the solution, until in the limit only feasible x vectors are accepted.

4 Numerical results of single objective optimization

4.1 *Cost optimization*

Cost saving can be a prime design aim of sandwich structures because the composite materials are very expensive. Table 1. shows the result of cost optimization of the analyzed structure based on the cost function (Eq. 1) and design constraints (Eqs. 3-9). The obtained continuous optimal number and geometries of the stiffeners and total costs for case of different numbers of layers (16-32 pieces) are as follows:

Table 1. Result of cost optimization

Number of layers	*Optimal discrete stiffener numbers and dimensions*			Cost
n [pieces]	h_{Al} *[mm]*	t_w *[mm]*	n_s *[mm]*	[€]
16	60	2.5	15	1730
18	60	2.5	14	1841
20	60	2.5	12	1919
22	55	2.5	11	2014
24	55	2.5	10	2126
26	60	2.5	8	2219
28	50	2.5	8	2340
30	45	2	8	2452
32	45	2	7	2570

It can be summarised based on the obtained results that the increasing number of deck layers causes significant increasing of total cost. The optimal structure is a laminated plate with 16 layers. After continuous optimization a secondary search is necessary to find discrete optimum sizes (standard geometries). The global cost optimum is obtained in case of laminate of 16 layers and 15 pieces of 60x60x2,5 mm stiffeners.

4.2 *Mass optimization*

Table 2. shows the result of mass optimization of the examined structure according to the mass function (Eq. 2) and design constraints (Eq. 3-9). The obtained continuous optimal number and geometries of the stiffeners for the case of different numbers of layers (16-32 pieces) of *CFRP* deck panels can be seen in Table 2.

Table 2. Result of mass optimization

Number of layers	*Optimal discrete stiffener numbers and dimensions*			Mass [kg]
n [pieces]	h_{Al} *[mm]*	t_w *[mm]*	n_s *[mm]*	
16	60	2.5	15	78.317
18	60	2.5	14	78.064
20	55	2.5	13	73.862
22	55	2.5	11	70.723
24	55	2.5	10	70.8
26	50	2.5	9	68.1
28	50	2.5	8	66.445
30	45	2	8	65.32
32	45	2	7	66.469

The global mass optimum is obtained in case of the 30 layered deck plate. This optimum is a global optimum only for the examined interval of n, but it is clear that the total stiffness of the examined structure can be increased by the continuous increase of the number of layers of the deck panel which causes the reduction of number and geometry of stiffeners. So a lighter structure can be constructed in this way, but the cost of it will be extremely high.

After discretization the optimal structure has 30 layers of deck plates and 8 pieces of 45x45x2 mm stiffeners.

5 Sensitivity analysis

We used sensitivity analysis to determine how sensitive the structure is to changes in the value of the parameters of the model and to changes in the structure of the model. Different values of many parameters were set to see how a change in the parameters causes a change in the optimal structural construction.

At first, design variables were analysed in aspect of sensitivity in case of the 20, 22, 24, 26 layered plate structures.

It can be realised that design variables have no significant effect on value of objective functions. After that we analyzed the other components of the objective functions. We have found that the optimal solution is very sensitive to changing of specific fabrication cost (k_f).

We completed the multiobjective optimization for case of different values (1; 2; 2,5; 3; 4 times higher value) of specific fabrication cost to present the effect of sensitivity.

Table 3. includes the result of Particle Swarm Optimization (PSO) (Farkas & Jármai 2003) completed for 26 layered deck plate structure. During the optimization the normalized weighting method were used to show the weight of the cost- and mass objective functions. The normalized objectives method solves the problem of the pure weighting method e.g. at the pure weighting method, the weighting coefficients do not reflect proportionally the relative importance of the objective because of the great difference on the nominal value of the objective functions. At the normalized weighting method we reflect closely the importance of objectives.

$$f(x) = \sum_{i=1}^{r} w_i f_i(x) / f_i^0 \quad \text{where } w_i \geq 0 \text{ and } \sum_{i=1}^{r} w_i = 1 \tag{10}$$

The condition $f_i^0 \neq 0$ is assumed.

Table 3. Result of Particle Swarm Optimization

	weights of objective functions	h_{Al} *[mm]*	t_w *[mm]*	n_s *[mm]*
k_f	100-0% weight	60	2.5	8
	0-100% weight	50	3	9
	50-50% weight	50	3	9
$2 \cdot k_f$	80-20% weight	50	3	9
	90-10% weight	55	3	8
	95-5% weight	60	3	8
	100-0% weight	70	3	7
$2.5 \cdot k_f$	100-0% weight	80	4	6
$3 \cdot k_f$	100-0% weight	85	4	6
$4 \cdot k_f$	100-0% weight	90	4	6

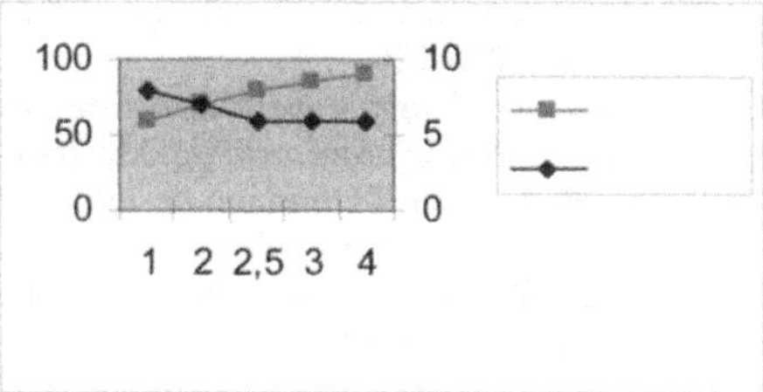

Figure 2. Specific fabrication cost in function of stiffener number and stiffener geometries.

Table 3 includes the optimal structure alternatives for 26 layered deck plate structure. Table summarises the optimal stiffener number and stiffener geometries in case of different value of specific fabrication cost and different weight of objective functions. The first number of weight (2. column of Table 3) represents the effect of cost function in percentage, the second number represents the weight of mass objective function in the multiobjective optimization.

Figure 2 shows the effect of changing of value of specific fabrication cost on optimal stiffener number and stiffener geometries. It can be summarised that the number of stiffeners (n_s) decreases and width of stiffeners (h_{Al}) increases when value of specific fabrication cost increases.

6 Conclusions

A new structural model of a sandwich plate riveted from two aluminium square hollow section rods and two *CFRP* deck plates is investigated by an optimization procedure. In an optimum design procedure the dimensions and number of stiffeners as well as number of layers of sandwich plates are determined, which fulfil the design constraints and minimize the cost and mass. It is shown that significant mass and cost savings can be achieved in the design stage through optimization.

It can be also summarized – based on the mass saving and the disadvantageous extra cost – that the application of fibre reinforced laminates is suggested in those applications where the mass saving is the prime design aim and the cost saving is only secondary. (e.g.: space flight, air-, water- and land vehicles, building parts etc.). Additional advantageous characteristics of these composite structures are the vibration damping and corrosion resistance. Due to the corrosion resistance the surface treatment and painting costs can be neglected which can reduce structural cost significantly.

Acknowledgements

The research work was supported by Öveges József scholarship of the National Office for Research and Technology, OMFB-01431/2006 project.

References

Barbero E. J. (1999) *Introduction to composite materials design*, USA: Taylor & Francis.

Farkas, J. & Jármai, K. (1997) *Analysis and optimum design of metal structure,* Balkema: Rotterdam-Brookfield.

Farkas, J. & Jármai, K. (1998) Minimum material cost design of five-layer sandwich beams. *Structural Optimization* **15** No.3-4, 215-220

Farkas, J. & Jármai, K. (2003) *Economic design of metal structures*. Rotterdam: Millpress.

Himmelblau, D.M. (1972): *Applied nonlinear programming*. McGraw-Hill, New York.

Jármai, K., Farkas, J. & Petershagen, H. (1999) Optimum design of welded cellular plates for ship deck panels. *Welding in the World* **43** No.1, 51-54

Noor, A. K., Burton,W.S. & Bert,C.W. (1996) Computational models for sandwich panels and shells. *Appl. Mech. Rev.* **49** No. 3, 155-199

Stamm, K. & Witte, H. (1974) *Sandwich-Konstruktionen,* Berlin: Springer.

Vinson, J. R. (2001) Sandwich structures, *Appl. Mech. Rev.* **54** No. 3, 201-214

Zenkert, D. (1995) *An introduction to sandwich construction*, W Midlands: EMAS Publ.

3.4 A Simple Function to Estimate Fabrication Time for Steel Building Rigid Frames

Kiichiro Sawada, Hitoshi Shimizu, Akira Matsuo, Takaichi Sasaki, Takashi Yasui and Atsushi Namba
Hiroshima University, Kagamiyama 1-4-1, Higashi-hiroshima, Japan
e-mail: kich@hiroshima-u.ac.jp

Abstract

This study presents a simple function to estimate fabrication time for steel building rigid frames derived from questionnaires given to fabricating workers of three companies in Japan. Next, it presents the coefficients of each function computed using the least squares method from the fabrication time data for four steel buildings of a fabricating company in Japan. Finally, the fabrication time of minimum weight steel building frames, considering different linking groups of design variables, is estimated by the presented function.

Keywords: *fabrication time function, steel building, linking groups, member depth*

1 Introduction

Structural weight does not necessarily predict the fabrication cost adequately. One reason is that the fabrication cost of steel members depends on the complexity of the connections rather than the structural weight. Some fabrication time functions have previously been proposed and applied to a welded stiffened plates (Jármai 2002) and steel frames (Pavlovcic et al.(2004))

This paper presents a simpler fabrication time function for steel building rigid frames, which has been derived from questionnaires given to managers of three fabricating companies in Japan. The proposed function is based on the following assumptions.

(1) The total fabrication time includes the preparatory process time, such as cutting and drilling bolt holes, the assembly time of columns and beams, the welding time and the time of preparing shop drawings. (2) The preparatory process time is proportional to the number of diaphragms and beams. (3) The assembly time of columns and beams is proportional to the number of steel structural parts, such as columns, beams and diaphragms. (4) The welding time is proportional to the jointed sectional area of columns, beams and diaphragms. (5) The time of preparing shop drawings is proportional to the structural weight and the number of columns and beams of different sizes. (6) The painting, transportation and erection times are not considered.

Next, the coefficients of each function are presented, which are computed from the fabrication time data for four steel buildings in Japan, using the least squares method. Finally, the fabrication time of minimum weight steel building frames, considering different linking groups of design variables, are estimated by this function. The minimum weight design of steel building frames is performed by the genetic algorithm, based on the ranking selection (Ohsaki (1995)). The constraints are based on the Japanese seismic design code. The design variables are the discrete cross-sectional depth and thickness. The effect of linking the member depth design variables on the fabrication times is discussed by analyzing these computational results.

2 Typical beam-to-column connection in Japan

Figure 1 shows typical H-beam-to-RHS-column connections in Japan. At the fabricating company, the box column connection is first welded to two through diaphragms by full-penetration welds, as shown in Fig. 2. The connection is welded to the flanges of the bracket by full-penetration welds and to the web of the bracket by fillet welds as shown in Fig. 3. Finally, the columns are welded to the connection by full-penetration welds as shown in Fig. 4. In the field, the bracket is connected to the beam by high strength bolts. This study deals with buildings having the beam-to-column connection shown in Fig. 1.

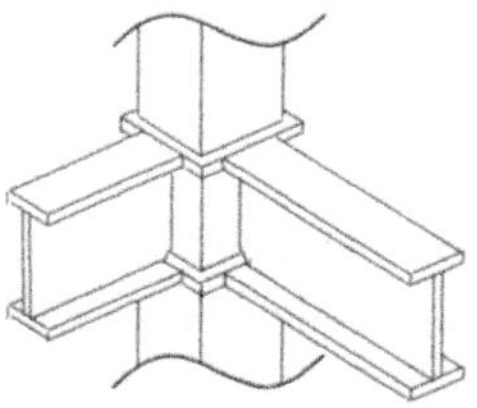

Figure 1. Typical Beam to column connection

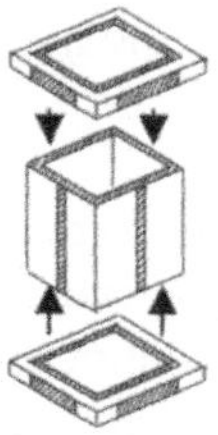

Figure 2.

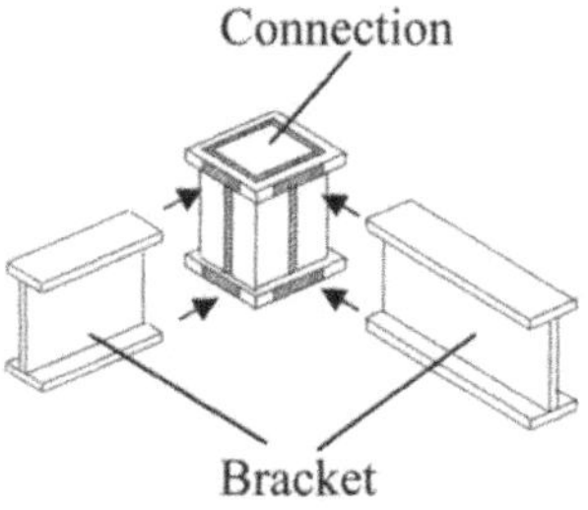

Figure 3.

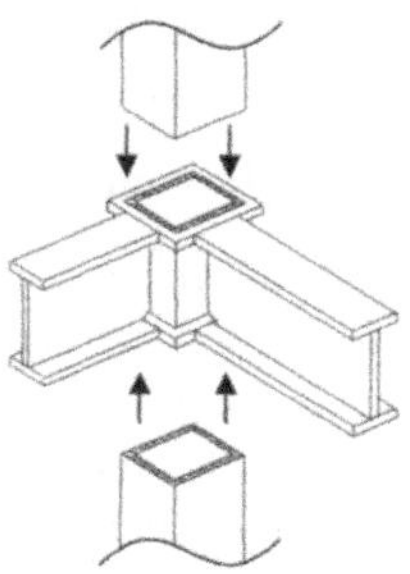

Figure 4.

3 Fabrication time functions (Sasaki et al. 2007)

In this study, the following function is used to predict the steel fabrication time:

$$TF = TP + TB + TW + TI \tag{1}$$

where TF represents the steel fabrication time, TP represents the preparatory process time, TB represents the assembly time, TW represents the welding time, and TI represents the time of preparing shop drawings.

The preparatory process consists of marking and drilling of diaphragms, marking, drilling and blasting of brackets and marking, drilling, blasting and flanging bevels of beams. The questionnaires indicated that the preparatory process time depends on the number of parts such as diaphragms, beams and brackets rather than structural weight. The following function is proposed to estimate the preparatory process time TP.

$$TP = KP \cdot \left(\sum_{i=1}^{nj} NP_{Di} + \alpha_{PB} \cdot NP_B\right) \tag{2}$$

where NP_{Di} represents the number of diaphragms for the beam to column connection i, nj represents the number of beam to column connections, NP_B represents the number of beams and brackets and α_{PB} and KP represent the coefficients for evaluating the preparatory process time.

The questionnaires indicated that assembly time also depends on the number of parts rather than the structural weight. Therefore, function TB to estimate the assembly is expressed as follows:

$$TB = KB \cdot (\sum_{i=1}^{nj} NB_{0_i} + \alpha_{BC} \cdot NB_C) \tag{3}$$

where NB_{0i} represents the number of parts consisting of connection panels and brackets for the beam to column connection i, NB_C represents the number of columns and α_{BC} and KB represent the coefficients for evaluating the assembly time.

The questionnaires showed that the welding time depends on the sum of jointed sectional areas. The following function is proposed to estimate the welding time TW.

$$TW = KW \cdot \left(\sum_{i=1}^{nj} A_{D_i} + \sum_{i=1}^{nbb} A_{BB_i} \right) \tag{4}$$

where A_{Di} represents the jointed sectional area between the column and the diaphragm for the beam to column connection *i*, and A_{BBi} represents the jointed sectinal area between the column and the bracket, *nbb* represents the number of brackets and *KW* represents the coefficient for evaluating the welding time.

Since the time of preparing shop drawings depends on the number of sheets of shop drawings, the following function, based on the number of columns and beams, is proposed:

$$TI = KIc \cdot NIc + KIb \cdot NIb + KIgAW \tag{5}$$

where NIc represents the number of shop fabricated column trees, *NIb* represents the number of beam groups having the same cross-sectional size, *W* represents the total structural weight of the frame and *KIc*, *KIb* and *KIg* represent the coefficients for evaluating the time of preparing shop drawings.

4 Coefficients to evaluate the fabrication time (Sasaki et al. (2007))

The values of α_{PB} and α_{BC} in Eqs.(2) and (3) were computed from questionnaires on fabrication time as follows.

$\alpha_{PB} = 2$, $\alpha_{BC} = 7$

The values of *KP, KB, KW, KIb, KIc* and *KIg* in Eqs. (2), (3), (4) and (5) were computed from the least squares approximation, based on recorded fabrication time data of the fabricating company, as follows:
KP = 0.085 (hours), *KB* = 0.50(hours), *KW* = 0.0071 (hours/cm^2), *KIb* = 1.54 (hours), *KIc*= 5.30 (hours), *KIg* = 0.044 (hours/kN)

5 Cost estimation of minimum weight steel frame

The minimum weight design problem of steel structural frames, shown in Fig.6, can be formulated as follows.

$$\begin{aligned}
&Find\ \ D_{idc}(idc=1,...,NDC), T_{ic}(ic=1,...,NC),\\
&H_{idb}(idb=1,...,NDB), B_{ib}, Tw_{ib}, Tf_{ib}(ib=1,...,NB)\\
&which\ \ \text{minimize}\ \ W=\rho\sum_{i=1}^{M}A_iL_i\\
&subject\ to
\end{aligned} \tag{6}$$

$$\begin{aligned}
&g_{S\,j}=\frac{N_j}{A_jf_{N\,j}}+\frac{M_j}{Z_jf_{M\,j}}\le 1\quad (j=1,2...,NM)\\
&g_{Dk}=\frac{\delta_k/H_k}{1/200}\le 1\qquad (k=1,2....,NF)\\
&g_P=\lambda_P\ge 1
\end{aligned} \tag{7a-c}$$

where W, ρ, A_i and L_i denote the structural weight, weight per unit volume of steel, the cross-sectional area, and member length, respectively; NM and NF denote the number of members and the number of stories, respectively. N_j, M_j, f_{Nj}, f_{Mj}, δ_k and H_k denote the axial force, bending moment, allowable stress for the axial force and bending moment, interstory drift of story k, and height of story k, respectively. λ_p is the collapse load factor. The design variables, depth of box-column section Didc, thickness of box-column section T_{ic}, depth of H-beam section H_{idb}, width of H-beam section B_{ib}, thickness of web T_{wib} and thickness of flange Tf_{ib} are chosen from the list of standard section sizes (Sawada et al. 2004). These constraints are based on the Japanese building standard law (1994) and the Japanese Design Standard for Steel Structures (2002).

One of the plane frames of a five-story building, shown in Fig. 5, are designed to minimise structural weight subjected to elastic and plastic constraints. Young's modulus E, yield stress F and steel weight per unit volume ρ are specified as follows: $E = 2.06\times10^5$ (N/mm^2), $F = 235$ (N/mm^2), $\rho = 76.93$ (N/cm^3)

The design load for the elastic constraints, Eq. (2a) and (2b), is shown in Fig. 10. The design horizontal load for the plastic constraints, Eq. (2c), is twice that shown in Fig. 6. Three different design variable linkings (Nos.1, 2 and 3) are given. The No.1 frame has a large number of independent design variables for minimizing the structural weight, the No.3 frame has a small number of independent design variables for realizing simple connections and the No.2 frame is the intermediate case.

The genetic Algorithm (GA) is applied for the minimum weight design. The GA control parameters are a population of 100, crossover probability of 1.0 and mutation probability of 0.01. The following computational results are the minimum weight of 10 solutions obtained using a different initial random number each time. Table 1 shows the minimum weight solutions of three frames. This table indicates that the structural weight decreases as the number of independent design variables increases. Figure 7 shows the fabrication time of three frames calculated from the proposed fabrication time function. The fabrication time for frame No.1 is the

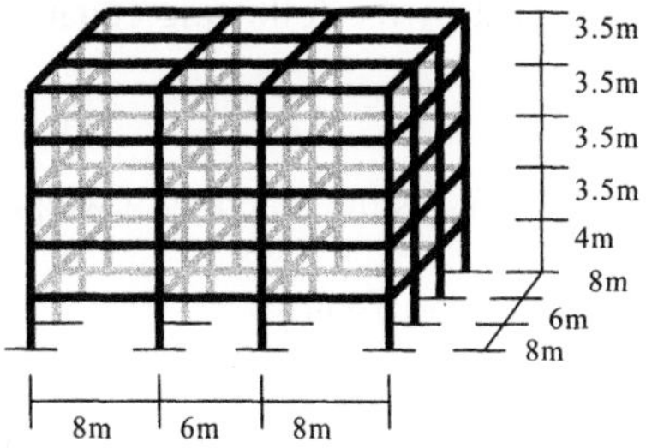

Figure 5. Five-story building

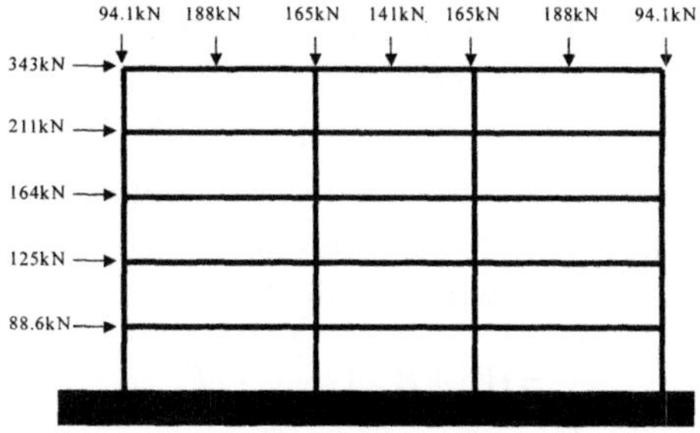

Figure 6. Five-story plane frame

Table 1. Minimum weight solutions

(A) No.1

	Sectional size (mm)	A(cm2)
Column		
5F	□ -400× 16	234.8
4F	□ -400× 16	234.8
3F	□ -400× 16	234.8
2F	□ -450× 16	266.8
1F	□ -450× 16	266.8
Outer girder		
RF	H-400× 200× 9× 16	98.57
5F	H-400× 200× 9× 22	121.5
4F	H-500× 200× 12× 19	132.9
3F	H-700× 200× 12× 22	169.5
2F	H-700× 300× 12× 19	196.2
Inner girder		
RF	H-500× 200× 9× 12	92.29
5F	H-400× 200× 9× 19	110.0
4F	H-600× 250× 12× 22	178.2
3F	H-500× 200× 9× 16	107.6
2F	H-500× 250× 9× 19	138.0
ND	12	
NT	12	
W (kN)	254.8	
W+WD (kN)	272.4	
TF (hours)	1031.1	

(B) No.2

	Sectional size (mm)	A(cm2)
Column		
5F	□ -450× 16	266.8
4F	□ -450× 16	266.8
3F	□ -450× 16	266.8
2F	□ -450× 16	266.8
1F	□ -450× 16	266.8
Outer girder		
RF	H-500× 200× 9× 19	119.0
5F	H-500× 200× 9× 19	119.0
4F	H-500× 200× 9× 19	119.0
3F	H-700× 250× 12× 19	177.2
2F	H-700× 250× 12× 19	177.2
Inner girder		
RF	H-500× 200× 9× 16	107.6
5F	H-500× 200× 9× 16	107.6
4F	H-500× 200× 9× 16	107.6
3F	H-700× 200× 12× 22	169.5
2F	H-700× 200× 12× 22	169.5
ND	3	
NT	6	
W (kN)	265.6	
W+WD (kN)	279.3	
TF (hours)	903.5	

(C) No.3

	Sectional size (mm)	A(cm2)
Column		
5F	□ -500× 16	298.8
4F	□ -500× 16	298.8
3F	□ -500× 16	298.8
2F	□ -500× 16	298.8
1F	□ -500× 16	298.8
Outer girder		
RF	H-600× 200× 12× 19	144.9
5F	H-600× 200× 12× 19	144.9
4F	H-600× 200× 12× 19	144.9
3F	H-600× 200× 12× 19	144.9
2F	H-600× 200× 12× 19	144.9
Inner girder		
RF	H-600× 200× 12× 19	144.9
5F	H-600× 200× 12× 19	144.9
4F	H-600× 200× 12× 19	144.9
3F	H-600× 200× 12× 19	144.9
2F	H-600× 200× 12× 19	144.9
ND	2	
NT	2	
W (kN)	288.1	
W+WD (kN)	304.8	
TF (hours)	937.0	

ND: number of depth design variables, NT: number of thickness design variables, W: structural weight without diaphragm, WD: diaphragm weight, TF: fabrication time

□-400 × 16 represents the box column whose depth is 400 (mm) and thickness is 16(mm).

H-400×200×9×16 represents the H beam whose depth is 400 (mm), width is 200 (mm), thickness of web is 9 (mm) and thickness of flange is 16(mm).

longest because a large number of independent variables lead to complex connections. The total cost function K can be expressed by the following equation (Jarmai & Farkas (1999)).

$$K = k_m \cdot (W + W_D) + k_f \cdot TF \tag{8}$$

where W is the structural weight without a diaphragm, W_D is the diaphragm weight and k_m and k_f are the corresponding cost factors.
Eq. (8) can be written in the following form (Jarmai & Farkas 1999).

$$\frac{K}{k_m} = (W + W_D) + \frac{k_f}{k_m} \cdot TF \tag{9}$$

Figure 8 shows the relationship between K/k_m and k_f/k_m for Nos.1, 2 and 3.

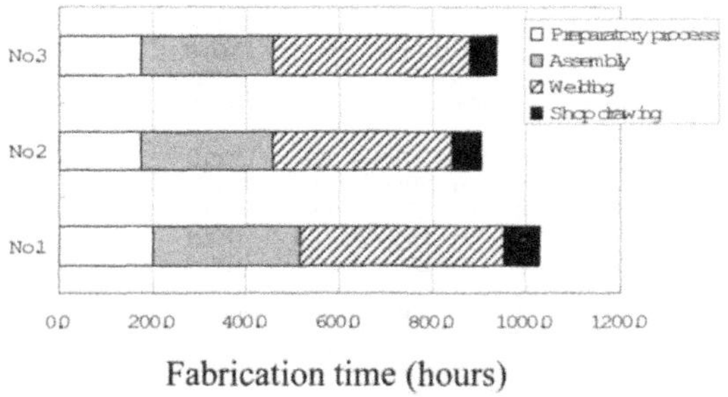

Figure 7. Fabrication time

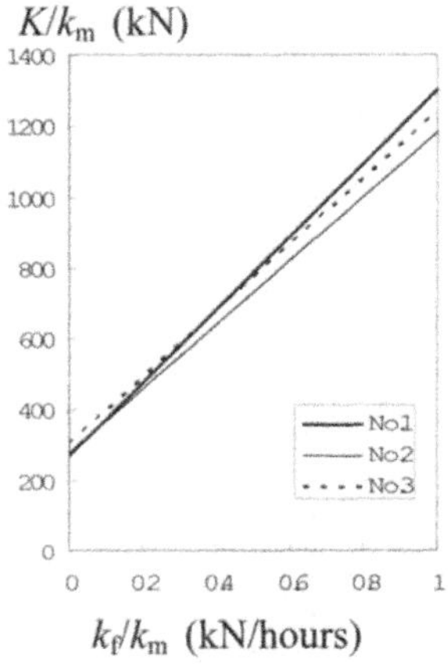

Figure 8. Total cost K/k_m

Frame No.1 gives the smallest total cost only when kf/km is less than around 0.06. Frame No.2 gives the smallest total cost when k_f/k_m is greater than around 0.06. This figure shows that the minimum weight frame (No.1) does not necessarily have the minimum cost, and that adequate design variable linking (No.2) leads to a lower total cost.

6 Conclusions

This paper presented a simple fabrication time function and its coefficients for steel building rigid frames. Next, using the function, the fabrication time for minimum weight steel building frames, considering different linking groups of member depth design variables was estimated. This study revealed that the minimum weight frame does not necessarily produce the minimum cost, and that adequate design variable linking leads to a lower total cost.

References

Architectural Institute of Japan (2002), *Design Standard for Steel Structures*, (In Japanese)

Jármai, K. (2002), Design, Fabrication and Economy. *European Integration Studies*, Miskolc, **1** No. 2, 91-107.

Jármai, K. & Farkas J. (1999), Cost calculation and optimisation of welded steel structures. *Journal of Constructional Steel Research* **50** 115–135.

Ohsaki, M. (1995) Genetic Algorithm for Topology Optimization of Trusses. *Computers & Structures* **57** No. 2. 219-225.

Pavlovcic, L., Krajnic, A. & Beg, D. (2004) Cost function analysis in the structural optimization of steel frames. *Struct. Multidisc. Optim.,* **28** 286–295.

Sasaki, T., Shimizu, H., Sawada K., Matsuo A. & Namba T. (2007) A Study on an Estimation Method of Steel Fabrication Cost Based on the Recorded Fabrication Times Data. *Journal of constructional steel*, Vol.15, (In Japanese)

Sawada, K. & Matsuo, A. (2004) An Exact Algorithm and Approximate Algorithms for Discrete Optimization of Steel Building Frames, *CJK-OSM3*, pp. 663-668.

The Ministry of Construction of Japan (1994), *The Building Standard Law of Japan*, (In Japanese)

3.5 Topology, Shape and Standard Sizing Optimization of Trusses Using MINLP Optimization Approach

Simon Šilih, Stojan Kravanja
University of Maribor, Faculty of Civil Engineering, Smetanova 17, SI-2000 Maribor, Slovenia, e-mail: simon.silih@uni-mb.si, stojan.kravanja@uni-mb.si

Abstract

The paper presents the simultaneous topology, shape and discrete/standard sizing optimization of steel trusses using Mixed-Integer Non-linear Programming (MINLP) approach. The discrete/continuous non-convex and non-linear optimization problems are solved by the Modified OA/ER algorithm. Two types of objective functions are introduced for the optimization, namely a volume/mass objective function and an economic objective function. The design constraints are defined in accordance to Eurocode 3. A numerical example is presented at the end of the paper in order to show the suitability of the proposed approach and to emphasize the importance of selecting a proper objective function.

Keywords: *Structural synthesis, Mixed-Integer Non-linear programming, Topology optimization, Discrete sizing optimization, Trusses.*

1 Introduction

The paper presents the Mixed-Integer Non-Linear Programming (MINLP) optimization approach to the synthesis of trusses. The solution of discrete/continuous and non-linear optimization problems is discussed with respect to the simultaneous topology, shape and discrete/standard dimension optimization of trusses. The MINLP synthesis of trusses is performed through three steps, Kravanja et al. (1998): the first one is the *generation of truss superstructure*, the second one the *development of a special MINLP model formulation* and the final step is the *solution of the defined MINLP problem.* All three steps are briefly described in the following sections. The discrete/continuous, non-convex and non-linear problems are solved using the Modified Outer-Approximation/Equality Relaxation (OA/ER) algorithm, Z. Kravanja & Grossmann (1994). Topology optimization of trusses is in general a high combinatorial/expansive problem and needs a high number of alternative arrangements of elements to be defined in the truss superstructure. The Hierachical Superelement Approach (HSA), Šilih & Kravanja (2005), was thus developed with the object of reducing the combinatorial expanse of the problem.

Beside many non-convexities and non-linearities, the main problem of truss synthesis is that a very high number of discrete/binary variables, particularly those defining discrete/standard dimensions can be included in the optimization. In order to be able to solve such comprehensive optimization problems, we apply multilevel strategies. Some of them, namely the Two-Phase (TP) MINLP strategy, Kravanja et al. (1992), and the prescreening of discrete dimensions are briefly described in the paper.

Beside the usual mass/volume objective function a simple cost objective function for the truss synthesis is also introduced. The importance of cost optimization and its possible influence on the final/optimal result is shown through a numerical example presented at the end of the paper.

2 MINLP synthesis of trusses

The MINLP synthesis of trusses is performed through three steps: the process is started with the *generation of truss superstructure*, followed by the *development of a special MINLP model formulation* and the final step is the *solution of the defined MINLP problem*.

2.1 *Generation of truss superstructure*

The first step of the truss synthesis is the generation of a MINLP truss superstructure, which includes all possible topology/structure alternatives to compete for a feasible and optimal solution. The discrete topology optimization problems of trusses generally comprise a high number of alternative structural elements and their configurations. A number of them will never be selected to compose the optimal topology/structure. In order to reduce the combinatorics of the discrete optimization, it is preferred for such elements to be removed from the set of the defined alternatives. For this reason, we propose a special topological formation, the so-called Hierarhical Superelement Approach (HSA), by which a truss superstructure is in a hierarchical manner composed from a number of different superelements, see Figure 1.

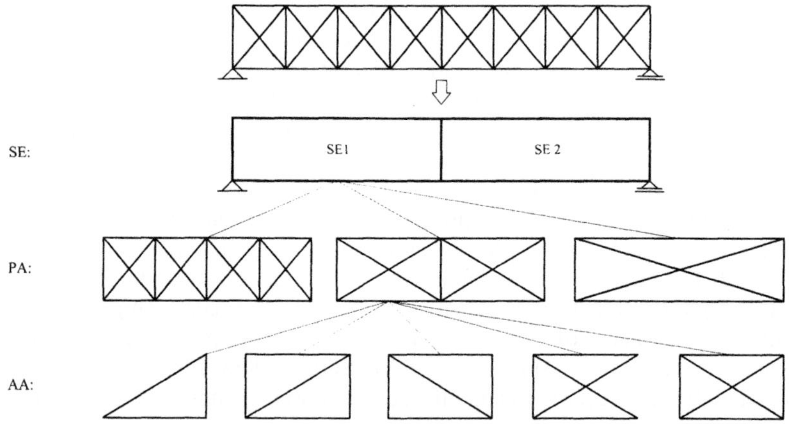

Figure 1. The Hierarchical Superelement Approach (HSA).

The highest level of superelements are segments (SE), which are then further partitioned into one of the pre-defined alternative numbers of internal partitions - panels (PA). The panels represent the medium level of superelements. Each alternative panel can additionally be formed from any possible alternative arrangement (AA) of bracing members which assure the kinematical stability of the structure. Alternative arrangements comprise the lower level of superelements.

While the upper levels, i.e. the segments and panels, represent the quantity of the alternative superelements incorporated in the superstructure, the lower level determines their quality. The number of segments is fixed and defined before the optimization. Though different topology alternatives (active/non-active nodes and elements) are independently determined and optimized inside each SE, the entire superstructure is optimized simultaneously, all the defined SE-s included.

2.2 *The MINLP model formulation for truss superstructure*

The general non-linear and non-convex discrete/continuous optimization problem can be formulated as an MINLP problem in the form:

$$
\begin{aligned}
\min \quad & z = \boldsymbol{c}^{\mathrm{T}} \boldsymbol{y} + f(\boldsymbol{x}) \\
s.t. \quad & \boldsymbol{h}(\boldsymbol{x}) = \boldsymbol{0} \\
& \boldsymbol{g}(\boldsymbol{x}) = \boldsymbol{0} \\
& \boldsymbol{B}\,\boldsymbol{y} + \boldsymbol{C}\,\boldsymbol{x} \le \boldsymbol{b} \\
& \boldsymbol{x} \in X = \{\, \boldsymbol{x} \in R^n : \boldsymbol{x}^{\mathrm{LO}} \le \boldsymbol{x} \le \boldsymbol{x}^{\mathrm{UP}} \,\} \\
& \boldsymbol{y} \in Y = \{\, 0,1 \,\}^m
\end{aligned}
\qquad \textbf{(MINLP)}
$$

where $\boldsymbol{x}$ is a vector of continuous variables specified in the compact set X and $\boldsymbol{y}$ is a vector of discrete, mostly binary 0-1 variables. Functions $f(\boldsymbol{x})$, $\boldsymbol{h}(\boldsymbol{x})$ and $\boldsymbol{g}(\boldsymbol{x})$ are non-linear functions involved in the objective function z, equality and inequality constraints, respectively. Finally, $\boldsymbol{B}\,\boldsymbol{y} + \boldsymbol{C}\,\boldsymbol{x} \le \boldsymbol{b}$ represents a subset of mixed linear equality/inequality constraints. The set of continuous variables $\boldsymbol{x}$ represents the continuous parameters (stresses, deflections, nodal coordinates etc.), while the set of discrete variables $\boldsymbol{y}$ contains parameters for discrete decisions (standard dimensions, elements).

The MINLP model formulation for truss superstructure was developed on the basis of the continuous NLP truss optimization model, which was previously successfully used for sizing and shape optimization of steel trusses, Šilih et al. (2002), composite trusses, Kravanja & Šilih (2003), Klanšek et al. (2006) and timber trusses by considering joint flexibility, Šilih et al. (2005). The main components of the nonlinear equality constraints are the finite element equations for the calculation of nodal displacements, reactions and member forces. The non-linear inequality constraints include the tensional and compressive/buckling resistance conditions of elements, as well as the displacement conditions. All the design conditions are defined in accordance with Eurocode 3 (1992).

Independent continuous variables include nodal coordinates and sizing variables. Each bar is defined as a truss element connecting the nodes i and j (in the further text noted as element i-j). The cross sections of bars are considered to be steel circular hollow sections. The cross section of each element i-j is thus defined by the variables $d_{i,j}$ and $t_{i,j}$, which represent the diameter and the wall thickness of the tube, respectively. The binary 0-1 variables are divided into topological binary variables and binary variables defining standard dimensions.

The topological binary variables subjected to the currently active truss elements take the value 1, while the rest of them are given the value zero. The binary variables subjected to standard dimensions are defined for each standard value that can be valued to any cross-section dimension of each element *i-j*.

Two different objective functions are proposed. The first one is a mass objective function, representing simply the mass of the truss structure (1), where ρ is the density of the material, while $A_{i,j}$ and $l_{i,j}$ are the cross-sectional area and the length of element *i-j*, respectively.

$$MASS = \rho \cdot \sum_{i,j} A_{i,j} \cdot l_{i,j} \tag{1}$$

Beside the mass objective function, a simple cost objective function (2) is also proposed in the form of:

$$COST = C_{st} \cdot \rho \cdot \sum_{i,j} A_{i,j} \cdot l_{i,j} + C_{ac} \sum_{i,j} A_{i,j}^{ac} + C_n \cdot n \tag{2}$$

In Eq.2, C_{st} defines the material costs of steel (in EUR/kg), C_{ac} represents the anti-corrosion protection (material and painting) costs (in EUR/m^2), while C_n represents the erection of one joint (including cutting and welding of steel profiles) in EUR. Further, $A_{i,j}{}^{ac}$ represents the exposed area for anti-corrosion protection of element *i-j*, and n represents the number of joints of the truss.

2.3 *Solution of the defined MINLP truss synthesis problem*

For the solution of non-linear and non-convex problems we used the Modified Outer-Approximation/Equality-Relaxation (OA/ER) algorithm. Beside many non-convexities and non-linearities, truss synthesis problems involve a high number of discrete/binary variables. Such problems are thus hard to solve in a single MINLP phase, where all the included binary variables are initialized in a single full set. For this reason we applied the Two-phase (TP) MINLP strategy. The TP approach performs topology, shape and standard dimension optimization separately in two phases. In the first phase, simultaneous topology, shape and sizing optimization is performed with dimensions of cross-sections being temporarily relaxed into continuous parameters.

When the optimal topology is obtained, the discrete dimensions are re-established and the process continues with the second phase, where the shape and standard dimension optimization is performed until the optimal solution is obtained. After the completion of the first phase (continuous sizing optimization), a prescreening procedure is additionally applied in order to reduce the number of active binary variables. This way only those binary variables are included in the second phase (discrete sizing optimization), which define the discrete/standard dimensions close to the optimal continuous dimensions, obtained in the first phase. The set of active binary variables becomes essentially smaller than the full initial set, which leads to shorter and solvable optimization.

The optimization model was developed using General Algebraic Modelling System (GAMS), Brooke et al. (1988). The optimization was performed by the MINLP computer package MIPSYN, the extension of PROSYN, Z. Kravanja & Grossmann (1994). GAMS/CONOPT, Drudd (1994) has been used to solve the NLP subproblems and GAMS/CPLEX to solve the MILP master problems.

3 Numerical example

As a numerical example the synthesis of a simply supported truss girder over the span of 20 m is presented. The simultaneous topology, shape and sizing optimization was performed by the use of the proposed Hierarchical superelement approach. The defined truss superstructure consists of two equal segments, each of them may be further partitioned into 1, 2, 3, 4 or 6 panels (see Figures 2 and 3). The truss is subjected to a vertical nodal load F_d of 100 kN (design value), acting on the mid-span joint of the bottom chord. The structure is designed in accordance with Eurocode 3. The buckling lengths of the truss elements are considered as being equal to the system lengths of the elements for both in-plane and out-of-plane buckling. The bars are designed from circular hollow sections made of S235 steel. The vectors of discrete/standard alternative values for the diameter d and wall thickness t of cross-sections are given as follows: $\mathbf{d}$ = {42.4, 48.3, 60.3, 76.1, 88.9, 108.0, 114.3, 133.0, 139.7, 159.0, 168.3, 219.1, 273.0, 323.9, 355.6, 406.4, 457.0, 508.0, 558.8, 609.6} [mm] and $\mathbf{t}$ = {2.0, 2.9, 3.2, 4.0, 5.0, 6.3, 7.1, 8.0, 10.0, 12.5, 14.2, 16.0} [mm]. With respect to the available standard cross sections, the lower/upper bounds of the wall thickness (t^{LO}/t^{UP}) for each individual value of diameter d of cross section are defined. The cross sections of the chords are forced into being constant through the entire span. The vertical coordinates of top chord joints were defined as geometric variables with their lower and upper bounds of 200 and 700 cm.

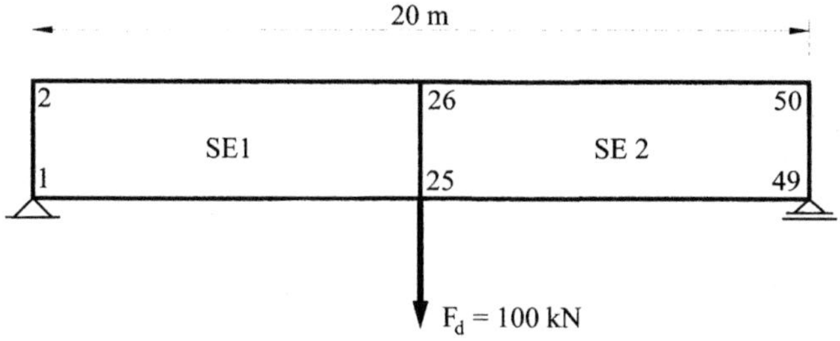

Figure 2. The superstructure for 20 m simply supported truss.

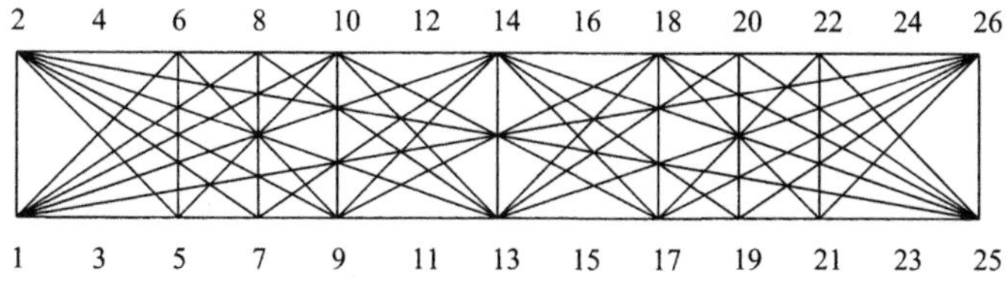

Figure 3. The proposed superelement (SE1) for 20 m simply supported truss.

Since the loading of the defined truss is symmetric, symmetry of topology with respect to the vertical axis through the midspan of the structure is requested. The topology is thus optimized within the segment SE1, while SE2 represents its mirror image.

Three separate optimizations were performed:

- Example A: Mass optimization – objective function (1), with $\rho = 7850$ kg/m^3
- Example B: Cost optimization – objective function (2), with C_{st} = 1.0 EUR/kg, C_{ac} = 20.0 EUR/m^2 and C_n = 0.0 EUR (only material and anti-corrosion protection costs considered)
- Example C: Cost optimization – objective function (2), with C_{st} = 1.0 EUR/kg, C_{ac} = 20.0 EUR/m^2 and C_n = 50.0 EUR/joint.

The optimal truss structures obtained from examples A, B and C are shown in Figures 4 to 6, while the optimal standard cross sectional dimensions are listed in Tables 1 to 3.

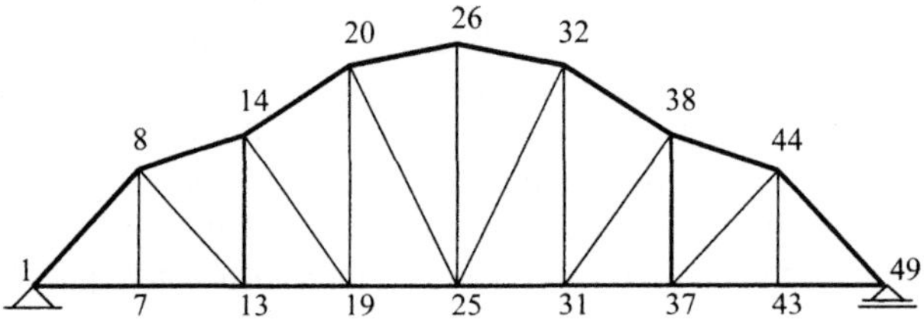

Figure 4. Optimal solution, Example A.

Table 1. Optimal standard cross-sections, Example A

Element	Diameter d [mm]	Wall thickness t [mm]
Top chord (8-44)	108.0	2.9
Bottom chord (1-49)	88.9	3.2
1-8, 44-49	108.0	2.9
7-8, 8-13, 14-19, 19-20, 20-25, 25-26, 25-32, 31-32, 31-38, 37-44, 43-44	42.4	2.0
13-14, 37-38	76.1	2.9
8-9, 9-12, 9-10	42.4	2.0
Height at mid-span: 5.55 m		
Optimal mass = 444.33 kg		

The obtained optimal results show considerable differences in the final solutions when different objective functions are included into the optimization model. Compared to Example A, where only the mass of the structure was optimized, the inclusion of anti-corrosion protection costs in Example B yielded a reduction in the number of bars.

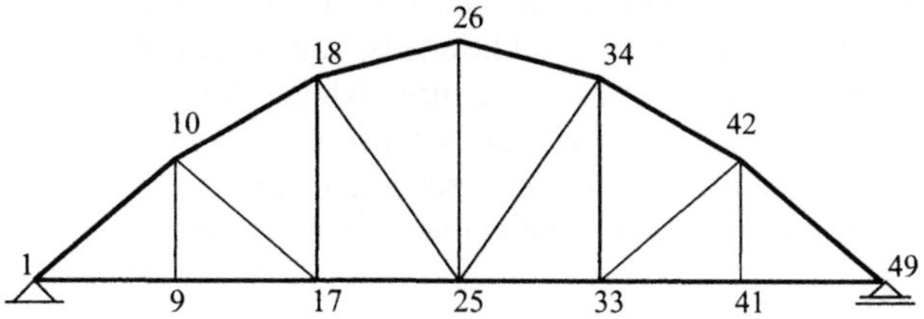

Figure 5. Optimal solution, Example B.

Table 2. Optimal standard cross-sections, Example B

Element	Diameter d [mm]	Wall thickness t [mm]
Top chord (10-42)	108.0	2.9
Bottom chord (1-49)	88.9	4.0
1-10, 42-49	114.3	3.2
9-10, 10-17, 25-26, 33-42, 41-42	42.4	2.0
17-18, 33-34	60.3	2.9
18-25, 25-34	48.3	2.0
Height at mid-span: 5.56 m		
Optimal costs = 862.68 EUR		

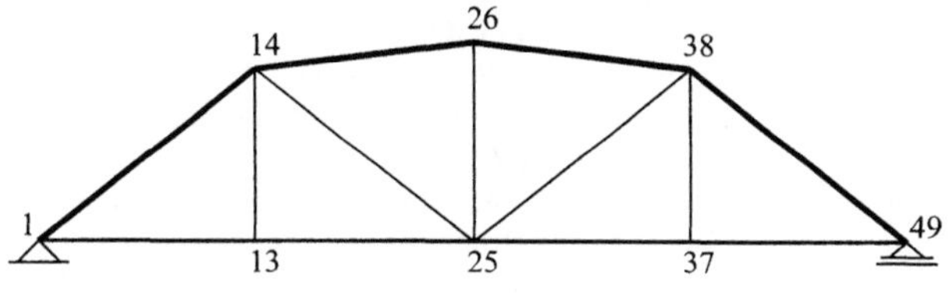

Figure 6. Optimal solution, Example C.

Table 3. Optimal standard cross-sections, Example C

Element	Diameter d [mm]	Wall thickness t [mm]
Top chord (14-38)	133.0	4.0
Bottom chord (1-49)	76.1	4.0
13-14, 37-38	42.4	2.0
14-25, 26-26, 25-38	48.3	2.9
Height at mid-span: 4.56 m		
Optimal costs = 1264.09 EUR		

A further reduction in the number of bars was obtained, when the costs of joints (cutting and welding of bars) were additionally considered. Although this example represents a simple case of truss structure, where the solutions are somewhat expected (i.e. solutions with predominantly tensioned diagonals), it still proves that defining a proper objective function is a crucial part of the optimization process and that cost objective functions should take precedence over the volume/mass ones.

4 Conclusions

The paper presents the Mixed-Integer Non-Linear Programming (MINLP) optimization approach to the synthesis of trusses. The topology, shape and discrete/standard cross-sectional dimensions of the trusses are optimized in a uniform optimization process. The non-convex and non-linear problems are solved by the Modified OA/ER algorithm. Two different objective functions are proposed, namely a mass objective function and a simple cost objective function. The importance of defining a suitable objective function is discussed through a numerical example.

References

Brooke, A., Kendrick, D. & Meeraus, A. (1988) *GAMS (General Algebraic Modelling System), a User's Guide*, Redwood City: The Scientific Press.

CPLEX User Notes, ILOG inc.

Drud, A.S. (1994) CONOPT – A Large-Scale GRG Code. *ORSA Journal on Computing* **6** No.2, 207–216

Eurocode 3 (1992) *Design of steel structures*, Brussels: European Committee for Standardization.

Klanšek, U., Šilih, S. & Kravanja, S. (2006) Cost optimization of composite floor trusses. *Steel & Composite Structures* **6** No.5, 435–457

Kravanja, S., Kravanja, Z. & Bedenik, B.S. (1998) The MINLP approach to structural synthesis, Part I: A general view on simultaneous topology and parameter optimization. *International Journal on Numerical Methods in Engineering* **43** 293-328

Kravanja, S., Kravanja, Z., Bedenik, B.S. & Faith, S. (1992) Simultaneous topology and parameter optimization of mechanical structures. In: *Numerical methods in engineering 92, First European conference on numerical methods in engineering, Bruxelles, Belgium,* pp. 487-495

Kravanja, Z. & Grossmann, I.E. (1994) New developments and capabilities in PROSYN – an automated topology and parameter synthesizer. *Computers in Chemical Engineering* **18** 1097 –1114

Kravanja, S. & Šilih, S. (2003) Optimization based comparison between composite I beams and composite trusses. *Journal of constructional steel research* **59** 609-625.

Šilih S. & Kravanja S. (2005) Synthesis of trusses using the MINLP optimization approach. In: *Proc. Int. Conf. on Computer Aided Optimum Design in Engineering, Skiathos, Greece,* pp. 221-233

Šilih, S., Kravanja, S. & Bedenik, B.S. (2002) Shape optimization of plane trusses, in: *Finite Elements in Civil Engineering Applications, Proceedings of the Third DIANA World Conference, Tokyo, Japan,* pp. 369-373

Šilih, S., Premrov, M. & Kravanja, S. (2005) Optimum design of plane timber trusses considering joint flexibility. *Engineering Structures* **27** 145-154.

3.6 The Influence of Dimensional Accuracy on Steel Structures Cost Calculation

Elżbieta Urbańska-Galewska
Gdańsk Technical University, Pl-80-952 Gdańsk, Poland, e-mail: ugalew@pg.gda.pl

Abstract

Relation between tolerance values and costs of steel structures fabrication is presented. Classification of structure dimensions from the point of view of the proper structure erection is described and assembly loop equations are formulated. The basis of the tolerance analysis and synthesis are shortly presented. The new formula of steel structures cost optimisation is proposed. Costs of dimensional accuracy are the main condition of the tolerance allocation. Optimisation methods are pointed.

Keywords: *steel structures, costs optimisation, dimensional accuracy, tolerances*

1 Introduction

Minimising the mass of the structures is still the most common method of optimisation. Moreover, most of the method considers only ideal cases, i.e. only nominal values of design variables. In fact, real values differ from those theoretical. For example all dimensions of the steel structures result from fabrication and erection processes and variations of their values depend on dimensional accuracy. The sensitivity of some state variables to small variations of certain designs variables increases rapidly for near-optimal structures (Bauer et al. 2003). In case of optimum metal structures, where one or even several constraints come to their limit values, this may limit the safety of an erected structure. So, dimensional tolerances should be incorporated in optimum design of steel structures (Gutkowski & Bauer – 1999). There are at least two different reasons to take dimensional tolerances into consideration. The first one concerns the case when safety of the optimal structure can be violated because of differences between assumed and real structure dimensions caused by manufacturing tolerances. The second one concerns the minimum manufacturing cost, while keeping the tolerances within design specification.

The paper concerns the minimum cost design of steel structures. The dimensional deviations of manufacturing steelwork are incorporated in optimum design of steel structures.

2 Dimensional accuracy as the optimisation criterion

2.1 *Cost-tolerance relation*

A design of modern steel structures demands an optimum design meaning an achievement as low cost of a construction as possible at the level a high enough quality. The adequate level of quality should ensure:

- safe erection of structures within the shortest possible time, without any additional works on site concerning structural elements fitting,
- safe using of structures from the point of view of the ultimate limit state,

- fulfilment of functional and operational requirements from the point of view of the serviceability limit state.

The quality of the newly constructed structures is the function of the manufacturing, fabrication and erection tolerances (Urbańska-Galewska 2002). The high quality is obtained when both the assembling of the structure and the fixing non-structural components can be performed in an easy and safe way. We cannot avoid dimensional deviations, which are recognized as a natural part of the technological process. We can only set the limits on the deviations. A value of the permitted deviations (tolerances) depends on both required quality and cost effectiveness. Due to enormous development of the new production technologies and measuring equipment, it is possible to achieve extreme accuracy. However, from the economical point of view this way is not acceptable.

There is a very close relation between tolerance values and costs of manufacturing. A typical cost-tolerance function is shown in Figure 1. It is basically a reciprocal function, which estimates a decrease in cost for an increase in tolerance. Tighten tolerances demand more effort, more labour, so they cause the costs increase. The cost for each satisfactory dimension is a function of the following components of the manufacturing operations: a material cost, a manufacturing cost (machining, rework and inspection costs) and a scrap cost (He 1991). The cost components are fixed (material and machining costs) or variable (rework, inspection and scrap costs). Examples of more adequate cost-tolerance relation curves for common production processes are presented in the Figure 2 (Dong et al. 1994). In practice, the empirical cost-tolerance data should be directly obtained from machine shops through experiments or observations.

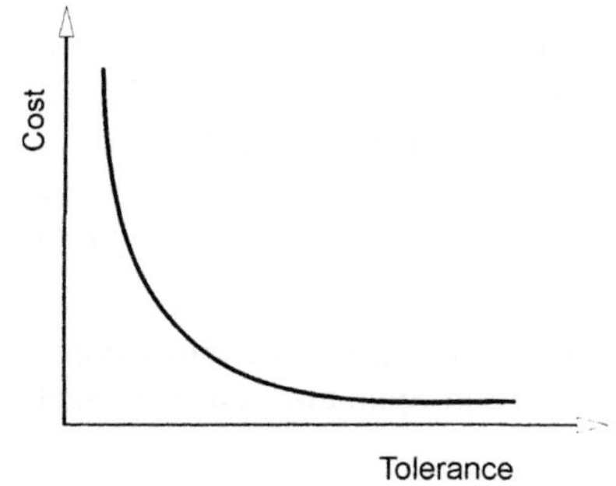

Figure 1. Typical cost-tolerance function.

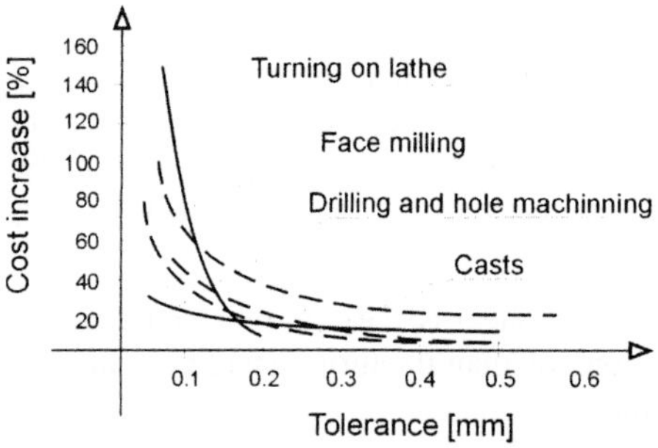

Figure 2. Common production processes cost-tolerance relations.

2.2 *Classification of structure dimensions*

Only dimensions, which are responsible for proper erection of structures, named assembly coordination dimensions (ACD), are important from the point of view of the dimensional accuracy (Urbańska-Galewska 2006). ACD consist of attachment dimensions (ATD) and axial coordination dimensions (AXD). The ATD are the chosen fabrication dimensions of the assembly members and influence the possibility of the site joining. The variation of the ATD influences the variation of the structure geometry. The AXD are the dimensions of the axial spans of the structure. They are set out in site.

The phenomenon of accumulation of ATD and AXD occurs at any steelwork assembly and unavoidable gaps (clearances) appear in the site connections as a result. However, lack of gaps in the site connections may mean incorrectly designed and/or manufactured structure. The phenomenon of deviations accumulation should be analyzed at the level of a single assembly loop, i.e. presented in the Figure 3.

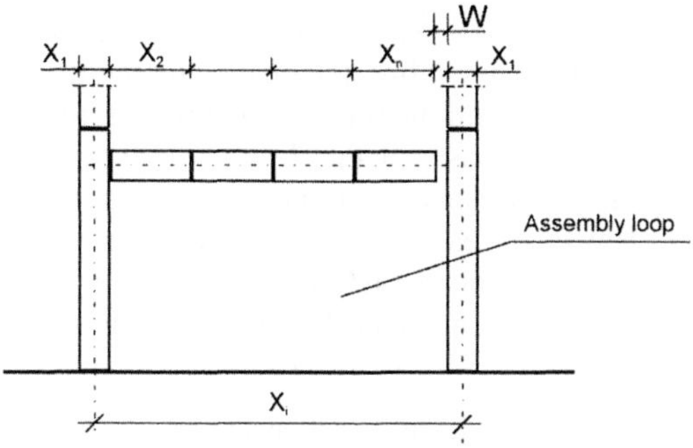

Figure 3. Example of assembly loop – erection of one bay of single storey.

Dimensional chains equations are applied to the mathematical description of the assembly loop. The component dimensions X_i (AXD, ATD) specified on workshop drawings are the independent variables. The assembly dimension W is also called the redundant or the resultant dimension and is the dependent variable (dimension of the gap between joined frame members). All dimensions mentioned above are components of the assembly loop equation:

$$W = F(X_1, X_{2,}......, X_n)\,. \tag{1}$$

2.3 *Tolerance analysis and synthesis*

Tolerance analysis determines the assembly tolerance T_W, meaning tolerance value of the resultant dimension of the dimensional chain. The two most commonly used tolerance analysis methods are Worst Case (WC or arithmetic sum) and Root Sum Squares (RSS or statistical sum). In the WC method, the assembly loop equation is the function of the independent variables as presented in Eq.1, where W – assembly dimension, dependable variable, X_i – component dimensions, undependable variables. The WC relation is shown in equation (2) and makes no assumptions of how the dimensions are distributed within the tolerance zone:

$$T_W = \sum_i \left| \frac{\partial W}{\partial X_i} \right| T_{X_i}\,. \tag{2}$$

The relation for the RSS method, shown in equation (3):

$$T_W = \sqrt{\sum_i \left(\frac{\partial W}{\partial X_i} \right)^2 T_{X_i}^2}\,, \tag{3}$$

is valid only if the assumption that all distributions of the random variable are the normal distributions.

A frame erection is possible to be carried out only when the value of the assembly tolerance T_W, determined at the tolerance analysis is not greater then its critical value $T_W < T_{Wcr}$. If not, the tolerance synthesis has to be carried out. It means that the values of the particular ACD tolerances have to be tightened and then the tolerance analysis has to be repeated.

The tolerance synthesis depends on the tolerance allocation which means the distribution of the specified assembly tolerance among the components of the assembly. In traditional design and manufacturing practice, tolerance allocations are performed sequentially on the basis of the designer's experience and non-optimal methods. The component tolerances can be distributed equally among all the assembly members. However, each component tolerance may have a different manufacturing cost according to the relation presented in Figures 1 and 2. By defining a cost-tolerance function for each component dimension, the component tolerances may be allocated to minimize the cost of production.

To take into account the erection of any steel structure, the assembly loop equations have to be formulated. The number of the assembly loops depends on a size and degree of the frame complexity. The equations of the dimensional chains are constructed from the nominal values of ACD, what has been presented by Urbańska-Galewska (2006). The nominal dimensions vector $\mathbf{X_k}$ ($X_1, X_2, \ldots, X_i, \ldots X_n$), where $k = 1, \ldots, m$ – number of assembly loop equations, $i = 1, \ldots, n$ – number of dimensions, is the design variable. The actual values of ACD may deviate from their nominal values by an allowable summand t_{Xi}, being the single side tolerance value of the dimension X_i. Therefore, the actual value of the ith dimension, in each assembly loop equation, stays within the range $< X_i - t_i \div X_i + t_i >$. The inequality constraints imposed on tolerances resulted from assembly loop equations and allowable deviations of design variables (tolerances) are described as follows:

in the deterministic model (WC method):

$$\sum_{i=1}^{n}\left(\frac{\partial W_k}{\partial X_i}\right) T_{Xi} \le T_k \quad \text{for each } k = 1,\ldots,m, \tag{4}$$

in the statistical model (RSS method):

$$\sum_{i=1}^{n}\left(\frac{\partial W_k}{\partial X_i}\right)^2 T_{Xi}^2 \le T_k^2 \quad \textit{for each } k = 1,\ldots.m, \tag{5}$$

where: W_k – assembly (resultant) dimension of the kth assembly loop, T_{Xi} – tolerance range of the X_i dimension, T_k – critical value of the assembly tolerance. For the symmetric tolerances $T_{Xi} = 2t_{xi}$.

3 Steel structures cost optimization

Due to the industrialization of the means of production, fabrication of the prefabricated steelwork becomes rather a manufacturing operation with the emphasis on accurate cutting, shaping, drilling and welding of components on a production line. Therefore, the cost of steel structure building with the significant contribution of labour cost becomes the essential optimization criteria (Jarmai & Farkas 1999).

The cost function recommended by Urbańska-Galewska can be expressed as follows:

$$K = \sum K_M + \left(\sum K_T + \sum K_N\right) + K_c, \tag{6}$$

where: K – total cost of steel structures, K_M – material cost, $(\Sigma K_T + \Sigma K_N)$ – total cost of structure manufacturing, K_T – toleranced dimensions (ACD) cost of manufacturing, K_N – other costs of manufacturing, K_C – cost of structure construction.

The material cost should not be minimized, as it was proved that minimum weight solutions might be up to 20% more expensive than solutions where also the manufacturing costs have been taken into consideration to optimize the design (Weynand et al. 1998). The frame producibility is the basic optimization criterion, which reflects the actual costs. As it is not possible to express this feature mathematically, only various manuals can be helpful at this stage of knowledge. The ACD manufacturing cost K_T depends on demanded accuracy of assembled members and cost optimization should be taken into account. Assembly elements with proper values of tolerances result with a structure construction without additional works connected with extra fitting of assembly members (lower construction cost K_C).

This paper concerns ACD cost optimization only, therefore the objective function has been reduced to the form:

$$f(\mathbf{T_X}) = \min \sum K_T \,. \tag{7}$$

According to the studies carried out by Dong (1994) and He (1991) the following function of the dimension manufacture cost has been introduced:

$$K_T = K_i(T_i) = a_i + \frac{b_i}{T_{Xi}}, \tag{8}$$

where: $K_i(T_i)$ – cost of ith dimension manufacturing at given tolerance range T_{Xi}; a_i, b_i – typical constants for given operation, which can be empirically obtained.

At last, the ACD manufacturing cost optimization problem has been formulated as follows:

$$\min_{T_X \in T_d} f(\mathbf{T_X}) = \sum K_T = \sum_{i=1}^{n} \left(a_i + \frac{b_i}{T_{Xi}} \right),$$
$$T_d = \left\{ \mathbf{T_X} \in R^n : g(\mathbf{T_X}) \le 0 \right\}, \tag{9}$$

where:

$$g(\mathbf{T_X}) = \sum_{i=1}^{n} \left(\frac{\partial W_k}{\partial X_i} \right) T_{Xi} - T_k \leq 0 \quad \text{(WC method)}$$

or

$$g(\mathbf{T_X}) = \sum_{i=1}^{n} \left(\frac{\partial W_k}{\partial X_i} \right)^2 T_{Xi}^2 - T_k^2 \leq 0 \quad \text{(RSS method)}.$$

Three methods of tolerance allocation for steel structure building purposes by: Lagrange Multipliers, combinatorial search procedure (the method of exhaustive search) and Monte Carlo simulation are elaborated by Urbańska-Galewska (2005).

4 Conclusions

To apply the new formula of the steel structures optimization in practice, the real value of the cost-tolerance relation should be obtained from steelwork workshops. This is the necessary condition to introduce a more effective system of tolerance selection proposed by Urbańska-Galewska (2006). The current system of tolerances selection does not meet the requirements and necessities of present day industrial methods of fabrication and erection of steel structures.

References

Bauer J., Gutkowski W. & Latalski J. (2003) Minimum structural weight with manufacturing tolerances constraints. In: *Proc. Int. Conf. on Metal Structures, Miskolc, Hungary*, 259-264.

Dong Z., Hu W. & Xue D. (1994) New production cost-tolerance models for tolerance synthesis. *Transactions of ASME. Journal of Engineering for Industry.* **116** 199-206.

Gutkowski W. & Bauer J. (1999) Manufacturing tolerance incorporated in minimum weight design of trusses. *Engineering Optimization* **31** 393-403.

He J. R. (1991) Tolerancing for manufacturing via cost minimization. *International Journal of Machining Tools for Manufacturing,* **31** No. 4, 455-470.

Jarmai K. & Farkas J. (1999) Cost calculation and optimisation of welded steel structures. *Journal of Constructional Steel Research,* **50** 115-135.

Urbańska-Galewska E. (2002) Tolerance analysis as the way to cost-effective structures. In: *Proc. Int. Conf. on Steel in Sustainable Construction*, Luxemburg, pp. 43-47.

Urbańska-Galewska E. (2005) *Tolerancje w budowlanych konstrukcjach stalowych łączonych na śruby (Tolerances in steel building structures with bolted connections*). Monografia **59**, Gdańsk, Wydawnictwo Politechniki Gdańskiej

Urbańska-Galewska E. (2006) Proposal of a new tolerances classification system. In: *Proc. Int. Conf. on Metal Structures, Rzeszów, Poland*, pp. 983-990.

Weynand K., Jaspart J. P. & Steenhuis M. (1998) Economy studies of steel building frames with semi-rigid joints. *Journal of Constructional Steel Research,* **46** No. 1-3, 85.

3.7 Effects of Residual Stresses on Optimum Design of Stiffened Plates

Zoltán Virág, Károly Jármai
University of Miskolc, H-3515 Miskolc, Hungary,
gtbvir@uni-miskolc.hu, altjar@uni-miskolc.hu

Abstract

In this overview of uniaxially compressed stiffened plates various types of loadings, and stiffener shapes are investigated. The global buckling strength of welded stiffened plates is calculated according to Mikami and Niwa. This method considers the effect of initial imperfection and residual welding stresses. The aim of the present study is to apply Okerblom's constraint to investigate the effect of the deflection of the plate due to longitudinal welds for the optimum design. The unknowns are the thickness of the base plate as well as the dimensions and number of stiffeners.

Keywords: *stiffened plates, plate buckling, deflection, residual stresses, optimum design*

1 Introduction

Welded stiffened plates are widely used in various load-carrying structures, e.g. ships, bridges, bunkers, tank roofs, offshore structures, vehicles, etc. They are subject to various loadings, e.g. compression, bending, shear or combined load. The shape of plates can be square rectangular, circular, trapezoidal, etc. They can be stiffened in one or two directions with stiffeners of L, trapezoidal or other shape.
In this overview of uniaxially compressed stiffened plates (Figure 1.) various types of loadings and stiffener shapes are investigated. The global buckling strength of welded stiffened plates is calculated according to Mikami & Niwa (1996). This method considers the effect of initial imperfection and residual welding stresses. Structural optimization of stiffened plates has been worked out by Farkas (1984), Farkas & Jármai (1997) and applied to uniaxially compressed plates with stiffeners of various shapes (Farkas &Jármai (2000), biaxially compressed plates (Farkas et al. (2001)).

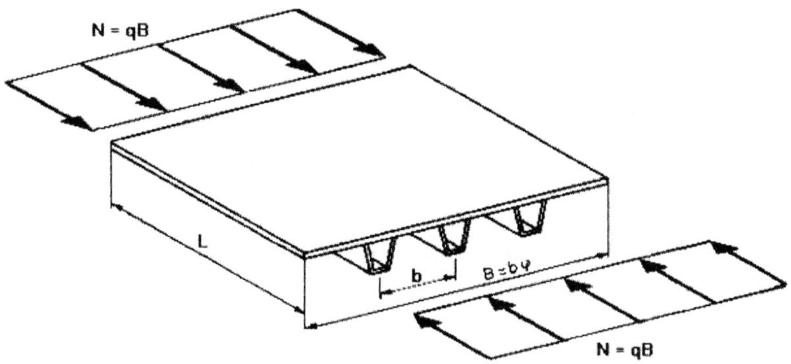

Figure 1. A uniaxially compressed longitudinally stiffened plate

The aim of the present study is to apply Okerblom's constraint to investigate the effect of the deflection of the plate due to longitudinal welds for the optimum design. In the minimum cost design the characteristics of the optimal structural version are sought which minimize the cost function and fulfil the design constraints. First, the special calculations of L- and trapezoidal stiffeners and design constraints are treated, then the general formulae for the cost function is described.

2 Geometric characteristics

Geometrical parameters of plates with L- and trapezoidal stiffeners can be seen in Figure 2.

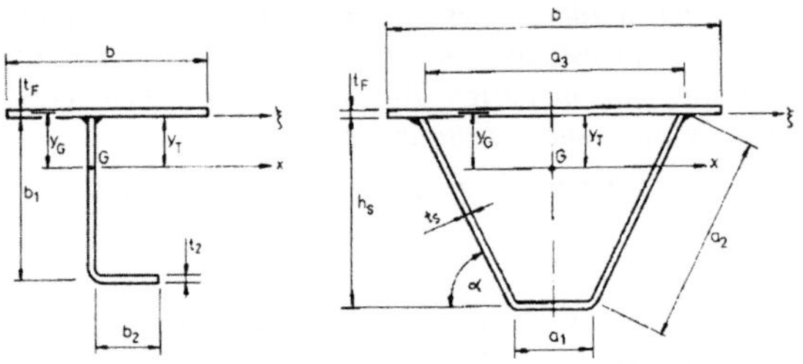

Figure 2. Dimensions of a L- and a trapezoidal stiffeners

The calculations of geometrical parameters of the L-stiffener are

$$A_s = (b_1 + b_2)t_s \tag{1}$$

$$b_1 = 30 t_s \varepsilon \tag{2}$$

$$b_2 = 12.5 t_s \varepsilon \tag{3}$$

$$y_G = \frac{b_1 t_s \dfrac{b_1 + t_f}{2} + b_2 t_s \left(b_1 + \dfrac{t_f}{2} \right)}{b t_f + A_s} \tag{4}$$

$$I_x = \frac{b t_f^3}{12} + b t_f y_G^2 + \frac{b_1^3 t_s}{12} + b_1 t_s \left(\frac{b_1}{2} - y_G \right)^2 + b_2 t_s (b_1 - y_G)^2 \tag{5}$$

$$I_S = \frac{b_1{}^3 t_s}{3} + b_1{}^2 b_2 t_s \tag{6}$$

$$I_t = \frac{b_1 t_s{}^3}{3} + \frac{b_2 t_s{}^3}{3} \tag{7}$$

The calculations of geometrical parameters of the trapezoidal stiffener are

$$A_S = (a_1 + 2a_2)t_S \tag{8}$$

$a_1 = 90$, $a_3 = 300$ mm, thus

$$h_S = \left(a_2^2 - 105^2 \right)^{1/2} \tag{9}$$

$$\sin^2 \alpha = 1 - \left(\frac{105}{a_2} \right)^2 \tag{10}$$

$$y_G = \frac{a_1 t_S \left(h_S + t_F / 2 \right) + 2 a_2 t_S \left(h_S + t_F \right) / 2}{b t_F + A_S} \tag{11}$$

$$I_x = \frac{b t_F^3}{12} + b t_F y_G^2 + a_1 t_S \left(h_S + \frac{t_F}{2} - y_G \right)^2 + \frac{1}{6} a_2^3 t_S \sin^2 \alpha + 2 a_2 t_S \left(\frac{h_S + t_F}{2} - y_G \right)^2 \tag{12}$$

$$I_S = a_1 h_S^3 t_S + \frac{2}{3} a_2^3 t_S \sin^2 \alpha \tag{13}$$

$$I_t = \frac{4 A_P^2}{\sum b_i / t_i} \tag{14}$$

$$A_P = h_S \frac{a_1 + a_3}{2} = 195 h_S \tag{15}$$

3 Global buckling of the stiffened plate

According to Mikami the effect of initial imperfections and residual welding stresses is considered by defining buckling curves for a reduced slenderness

$$\lambda = \left(f_y / \sigma_{cr} \right)^{1/2} \tag{16}$$

The classical critical buckling stress for a uniaxially compressed longitudinally stiffened plate is

$$\sigma_{cr} = \frac{\pi^2 D}{h B^2} \left(\frac{1 + \gamma_S}{\alpha_R^2} + 2 + \alpha_R^2 \right) \quad \text{for} \quad \alpha_R = L / B < \alpha_{R0} = (1 + \gamma_S)^{1/4} \tag{17}$$

$$\sigma_{cr} = \frac{2 \pi^2 D}{h B^2} \left[1 + \left(1 + \gamma_S \right)^{1/2} \right] \quad \text{for} \quad \alpha_R \geq \alpha_{R0} \tag{18}$$

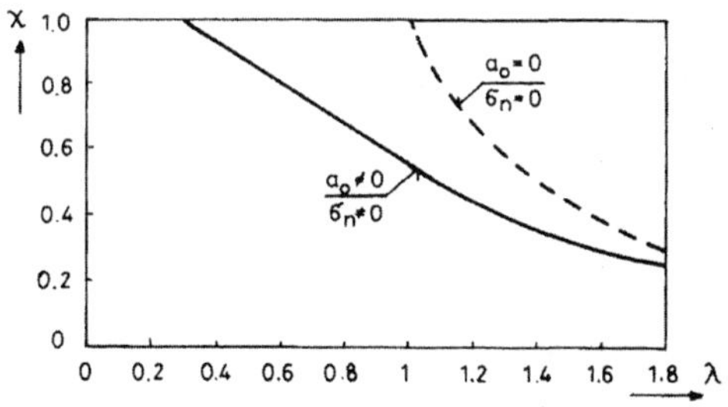

Figure 3. Global buckling curve considering the effect of initial imperfections $(a_0 \neq 0)$ and residual welding stresses $(\sigma_R \neq 0)$

Knowing the reduced slenderness the actual global buckling stress can be calculated according to Mikami as follows (Figure 3)

$$\sigma_U / f_y = 1 \quad \text{for} \quad \lambda \leq 0.3 \tag{19}$$

$$\sigma_U / f_y = 1 - 0.63(\lambda - 0.3) \qquad \text{for} \qquad 0.3 \le \lambda \le 1 \tag{20}$$

$$\sigma_U / f_y = 1/\left(0.8 + \lambda^2\right) \qquad \text{for} \qquad \lambda > 1 \tag{21}$$

The global buckling constraint is

$$\frac{N}{A} \le \sigma_U \frac{\rho_P + \delta_S}{1 + \delta_S} \tag{22}$$

where

$$A = Bt_f + (\varphi - 1)A_S \tag{23}$$

$$\delta_S = \frac{A_S}{bt_f} \tag{24}$$

and the ρ_P factor is

$$\rho_P = 1 \qquad \text{if} \qquad \sigma_{UP} \rangle \sigma_U \tag{25}$$

$$\rho_P = \sigma_{UP} / f_y \qquad \text{if} \qquad \sigma_{UP} \langle \sigma_U \tag{26}$$

4 Single panel buckling

This constraint eliminates the local buckling of the base plate parts between the stiffeners. From the classical buckling formula for a simply supported uniformly compressed in one direction

$$\sigma_{crP} = \frac{4\pi^2 E}{10.92}\left(\frac{t_F}{b}\right)^2 \tag{27}$$

the reduced slenderness is

$$\lambda_P = \left(\frac{4\pi^2 E}{10.92 f_y}\right)^{1/2} \frac{b}{t_F} = \frac{b / t_F}{56.8\varepsilon}; \qquad \varepsilon = \left(\frac{235}{f_y}\right)^{1/2} \tag{28}$$

and the actual local buckling stress considering the initial imperfections and residual welding stresses is

$$\sigma_{UP} / f_y = 1 \qquad \text{for} \qquad \lambda_P \le 0.526 \tag{29}$$

$$\frac{\sigma_{UP}}{f_y} = \left(\frac{0.526}{\lambda_P}\right)^{0.7} \qquad \text{for} \qquad \lambda_P \ge 0.526 \tag{30}$$

The single panel buckling constraint is

$$\frac{N}{A} \le \sigma_{UP} \tag{31}$$

5 Local and torsional buckling of stiffeners

These instability phenomena depend on the shape of stiffeners and will be treated separately for L stiffener.

The torsional buckling constraint for open section stiffeners is

$$\frac{N}{A} \leq \sigma_{UT} \tag{32}$$

The classical torsional buckling stress is

$$\sigma_{crT} = \frac{GI_T}{I_P} + \frac{EI_\omega}{L^2 I_P} \tag{33}$$

where $G = E/2.6$ is the shear modulus, I_T is the torsional moment of inertia, I_P is the polar moment of inertia and I_ω is the warping constant. The actual torsional buckling stress can be calculated in the function of the reduced slenderness

$$\lambda_T = \left(f_y / \sigma_{crT}\right)^{1/2} \tag{34}$$

$$\sigma_{UT} / f_y = 1 \quad \text{for} \quad \lambda_T \leq 0.45 \tag{35}$$

$$\frac{\sigma_{UT}}{f_y} = 1 - 0.53\left(\lambda_T - 0.45\right) \quad \text{for} \quad 0.3 \leq \lambda_T \leq 1.41 \tag{36}$$

$$\frac{\sigma_{UT}}{f_y} = \frac{1}{\lambda_T^2} \quad \text{for} \quad \lambda_T \geq 1.41 \tag{37}$$

6 Distortion constraint

In order to assure the quality of this type of welded structures large deflections due to weld shrinkage should be avoided. It has been shown that the curvature of a beam-like structure due to shrinkage of longitudinal welds can be calculated by relatively simple formulae (Farkas & Jármai 1998). The allowable residual deformations f_0 are prescribed by design rules. For compression struts Eurocode 3 (1992) prescribes $f_0 = L/1000$, thus the distortion constraint is defined as

$$f_{\max} = CL^2 / 8 \leq f_0 = L / 1000 \tag{38}$$

where the curvature is for steels

$$C = 0.844x10^{-3} Q_T y_T / I_x \tag{39}$$

Q_T is the heat input caused by welding for Submerged Arc Welding (SAW) is

$$Q_T = 1{,}3 * 59{,}5 a_W{}^2 \tag{40}$$

and y_T is the weld eccentricity

$$y_T = y_G - t_F / 2 \tag{41}$$

I_x is the moment of inertia of the cross-section containing a stiffener and the base plate strip of width b in the case of trapezoidal stiffeners, instead of b the larger value of $a_3 = 300$ [mm] and $b_3 = b$ - 300 [mm] should be considered.

7 Cost function

The objective function to be minimized is defined as the sum of material and fabrication costs

$$K = K_m + K_f = k_m \rho V + k_f \sum T_i \tag{42}$$

or in another form

$$\frac{K}{k_m} = \rho V + \frac{k_f}{k_m}(T_1 + T_2 + T_3) \tag{43}$$

where ρ is the material density, V is the volume of the structure, K_m and K_f as well as k_m and k_f are the material and fabrication costs as well as cost factors, respectively, T_i are the fabrication times as follows:

Time for preparation, tacking and assembly

$$T_1 = \Theta_d \sqrt{\kappa \rho V} \tag{44}$$

where Θ_d is a difficulty factor expressing the complexity of the welded structure, κ is the number of structural parts to be assembled;

T_2 is time of welding, and T_3 is time of additional works such as changing of electrode, deslagging and chipping. $T_3 \approx 0.3T_2$, thus,

$$T_2 + T_3 = 1.3 \sum C_{2i} a_{wi}^n L_{wi} \tag{45}$$

where L_{wi} is the length of welds, the values of $C_{2i} a_{wi}^n$ can be obtained from formulae or diagrams constructed using the COSTCOMP (1990) software, a_w is the weld dimension.

For SAW (Submerged Arc Welding) welded fillet welds

$$C_2 a_w^n = 0.2349x10^{-3} a_w^2 \tag{46}$$

8 The method of optimization

Rosenbrock's hillclimb (Rosenbrock 1960) mathematical method is used to minimize the cost function. This is a direct search mathematical programming method without derivatives. The iterative algorithm is based on Hooke & Jeeves searching method. It starts with a given initial value and it takes small steps in direction of orthogonal coordinates during the search. The algorithm is modified that secondary searching is carried out to determine discrete values. The procedure finishes in case of convergence criterion is satisfied or the iterative number reaches its limit.

9 Numerical example

The given data are width B = 4200 [mm], length L = 9000 [mm], compression force N = 1.974x10^7 [N], Young modulus E = 2.1 x 10^5 [MPa], density ρ = 7.85x10^{-6} [kg/mm^3] and the yield stress is f_y = 355 [MPa]. The plate is simple supported on four edges. The unknowns – the thicknesses of the base plate and the stiffener and the number of the ribs - are limited in size as follows:

$$3 \le t_f \le 40 \text{ [mm]} \tag{47}$$

$$3 \le t_s \le 10 \text{ [mm]} \tag{48}$$

$$3 \le \varphi \le 15 \tag{49}$$

9.1 *Results for L- stiffeners*

Table 1. Optimization results with L- stiffeners without distortion constraint

k_f/k_m	t_f *[mm]*	t_s *[mm]*	φ	K/k_m *[kg]*
0	28	10	15	11728
1	40	10	4	13687
2	40	10	4	14773

Table 2. Optimization results with L- stiffeners with distortion constraint

k_f/k_m	t_f *[mm]*	t_s *[mm]*	φ	K/k_m *[kg]*
0	28	10	15	11728
1	37	10	6	13699
2	37	10	6	15198

9.2 *Results for trapezoidal stiffeners*

Table 3. Optimization results with trapezoidal stiffeners without distortion constraint

k_f/k_m	t_f *[mm]*	t_s *[mm]*	φ	K/k_m *[kg]*
0	27	10	7	11014
1	31	10	5	12987
2	37	10	3	14194

Table 4. Optimization results with trapezoidal stiffeners with distortion constraint

k_f/k_m	t_f *[mm]*	t_s *[mm]*	φ	K/k_m *[kg]*
0	27	10	7	11014
1	29	10	6	13228
2	29	10	6	15349

10 Conclusions

Cost comparisons of structural versions obtained for a given numerical example by minimum cost design show the following:
The results show that the trapezoidal stiffener is the most economic one.
The distortion constraint in this case is significant, since the weld length is relatively large. It needs higher strength, so the number of the stiffeners is higher. Therefore, the cost is higher too.

The cost difference between the lower and higher fabrication cost is significant, which emphasizes the necessity of optimization.

Acknowledgement

The research work was supported by the Hungarian Scientific Research Found grants OTKA 72386.

References

COSTCOMP (1990) *Programm zur Berechnung der Schweisskosten.* Deutscher Verlag für Schweisstechnik, Düsseldorf.

Eurocode 3. (1992) *Design of steel structures. Part 1.1. General rules and rules for buildings.* European Prestandard ENV 1993-1-1. CEN European Committee for Standardisation, Brussels.

Farkas, J. (1984) *Optimum design of metal structures.* Budapest, Akadémiai Kiadó, Chichester, Ellis Horwood.

Farkas, J., Jármai, K. (1997) *Analysis and optimum design of metal structures.* Balkema, Rotterdam-Brookfield.

Farkas,J.,Jármai,K. (1998) Analysis of some methods for reducing beam curvatures due to weld shrinkage. *Welding in the World* **41** No.4. 385-398.

Farkas, J., Jármai, K. (2000) Minimum cost design and comparison of uniaxially compressed plates with welded flat-, L- and trapezoidal stiffeners. *Welding in the World 44:3, 47-51.*

Farkas, J., Simoes, L.M.C., Jármai, K. (2001) Minimum cost design of a welded stiffened square plate loaded by biaxial compression. *WCSMO-4, 4th World Congress of Structural and Multidisciplinary Optimization, Dalian China, Extended Abstracts, pp. 136-137.*

Mikami, I., Niwa, K. (1996) Ultimate compressive strength of orthogonally stiffened steel plates. *J. Struct. Engng ASCE* **122** No. 6, 674-682.

Rosenbrock, H.H. (1960) An automatic method for finding the greatest or least value of a function, *Computer Journal,* **3** 175-184.

Section 4
Fatigue design

4.1 Numerical Fatigue Analysis of Welded Orthotropic Bridge Decks

Anikó Alleram, László Dunai, Attila László Joó
Budapest University of Technology and Economics, H-1111 Budapest, Hungary
anikoalleram@gmail.com

Abstract

The paper deals with the numerical fatigue analysis of the steel orthotropic bridge decks. The main focus is on the welded connection of the longitudinal rib and the deckplate. Advanced numerical model is developed to predict the fatigue-life of this structural detail on the basis of hot-spot-stresses, using the recommendations of the IIW and Eurocode. By the developed procedure the effect of weld geometry on the fatigue life is analyzed.

Keywords: *fatigue, orthotropic deck, weld, finite element analysis*

1 Introduction

Due to the small self-weight and the easy manufacturing of the steel orthotropic decks they are widely used in the bridge construction. The biggest disadvantage of these systems is the necessity of the welded joints which are sensitive to fatigue. The national requirements and the Eurocode 3 (2005) (EC3) give instruction for detailing this construction.

1.1 *Structural configuration*

According to EC3 the thickness of the deckplate should be selected by the traffic category, the effects of composite action of the deckplate with the surfacing and the spacing of the supports of the deckplate by webs of stiffeners. The National Annex may give information on the plate thickness to be used.

The deckplate-stiffener connection is typically solved by a butt weld. The EC3 recommendation for the weld details of stiffeners are shown on Figure 1.

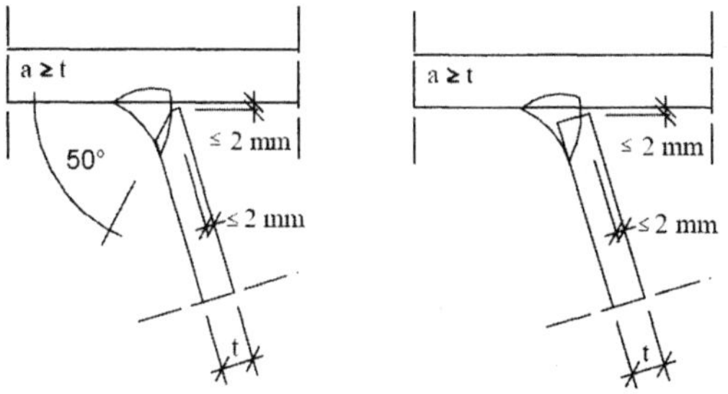

Figure 1. Weld preparation for rib-deck connection

The edges of the stiffeners should be chamfered. The requirements for the butt welds are as follows:

- the seam thickness „a" cannot be smaller than 0,9 x stiffener thickness;
- unwelded gap at root should be smaller than 0,25 x $t_{stiffener}$ or smaller than 2 mm.

The surface of the weld shape is very important as the welded structures are more sensitive to the external weld finishing than the properties of the material or defects deep under the weld surface. The bulge of the surface at the weld transition may easily lead to cracks, therefore the hollow surface should be preferred. The fatigue strength of welded joint is independent from the tensile and yield strength of steel material so the fatigue curves are valid for the different strength group. The butt welds should be preferred rather than fillet welds for fatigue sensitive structures.

The Hungarian National Standard, MSZ EN ISO 5817 (2003) determines the geometry of welded joint through limits for imperfections. The Standard classifies the welded joint into 3 quality level (B,C,D). Figures 2 and 3 shows these requirements for T-joints with butt weld.

Incorrect root gap (Figure 2):

B: $h \leq 0{,}5mm+0{,}1a$ but max. 2 mm

C: $h \leq 0{,}5mm+0{,}2a$ but max.3 mm

D: $h \leq 0{,}5mm+0{,}3a$ but max. 4 mm

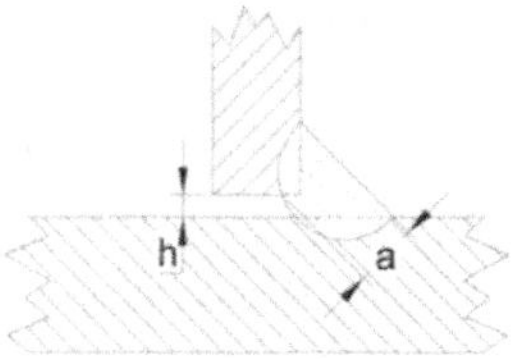

Figure 2. Incorrect root gap

Lack of penetration (Figure 3):

B and **C:** not permitted

D: $h \leq 0{,}5s$ but max. 2 mm

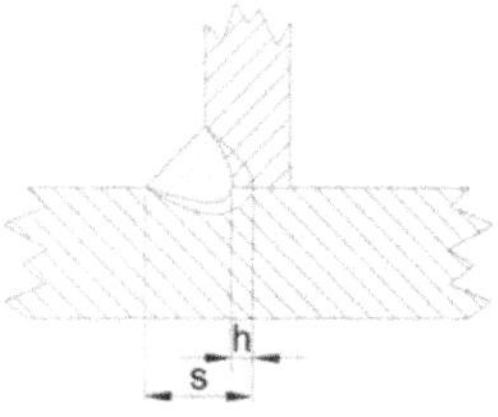

Figure 3. Lack of penetration

1.2 *Fatigue design by EC3*

The EC3 recommends a series of S-N curves, which corresponds to typical detail categories. Each detail category is designated by a number which represents, in N/mm^2 the reference value $\Delta\sigma_C$ for the fatigue strength at $2x10^6$ cycles. In the EC3 two procedures are recommended to perform the fatigue analysis. The first assessment calculates the design stress from the simplified fatigue load model with damage equivalence factors. The second method determines the stress spectrum by standard vehicles and traffic data. The fatigue check is performed on the basis of Palmgren-Miner rule. Table 1 presents the main steps of this method. Stress histories may be evaluated by cycle counting to determine stress range spectrum. The generated spectrum is characterized by the $\Delta\sigma_i$ stress ranges and n_i cycle numbers.

To calculate the D_d cumulate damage the N_i endurance values should be determined from the design fatigue strength curve. The cumulate damage is:

$$D_d = \sum_i^n \frac{n_{Ei}}{N_{Ri}}$$

The fatigue strength of the structural detail is checked by the cumulate damage (D_d<1,0).

Table 1. Steps of fatigue analysis

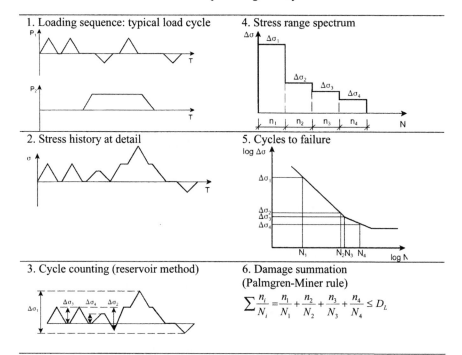

1. Loading sequence: typical load cycle	4. Stress range spectrum
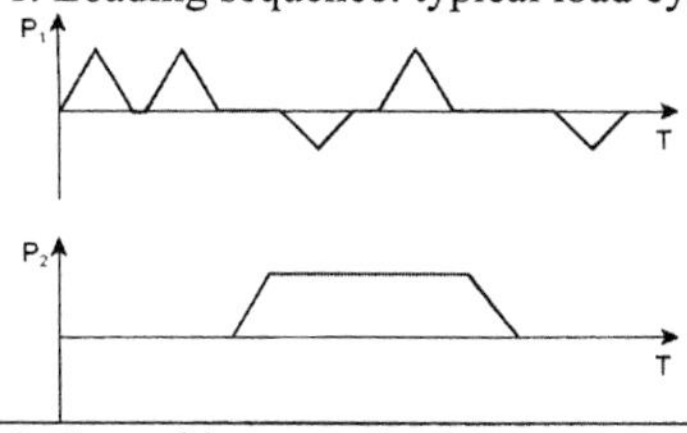	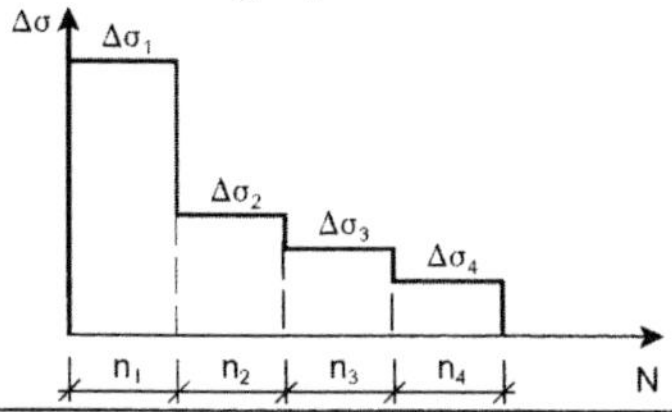
2. Stress history at detail	5. Cycles to failure
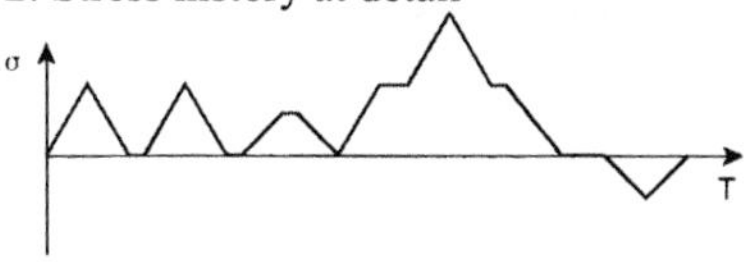	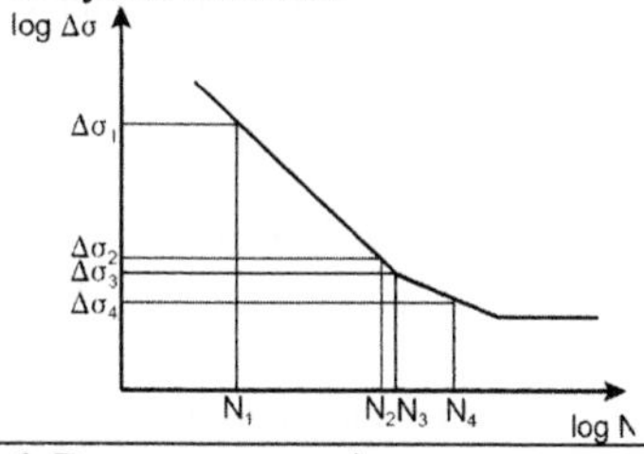
3. Cycle counting (reservoir method)	6. Damage summation (Palmgren-Miner rule)
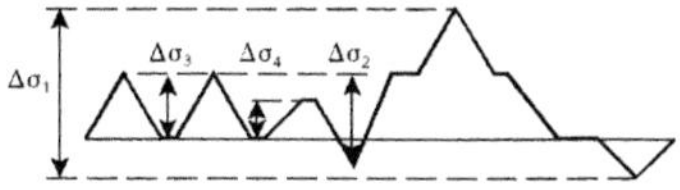	$\sum \frac{n_i}{N_i} = \frac{n_1}{N_1} + \frac{n_2}{N_2} + \frac{n_3}{N_3} + \frac{n_4}{N_4} \leq D_L$

1.3 *Nominal and geometrical stress*

The design fatigue strength curves can be applied for structural detail or structural point. Accordingly, the nominal stress (excluding all stress concentration) or geometric stresses (structural hot spot) are to be used to predict the fatigue life, for detail or point, respectively. The geometric stress takes into account the stress raising effect of a structural detail excluding all stress concentration due to local weld profile, as shown in Figure 4.

EC3 allows to use geometrical stress and recommends fatigue strength curves but does not give assessment to calculate stress concentration. According to the recommendations of the International Institute for Welding (IIW 2003) this stress can be analyzed by appropriate finite element (FE) model. The aim is to calculate the principal stress close to the weld toe, as illustrated in Figure 4. The IIW recommendation determines the stress by 2 or 3 reference points and calculates the hot spot stress with extrapolation.

The stress spectrum depends on the geometry of the vehicles, the axle loads, the vehicle spacing, the composition of the traffic and its dynamic effects. The Eurocode 1 (2002) (EC1) applies fatigue load model to analyse the effect on deckplates. The models use traffic categories which are defined by number of slow lanes, and the number of heavy vehicles. In this research the first category with $N_{obs}=2x10^6$ per year in the slow lane are assumed and used in the analyses.

Firstly, the details of the finite element models are presented; then its application is shown for different weld detailing. Finally, the results of the fatigue check are summarized.

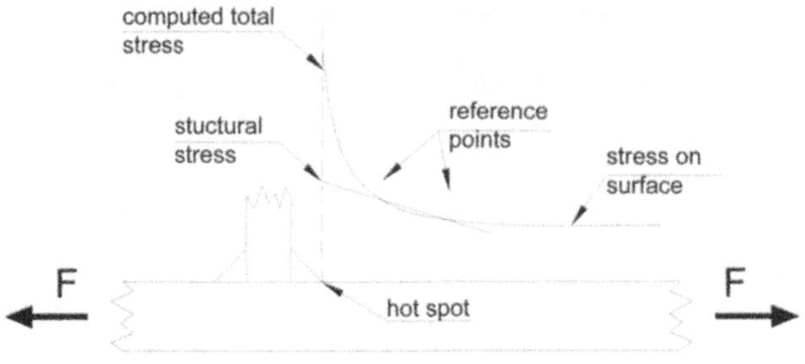

Figure 4. Definition of structural hot spot stress

2 The finite element models

To perform the fatigue analysis two FE models are developed with the same geometry by the Ansys finite element program (Ansys 2002). The *first* is a two-level-model consists of a global and local model. The global shell model carries the vehicle loads, meanwhile the local plane model contains the weld geometry and uses the deflections of global model as prescribed displacement loading. In the local model the geometrical stress for fatigue analysis is calculated. In the *second* solid-shell model the weld area is modelled by solid finite element, while the other parts of the structure is modelled with shell elements.

2.1 *The two-level model*

The *global model* is a 3000mm wide and 10800mm long orthotropic deck with longitudinal trapezoid ribs. The distances between the cross girders are 3600 mm (Figure 5.). The position and geometry of vehicle loads follow the orders of EC1 and also takes into account the transverse locations of vehicles and the different types of wheels.

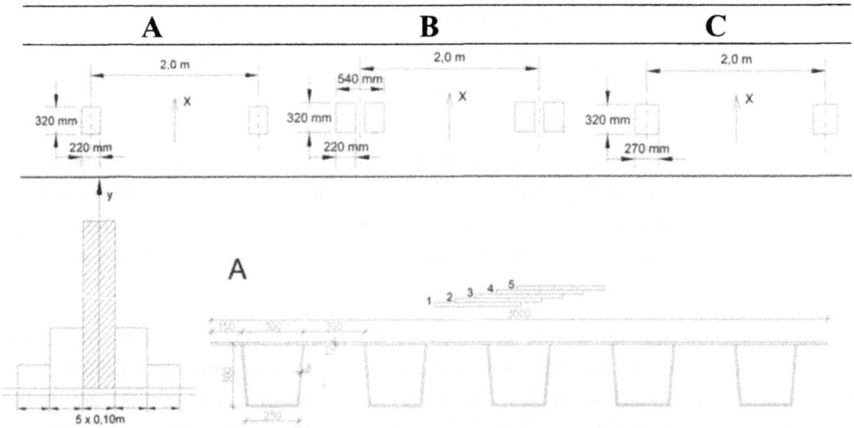

Figure 5. Wheel loads, distribution of transverse location of centre lines of vehicle

The *local model* is a plane model with unit thickness in plane strain state. The weld geometry is generated on the middle rib of the local model, which is able to take into account the effective geometry with the following parameters (Figure 6):

- lack of penetration;
- the „a" size of weld;
- gap between deck and rib;
- the contour of weld.

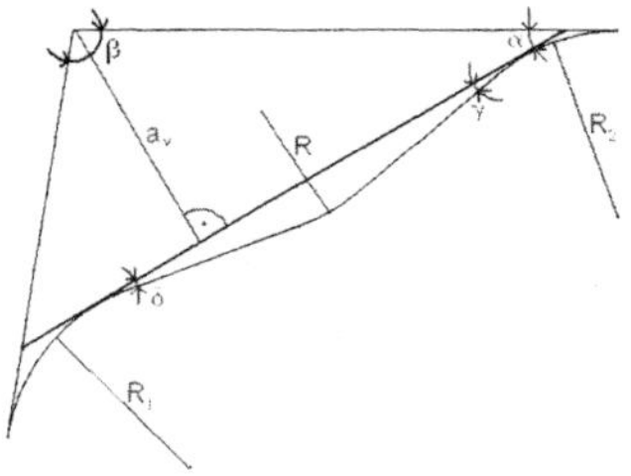

Figure 6. The geometry parameters of the studied welds

In the investigations six realistic weld shapes (1-6) are considered. The parameters of them are summarized in Table 2, and Figure 7 shows the image and the part of the local FE model of weld no. 4.

Table 2. The geometry parameters of the welds (the units are in mm),

Parameters	**1**	**2**	**3**	**4**	**5**	**6**
a_v	3,00	3,40	6,10	4,24	4,93	4,40
α	53,00	53,00	38,00	55,00	41,00	53,00
γ	4,00	12,00	9,00	6,00	8,00	6,00
δ	2,00	4,00	7,00	11,00	9,00	8,00
r_1	4,70	6,10	6,50	0,78	2,43	4,60
r_2	0,50	0,40	0,15	0,84	1,00	2,30
r	15,00	6,00	20,00	25,00	4,00	10,00
Penetration	2,00	2,00	0,00	0,86	1,14	1,30
Gap	0,60	0,30	0,30	0,60	0,40	0,20

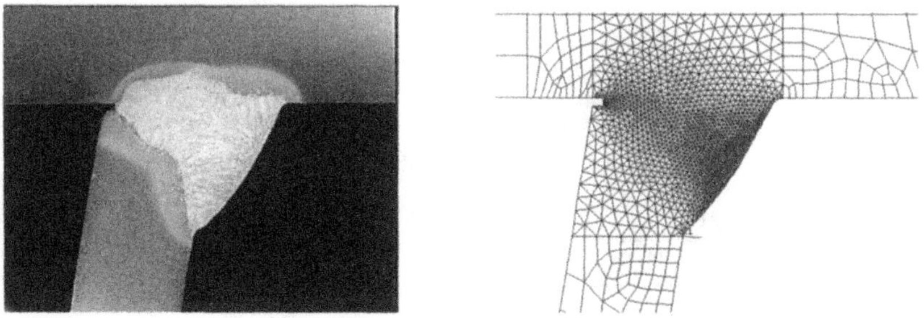

Figure 7. Studied weld no. 4

2.2 *The solid–shell model*

The weld area in the solid-shell model is built up with solid elements, the other parts with shell elements The stress and strain continuity is ensured between the two types of elements: Figure 8 shows the strain of middle-loaded trapezoid rib. In this model the weld stresses can be determined directly.

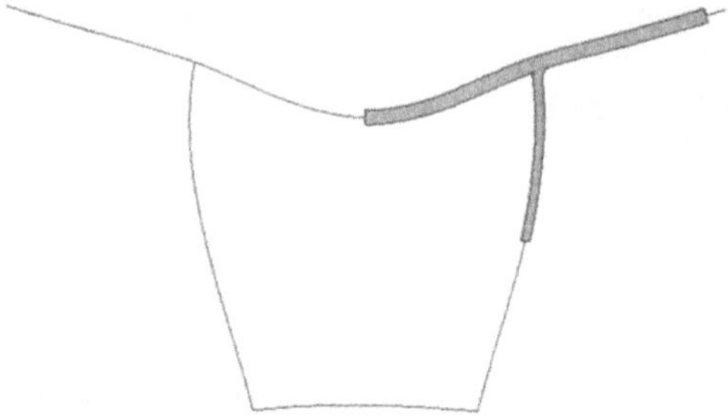

Figure 8. The deformation of middle rib in the solid-shell model

3.3 *Calculation of geometrical stresses*

According to the IIW recommendations the design value of geometrical stress can be calculate considering the size of finite element, as follows (Figure 9).

1) Fine mesh with element length not more than **0,4t** at the hot spot: evaluation of nodal stresses at two reference points **0,4t** and **1,0t**, and linear extrapolation:

$$\sigma_{hs} = 1{,}67 \cdot \sigma_{0,4t} - 0{,}67\sigma_{1.0t}$$

2) Fine mesh as defined above: evaluation of nodal stresses at three reference points **0,4t, 0,9t** and **1,4t**, and quadratic extrapolation; this method is recommended in cases with pronounced non-linear structural stress increase to the hot spot.

$$\sigma_{hs} = 2{,}52 \cdot \sigma_{0,4t} - 2{,}24 \cdot \sigma_{0,9t} + 0{,}72 \cdot \sigma_{1,4t}$$

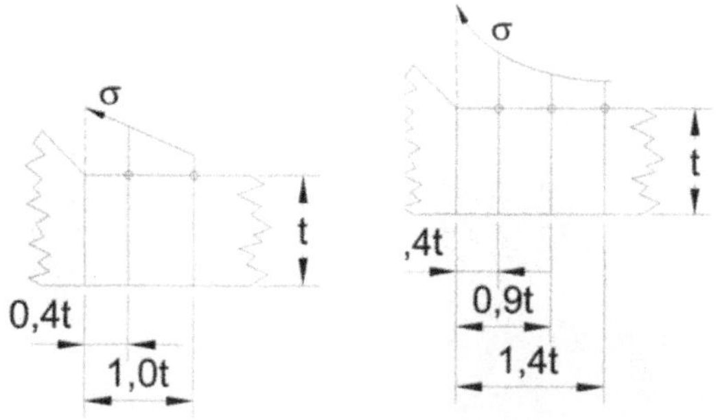

Figure 9. Determination methods of geometrical stress

3 Results of FE analysis

According to the two-level analysis the largest stress for all the welds evolved at the point signed as „maximum" on Figure 10. The linear and quadratic extrapolation nearly gave the same results for the different welds, as shown in Figure 11. Comparing the stresses at weld toe with the stresses calculated by quadratic extrapolation, the biggest difference occurs at weld no. 4 (~30%), where the bellow rounding (r_1) is the smallest. This result shows that the rounding at weld toe has significant effect on stresses.

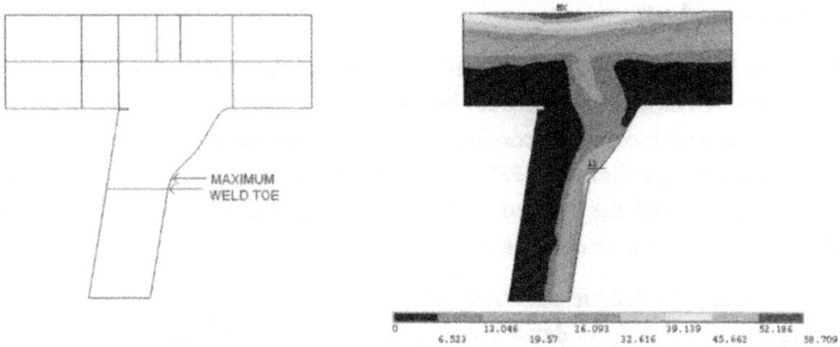

Figure 10. Principal stress distribution in weld no. 4

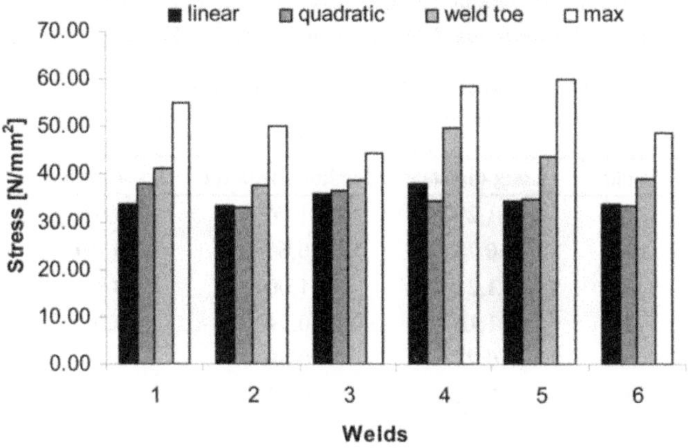

Figure 11. The numerical and the geometrical stresses for the different welds

The results from solid-shell model show the same tendency but the values, for example at weld toe, are less by 30-40 per cent (Figure 12): the assumed plain strain state at the local model resulted in stiffener structure with higher stresses.

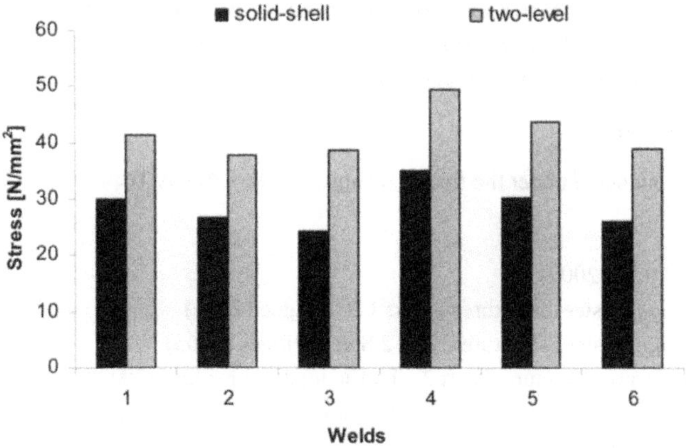

Figure 12. The weld toe stresses for the different welds

4 The fatigue analysis

To perform the fatigue analysis the detail category of the weld should be determined according to EC3. All of the analysed weld details belong to category $\Delta\sigma_c$=100 N/mm^2, which appoint a fatigue strength curve. According to EC3, if all of the stress values are bellow the constant amplitude fatigue limit (0,737*$\Delta\sigma_c$), the structure detail is not sensitive to fatigue. For the studied cases and the applied loads all of the calculated geometrical stresses are bellow this limit, so the fatigue check is not necessary.

In order to compare the fatigue sensitivity of the different weld geometries, however, the fatigue strength curve belong to detail category $\Delta\sigma_c$=45 N/mm^2 is used. In the first step the amplitude of stress values ($\Delta\sigma_i$) from different wheel types are calculated. In the next step the effective cycle numbers (n_i) are determined for the stress amplitudes. Finally the endurance values (N_i) are taken from the fatigue strength curve. On the basis of the cumulate damages the fatigue sensitivity of the welds are determined, as detailed in Table 3.

Table 3. Cumulate damages for the different welds

	Traffic type		
Weld	Long distance	Medium distance	Local traffic
1	1,29	1,00	0,67
2	0,78	0,60	0,39
3	1,21	1,00	0,82
4	1,15	0,94	0,70
5	0,92	0,68	0,38
6	0,82	0,63	0,41

5 Conclusions

Advanced finite element model based fatigue-life prediction method is developed for the welded details of steel orthotropic bridge deck. The process is based on the geometric stress and may produce a more established and more accurate results than the methods used before. The two different level of FE models are verified and applied for the geometrical stresses calculations of different weld details. In the models the real geometry of the welds are considered on the basis of fabricator's weld data. The application is illustrated by the fatigue analysis of the weld details, with assumed traffic data. On this basis the damage calculation the fatigue sensitivity of the different details are determined.

Acknowledgement

The research is conducted under the financial support of the OTKA T049305 project.

References

Ansys, Release 9.0A1 (2002)

Eurocode 3: Design of steel structures – Part 1-9: Fatigue (2005)

Eurocode 3: Design of steel structures, Part 2 Steel Bridges (2005)

Eurocode 1: Actions on structures, Part 2 : Traffic loads on bridges (2002)

IIW (2003) *Recommendations for fatigue design of welded joints and components.* International Institute of Welding.

MSZ EN ISO 5817 (2002) *Welding. Fusion-welded joints in steel, nickel, titanium and their alloys, Quality levels for imperfections.*

4.2 Fatigue Strength Investigations of Connections between Primary Ship Structural Components

Wolfgang Fricke, Anatole von Lilienfeld-Toal, Hans Paetzold
Hamburg University of Technology, Hamburg, Germany

Abstract

Primary structural components in ships are usually built-up I-beams with deep webs and flanges formed either by flat bars or by water-tight stiffened plating of the ship hull, decks or bulkheads. Typical examples are stiffened decks with large T-shaped girders beneath the decks or the floors and girders of the double bottom. Highly stressed are often the end connections between horizontal and vertical members which can be designed in different ways. One way is to apply large triangular or curved brackets which additionally support the connection. An alternative is offered by widening the depth of the component towards the connection with knuckled flange or plating. A further alternative particularly for ro/ro ships is a simple intersection of the I-beams which is well-known from steel structures. The different alternatives are discussed in the paper. As all designs contain fatigue-critical details, several numerical and experimental investigations have been carried out in the past, which are reviewed. Recent investigations on the third alternative, i.e. simple intersections of I-beams, are described in more detail. The stress analysis was performed with an extensive finite element model using the structural hot-spot stress approach for the assessment of the critical weld toes. The fatigue tests with large-scale models showed a fatigue behaviour which differed from the numerical assessment. Possible reasons for the discrepancies are discussed and conclusions are drawn.

Keywords: *Fatigue, girder, connection, ship structure*

1 Introduction

Ship structures are usually typical steel structures mainly consisting of stiffened plate fields. These plate fields are supported by girders (in longitudinal direction) and/or web frames or side transverses, deck transverses and double bottom floors (in transverse direction). These structural members, which may have a depth of 1 – 3 m in large ships, are called primary members of a ship structure. Fig. 1 shows these structural members in a transverse section through a large tanker.

In the double bottom and double side, the primary structural members consist of floor and web plates, which form a beam with two flanges, i.e. the plating of the inner and outer hull. Below deck and along the longitudinal bulkheads, the primary members consist of T-shaped beams with deep webs, forming a ring structure together with the web plates in the double hull, see Fig.1.

The loading is mainly due to pressure loads on the plates which are transmitted to the primary members by the stiffeners, but also due to global loads and deformations of the ship, e.g. relative vertical deflections between the double side and the longitudinal bulkheads. These cause high bending and shear stresses particularly at the ends of the members, requiring special attention and measures by the designer.

Frequently, large triangular brackets are arranged at the members' ends supporting the end connection and reducing the span of the primary members. Here, the toes of the brackets create a discontinuity which is prone to fatigue under cyclic loads (Munse, 1981; Jordan & Krumpen, 1984; Mizukami et al., 1994; TSCF, 1995). In double hull structures, the transition at the ends is often designed with a sloped inner hull, as shown in Fig. 1 for the connection between double bottom and side structure. Here, the knuckle of the inner plating is well-known to be the critical point, where the stress is increased due to the reduced effective width of the plating.

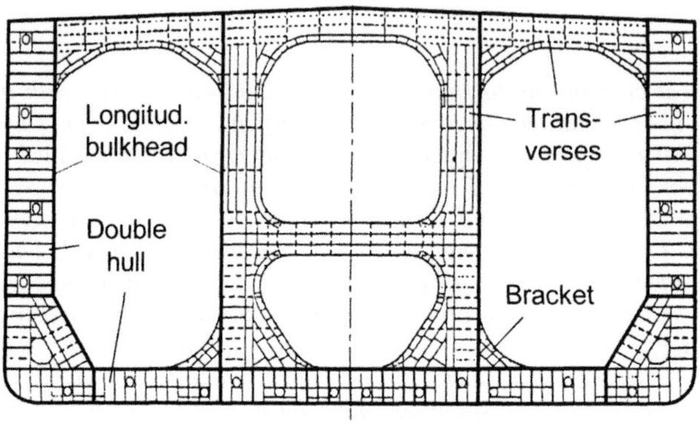

Figure 1. Midship section of a large tanker (Gutierrez-Fraile et al., 1994)

In some ship types such as roll-on/roll-off ships, a connection between primary structural members without sloped transition and without any brackets is desired in order not to loose space for trucks. Here, the connection is similar to steel structures of buildings, where the node has to be properly designed for the transfer of the bending moments.

The connections between primary members have been subject of several investigations during the past, with the focus being mainly on the fatigue strength. In the following, some of the investigations on the different types mentioned are reviewed and conclusions drawn with respect to appropriate designs.

2 End connections of primary structural members with brackets

At the ends of primary structural members, i.e. T-shaped beams and webs in double hulls, relatively large brackets are usually applied, which have to be reinforced by stiffeners at least along their free edge. The arrangement and size of this stiffener affects the local geometry and the stress concentration at the bracket toe.

Fig. 2 shows the effect of a stiffener placed on the surface of a bracket close to the free edge and of a soft 'nose' at the bracket toe on the stress concentration factor K_s for different load components acting on the primary structural member. The load components are the internal nominal force N, the shear force Q and the bending moment M in the beam section at the bracket toe. K_s is the ratio between the structural hot-spot stress computed with the finite element method (Fricke et al., 1991) at the weld toe and the nominal stress component. The figure shows that the stress increase due to the stiffener and the stress reduction due to the soft nose are more or less pronounced in the different load cases. Also shown is the superimposed structural hot-spot stress σ_s for a given load combination.

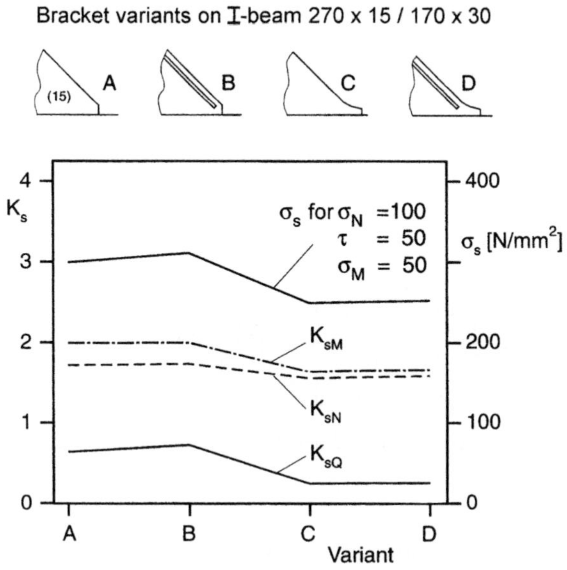

Figure 2. Stress concentration factor K_s and structural hot-spot stress σ_s for a load combination at different bracket toes

Another investigation (Paetzold et al., 2001) showed that the stress concentration becomes very high if the stiffener is arranged as face plate on the free edge of the bracket. Fig. 3 illustrates the three versions of brackets on an I-beam, which were investigated experimentally and numerically for a load case with predominant bending in the I-beam.

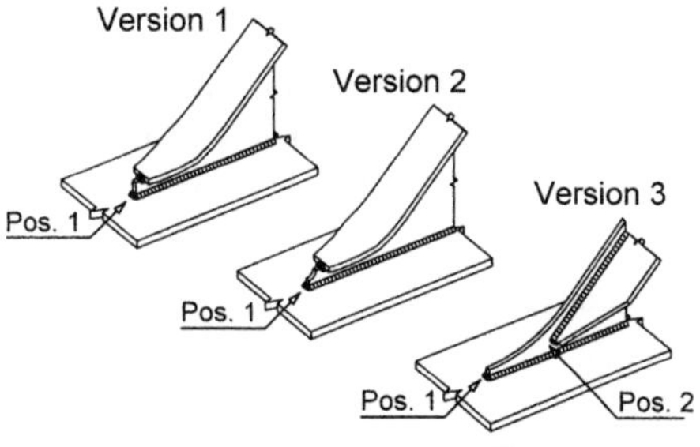

Figure 3. Versions of bracket toes investigated by Paetzold et al. (2001)

The stress plot in Fig. 4 shows a very high stress peak on the flange of the I-beam at the bracket toe (Pos. 1) of version 1. Also highly stressed is the end of the flange. In cyclic loaded structures it is recommended to taper the width as well as the thickness very softly (TSCF, 1995). A soft nose (version 2) can reduce the stresses particularly at the end of the flange. A great reduction of the hot-spot stresses at the bracket toe can be achieved by the arrangement of the stiffener as shown for version 3 in Fig. 3, which is quite common in shipbuilding. The fatigue tests have shown that Pos. 2 is less prone to fatigue compared to Pos. 1.

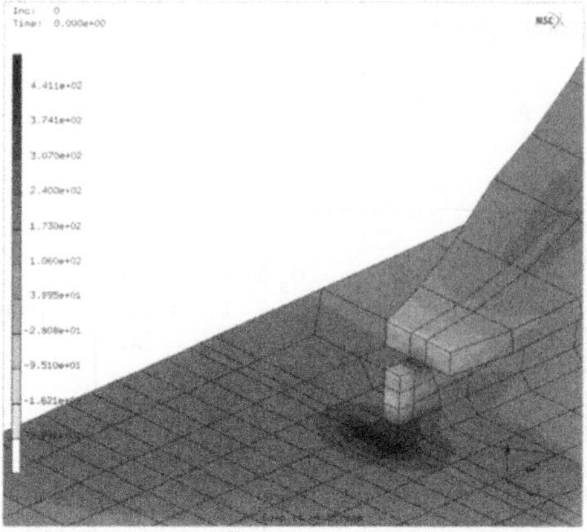

Figure 4. Distribution of longitudinal stresses at the bracket toe of version 1

The fatigue tests performed showed also that full-penetration welding is very important at highly stressed bracket toes in order to avoid cracks from the weld root. Due to the high force flow through the weld, the fatigue assessment should be performed using the fatigue class for load-carrying welds, i.e. FAT90 according to the IIW fatigue design recommendations (Hobbacher, 2007).

3 Knuckled transitions of primary structural members

Knuckled transitions are quite usual at the connections between horizontal and vertical primary structural members, as shown in Fig. 1 for the connection between the double bottom and the side structure. But also T-shaped members are often accordingly knuckled and widened at their ends in order to reduce the stresses at the connection.

As mentioned before, the knuckle of the plate or flange is the critical point, where an effective support is essential to avoid excessive secondary bending stresses. Fig. 5 shows different structural arrangements which are usual for the double bottom and side of ships.

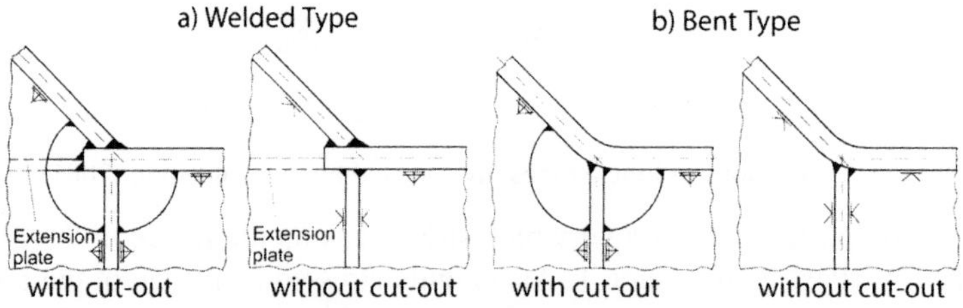

Figure 5. Typical structural arrangements for knuckled transitions of primary structural members

The knuckle of the plating or flange can either be realised by arranging a welded joint or by plate bending. Usually, cut-outs or scallops are arranged in order to allow proper welding, however it is also possible to omit the scallops, which requires very accurate fabrication in connection with full penetration welding.

The structural hot-spot stress can be rather large in the knuckle. Under axial loading of the plating, the angle of the knuckle as well as the reduced effective width of the plating contribute mainly to the stress increase but also the geometry of supporting structures and possible local secondary bending stresses. Fricke & Weissenborn (2004) showed by parametric finite element calculations structural stress concentration factors which are typically between 2 and 10 but can even be larger for a weak support of the knuckle.

Not only the knuckled plating or flange is prone to fatigue, but also the welded joints with the plates behind and particularly the ends of the scallops. Several experimental and numerical investigations have been performed in the recent past on this detail. Fatigue tests of different types of knuckled transitions within the EU-funded project FatHTS (Dijkstra et al., 2001) showed that fatigue cracks appeared relatively early in the welded type (upper left part in Fig. 5), however, the remaining lifetime was very long because the cracks ran into areas with lower stresses. The bent type (lower left part in Fig. 5) showed a higher fatigue strength and cracks appeared at first at the scallops, which can be regarded as less relevant for the safety because there is no leak in the beginning. Sometimes these cracks are arrested.

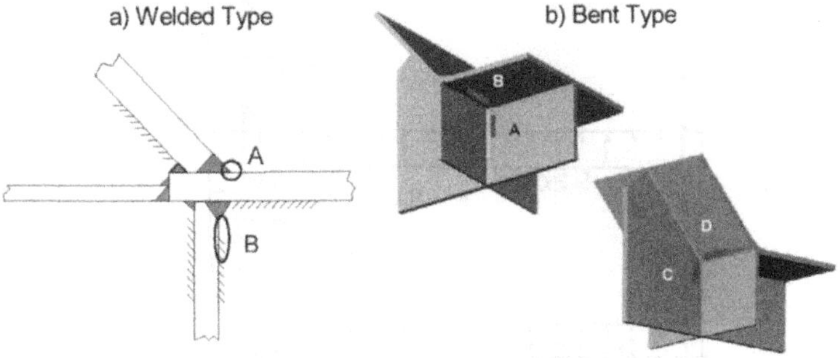

Figure 6. Potential hot spot for the welded and bent type of knuckled transition (Kang et al., 2004)

Gimperlein (1990) showed for the bent type that the fatigue strength is strongly affected by the bending of the supporting component. Extensive computations and fatigue tests of the welded and bent type without scallops were performed by Kang et al. (2004). Full penetration welding was necessary for the connections with the transverse members to avoid cracks originating from the weld root. The potential hot spots are shown in Fig. 6. The tests confirm that pos. A is the most critical point of the welded type. Finite element calculations showed that for small bending radii pos. A and C are more critical, while for larger bending radii pos. D becomes critical. The tests of realistic models with small bending radius revealed early cracks at pos. C. Probably, unfavourable manufacturing conditions and high residual stresses may have played a role here.

The determination of the structural hot-spot stress may be also problematic here because the stress has to be extrapolated in two directions into the corner. In ship structural design, a further problem arises due to the preferred use of shell models, where the stress extrapolation to the intersection point gives too large stresses.

Lotsberg et al. (2007) have developed a method to correct these stresses for the applications shown here.

4 Intersections of I-beams

As mentioned in the beginning, intersections if I-beams without any brackets or widened webs are typical on ro/ro ships, see Fig. 7. In such intersections, local stresses at the welded joints are not only increased due to reduced effective width but also due to the transfer of bending moments by shear in the nodal plate, which is illustrated in Fig. 8.

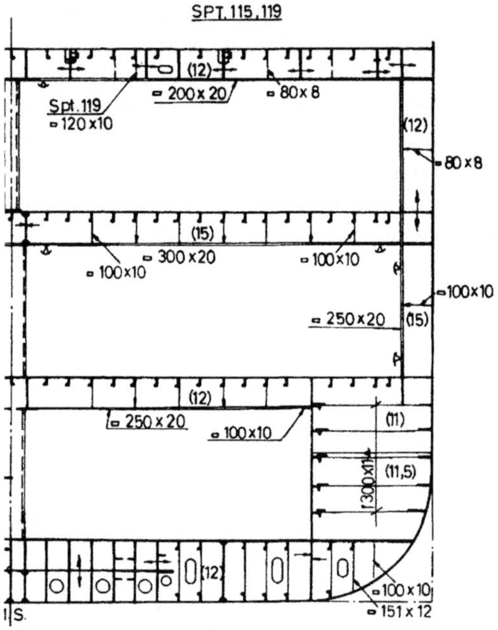

Figure 7. Midship section of a ro/ro ship (Schidlowski, 1977)

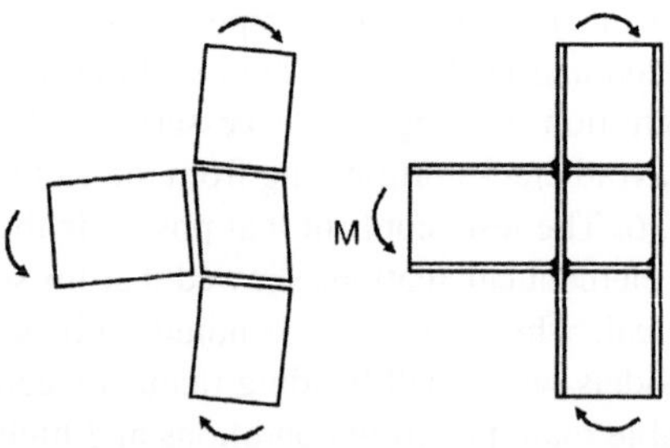

Figure 8. Intersections of I-beams under bending load

In a recent research project, a similar connection forming the upper corner of the side structure of a ro/ro ship, has been investigated experimentally and numerically. Fig. 9 shows the almost 4 m long test model with two parallel T-bars being mainly subjected to bending by the diagonally acting hydraulic cylinder. The T-bars have a web of 600 x 10 and a flange of 250 x 20 in the horizontal continuous beam and 200 x 20 in the vertical non-continuous beam, respectively. The thickness of the plate is 11 mm.

Figure 9. Test model for corner connection of I-beams

A cut-out is arranged at the end of the vertical non-continuous beam to ensure sound full-penetration welding of the vertical flange, see Fig. 10. The weld toes showing the highest structural hot-spot stresses in a finite element analysis according to the IIW guidelines (Niemi et al., 2006) are denoted HS1 – HS4. The highest structural hot-spot stress was found for HS2 from both, the finite element analysis and strain measurement, while HS1 and HS3 showed similar, burt smaller results, and HS4 the smallest stresses.

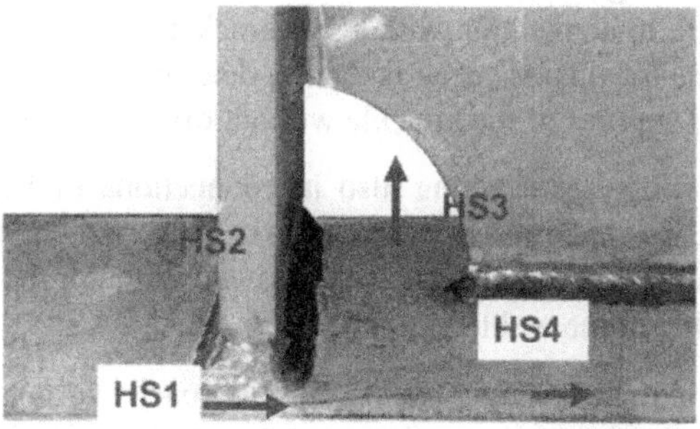

Figure 10. Critical weld toes at the connection of the I-beam

In the fatigue tests, first cracks appeared at HS3 and close to HS4, however, here from the root of the fillet weld. This crack stopped after reaching the width of the weld, while the crack at HS3 penetrated through the flange and dominated the fatigue failure. Fig. 11 shows the propagation of the cracks over the load cycle number.

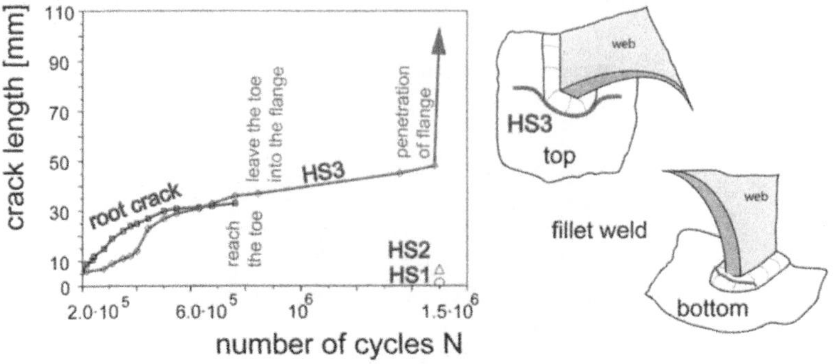

Figure 11. Propagation of cracks at HS3 and HS4

It is interesting to note that the same failure behaviour was observed at both models tested and that no crack was observed at the most highly stressed HS1 nor at HS2. Apart from the weld quality, different residual stresses are considered to be the main reason for this.

5 Conclusions

The end connection of primary structural members in ship structures is designed in different ways, usually being supported by large brackets or with widened depth to reduce the span and the stresses. However, local stress concentrations occur, either at the bracket toes or at the knuckles of the flange or plating, which require a fatigue strength assessment in the presence of cyclic stresses.

Different structural configurations have been investigated experimentally and numerically in the past, from which conclusions can be drawn for structural design. In brackets, the stress concentration is mainly determined by the stiffener or flange arranged at the free edge of the bracket. A stiffener placed at the side shows an improved fatigue life in connection with a soft bracket toe. In the case of a knuckled flange or plating, the local stress peak is mainly determined by the reduced effective width. Here, a good support of the knuckle without offset is essential.

Rather high local stresses are acting also in connections of I-beams without any bracket or knuckel. The fatigue-critical point may be here the cut-out for welding – as also for knuckled flanges –, as tests have shown, although the structural hot-spot stress might be higher at the welded joint of the flange.

Nevertheless, the structural hot-spot stress approach is considered to be an appropriate tool for the fatigue strength assessment of the critical welds. It should be emphasized that full-penetration welds are required here in order to avoid early fatigue cracks from non-welded root faces.

6 Acknowledgements

The investigations have been performed in different research projects, the last mentioned in the project 'Stiffened plate structures in Shipbuilding', funded by AiF (Arbeitsgemeinschaft industrieller Fördervereinigungen) through the Center of Maritime Technologies (CMT) in Hamburg. The test models have been provided by Flensburger Schiffbau-Gesellschaft.

References

Dijkstra,O., Janssen,G.T.M. & Ludolphy,J.W.L. (2001) Fatigue Tests of Large Scale Knuckle Specimens. *Practical Design of Ships and Other Floating Structures* (Ed.: Y.-S. Wu, W.-C. Cui und G.-J. Zhou), Elsevier.

Fricke,W., Borchardt,H., Gritl,D. & Pohl, S. (1991) Bewertung der Betriebsfestigkeit schiffbaulicher Strukturdetails mit Hilfe von Formzahlen. *Jahrb. Schiffbautech.* Ges., **85** Vol., Springer-Verlag, Berlin, Heidelberg, New York.

Fricke,W. & Weißenborn,C. (2004) Vereinfachte Formzahlen für ermüdungskritische Strukturdetails in der Schiffskonstruktion. *Schiffbauforschung* **43** 2/2004, pp. 39-50.

Gimperlein,D. (1990) Tragverhalten von Rahmenecken mit geknickten Gurten. *Schweißen und Schneiden* **42** S. 234-240, with English version.

Gutierrez-Fraile,R., Rosenberg,H., Person,P., Cumin,A. & Paetow,K.-H. (1994) The European E3-Tanker: Development of an Ecological Ship. *Trans. SNAME*, 102.

Hobbacher,A., Ed. (2007): Recommendations for Fatigue Design of Welded Joints and Components. IIW-Doc. XIII-1965-03 / XV-1127-03, International Institute of Welding.

Jordan,C.R. & Krumpen,R.P. (1984): Performance of structural details. *Welding Journal*, Jan. 1984, 18-28.

Kang,J.-K., Kim,Y. & Heo,J.-H. (2004): Fatigue Strength of Bent Type Hopper Corner Detail in Double Hull Structure. *Proc. of OMAE Specialty Conf. on FPSO Integrity*, Houston.

Lotsberg,I., Rundhaug,T.A., Thorkildsen,H., Bøe,A. & Lindmark,T. (2007) Fatigue design of web stiffened cruciform connections. *Proc. PRADS'2007*, pp. 1003-1011, ABS, Houston.

Mizukami,T, Ishikawa,I. & Yuasa,M. (1994) Trends of recent hull damage and countermeasures. *NK Tech. Bulletin*, Tokyo.

Munse,W.H. (1981) Fatigue criteria for ship structure details. *Proc. Extreme Loads Response Symp., SNAME*, Arlington.

Niemi,E., Fricke,W. & Maddox,S.J. (2006) Fatigue Analysis of Welded Components - *Designer's Guide to the Hot-Spot Stress Approach.* Woodhead Publ., Cambridge.

Paetzold,H., Doerk,O. & Kierkegaard,H. (2001) Fatigue behaviour of different bracket connections. In: *Practical Design of Ships and Other Floating Structures* (Ed. Y.-S. Wu, W.-C. Cui and G.-J. Zhou), Elsevier.

Schidlowski,D. (1977) Roll-on/Roll-off-Motorschiff „Reichenfels". *Hansa* **114** 22/1977, pp. 1993 – 1998.

TSCF (1995) Tanker Structure Co-operative Forum: *Guidelines for the Inspection and Maintenance of Double Hull Tanker Structures*, Witherby & Co. Ltd., London

4.3 Structural Stress Concentration Due to Axial Misalignment in Monorail Beam

Tapani Halme
Lappeenranta University of Technology, PO 20 Lappeenranta, Finland,
tapani.halme@lut.fi

Abstract

Stress and displacement analysis of a beam used in so-called monorail hoist system is analyzed in the context of the first-order Generalised Beam Theory, which takes into account the distortional effects of thin-walled structural members. In the analysis the structural stress in the loading point is calculated using GBT and the results are verified using finite element analysis. These results are then used in fatigue analysis to estimate the effect of plate misalignment in a butt weld joint in the beam and a life estimate based on stress concentration due to misalignment is presented.

Keywords: *Generalised Beam Theory, structural stress, weld misalignment, fatigue*

1 Introduction

Generalized Beam Theory (GBT), developed by Prof. Schardt and his associates in Darmstadt, Schardt (1989) is a tool for the analysis of prismatic thin-walled structures based in the theory of Vlasov (1961). GBT unifies the concept of warping in Vlasov's theory into all the deformation modes. GBT handles the different deformation modes, i.e. extension, bending about the principal axes, torsion and the distortional modes independently according to the first-order theory. They are thus uncoupled and can be analyzed separately before their effects are combined with the simple procedure of superposition, as in the classical beam bending analysis. In this paper a stress and displacement analysis of a monorail beam is performed as a case study using the first-order GBT. The resulting structural stress at a butt weld joint is used as basis to qualitatively estimate the design life of the structure.

2 Theoretical background of GBT

Prismatic structures have two main directions: the longitudinal direction, x, and the transverse direction, s. Thus the total deformation of the structure can thus be expressed as the sum of product functions $F(s)\cdot V(x)$ where $F(s)$ expresses the relative cross-sectional deformation, which also includes distortion, and $V(x)$ (generalised displacement), which is the amplitude function along the x-axis. Both functions must fulfil the boundary conditions of the structure, $F(s)$ transversally and $V(x)$ longitudinally. The generalized displacement $V(x)$, i.e. transverse deformation along the length of a prismatic beam, can be solved from the basic equation of GBT, which is written in a form

$$E\,{}^{k}C\,{}^{k}V'''' - G\,{}^{k}D\,{}^{k}V'' + {}^{k}B\,{}^{k}V = {}^{k}q\,. \tag{1}$$

The index k refers to the deformation modes or eigenmodes of a prismatic cross-section, which are obtained from the eigenvalue analysis of the cross-section. As an example the first fifteen deformation modes of a monorail hoist beam profile is presented in Fig. 1. The first deformation mode is the axial displacement. The next two modes are the bending deformation about the principal axes and the fourth one is the torsion mode, i.e. rotation about the shear centre. The following are the distortional modes.

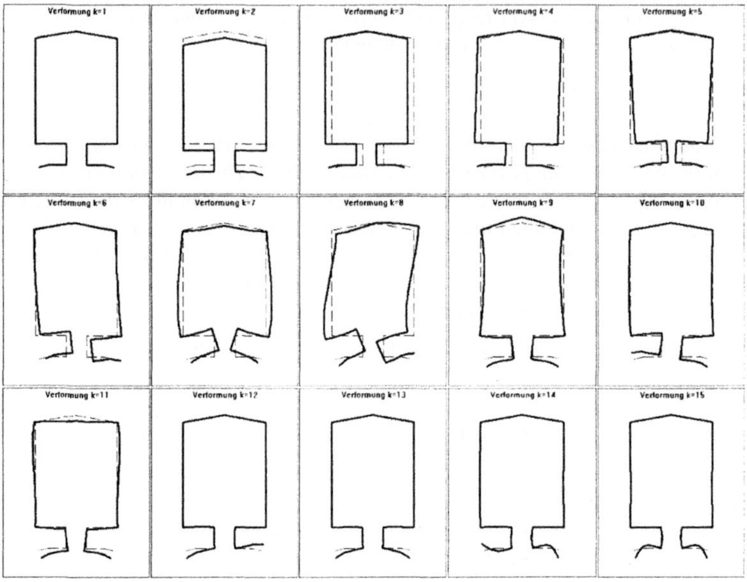

Figure 1. The first fifteen deformation modes of a monorail beam cross-section.

In GBT three different degrees of freedom or displacements are required to define displacement in one plate strip. The three displacements are the axial displacement, $u(s)$, which is the warping function of the cross-section, the in-plane displacement function, $f_s(s)$ and the out-of-plane displacement function, $f(s)$. When deriving the basic equation, these three displacements are the basis when defining the following cross-sectional constants (tilde above the displacement functions refers to normalised values):

$$^{k}C = \int_A {^k\tilde{u}^2} \mathrm{d}A + \frac{K}{E}\int_s {^k\tilde{f}^2} \mathrm{d}s\,, \quad \text{compare} \quad I_z = \int_A y^2 \mathrm{d}A,\, I_y = \int_A z^2 \mathrm{d}A,\; I_\omega = \int_A \omega^2 \mathrm{d}A$$

$$^{k}D = \frac{t^3}{3}\int_s {^k\dot{\tilde{f}}^2}\, \mathrm{d}s\,, \quad \text{compare} \quad I_v = \frac{1}{3}\sum bt^3 \tag{2}$$

$$^{k}B = K\int_s {^k\ddot{\tilde{f}}^2}\, \mathrm{d}s, \qquad K = \frac{E\,t^3}{12(1-\nu^2)}.$$

The first two cross-sectional properties in the basic equation are analogous the well known cross-section properties of the classical beam theory. The warping constant ^{k}C is analogous to the flexural moment of inertia in bending or sectorial moment of inertia in torsion and ${}^{k}D$ is the torsion constant. The third property, ${}^{k}B$, is the transverse bending stiffness, which defines the stiffness of the cross-section in distortion. The cross-section deformation modes can be divided into two basic modes. The first is rigid section modes, which are extension, bending about principal axes, and torsion. The flexible section modes are the distortional modes starting from mode $k = 5$.

Solving the basic cross-sectional properties is the first phase in the analysis of GBT. In the next phase the deformation function ${}^{k}V$ is solved from Eq. 1 for each individual mode as in the classical bending theory.

3 Case study of a monorail beam

A stress and displacement analysis of a monorail beam was performed using the first -order GBT. The beam is simply supported at the ends with a point load at the middle. The length of the beam is L = 6000 mm and the load is n F = 4 kN. The cross-section of the beam is shown in Fig. 2.

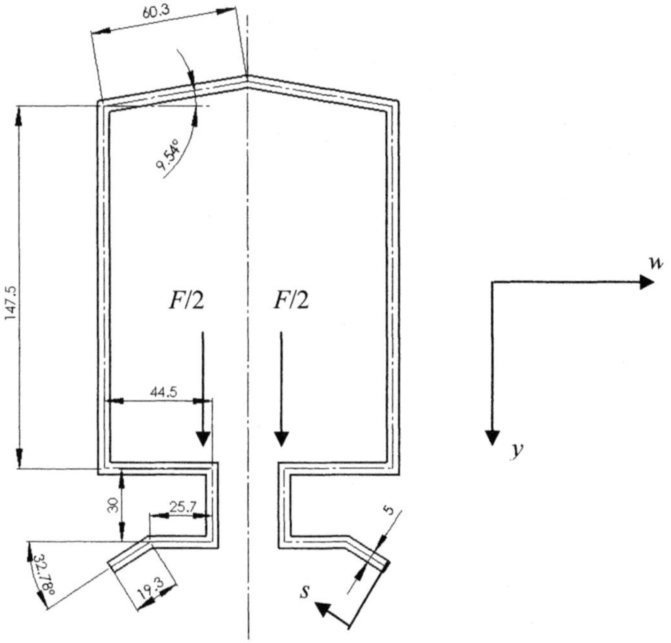

Figure 2. Cross-section of the beam showing the loading points and coordinate systems.

Thickness of the plate strips is constant t = 5 mm. Material is steel with material properties E = 210000 MPa, G = 80000 MPa and ν = 0,3. The axial membrane normal stress at the free end of the beam (s = 0) was calculated to obtain the structural or hot spot stress at butt weld joint. The first fifteen deformation modes of the cross-section are shown in Fig. 1. These modes and corresponding cross-section properties were calculated using program VTB, Schardt (1996). As the loading is symmetrical, only the symmetrical modes are required in the analysis. Thus six modes, i.e. modes 2, 5, 7, 9, 11 and 13 were chosen for the analysis. The main cross-sectional properties of the chosen six modes are presented in Table 1:

Table 1. Cross-sectional properties of modes 2, 5 and 7

k	${}^kC$ [cm^4]	${}^kD$ [cm^2]	${}^kB$ [kN/cm^2]
2	1520,63	0	0
5	3,7187	8,8068E-4	0,013324
7	6,4567	6,9248E-4	0,463127
9	2,711	0,04567	34,27
11	2,610	0,05871	38,33
13	0,810	0,2642	123,5

The loading term of the differential equation in GBT is based on the virtual work done by the external forces, resulting for a point load the following equations:

$$\begin{aligned} &E\,{}^2C\,{}^2V''''(x) = F\,{}^2\tilde{v}_F \\ &E\,{}^5C\,{}^5V''''(x) - G\,{}^5V''(x) + {}^5B\,{}^5V(x) = F\,{}^5\tilde{v}_F \\ &E\,{}^7C\,{}^7V''''(x) - G\,{}^7V''(x) + {}^7B\,{}^7V(x) = F\,{}^7\tilde{v}_F \\ &E\,{}^9C\,{}^9V''''(x) - G\,{}^9V''(x) + {}^9B\,{}^9V(x) = F\,{}^9\tilde{v}_F \\ &E\,{}^{11}C\,{}^{11}V''''(x) - G\,{}^{11}V''(x) + {}^{11}B\,{}^{11}V(x) = F\,{}^{11}\tilde{v}_F \\ &E\,{}^{13}C\,{}^{13}V''''(x) - G\,{}^{13}V''(x) + {}^{13}B\,{}^{13}V(x) = F\,{}^{13}\tilde{v}_F. \end{aligned} \tag{3}$$

Table 2 shows the normalised eigendisplacements in the direction of the load at the loading points for the modes used in the analysis.

Table 2. The normalised vertical eigenvectors at the loading point of the chosen modes

${}^2\tilde{v}_F$	${}^5\tilde{v}_F$	${}^7\tilde{v}_F$	${}^9\tilde{v}_F$	${}^{11}\tilde{v}_F$	${}^{13}\tilde{v}_F$
1,000	0,125	0,220	0,227	0,550	0,007

The corresponding generalised force component is called the stress resultant or warping moment ${}^kW(x)$

$${}^kW(x) = -E\,{}^kC\,{}^kV''(x)\ \text{, compare } M = -EIv''. \tag{4}$$

The longitudinal stress distribution is calculated from the superposition of stresses of the individual modes $k = 1 \ldots l+1$ (l is the number of folded plate strips):

$$\sigma_x(x,s) = -\sum_{k=1}^{l+1} \frac{{}^kW(x)\,{}^k\tilde{u}(s)}{{}^kC}\text{, compare } \sigma_x(x,s) = \frac{M_1 e_1(s)}{I_i} + \frac{M_2 e_2(s)}{I_2}. \tag{5}$$

Table 3 shows the warping function value at the free end of the cross-section where the axial normal stress is calculated in post-processing phase using Eq. 5.

Table 3. Warping function value at the free end of the cross-section of the chosen modes [cm]

${}^{2}\tilde{u}(0)$	${}^{5}\tilde{u}(0)$	${}^{7}\tilde{u}(0)$	${}^{9}\tilde{u}(0)$	${}^{11}\tilde{u}(0)$	${}^{13}\tilde{u}(0)$
-9,5978	-1,0000	-1,0000	2,44E-4	0,1209	-1,0000

Deformation mode k = 2, i.e. bending about the main axis, can be directly solved from textbook tables as the differential equations for the GBT and for the classical beam theory are analogous. The distortional deformation modes can be solved using some numerical method or the analytical solutions obtained from the theory of the beam on elastic foundation (BEF). The general expression of the differential equation of an axially loaded beam on elastic foundation, Fig. 3, is

$$E\,I\,v''''(x) - N\,v''(x) + k\,v(x) = q(x)\,. \tag{6}$$

The equation is clearly analogous with the differential equation of the first-order GBT. The following conversions must be made when using the available analytical solutions either from the classical beam theory or from the BEF:

Moment of inertia = warping constant	$I = {}^{k}C$
Axial load = torsional stiffness	$N = G\,{}^{k}D$
Foundation modulus = transverse bending stiffness	$k = {}^{k}B$
Bending moment = generalised warping moment	$M = {}^{k}W$
Point load = virtual work done by point load	$P_y = F\,{}^{k}\tilde{v}_F$

The distortional deformation modes due to point loads typically dampen quite rapidly. As the beam is long compared to the cross-section dimensions, the analytical results for an infinite beam loaded by a point load P_y, can be used to solve the distortional modes with sufficient accuracy. The bending moment is, Hetenyi (1946)

$$M(x) = \frac{P_y}{4\alpha\beta} e^{-\alpha x}\left(\beta\cos(\beta x) - \alpha\sin(\beta x)\right)$$

$$\alpha = \sqrt{\sqrt{\frac{k}{4EI}} + \frac{N}{4EI}}, \quad \beta = \sqrt{\sqrt{\frac{k}{4EI}} - \frac{N}{4EI}}, \; N > 2\sqrt{kEI}\,. \tag{7}$$

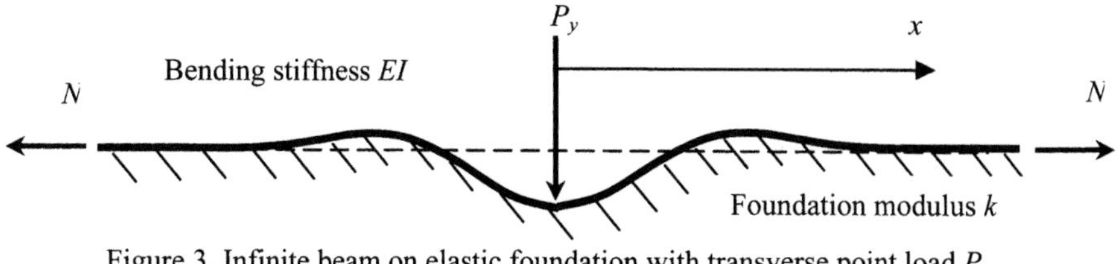

Figure 3. Infinite beam on elastic foundation with transverse point load P_y

The results for the axial membrane normal stress at the free end of the cross-section is given in Fig. 4. The curve clearly shows the rapid dampening of the stress due to distortion as the curve is almost linear already a meter from the point of loading. The comparison with the results between FEM and GBT shows excellent agreement with the two methods.

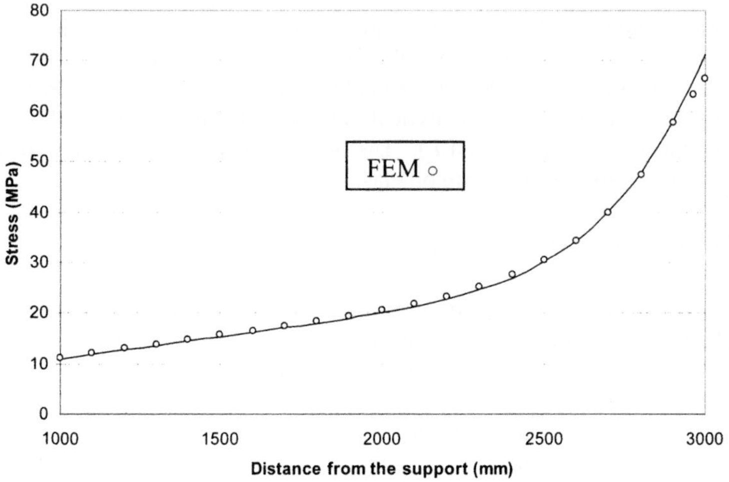

Figure 4. Normal stresses at the free end of the profile

The obtained structural or hot spot stress is used in the fatigue analysis of the butt weld joint of the hoist beam. A joint is assumed to be located at the centre of the beam where the maximum stress occurs. In the analysis the axial misalignment of the butt weld joint, Fig. 5, is included in the analysis.

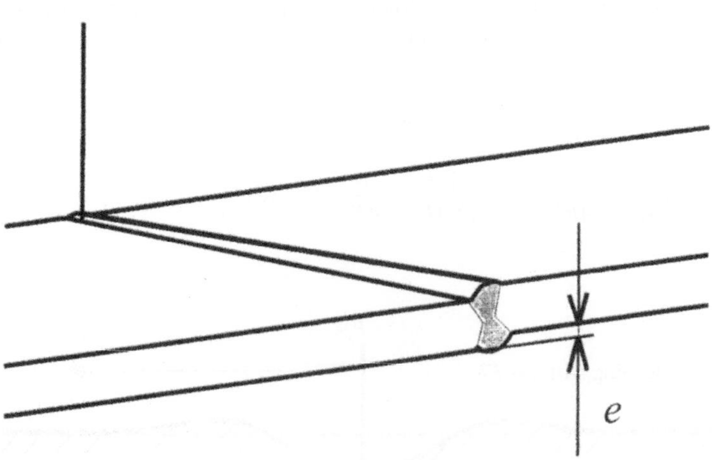

Figure 5. Axial misalignment of plates in a butt weld joint.

The structural stress concentration factor due to axial misalignment is

$$K_m = 1 + 3\frac{e}{t}, \tag{8}$$

where e is the eccentricity of the plates and t is the thickness of the plates.

The structural or hot spot stress range can then be calculated as

$$\Delta\sigma_{hs} = K_m \,\Delta\sigma \; = K_m \, K_s \,\Delta\sigma_{nom} \tag{9}$$

In the equation the structural stress concentration due to distortion of the cross-section is taken into account using stress concentration factor K_s. It can be obtained from the analysis of GBT by normalising the structural stress due to distortion by global stress, i.e. nominal bending normal stress:

$$K_s = 1 + \frac{{}^{5}\sigma_x}{{}^{2}\sigma_x} + \frac{{}^{7}\sigma_x}{{}^{2}\sigma_x} + \frac{{}^{9}\sigma_x}{{}^{2}\sigma_x} + \frac{{}^{11}\sigma_x}{{}^{2}\sigma_x} + \frac{{}^{13}\sigma_x}{{}^{2}\sigma_x}. \tag{10}$$

The design life of the structure can then be calculated using S-N –curve

$$\Delta\sigma_{hs}^{m} \, N = C_d \tag{11}$$

where $C_d = 2\cdot10^{12}$ is the fatigue capacity in the hot spot method and $m = 3$ (IIW 2007). Constant C_d is valid when $N < 10^7$. The resulting design life as a function of the misalignment normalized with the maximum design life is presented in Fig. 6. The upper curve represents the design life without distortional effects.

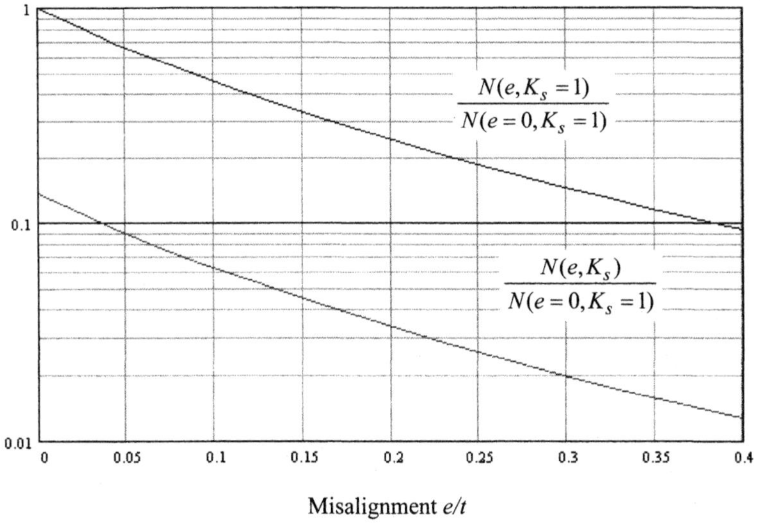

Figure 6. Normalised design life of the joint as a function of misalignment.

The result clearly shows the influence of the misalignment; e.g. when misalignment is less than one tenth of the thickness of the plate, the design life is halved.

4 Conclusions

Generalized beam theory offers a unique system to obtain structural stresses due to distortion of a prismatic beam. It offers the possibility to derive parametric solution methods to investigate different phenomena in thin-walled structural members. In this paper a monorail hoist beam is used as a case study to show that GBT can be used instead of FEM to the estimate the design life of a butt weld joint in the beam. The results can then be used to define the weld quality specifications for the manufacturing process.

References

IIW (2007) *Recommendations for fatigue design of welded joints and components*, International Institute of Welding, IIW document XIII-2151-07/XV-1254-07, May 2007.

Hetenyi, M. (1946) *Beams on Elastic Foundation*, University of Michigan Press, Ann Arbor, MI.

Schardt, R. (1989) *Verallgemeinerte Technische Biegetheorie*. Springer-Verlag, Berlin. ISBN 3-540-51339-6.

Schardt, R. (1996) *Program VTB*, User's Guide.

Vlasov, V.Z. (1961) *Thin-walled Elastic Beams*. Israel program for scientific calculations. Translation of *Tonkostennye uprugie sterzhni*, Jerusalem.

4.4 Estimation of the Applicability of Thin Steel Plate as Sacrificial Test Piece

Yoshihiro Sakino[1], You-Chul Kim[1] and Kohsuke Horikawa[2]
[1]Joining and Welding Research Institute, Osaka University
11-1, Mihogaoka, Ibaraki, Osaka, Japan, 567-0047
sakino@jwri.osaka-u.ac.jp
[2]Professor Emeritus, Osaka University

Abstract

Thin steel plates, which have initial cracks at the center, are used as the Sacrificial Test Pieces in this study. "The Sacrificial Test Piece" is attached to the member of a main structure in order to evaluate the damage before the appearance of a crack in the member of the main structure. The purpose is to show the practical applicability of "the thin steel plate with crack as the Sacrificial Test Piece" for monitoring the fatigue damage parameters on bridge members. In this research, it is decided that the applicable range of crack length and the crack propagation properties of the Sacrificial Test Pieces to evaluate the fatigue damage parameter are obtained. And it is suggested that the fatigue damage parameter can be estimated by the Sacrificial Test Pieces with practical accuracy by the test under constant amplitude loading.

Keywords: *Sacrificial Test Piece, Fatigue Damage Parameter, Thin steel plate, Crack Growth*

1 Introduction

"The Sacrificial Test Piece" is used as a specimen attached to the member of a main structure in order to evaluate the damage before the appearance of a crack in a member of the main structure. The Sacrificial Test Piece is designed so that it is damaged earlier than the main members under the same loads because of crack and stress magnification. The damage to the bridge members can be estimated by the observation of the Sacrificial Test Piece. If the fatigue damage parameter can be made clear by the behaviour of the Sacrificial Test Piece, the maintenance management of the structure can be determined. Some types of the Sacrificial Test Piece are proposed and investigations to apply these to the structures are going on. (Taguchi et al. (1999), Fujimoto et al. (1999), Komon et al. (1999), Matuda et al. (2004))
As shown in Figure 1, thin steel plates, which have initial cracks at the center, are used as the Sacrificial Test Pieces in this study (Sakino et al. 2002, Sakino et al.

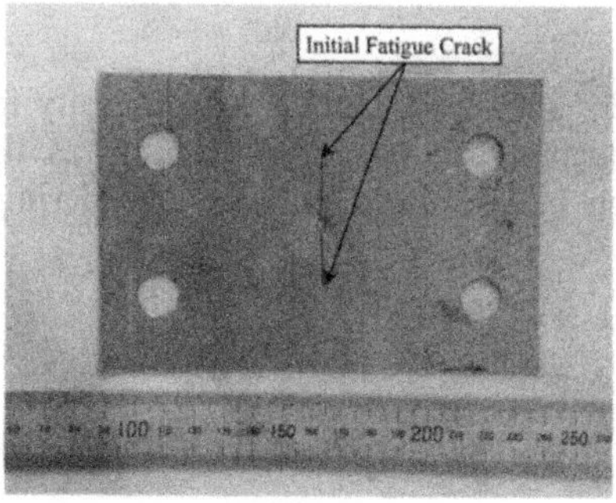

Figure 1. Sacrificial test pieces

2006). When strains are applied to the main member, these are transmitted from the main member to the thin steel plate and the crack in the thin steel plate will grow as a result. Therefore, the monitoring of fatigue damage parameters on the bridge can be carried out by the observation of the crack growth in the thin steel plate. If the thin steel plate can be used as the Sacrificial Test piece, it seems that fatigue damage on a bridge can be monitored at a low price and widely. Because the thin steel plate is cheap, everyone can obtain it easily.

In this paper, the estimating method for the fatigue damage parameter by crack growth of the thin steel plate as the Sacrificial Test Pieces is investigated and the crack propagation properties and applicable range of crack length of the thin steel plate as the Sacrificial Test Pieces to evaluate the fatigue damage parameter are obtained through a numbers of tests. The applicability of this method under constant amplitude loading is also investigated.

2 Monitoring of fatigue damage parameter

According to the Miner's law, damage of the bridge member by forced fluctuating amplitude loading can be written as follows;

$$\sum\left(\sigma_i^{\mathrm{m}} n_i\right) \tag{1}$$

where σ is stress amplitude, n is number of cycles and lower suffix i is operation number.

Eq. 1 is termed "the fatigue damage parameter".(Mori 1997, Ookura 1994). We propose a method for measuring these fatigue damage parameter by the crack growth of the Sacrificial Test Piece. The basic theory and assumptions are shown as follows:

1) The crack at the center of the Sacrificial Test Piece grows by the strain that is transmitted from the member to the Sacrificial Test Piece.

2) The relationship between a stress component of the live load and the crack growth, which is generated by the stress component, is expressed by Paris' law as follows:

$$\mathrm{d}a_i/\mathrm{d}n_i = \mathrm{A}\left(\Delta K_i\right)^{\mathrm{m}} \tag{2}$$

where a is the crack growth, A and m are constants, and K is the stress intensity factor.

3) The stress intensity factor coefficient under constant displacement amplitude can be expressed as follows;

$$K_i = \mathrm{B}\sigma_i \tag{3}$$

where B is a constant. Eq. 3 shows that the stress intensity factor for the constant displacement amplitude testing can be expressed only as the function of stress amplitude "σ", and can be expressed without considering the effect of crack length "a".

4) Substituting Eq. 3 in Eq. 2, produces Eq. 4;

$$\mathrm{d}a_i/\mathrm{d}n_i = \mathrm{A}\left(\mathrm{B}\Delta\sigma_i\right)^{\mathrm{m}} \tag{4}$$

It is assumed that m is approximately 3 for steel. It follows from Eq. 4, that;

$$a_i = \mathrm{A}\,\mathrm{B}^{\mathrm{m}}\left(\sigma_i^{\mathrm{m}} n_i\right) \tag{5}$$

5) The crack growths due to each stress component of live load do not affect each other and can be summed simply. Thus, the total crack growth can be written as follows:

$$\sum\left(\sigma_i^{\mathrm{m}} n_i\right) = a/\mathrm{AB}^{\mathrm{m}} \tag{6}$$

where a is the total crack growth.

The constant A, B and m can obtain by examination or theoretical calculation in advance. So by

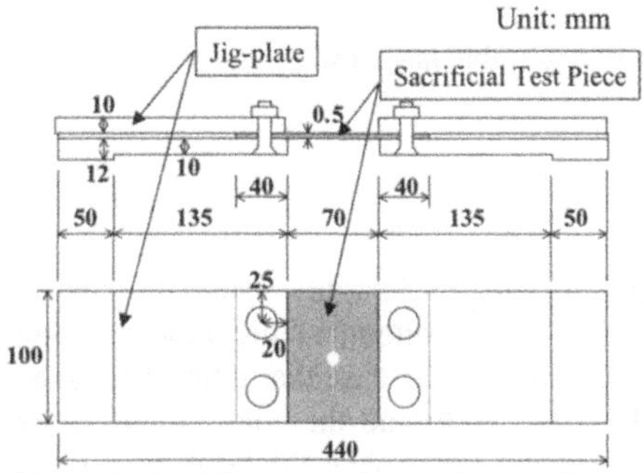

Figure 2. Sacrificial test pieces with jig-plate

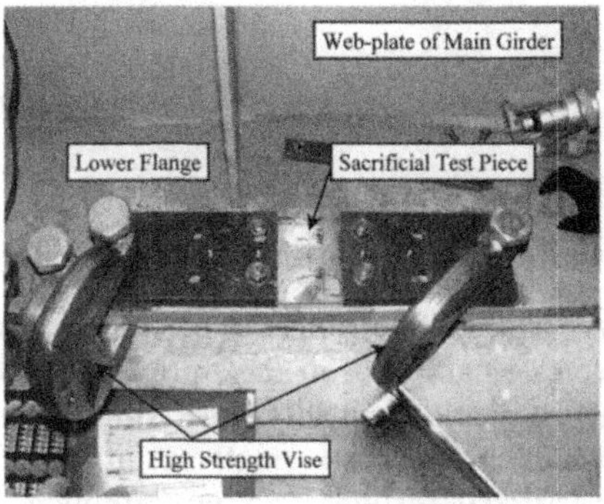

Figure 3. Example of attachment to lower flanges of a highway bridge

these assumption, if a is measured, the fatigue damage parameter (Eq. 1) can be obtained via Eq. 6.

3 Application to Bridge Members

The Sacrificial Test Piece has been attached to four steel jig-plates by some bolts. The shape and the dimension of the jig-plates are shown in Figure 2. The thickness of the Sacrificial Test Piece is 0.5 mm and the thickness of one side edge of the jig-plate is 12mm and other part of the jig-plate is 10mm. Using the jig-plates, a strain between the connected points is concentrated at the Sacrificial Test Piece by the difference in stiffness between thin plate and jig-plate. Strain in the Sacrificial Test Piece is concentrated more than about 3 times that of the flange by theoretical calculation. This strain concentration makes the crack growth faster and the measurement in bridge members can be carried out in short period.

To avoid compression loading on the Sacrificial Test piece by uplift of the bridge member, pre-tensile stress is applied to the Sacrificial Test Piece by heating the specimen before attached to the member. After the specimen is attached, the temperature of the specimen falls to room temperature and pre-tensile stress will be forced into the Sacrificial Test Piece because of thermal deformation.

The specimen is attached on the lower flange of bridge members by high strength vices at the edge of the jig-plates, as shown in Figure 3. The high strength vices are often used on site for rigid fixing and the vice is tightened up using a torque wrench.

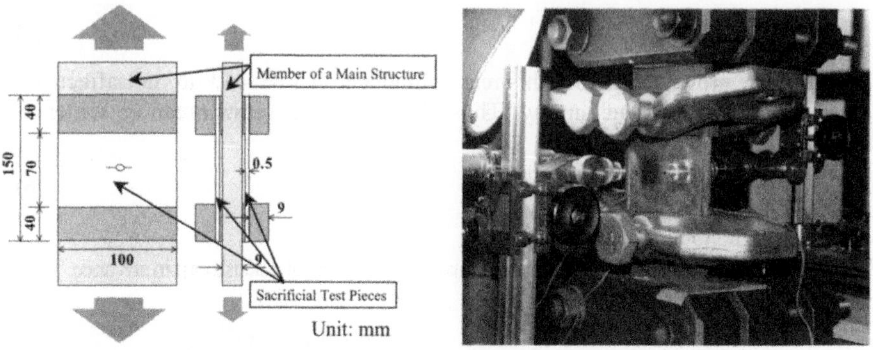

Figure 4. Figure and photo of applicable range test

4 Applicable Range of Crack Length

As one can see from Eq. 4, crack propagation velocity “da/dn” is not affected by crack length “a”. So crack propagation velocity should remain stable under constant stress amplitude in all ranges. But Eq.3 is valid only in the case that both of the plate width and the crack length are infinity. The plate width and the crack length in the Sacrificial Test Piece are not infinity, so we should make clear the applicable range of crack length that Eq. 3 and Eq. 4 can be valid.

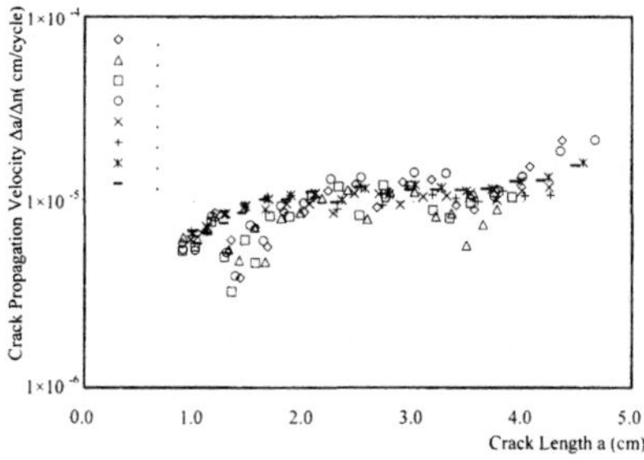

(Δσ=120MPa)

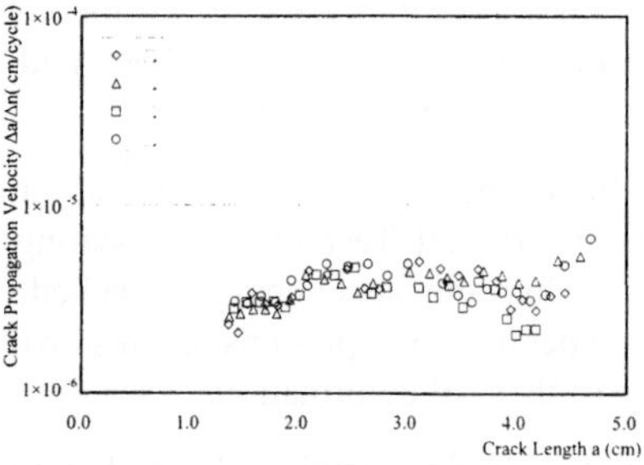

(Δσ=80MPa)

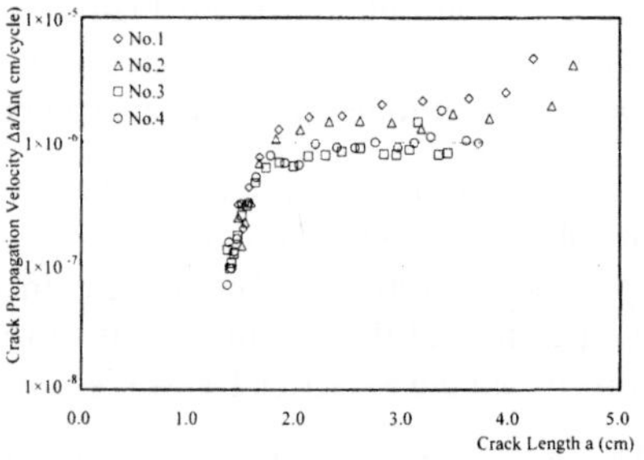

(Δσ=60MPa)

Figure 5. Δa/ΔN-a relationship

Figure and photo of applicable range test are shown in Figure 4. A couple of the Sacrificial Test Pieces are fixed to front and back of the main member whose thickness is 9mm without jig-plates. Backing plates are put between the main member and the Sacrificial Test Pieces and then the Sacrificial Test Pieces and backing plates are fixed by the high strength vices. Material of the main member and backing plates are mild steel. The crack length of the Sacrificial Test Pieces was measured every 10,000 cycle of the constant amplitude loading by uniaxial fatigue machine. Scale-readout microscope was used to measure the length of crack. Three ranges of stress amplitude, 60MPa, 80 MPa and 120 MPa, were loaded and stress ratios were 0.33, 0.27 and 0.2, respectively. One series of test (two thin plates) in case of 60MPa and 80 MPa, and two series of test (four thin plates) in case of 120MPa were run.

Figure 5 show relationship between the crack propagation velocity and the crack length under constant stress amplitude. As mentioned above, the crack propagation velocity should remain stable under the constant stress amplitude because the crack propagation velocity is not affected by crack length. So it can be said that the stable range of the crack propagation velocity is the applicable range of the thin steel plate as the sacrificial test piece. According to the results, the crack propagation velocity from 2cm to 4cm of the crack length remains approximately stable regardless of the stress amplitude. A cause of unstability of the crack propagation in the area under 2cm seems that the crack length is too short to satisfy the qualification of semi-infinite crack length. A cause of unstablility of the crack propagation in the area over 4cm seems that the ligament length is too short to satisfy of qualification of semi-infinite ligament length.

So in the case of the steel plate of 10 cm in width that used as the Sacrificial Test Piece in this research, it can be said that the applicable range of the crack length in the Sacrificial Test Pieces is from 2cm to 4cm.

5 Crack propagation properties under constant amplitude loading

In case of calculating the fatigue damage parameter from the crack length a by Eq. 6, the constant A and m should be decided besides the restraint coefficient B that can be decided by theoretical calculation. These constants are decided by measuring

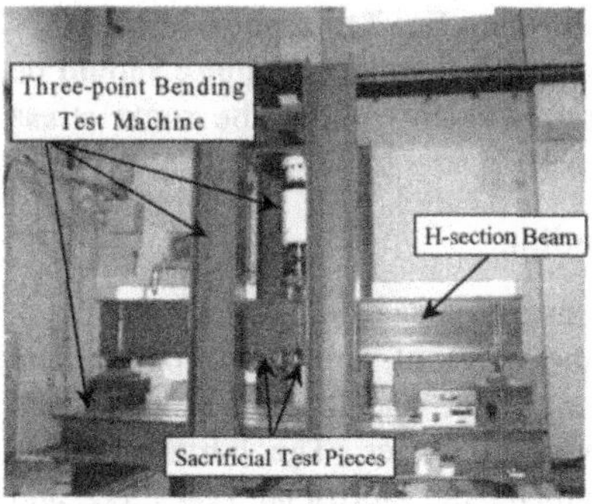

Figure 6. Fatigue test by three point bending test machine

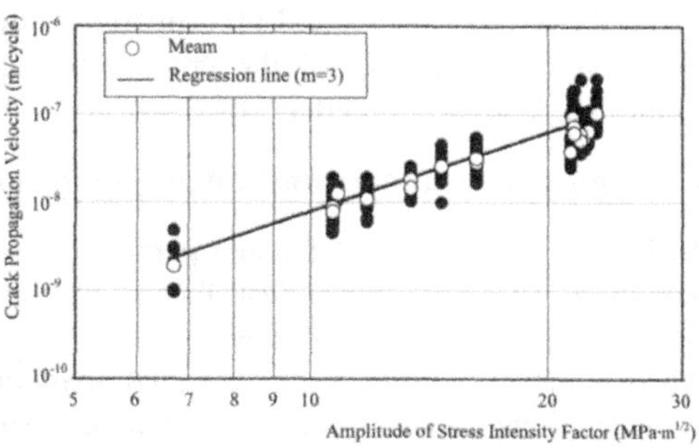

Figure 7. Crack propagation velocity – amplitude of stress intensity factor relationship

the crack propagation velocity under some stress amplitude. So the constant A and m were decided by experiments.

The three-point bending fatigue test machine, shown in Figure 6, was used. The Sacrificial Test Pieces with jig-plates are fixed by the way proposed in chapter 3 on the lower flange of an H-section beam modelled from the highway bridge member. The section size and length of H-section beam were 600×300×12×19mm and 4,000mm. Strain gauges were pasted on the center of the Sacrificial Test Pieces and on the same place of the H-section beam. To investigate of the effect of the mechanical property, two sorts of the thin steel plates that were made in another lot were used. Four set of the Sacrificial Test Pieces with the jig-plate were examined in one series and a total of five series of experiment were run. Series 1, 3, 4 are same lot and series 2, 5 are another lot of the thin steel plate.
In the series 1 ~ 3, the crack lengths were measured by microscope after every 10,000 cycle of the constant amplitude loading for 10 ~ 15 times. And then same measurements were done 2 or 3 times after changing the stress amplitude of the bridge member $\Delta\sigma_B$. In the series 4 and 5, the crack lengths were measured by crack gauges under the constant amplitude loading of about 30MPa.

Figure 7 shows relationships between the crack propagation velocity and the stress intensity factor in the Sacrificial Test Pieces. Data of specimen No.2 in series 2 were disregarded because the Sacrificial Test Pieces was damaged when it was fixed to the H-section beam. Difference in the crack propagation between the two sorts of the thin steel plates was not observed. So plots are not distinguished in Figure 6 by each sort. It seems that the same value of constants can be used if steel type and size are same.

In most of "fatigue design recommendations for steel structures" including IIW and JSSC, m=3 is widely used. Substituting m=3, following regression equation is obtained.

$$\log(da/dn) = 3.0 \cdot \log(\Delta K) - 11.1 \tag{7}$$

The solid line in Figure 6 is a regression line with the value of m is fixed as 3. The plots and the solid line agree well. From these results, it can be said that the Paris' law, Eq. 2, and m=3 can also apply to the Sacrificial Test Pieces. And we also obtain that the values of A is 7.94×10^{-12} by transforming Eq. 7 to shape of Eq. 2.
From these experiments and investigations, the constants m and A, those should be obtained by measuring of the crack propagation velocity under the same stress amplitude, can be decided as $m = 3$ and $A = 7.94\times10^{-12}$.

6 Applicability under constant amplitude loading

In Figure 8, the fatigue damage parameters measured and calculated by the crack length of the Sacrificial Test Pieces under the constant amplitude loading are compared with those calculated by the stress amplitude and loading times (Stress Measurement). Figure 9 shows a comparison of all series. The horizontal axis represents the value of the fatigue damage parameter by Sacrificial Test Piece and the vertical axis represents those of Stress Measurement. estimation of the fatigue damage parameter. Especially the means of 4 set of the Sacrificial Test piece agree well in all ranges of the fatigue damage parameter as shown in Figure 9.

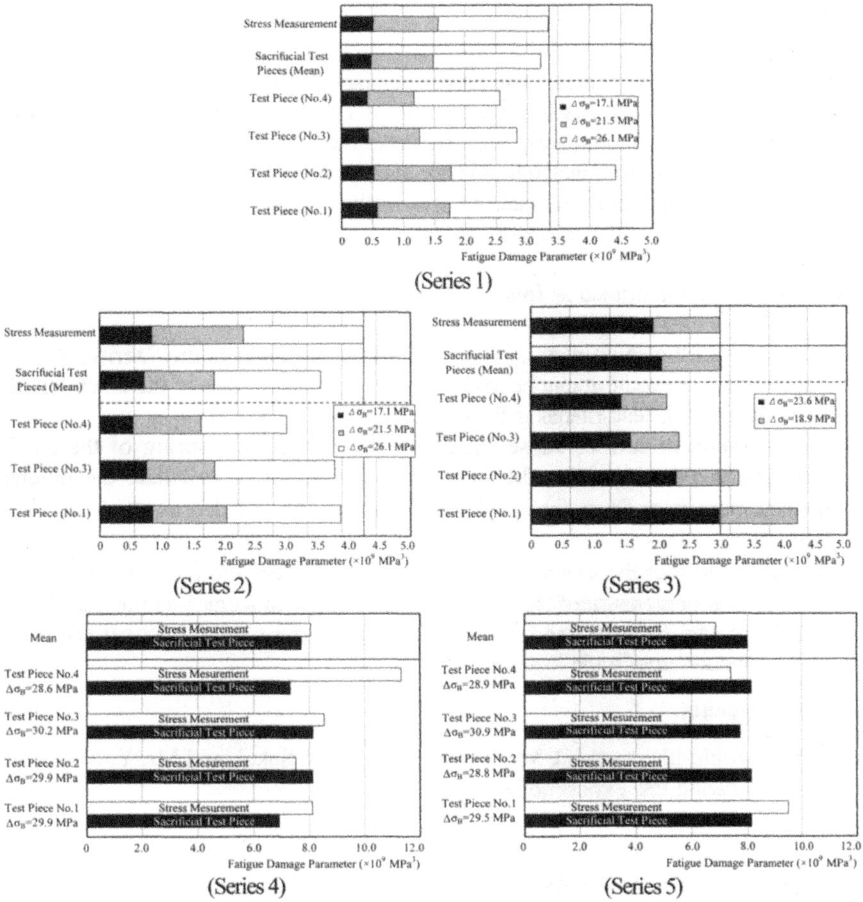

Figure 8. Comparisons of fatigue damage parameter

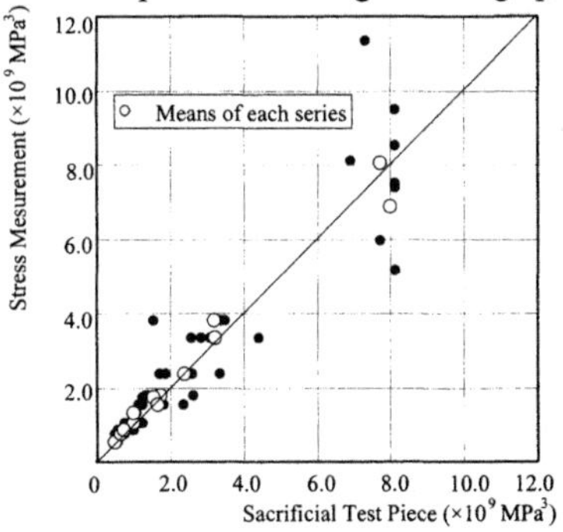

Figure 9. Comparisons of fatigue damage parameter

So it can be said that the proposed method is valid under the constant amplitude loading. It is demonstrated that the thin steel plate as the Sacrificial Test Piece can estimate the fatigue damage parameters with practical accuracy under the constant amplitude loading.

7 Conclusions

We propose a method to monitor the fatigue damage parameters on bridge members by the thin steel plate with crack as the Sacrificial Test Piece. By using this method, the fatigue damage parameters can be estimated with lower cost than by conventional methods. In this study, the applicable range of the crack length and the constants that are needed to estimate by this method were obtained by experiment. The applicability of this method under constant amplitude loading is also investigated.
Main results are summarized as follows.

(1) In the case of the steel plate of 10 cm in width that used as the Sacrificial Test Piece in this research, it can be said that the applicable range of the crack length in the Sacrificial Test Pieces is from 2cm to 4cm.
(2) The constants m and A, those should be obtained by measuring of the crack propagation velocity under the same stress amplitude to calculate the fatigue damage parameter from the crack length a, can be decided as m=3 and A = 7.94×10^{-12}.
(3) It can be said that the estimating method is valid under the constant amplitude loading. It is demonstrated that the thin steel plate with a crack as the Sacrificial Test Piece can estimate the fatigue damage parameters with practical accuracy under constant amplitude loading.

Acknowledgments

The authors would like to thank Mr. T. Taguchi., Mr. K Sakai, and Mr. Y. Nakatsuji for their help during the experiments.

References

Fujimoto,Y., Hamada,K. & Shintaku,E. (1999) *Proc. of Welding Structures Symp. '99*, 611-618.
Fujimoto,Y., Ito,H. & Shintaku,E. (2002) *Proc. of Welding Structures Symp. 2002*, 509-516
IIW (1994) *Fatigue Design Recommendations*, XIII-1539-94 /XV845-94,
JSSC (1995) *Fatigue design recommendations for steel structures*, Japanese Society of Steel Construction Technical Report, 32.
Komon,K., Abe,M., Marumoto,A., Sugidate,M. & Miki,C. (1999) *Journal of Constructional Steel*, No.7, 189-184.
Matsuda,H., Muragishi,O. & Nihei,K. (2004) *Proc. of Welding Structures Symp. 2004*, 105-116.
Mori,T. (1997) *J. Japan Welding Soc.*, **8** No. 66. 6-10.
Ookura I. (1994) Fatigue of Steel Bridge: TOYOSHOTEN
Taguchi,T., Horikawa,K. & Sakino,Y. (1999), *Trans. of JWRI* No.28-2, 77-79.
Sakino Y., Horikawa K. & Taguchi T. (2002) *Proc. of Welding Structures Symp. 2002*, Japan, 714-719.
Sakino Y, Kim Y-C & Horikawa K(2006), *Trans. of JWRI* No.35-1, 63-70.

4.5 Possibilities of Predicting the Fatigue Life of Resistance Spot Welded Joints

Péter Szabó[1], István Borhy[2]
[1]*Air Liquide Hungary Kft.* [2]*TÜV Rheinland InterCert Kft.*

Abstract

Examinations performed in association with the planning and manufacturing issues of road and railway vehicle structures are very current nowadays. There are major international researches conducted in order to develop the apparatuses that can be used for the planning of welded vehicle structures, as well as for the definition of the technology of the required welding procedures. This presentation aims to give an overview on the current issues arising from our research work that focuses on the optimization of spot welded railway vehicle structures and the computer-aided planning of welding processes with special attention paid to the way how the fatigue life of spot welded structures can be determined.

Keywords: *fatigue, resistance spot welds, railway vehicle structures, welding technology*

1 Introduction

The attitude to any product in the market (the success of the product) is basically determined by its competitiveness. Such competitiveness is primarily estimated with respect to the extent of costs incurred during the entire lifecycle of the product – i.e. in the period from the planning of the product through manufacturing and operation up to the time of disposal. In the course of planning and manufacturing, several aspects and criteria are to be taken into account and considered in order to satisfy all and nay – often rather contradicting – customer demands in relation to the product. It is thus a difficult task the achieve any optimal result in the light of the various demands because in addition to the strength, reliability, aesthetic, etc. requirements posed against the structure particular attention is to be paid to the fulfilment of manufacturability and controllability criteria. Therefore, today the goal is not simply to develop welded joints being free from the discrepancies that can be detected by means of non-destructive testing procedures but also to ensure the best possible joint quality and longest lifetime as regarding the given economic, lifecycle and environmental requirements.

An essential precondition of fulfilling such goals is the close cooperation of design engineers, technologists and experts involved in testing as to be headed for in the phases of product design and technology planning. At the same time, competitiveness calls for the reduction of time used for product design and technology planning. With respect to the high rate of complexity of the task to be attended, the narrow time frame being available, as well as the enhanced requirements against the experts involved in the attendance of the task (well-grounded theoretical knowledge and outstanding practical experience), expert systems to be applied in the field of planning tasks are also to be developed in order to meet the objectives set.

2 Compliance of welded railway vehicle structures

The underframes and bodies of road and railway vehicles are designed as following the so-called lightweight construction principles (Sostarics 1991) (Figure 1).

Requirements posed against such vehicle structures delineate a structure that has appropriate strength, withstands fatigue loading and at the same time features the lowest possible weight and moderate manufacturing costs. One of the most common method to manufacture such structures is the application of resistance spot welding with respect to the several advantages of the procedure (productivity, mechanizability, even joint quality, etc.). The examinations performed in association with the planning and manufacturing issues of these vehicle structures have become widely used today.

The main reason behind this heightened attention is that this industrial sector can be characterized by large-scale series where economic issues tend to be in the foreground. Such goals can only be achieved by enhancing the quality and reliability of these vehicle structures, as well as by reducing planning and manufacturing, operating and maintenance costs. Thus, market competition calls for the ongoing development of the materials used, the design together with the related manufacturing and control procedures, as well as the application of state-of-the-art design and manufacturing methods. Without these factors, no market success – gaining of additional market positions and the retention of the market positions having been occupied – can be ensured. With a view to the fact that vehicle-manufacturing industry is one of the drives behind global economy, ceaseless developments are definitely required.

The second reason is that the above enhanced quality requirements against the vehicles and vehicle parts can be guaranteed exclusively by exercising proper control over the entire manufacturing process from design through the selection of adequate materials to any follow-up supervision. Therefore, it is reasonable to draft strict regulations on the design, manufacturing and control of vehicle structures, including the enhancement of the reliability of welded joints (Borhy 2000). The heightened demand for proper product safety also calls for continuous developments.

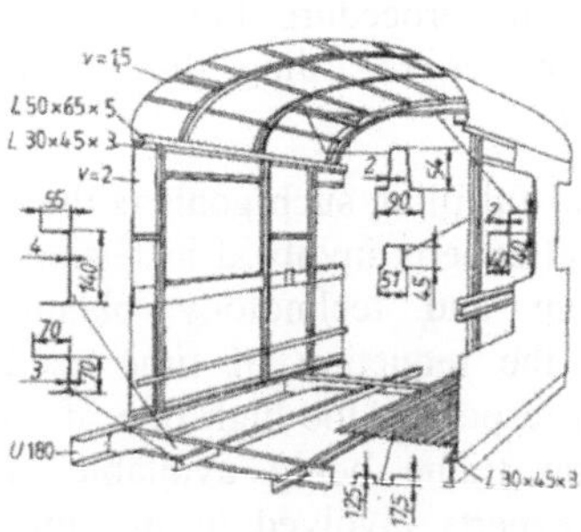

Figure 1. Cross-sectional drawing of a lightweight construction railroad carriage body

In the course of designing and manufacturing spot welded vehicle structures we are to strive for developing such constructions that feature the most optimal parameters, as focusing on compliance with customer demands, as well as analyzing and quantifying (transforming demands into measurable indicators) the same. In other words, the fulfilment of customer demands are to be ensured by means of solving a quality-oriented optimization task that:

- Guarantees the observation of the strict quality (reliability) requirements posed against vehicle structures;
- Guarantees proper control over the entire planning and manufacturing process (from design through the selection of adequate materials to any follow-up supervision);
- Guarantees enhanced reliability for welded joints.

In the case of vehicle structures it is not sufficient to concentrate on the optimization of the quality of the individual spot welded seams but the compliance of the entire welded structure is to be taken into account. Structure optimization is in fact a synthesis wherein design, manufacturing, control, aesthetic and economic aspects equally count. It can thus be stated that the enhanced quality requirements posed against vehicle structures can be fulfilled exclusively by means of exercising proper control over the entire planning and manufacturing process.

3 Framing of the scientific objective

For several decades, the associated departments of the Budapest University of Technology and Economics and the University of Miskolc have been conducting notable research in the field of the optimization of welded structures, as well as computer-aided welding processes (Brenner & Palotás 1989, Jármai & Iványi 2001, Szabó 2003). The activities associated with this research process have a double goal:

I.) To establish a structure-optimization method in connection with the design of spot welded vehicle structures with respect to the fatigue life and restrictive conditions;

II.) On the other hand, to improve the quality-oriented optimization method of the spot welding technology with special attention paid to the indentation limit.

In order to fulfil these goals, we are to:

- Analyze and quantify the compliance requirements posed against spot welded vehicle structures;
- Have an overview on the methods suitable for the prediction of the fatigue life of spot welded vehicle structures;
- Perform examinations to determine the fatigue life of spot welded structures;
- Establish a complex evaluation procedure for the optimization of spot welded vehicle structures;
- Carry out the quality-oriented optimization of technological parameters with proper respect to the indentation limit;

The practical applicability of the outcomes of such researches is intended to be confirmed in the course of the planning and manufacturing actual vehicle structures. In this presentation, we would like to demonstrate the current issues of this multi-faceted research work (Borhy & Szabó 2005, Borhy & Kovács 2003, Borhy & Schwartz 2002, Borhy & Palotás 2003) with special attention paid to the determination of the fatigue lifes of spot welded structures.

4 Fatigue life of resistance spot welded joints

The compliance of road and railway vehicle structures produced with resistance spot welding is basically affected by the fatigue life of spot welded seams. While earlier researches tended to focus on the resistance of spot welded joints against static loading, recent years have witnessed the fatigue life of spot welded joints coming to the foreground of interests. In order to fulfil the above goals, we are to:

- Have an overview what parameters influence the fatigue life of spot welded joints;
- Consider the theoretical background of methods used for the prediction of fatigue life of spot welded joints;
- Establish a finite-element model for the prediction of the fatigue life of spot welded joints;
- Control the results derived from the model by means of fatigue testing.

5 Factors influencing the fatigue life of resistance spot welded joints

Fatigue is the most critical failure type of spot welded joints wherein joints become unserviceable in various ways. Cracking usually starts out from around the nugget, the contact plane of the plates. According to the experience obtained in vehicle-manufacturing industry, joints generally become defective not in the weld nugget, but the most frequently via within the heat-affected zone as starting out from the crack tip (Figure 2). Thereafter, such cracks tend to extend either in the base material towards external surfaces (Case I, Figure 2), or first radially in the welded joint, and then as deflected towards the external surfaces of the plate (Case II, Figure 2). In the event of any small-cycle fatigue, cracks can as well originate from the base metal (Case III, Figure 2). In the light of the related experience, vehicle structures can be rather characterized by defects involving cracks that increase via the width of the plate or feature large diameters. The location of fatigue cracks and the spread of such cracks are described in more details in publications (Yuuki et al. 1985, Satoh et al. 1991, Rui et al. 1993).

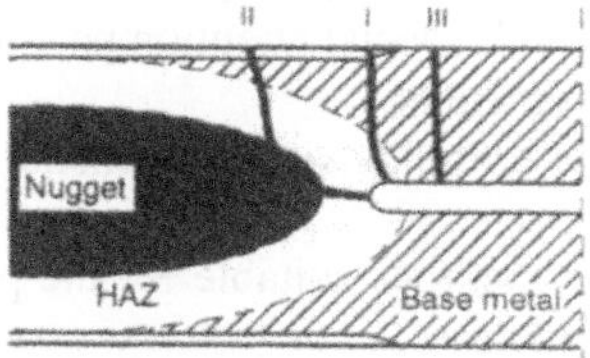

Figure 2. Typical failure types of spot-welded joints in the case of fatigue loading (Henrysson 1998)

The literature focusing on the fatigue of spot welded joints specifies the factors influencing fatigue life, as well as the actual effects of such factors in details (Swellam et al. 1994, Satoh et al. 1996). Accordingly, factors influencing the fatigue life of any spot welded seam are as follows:

- diameter of the seam;
- plate thickness;
- type of loading;
- mode of loading ($R = F_{min} / F_{max}$);
- material quality;
- welding parameters;
- surface quality;

In general, it can be stated that fatigue life will extend with the increase of the diameter of the spot welded seam or plate thickness. The effects of these geometrical parameters on fatigue life can be assessed with the stress measured at the spot welded seam or the stress intensity factor. Examinations on the mode of loading have indicated that fatigue life will double in case $R=-1$ as compared to $R=0$, while it is almost halved in case $R=0.5$.

6 Prediction of the fatigue life of spot welded joints

Procedures to predict the fatigue life of spot welded joints can be classified to the following method categories (Radaj & Sonsino 1998):

- stress-life method ($S-N$ curve);
- strain-life method ($\varepsilon-N$ curve);
- method based on linear elastic fracture mechanics ($\Delta K_{eq}-N$ curve);

All the three methods consist of similar phases:

Phase 1: Determination of constant-amplitude and variable-amplitude loading as relies on actual measurements or finite-element analyses;

Phase 2: Establishment of the $S-N$ curve, $\varepsilon-N$ curve, $\Delta K_{eq}-N$ curve of $da/dn - \Delta K$ curve belonging to the constant-amplitude loading on the basis of the results of the experiments;

Phase 3: Prediction of fatigue life as using any cumulative damage law (e.g. Miner's rule).

With respect to Figure 3, the equivalent stress intensity factor can be calculated by using the following relation (Pan & Sheppard 2003):

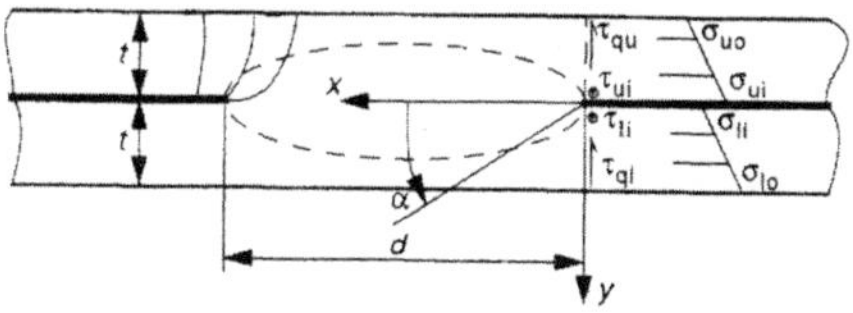

Figure 3. Stresses generated in spot welded joints

$$K_{eq}=K_I\cos^3\left(\frac{\alpha_{max}}{2}\right)-3K_{II}\cos^2\left(\frac{\alpha_{max}}{2}\right)\sin\left(\frac{\alpha_{max}}{2}\right) \tag{1}$$

wherein:

$$K_I=\frac{1}{6}\left[\frac{\sqrt{3}}{2}\left(\sigma_{ui}-\sigma_{uo}+\sigma_{li}-\sigma_{lo}\right)+5\sqrt{2}\left(\tau_{qu}-\tau_{ql}\right)\right]\sqrt{t} \tag{2}$$

$$K_{II}=\left[\frac{1}{4}\left(\sigma_{ui}-\sigma_{li}\right)+\frac{2}{3\sqrt{5}}\left(\tau_{qu}+\tau_{ql}\right)\right]\sqrt{t} \tag{3}$$

$$\alpha_{max}=2\tan\left(\frac{1}{4}\left(\gamma\pm\sqrt{\gamma^2+8}\right)\right) \tag{4}$$

$$\gamma=\frac{K_I}{K_{II}} \tag{5}$$

Fatigue life can be determined on the basis of the Paris law as follows:

$$\overline{\Delta K} \approx C^{-1/m}\left(\frac{N}{t}\right)^{-1/m} \tag{6}$$

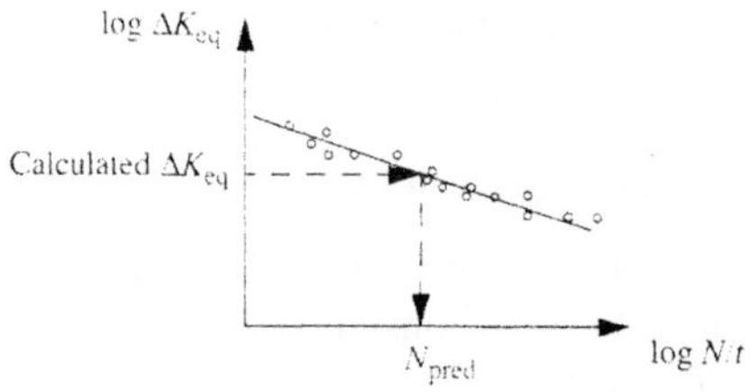

Figure 4. Determination of fatigue life on the basis of the Paris law

7 Examinations to determine fatigue life

To predict the fatigue life of spot welded vehicle structures, the calculation method suggested by Rupp et al. (1995) and then improved by Henrysson (2000) is to be applied. Fatigue life is assessed by using the equivalent stress intensity factor, which can be established from structural stresses. Stresses originating from the individual spot welded seams can be determined with the finite-element model.
To confirm the compliance of the applied calculation method, control examinations are to be also performed. We have prepared the finite-element models of the specimens presented in Figure 5 and 6 and intend to confirm the results of such calculations by means of carrying out actual fatigue testing.

Figure 5. Peel loading specimen (HP)

Figure 6. Shear loading specimen (HS)

The typical loading affecting the spot welded joints of the specimen shown in Figure 5 is the loading force being perpendicular to the joint, while in the case of the specimen demonstrated in Figure 6 spot welded joints are subject to shear loading.
The finite-element models of the individual specimens are presented in Figure 7 and

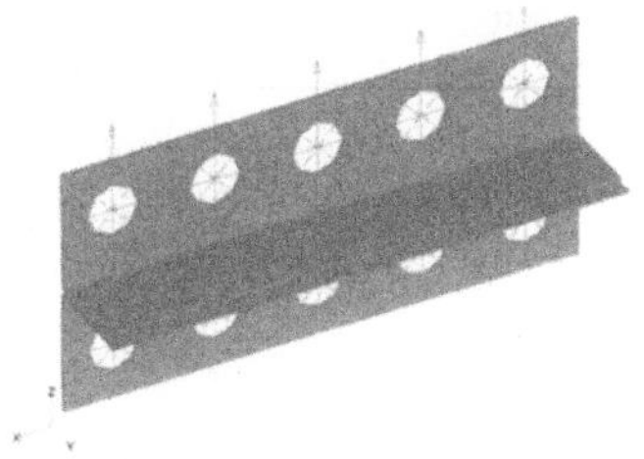

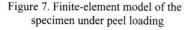

Figure 7. Finite-element model of the specimen under peel loading

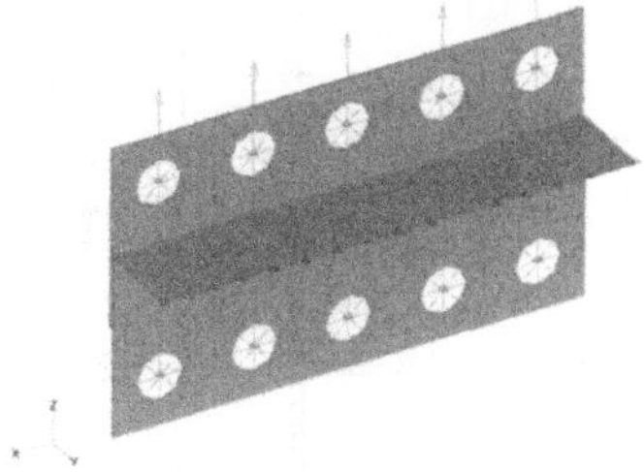

Figure 8. Finite-element model of the specimen under shear loading

Finite-element examinations have been performed with Suite MSC.Nastran/Patran v. 70.5. The plates of the specimens and spot welded joints have been modelled with CQUAD4 four-node general shell element and CBAR two-node general rod-element, respectively. The collars of the bores (for the fixing screws of the fatigue device) are linked with the center by means of RBE2 interlocking elements. For the lower element of the H-panel, bore centers have been subjected to clamp connections, while in the case of the upper element vertical shifts have been allowed. The symmetry of the model has been ensured by the clamping of shifts as perpendicular to the mid-plane, as well as of the twist of the complementary element.

Identical extents of loading affect the location of the fixing screws of the fatigue device vertically, as shown in the figure. The thickness of specimens is $t = 2$ mm with their material quality being S235 J2C+N under Standard EN 10025-2:2004. Figure 9 and 10 demonstrate the equivalent stress distribution and deformation as generated by $F_i = 100$ N loading.

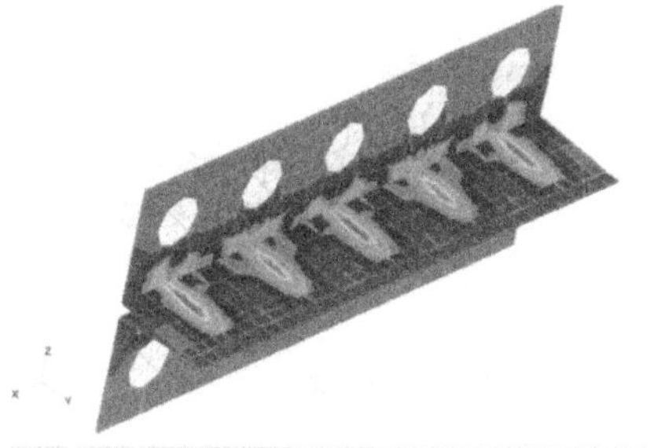

Figure 9. Equivalent stresses and deformations in the case of the HP-type specimen ($F_i = 100$ N)

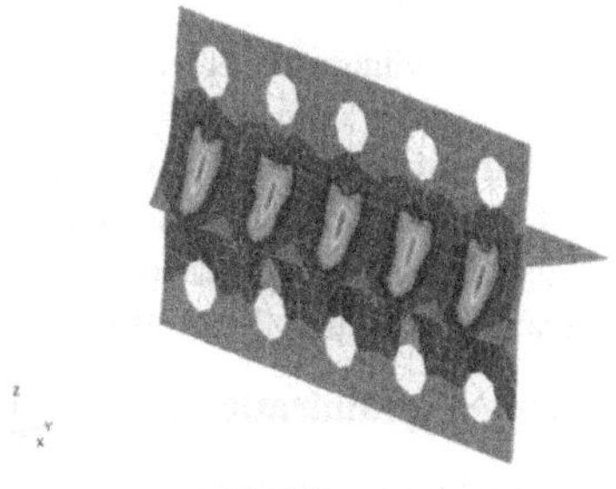

Figure 10. Equivalent stresses and deformations in the case of the HS-type specimen ($F_i = 100$ N)

Fatigue life can be established on the basis of (1), as knowing the forces loading the individual spot welded seams. The control the results of the finite-element modelling, fatigue testing is to be performed. Figure 11 shows the geometrical arrangement of HP-type specimens, while Figure 12 demonstrates the 3D model of the clamping device.

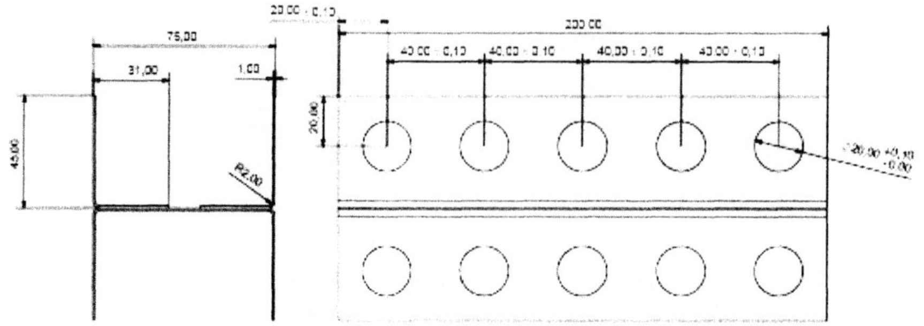

Figure 11. Geometrical arrangement of the HP-type specimen

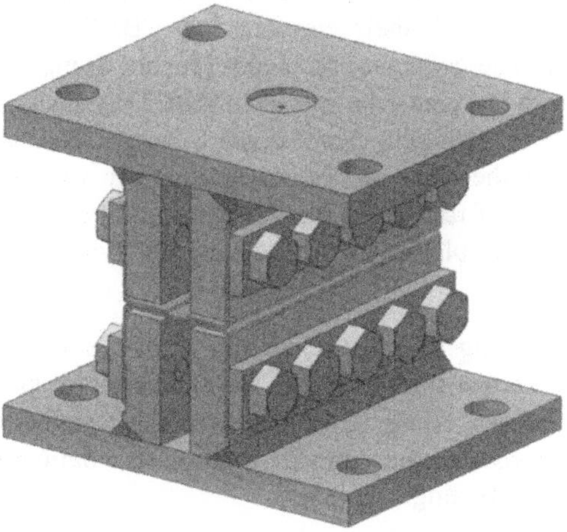

Figure 12. 3D model of the clamping device

The above examinations are due to be performed in the Test Laboratory of the Department of Railway Vehicles, Budapest University of Technology and Economics. Any further outcomes of the researches, as well as the potentials of practical applications will also be continuously presented.

8 Quality-oriented optimization of the spot-welding technology

Quality target function proposed to be used for the determination of the technological parameters of welded joints having optimal characteristics:

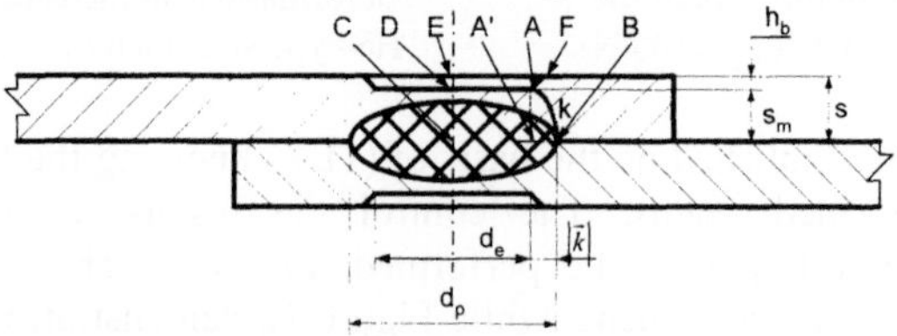

Figure 13. Characteristic parameters of the quality target function [5]

$$Y = \frac{k \cdot T_a}{s}, \quad (7)$$

wherein: Y: quality target function
k: length of the fracture curve
T_a: shape factor
s: material thickness of the piece

As concerning the value of the target function, the following statements can be made:

- if $Y<0$, the joint is sheared in the plane of jointing;
- $Y=0$ is the lower limit of the unbuttoning of the joint;
- In the case of $0<Y<1$ the joint is unbuttoned from the base material;
- $Y=1$ is the upper limit of the unbuttoning of the joint;
- if $Y>1$, the rupture occurs in a distance from the edge of the seam, within the heat-affected zone.

The examinations having been carried out have suggested that any welded joint in a quality complying with customer demands – with respect to the indentation limit – is arrived at when reaching the Y=0.6 value.

As using the multivariable finite-element model established to study the development process of spot-welded joints, non-linear, transient and linked analyses have been implemented with the related outcomes (temperature field, stress and deformation field) to be utilized as starting data for an expert system assisting the technological planning of spot welding (Borhy & Palotás 2003).

9 Conclusion

This presentation have aimed at highlighting the current issues of the research work that is connected to the optimization of spot welded railway vehicle structures, as well as to the improvement of the quality-oriented optimization method of the spot welding technology with special attention paid to the determination of the fatigue life of spot welded structures. We have described our research objectives and presented what questions are to be answered in order to achieve the contemplated results. We have described the factors influencing the fatigue life of spot welded structures in details together with the methods used for the determination of fatigue life. We have also presented the results of the research having been carried out and being in progress. We have provided information on the results of the finite-element modelling established for the prediction of fatigue life and described the main characteristics of the examinations to be performed with a view to the control of such results. We have also given an overview on the tasks to be attended in the near future, as well as on prospective examinations and issues to be still answered.

References

Borhy, I. (2000) Qualitätsfragen der Produktion von Schienenfahrzeugen und –fahrzeugteilen im Spiegel des Beitritts zur Europäischen Union, In: *Proceedings of 3rd GTE-MHtE-DVS* International Conference on Welding, pp.: 263.-268., Budapest

Borhy, I. & Schwartz, I. (2002) Practical application of FEM in the technology design of resistance spot welding in railway production, In: *Proceedings of 3rd Conference on Mechanical Engineering*, pp. 479-484. Springer-Verlag, Budapest

Borhy, I. & Kovács, L. (2003) Application of FEM in the production of welded railway vehicle parts, In: *Proceedings of International Conference on Metal Structures*, pp. 311-315, University of Miskolc

Borhy, I. & Szabó, P. (2005) Topical issues in the development of expert systems for use in the process planning of resistance spot welded railway vehicle structures, In: Mathematical Modelling of Weld Phenomena 7, pp. 1099-1109, Verlag der Technischen Universität Graz

Brenner, A. & Palotás, B. (1989) *Computer aided welding* (in Hungarian), BME Mérnöktovábbképző Intézet, Budapest

Henrysson, H.-F. (1998) Short fatigue crack propagation at spot welds: experiments and simulations, In: *Proceedings of Small Fatigue Cracks Mechanics*, Mechanisms and Applications Conference, pp. 483-490, Elsevier Science, Oxford (UK)

Henrysson, H.-F. (2000) Fatigue life predictions of spot welds using coarse FE meshes, Fatique Fract. Engng. Mater. Struct., **23** 737-746

Jármai, K. & Iványi, M. (2001) *Analysis and design of economic metal structures* (in Hungarian), Műegyetem kiadó, Budapest.

Pan, N. & Sheppard, S. D. (2003) Stress intensity factors in spot welds, *Eng. Fract. Mech.*, **70**, 671-684

Radaj, D. & Sonsino C. M.(1998) *Fatigue assessment of welded joints by local approaches*, Abington Publishing, Cambridge (UK)

Rui, Y.; Borsos, R. S.; Golpalakrischnan, R.; Agrawal, H. N. & Rivard, C.(1993) The fatigue life prediction method for multi-spot-welded structures, *SAE Technical Paper* 930571, Society of Automotive Engineers, Warrendale (PA, USA)

Rupp, A.; Störzel, K. & Grubisic, V.(1995) Computer aided dimensioning of spot welded automotive structures, *SAE Technical Paper* 950711, Society of Automotive Engineers,Warrendale (PA, USA)

Satoh, T.; Abe, H; Nishikawa, K. & Morita, M.(1991) On three-dimensional elastic-plastic stress analysis of spot-welded joint under tensile shear load, *Transactions of the Japan Welding Society*, **22** 46-51.

Satoh, T.; Abe, H.; Nakaoka, T. & Hayata, Y.(1996) The fatique life of the spot-welded joint under a repeated load of R=-1: Comparison of mild steel and high steel, *Welding in the Word*, **37** 12-15.

Sostarics, Gy.(1991) *Railway vehicles* (in Hungarian), Tankönyvkiadó, Budapest

Swellam, H.; Banas, G. & Lawrence, F. V.(1994) A fatigue design parameter for spot welds, *Fatigue Fract. Engng. Mater. Struct.*, **17**, 1197-1204.

Szabó, P. (2003) *Optimization of resistance spot welded joints of thin plates* (in Hungarian), Ph.D. Thesis, University of Miskolc

Yuuki, R.; Ohira, T.; Nakatsukasa, H. & Yi, W.(1985) Fracture mechanics analysis and evaluation of the fatigue strength of spot welded joints, Trans. *Japan. Soc. Mech. Engrs.*, 1772-1779.

4.6 Fatigue Behaviour of Friction Stir Spot Welded Joints with Re-filled Probe Hole in Al-Mg-Si Alloy

Yoshihiko Uematsu[1], Keiro Tokaji[1], Yasunari Tozaki[2], Tatsuo Kurita[3] and Shunsuke Murata[4]

[1]*Gifu University, 1-1 Yanagido, Gifu 501-1193, Japan,*
yuematsu@gifu-u.ac.jp,
[2]*Gifu Prefectural Research Institute for Machinery and Material, Oze, Seki 501-3265, Japan,*
[3]*Pacific Industrial Co. Ltd., Gohdo-cho, Anpachi-gun, Gifu 503-2397, Japan,*
[4]*Matsushita Electric Industrial Co. Ltd., Kadoma, Osaka 571-8501, Japan*

Abstract

In this study, fatigue behaviour of FSSW (friction stir spot welding) joints of Al-Mg-Si aluminium alloy was investigated. FSSW was performed using specially designed double-acting tool, which could re-fill probe hole of the joints. Fatigue tests were conducted using lap-shear specimens with probe hole and re-filled one at a stress ratio of $R = 0.1$. The re-filling process was beneficial to tensile strength, while detrimental to the fatigue strength at high applied loads. The dependence of fracture mechanism on FSSW tool geometry was discussed based on the macroscopic and microscopic structures near the weld zone.

Keywords: *Friction stir spot welding, Fatigue, Fracture mechanism*

1 Introduction

Aluminium alloy sheets are increasingly being used in the automotive industry, in which weight saving is extremely important. Resistance spot welding (RSW) has been widely used for decades for fabricating sheet metal assemblies. However, the conventional RSW technique is unsuitable for joining aluminium alloys because of its disadvantages such as high operating and investment costs due to high thermal and electrical conductivity of aluminium alloys, consumption of RSW probe during joining and large heat distortion.

A spot welding process using friction stir welding (FSW) technique has been newly developed, which is called friction stir spot welding (FSSW). This method is expected to apply to joining for aluminium sheets because of its advantages such as low heat distortion, excellent mechanical properties and little waste or pollution compared with conventional RSW. For future applications of FSSW to joining of load-bearing components, it is significant to elucidate the fatigue strength and the fracture mechanisms of FSSW joint. In the previous reports, therefore, authors investigated the static fracture and fatigue behaviour of FSSW joints welded by conventional concave tool: Tozaki et al (2007), Uematsu et al (2006). However, one of the disadvantages of FSSW joint is that probe hole inevitably remains at the centre of the weld nugget. Therefore, the double-acting FSSW tools, consisting of outer shoulder and inner probe, were developed in order to re-fill probe hole: Allen & Arbegast (2005). However, the fatigue behaviour of the joints with re-filled probe hole is unclear.

In the present study, fatigue tests were conducted using lap-shear specimens of friction stir spot welded Al-Mg-Si aluminium alloy with probe hole and re-filled one. The fatigue strength was evaluated and the fracture mechanisms were discussed based on experimental observation of the weld zone.

2 Experimental details

The material used is T4 treated Al-Mg-Si aluminium alloy, whose chemical composition is listed in Table 1. Specimens were made by using two 30 mm by 100 mm sheets with a 30 mm by 30 mm overlap area. The thickness of the sheet is 2 mm. The welding was performed by a double-acting tool consisting of outer shoulder and inner probe, which could re-fill probe hole as schematically shown in Fig.1. The tool is plunged into work pieces similar to conventional FSSW process, while the inner probe is retracted into the outer shoulder after joining and then the flat face of the tool is again plunged in order to re-fill probe hole. In this study, FSSW was performed with and without re-filling process.
Fatigue tests were conducted using an electro-hydraulic fatigue testing machine at a frequency of 10 Hz and a stress ratio of $R = 0.1$.

Table 1. Chemical composition of material

Mg	Si	Fe	Mn	Cr	Zn	Ti	Al
0.6	1	<0.2	0.05	<0.05	<0.3	<0.05	Bal.

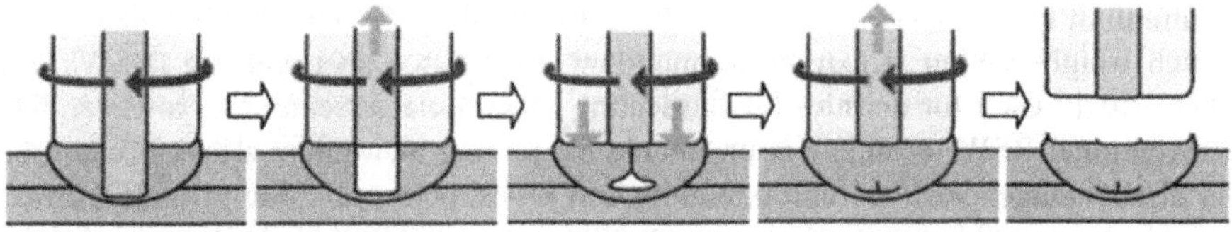

Figure 1. Schematic illustration of FSSW and re-filling processes.

3 Result

3.1 *Microstructures in weld zone*

Figure 2 shows the top view and the cross-section of the weld zone in a joint with probe hole, i.e. without re-filling process. Figures 3 (a) and (b) indicate the microstructures in the regions denoted as SZ and MZ in Fig.2, respectively. The classification of these regions will be mentioned later. Figure 3(c) is the microstructure of the parent metal (PM). As shown in Figs.3 (a) and (b), the grains in the regions SZ and MZ are much finer than those in PM (Fig.3 (c)) due to dynamic recrystallization during joining process. The average grain sizes in PM, SZ and MZ are 35μm, 5μm and 3μm, respectively. The grain refinement was recognized in the region inside the solid line in Fig.2, so that this region could be referred as stir zone (SZ). Furthermore, in SZ, the ring shape zone indicated by the dotted line in Fig.2 was recognized, where grains were slightly finer than SZ. It is believed that the material in this region was stirred more severely than the surrounding SZ. Such region is defined as the mixed zone (MZ) in the present study, in which the upper and lower sheets are mixed severely, and MZ is considered to be the effective nugget of FSSW joint. As a result, the microstructure could be classified into three regions, MZ, SZ and PM. The boundary between the upper and lower sheets is also indicated by an arrow in Fig.2, which results from the upward material flow of the lower sheet. The shape of the boundary could be characterized by rather vertical material flow around the probe hole.

Figure 4 shows the top view and the cross section of the weld zone in a joint with re-filled probe hole. It is clear that probe hole was successfully re-filled by double-acting tool. The microstructures in SZ and MZ are revealed in Figs.5 (a) and (b), respectively. The average grain size is 4μm in SZ and 8μm in MZ, indicating that the grains in MZ are slightly coarser than those of the joint with probe hole, which can be attributed to the heat input into MZ during re-filling process. The boundary between the upper and lower sheets is indicated by an arrow in Fig.4. The boundary on the re-filled hole is almost horizontal due to the squeezing of material by the flat face of the tool.

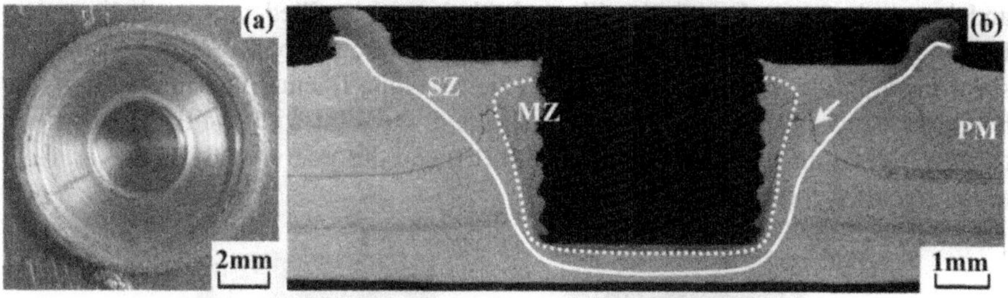

Figure 2. Macroscopic appearance of FSSW joint with probe hole:
(a) Top view of weld zone, (b) Cross section of weld zone.

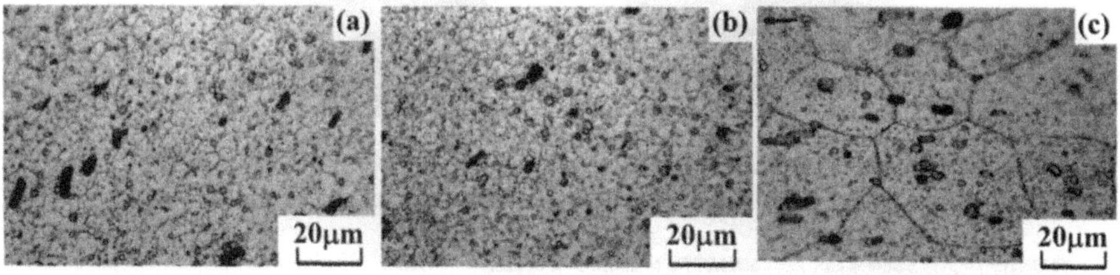

Figure 3. Microstructures in FSSW joint with probe hole: (a) SZ, (b) MZ, (c) PM.

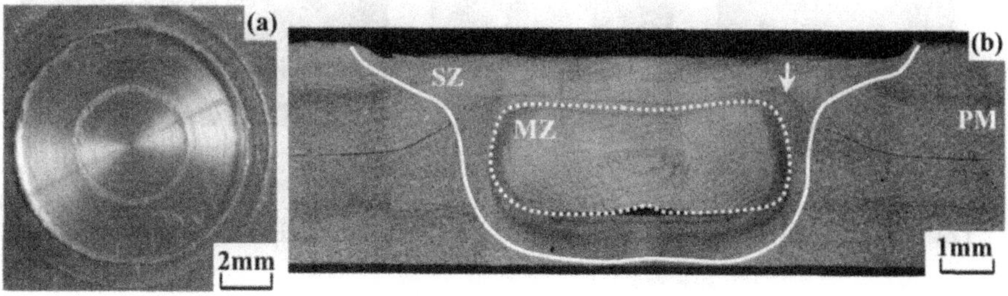

Figure 4. Macroscopic appearance of FSSW joint with re-filled probe hole:

(a) Top view of weld zone, (b) Cross section of weld zone.

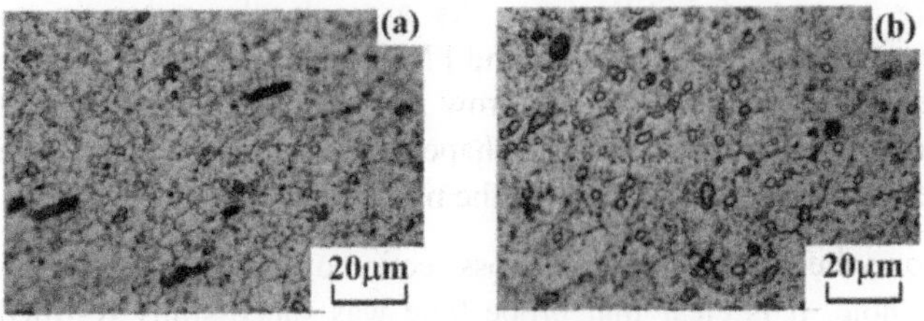

Figure 5. Microstructures in FSSW joint with re-filled probe hole: (a) SZ, (b) MZ.

3.2 *Tensile strength*

The tensile strengths of the joints with probe hole and re-filled one are 2654N and 3458N, respectively. Since the tensile strength of the FSSW joint fabricated by a conventional concave tool was 2948N: Uematsu et al (2006), it is seen that tensile strength can be improved by re-filling process. The fracture surfaces of the upper and lower sheets in the joints with probe hole and re-filled one are indicated in Figs.6 and 7, respectively. It is clear that tensile fracture occurred though MZ in shear mode in the joint with probe hole (Fig.6), while fracture took place along the boundary between the upper and lower sheets in the joint with re-filled probe hole, as seen in fracture surfaces (Figs.7 (a) and (b)) and the side view of the lower sheet of the fractured specimen (Fig.7 (c)).

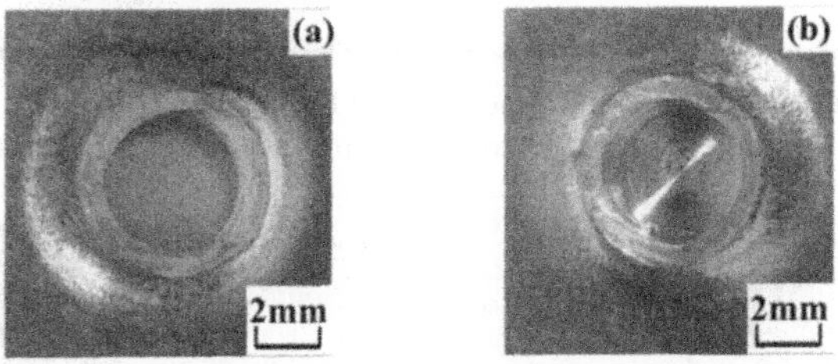

Figure 6. Static fracture surfaces of FSSW joint with probe hole: (a) Upper sheet, (b) Lower sheet. The loading direction is the horizontal direction.

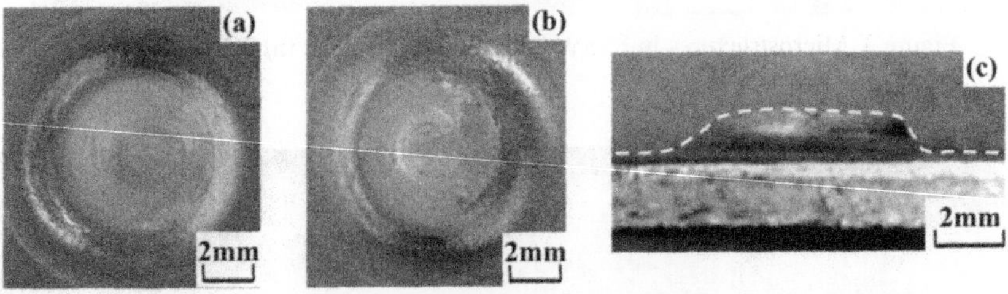

Figure 7. Static fracture surfaces of FSSW joint with re-filled probe hole: (a) Upper sheet, (b) Lower sheet, (c) Side view of fractured specimen. The loading direction is the horizontal direction.

3.3 *Fatigue strength*

Figure 8 shows the relationship between maximum load, P_{max}, and number of cycles to failure, N_f, in the joints with probe hole and re-filled one. The test results of the FSSW joints fabricated by conventional concave tool are also indicated in the figure for comparison: Uematsu et al (2006). Although the fatigue strengths of both joints are almost the same at low applied loads, the joint with probe hole has slightly higher fatigue strength than the joint with re-filled one. It should be noted that the joint with re-filled hole, which has higher tensile strength, has lower fatigue strength at high applied loads.

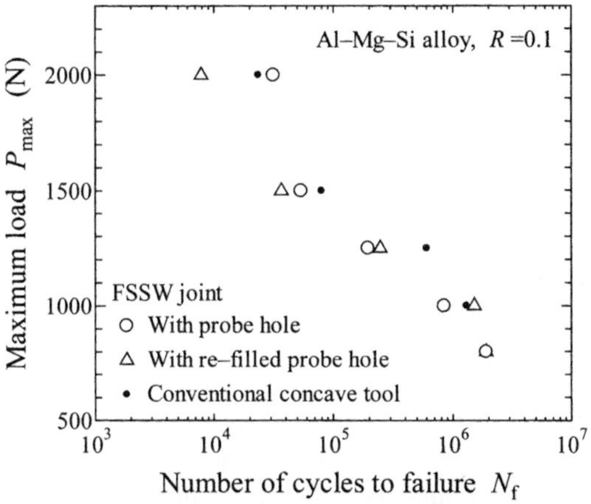

Figure 8. Relationship between maximum load, P_{max}, and number of cycles to failure, N_f.

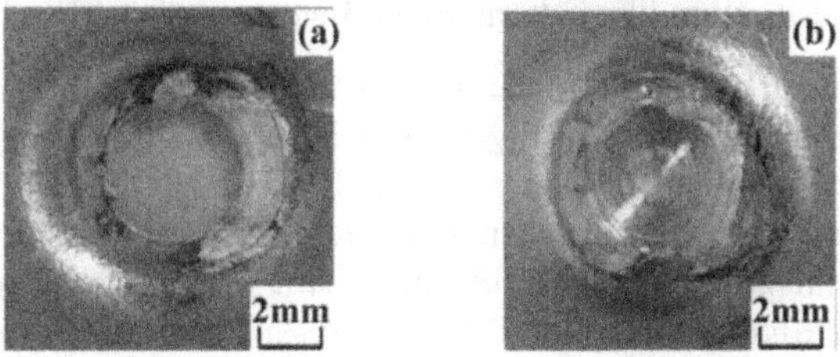

Figure 9. Fatigue fracture surfaces of FSSW joint with probe hole (P_{max} = 2000N): (a) Upper sheet, (b) Lower sheet. The loading direction is the horizontal direction.

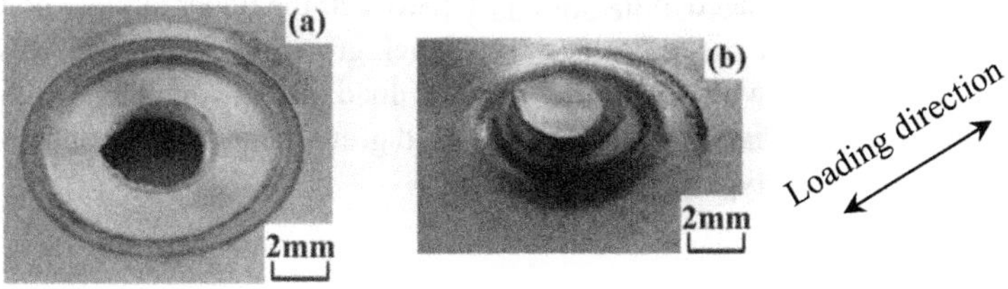

Figure 10. Fatigue fracture surfaces of FSSW joint with re-filled probe hole (P_{max} = 2000N): (a) Upper sheet, (b) Lower sheet.

3.4 *Macroscopic fatigue fracture morphology*

Typical fracture surfaces of the joint with probe hole tested at P_{max} of 2000N are revealed in Fig.9. As can be seen in the figure, fatigue fracture took place through MZ, whose fracture mode was similar to that of static fracture (Fig.6), and could be denoted as shear type fracture. Figure 10 is the oblique views of fracture surfaces of the joint with re-filled probe hole tested at P_{max} of 2000N. It was found that fatigue crack grew around the nugget and final fracture took place due to the pull out of the nugget, which could be denoted as plug type fracture. It is noteworthy that fatigue fracture mode is different form that of static fracture in the joint with re-filled probe hole.

3.5 *Microscopic fatigue fracture morphology*

Figure 11 indicates SEM micrographs of fracture surface observed on the upper sheet of the joint with probe hole, where P_{max} is 2000N. Figures 11 (b) and (c) are the magnified views at the points "A" and "B" in Fig.11 (a), respectively. The typical fatigue fracture surface with striation like patterns is seen at the point "A", while static fracture surface with dimples is recognized at the point "B". The presence of the striation like patterns in Fig.11 (b) suggests that the crack initiated at the edge of the nugget and grew in the direction parallel to the longitudinal direction through MZ as shown by an arrow in Fig.11 (a). In order to figure out fatigue fracture mechanism, fatigue test conducted at P_{max} = 800N was interrupted at N/N_f = 0.36, and the longitudinal section (LS) of the weld zone was observed. Figure 12 (a) indicates the LS view of the weld zone in the interrupted specimen. As seen in the figure, crack initiated at the boundary between the upper and lower sheets and grew through MZ. This crack growth path coincides well with the crack growth direction expected from the striation like pattern on the fracture surface (Fig.11 (b)).

SEM micrographs of fracture surface observed on the lower sheet of the joint with re-filled probe hole are shown in Fig.13, where P_{max} is 2000N. Figs.13 (b) and (c) are the magnified views at the points "A" and "B" in Fig.13 (a), respectively. The typical fatigue fracture surface is seen at the point "A", while static fracture surface is recognized at the point "B". This indicates that crack grew around the nugget and final fracture occurred at the point "B". Fatigue tests conducted at P_{max} = 800N and 1500N were interrupted at N/N_f = 0.36 and 0.63, respectively, and the LS view of the weld zone was observed. Figures 12 (b) and (c) indicate the LS views of the weld zone in the interrupted specimens tested at P_{max} = 800N and 1500N, respectively. Crack initiated at the boundary between the upper and lower sheets and grew through the upper sheet. Furthermore, crack growth path approaches vertical to the loading direction with increasing applied load. Based on these findings, it is believed that crack initiated at the boundary and grew around the re-filled probe hole and finally led to plug type fracture.

4 Discussion

4.1 *Effect of re-filling process on tensile strength*

The re-filling process improved static strength by about 30%. As indicated in Fig.6, static fracture mode was characterized by shear fracture through the nugget.

Therefore, the re-filling of prove hole resulted in the increase of effective sectional area of the nugget and consequently the improvement of tensile strength was achieved.

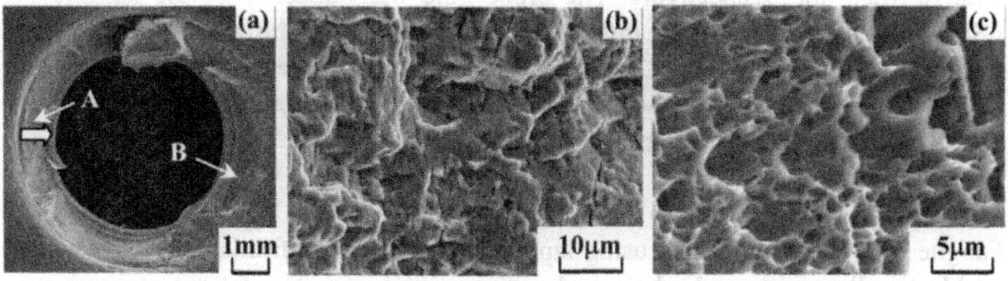

Figure 11. SEM micrographs of fracture surfaces in FSSW joint with probe-hole: (a) P_{max} = 2000N, N_f = 3.2×10^4, (b) and (c) Magnified views at A and B in (a), respectively.

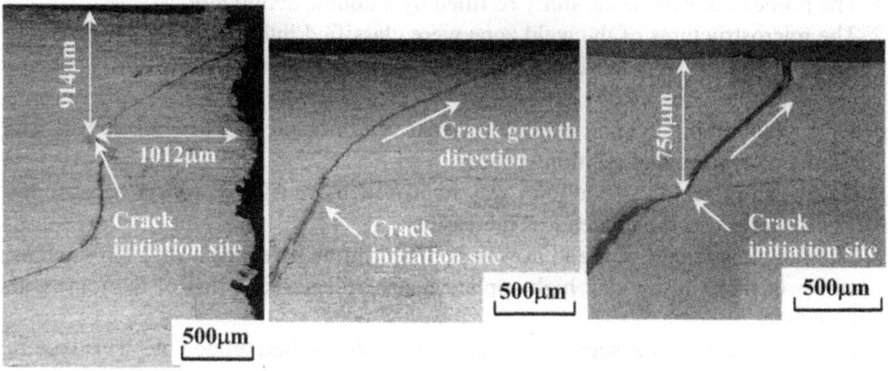

Figure 12. Optical micrographs showing LS view of interrupted specimen, (a) Joint with probe hole at P_{max}=800N, N/N_f=0.36, (b) and (c) Joint with re-filled probe hole at P_{max} = 800N, N/N_f = 0.36 and P_{max} = 1500N, N/N_f=0.63, respectively

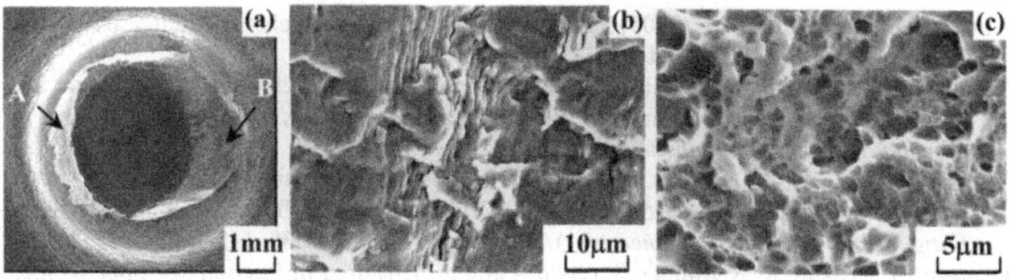

Figure 13. SEM micrographs of fracture surfaces in FSSW joint with re-filled probe-hole: (a) P_{max} = 2000N, N_f = 3.2×10^4, (b) and (c) Magnified views at A and B in (a), respectively.

4.2 *Effect of re-filling process on fatigue strength*

In the joint with re-filled probe hole, plug type fracture was dominant under fatigue loading, while shear type one under monotonic loading. It is believed that this transition of fracture mode is responsible for the lower fatigue strengths of the joint with re-filled probe hole at high applied loads. As shown in Fig.12 (b), crack initiated at the boundary between the upper and lower sheets and tended to grow through the upper sheet under fatigue loading. The thickness of the upper sheet was about 914μm (Fig.12 (a)) and 750μm (Fig.12 (c)) for the joints with probe hole and with re-filled one, respectively. The probe hole was squeezed by the shoulder of FSSW tool during re-filling process, which resulted in the thinner upper sheet thickness. Furthermore, crack tended to grow vertical to the loading direction with increasing applied load as shown in Figs.12 (b) and (c), leading to shorter crack growth path and thus resulting in the lower fatigue strength at high applied loads.

5 Conclusion

Fatigue tests were performed using lap-shear specimens of friction stir spot welded Al-Mg-Si alloy with probe hole and with re-filled one. The fatigue strength was evaluated and the fracture mechanisms were discussed based on experimental observation of the weld zone.

1. The probe hole was successfully re-filled by a double acting tool.
2. The microstructures of the weld zone were classified into MZ and SZ, where fine equiaxed grains were observed due to dynamic recrystallization during FSSW process.
3. Under monotonic loading condition, shear type fracture through the nugget was dominant regardless of re-filling process.
4. The tensile strength of the joint was improved by re-filling process because the effective cross sectional area was increased.
5. The fatigue strength of the joint with re-filled probe hole was nearly the same as, but lower than, at low and high applied loads, respectively, that of the joint with probe hole.
6. Plug type fracture was seen only in the joint with re-filled probe hole. This type of fracture mode was responsible for the observed lower fatigue strength of the joint at high applied loads.

References

Allen, C.D. & Arbegast, W.J. (2005) Evaluation of Friction Spot Welds in Aluminium Alloys. *SAE Technical Paper* 2005-01-1252

Tozaki, Y., Uematsu, Y. & Tokaji, K. (2007) Effect of Tool Geometry on Microstructure and Static Strength in Friction Stir Spot Welded Aluminium Alloys. *International Journal of Machine Tools and Manufacture* **47** 2230-2236

Tozaki, Y., Uematsu, Y. & Tokaji, K. (2007) Effect of Processing Parameters on Static Strength of Dissimilar Friction Stir Spot Welds between Different Aluminium Alloys. *Fatigue and Fracture of Engineering Materials and Structures* **30** 143-148

Uematsu, Y., Tokaji, K. & Murata, S. (2006) Fatigue Behaviour of Friction Stir Spot Welded Joints in Al-Mg-Si Alloy. *Proc. 6th Int. Symposium on Friction Stir Welding, Saint-Sauveur, Canada* (CD-ROM)

4.7 Applicability of Different Approaches for the Fatigue Assessment of Welded Components in Vehicle Construction

Martin Vogt[1], Ekke Hanssen[1], Tim Welters[1] and Klaus Dilger[1]
[1]*Institute of Joining and Welding Technique, Technical University of Braunschweig, Langer Kamp 8, 38106 Braunschweig, Germany, m.vogt@tu-bs.de*

Abstract

Within the scope of the research results presented here, the applicability of the structural and notch stress approach is under examination on the basis of different arc welded and dynamically loaded steel structures, taken from the automotive and railway sector. In detail these are a transverse control arm and a crossbeam connection from the underframe of a railcar body. Components and specimens with critical regions of failure are tested under dynamic loading with constant amplitude. With the help of strain gauges, the technical incipient crack is determined. The specimens are the basis for application and evaluation of the different concepts for the assessment of fatigue life. The numerical determination of the nominal, structural and notch stresses is performed with finite-element models. Finally the experimental and computational results allow the derivation of structural and notch Woehler-lines (S-N-curves).

Keywords: *arc welded steel structures, local stress approaches, fatigue assessment*

1 Introduction

For thin-walled steel structures the fatigue strength of weld seams is in many cases essential for the overall component strength because the welded joints often turn out to be the weak spots of structures under dynamic loading with constant or variable amplitudes. The assessment of fatigue life allows, in an early phase of design, a dimensioning of components appropriate for the expected working loads. Therefore, well-defined and in terms of their prediction accuracy reviewed strength concepts are necessary.

The notch stress approach with a fictious notch radius r_f=1mm, is only defined in the IIW-Recommendation (Hobbacher, 1996) and in the FKM-Guideline (2002) and only applicable for thick steel plates with thicknesses of 5 mm or greater. Additionally components are partial loaded with variable amplitudes, which results in the well known problems concerning the existing cumulative damage hypotheses with D=1 and the applicable strength criterion for multiaxial stress states (Sonsino & Maddox, 2001). So the different existing approaches have to be investigated.

The results presented here are developed in the context of the research cluster "applicability of fatigue design concepts for dynamic loaded weld structures", which is promoted by the AiF (see the acknowledgements). The applicability of the structural and notch stress approach on the basis of different arc welded steel structures is under examination. These are a transverse control arm (automotive sector) and a crossbeam connection from the underframe of a railcar body (railway sector). The derivation of structural and notch Woehler-lines (S-N curves) is the aim of the experimental and computational results.

2 Materials, specimen shapes and testing method

The crossbeam connections are the critical areas, concerning the underframe of the car body. The construction is build of u-profiles with different cross-sections and different sheet thicknesses. This leads within the region of the weld joints to an increase of stress. The construction steel S355J2G3 (1.0116) in a plate thickness of 3.0 mm and the fine grain cold pressure steel S500MC (1.0984) in a plate thickness of 4.0 mm are used for the investigation. To avoid testing a whole railcar body, different types of components and specimen are designed. The critical area of the crossbeam connection is the master for the different types of specimens and components (Figure 1). All specimens were manufactured by manual MAG-welding by the company ALSTOM LHB (Salzgitter, Germany). Component C (Figure 1) is a complex structure near the real railcar body, which will be tested under bending. A simplification of the crossbeam connection from the underframe of a railcar body with the load of component C, is the specimen B. It includes the different sheet thickness and the real welding seams, including the beginning and the end of the weld seam, as a combination of a butt and fillet joint. A further simplification of the critical area is built up in specimen A. The butt joint includes a misalignment of weld edges concerning the different sheet thicknesses (Figure 2). Specimen E is characteristic for a tensile loading of the longitudinal solebar, which is typical for the load in a railcar body.

In order to determine the fatigue strength behaviour, load-controlled test with constant amplitude and pulsating load (R = 0) were carried out with specimen A and B. For each specimen type 16 – 29 tests were carried out with a frequency of f = 30 Hz at room temperature in air. With the help of strain gauges, the technical incipient cracks were determined. The end of the test is defined by the fracture of the specimen or by 10^7.numbers of cycles. These results are the basis for application and evaluation of the different concepts for the assessment of fatigue life.

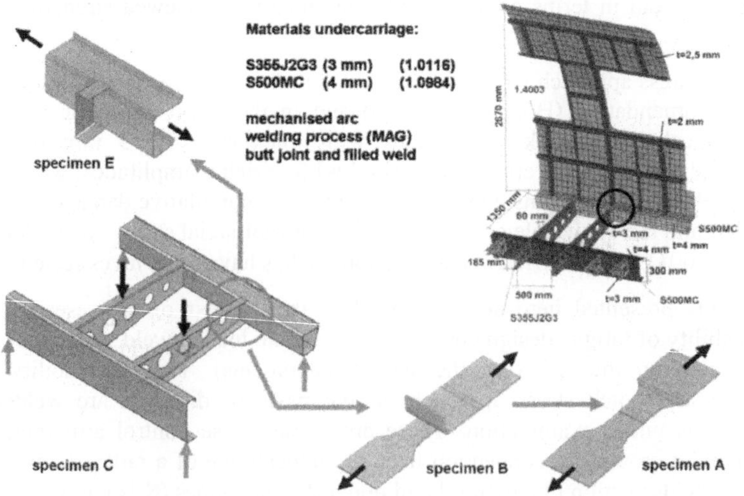

Figure 1. Specimens and components for fatigue tests

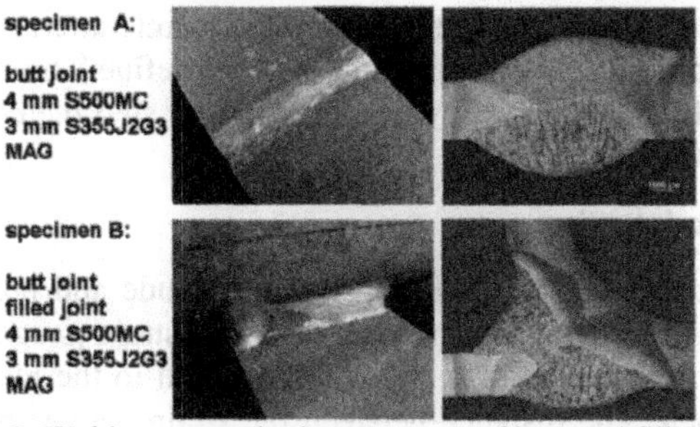

Figure 2. Weld seams and microstructure of butt and filled weld

The parts of the transverse control arm consist of the S460MC (1.0982), S355MC (1.0976) and DD13 (1.0335) with sheet thicknesses between 2.5 and 7 mm. The critical region of failure and the resulting detail are located at the rear bearing tube (Figure 3). The assembly is joined with an automatic MAG welding process. The stiffness of the coarse FE model is compared with the corresponding static loaded experiment by means of optical three dimensional deformation analysis and local strain measurements with strain gauges.

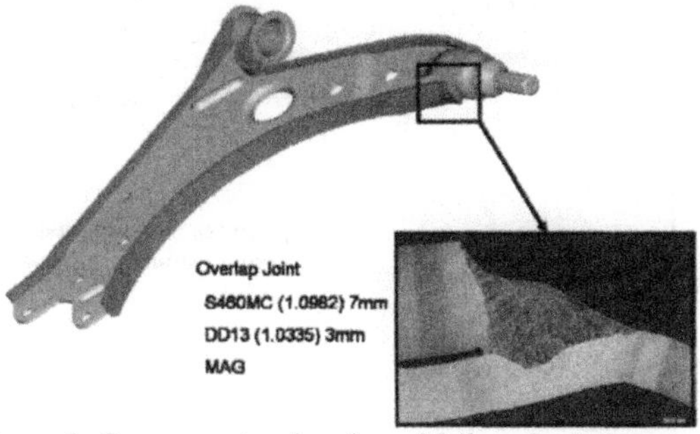

Figure 3. Component and region of failure for fatigue tests

After performing the stiffness alignment of the coarse model, a submodel of the region of failure is created. The real weld seam geometry is measured by means of laser-triangulation and used for modelling the weld seam in the submodel. The fictious notch rounding is performed with a notch radius r_f = 0.05mm (Figure 4). Because of the presence of a heavily multiaxial stress state acting on the weld seams of the control arm the Mises strength criterion is used to determine the equivalent notch stress.

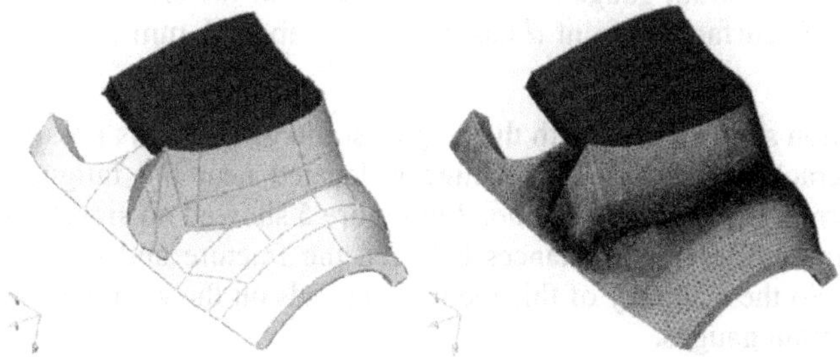

Figure 4. Submodel of the overlap joint with fictious notch radius r_f=1mm and r_f=0.05

First, the transverse control arm is tested under dynamic alternating loading (R=-1) with constant amplitude. The end of the test is defined by the fracture of the component or by 10^7 numbers of cycles. In a further step the control arm is tested under variable amplitudes.

3 Incipient crack detection

During the load-controlled test with constant amplitude and pulsating load (R=0), the signal of three strain gauges were recorded. The strain gauges were located in a distance of 3 mm from the weld seam, witch is equal to the plate thickness in this area of the specimen. The distance between the strain gauges perpendicular to the load direction was 10 mm and the total width of the specimen was 41 mm. In Figure 5 is the strain amplitude versus the number of cycles shown. The strain gauge no. 3 shows a significant change in the strain amplitude after 200.000 cycles. The experiment was stopped after a 10 % change in strain amplitude.

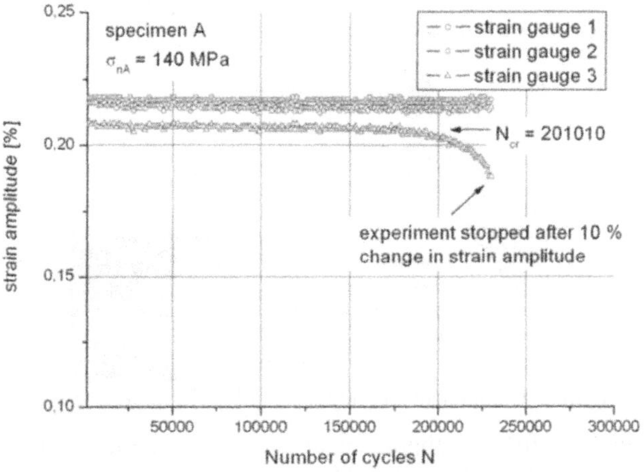

Figure 5. Incipient crack detection by strain gauges

In Figure 6 the fracture surface of specimen A after a 10 % change of the local strain amplitude is shown. For this purpose, the specimen was cooled with liquid nitrogen to produce a brittle fracture by load. Four different areas of fatigue crack surfaces were identified. The strain gauge no. 3 was located near the crack surfaces *c* and *d*. The fatigue crack surface at point *d* has a depth of about 1 mm at this time of the experiment.

This investigation showed that, with the help of strain gauges, it is possible to detect the incipient crack, when the strain gauge is located near the fatigue crack. For example, the strain gauges no. 1 and no. 2 in Figure 5 showed no significant changes in the strain amplitudes. The distances between the fracture and the strain gauges were too high. So the accuracy of this method depends on the geometry between the crack and the strain gauges.

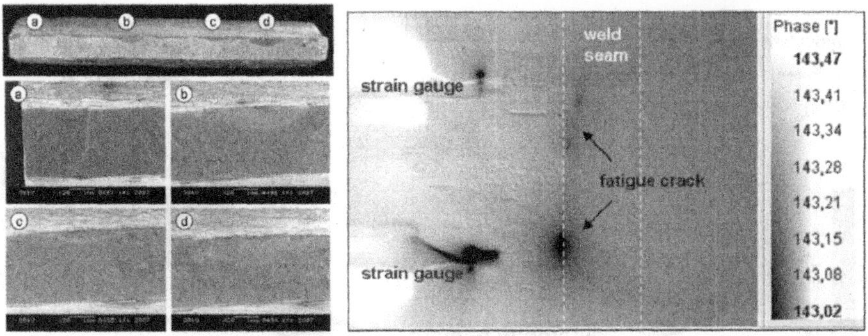

Figure 6. a) Fracture surface after 10 % change of local strain amplitude under dynamic loading, b) Thermography investigation of incipient crack detection

A further possibility to get information about the incipient crack is given by the ultrasonic-burst-phase-thermography. In this research, a short outlook of this method is given. Figure 6 shows a phase-image of a butt joint specimen after a 10 % change of strain amplitude. The dimension perpendicular to the load direction of the fatigue crack surface is identifiable. In contrast to the strain gauge method, the experiment must be stopped. For the ultrasonic-burst-phase-thermography an external experimental setup is necessary.

4 Finite Element Analysis (FEA)

The numerical determination of the nominal, structural and notch stresses is performed with proper finite-element models. For a local approach, according to the notch stress analysis, a submodelling technique is used. Existing weld seam and specimen tolerances are included by the usage of parametric models. The FE models are compared with the corresponding experiment by means of an optical three dimensional deformation analysis and strain measurements with strain gauges. Figure 7 shows the results (equivalent stress v. Mises) of the global FE-models of the tested specimens by a nominal tensile load of $\sigma_a = 100$ MPa. These solid models were the basis for the determination of nominal stresses and hot-spot stresses.

The models include the misalignment of weld edges concerning the different sheet thicknesses, which leads to an increase of load in one region of the weld seam. To calculate the hot-spot stresses, a linear extrapolation with the evaluation points 0.4 t and 1.0 t is chosen. The notch stresses are determined with the help of a submodelling technique. Figure 8 shows the equivalent notch stresses (v. Mises) with the fictitious radius $r_f = 1.0$ mm. In course of the different geometry of the specimens and the different weld geometry, the notch stresses of specimen B are higher than that of specimen A.

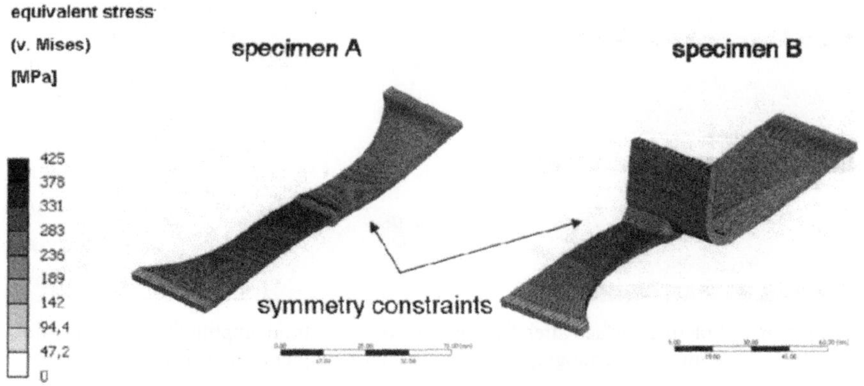

Figure 7. Results of the FE-global models of butt (specimen A) und fillet (specimen B) joint

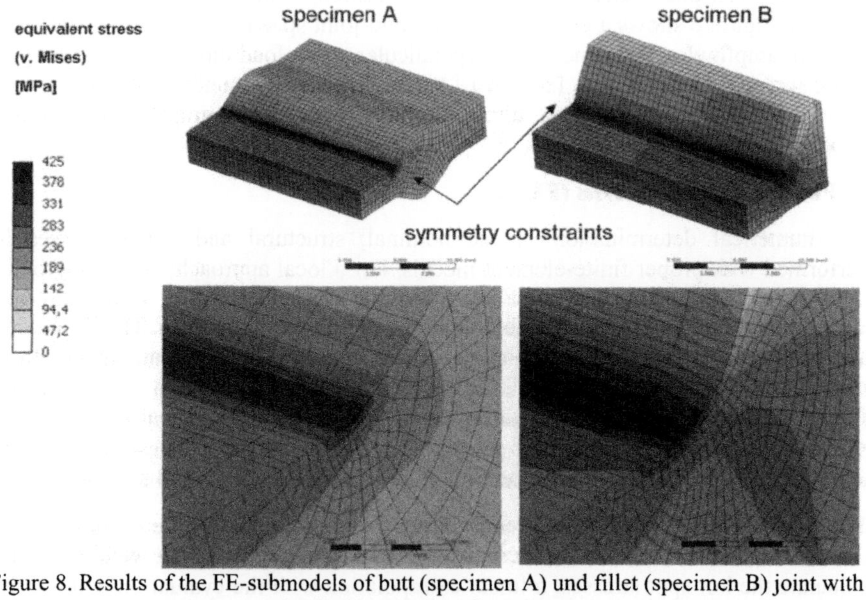

Figure 8. Results of the FE-submodels of butt (specimen A) und fillet (specimen B) joint with the fictitious radius rf = 1.0 mm

In the comparison between the FE-models and the failure of the specimens in the experiments, the hot-spots lie at the same places.

5 Results of the fatigue strength tests

The test results for the specimens with butt joint and with the combination of butt und fillet joint are shown in the nominal stress system in Figure 9. The misalignment of weld edges leads to an additional bending load. In this examination, a distinction was made between the fatigue life to crack initiation (technical crack depth a = 1.0 mm) and to final rupture. The results permitted an evaluation according to the concept of uniform SN-curve (Hobbacher, 1996) with an appointed knee point at $N_k = 10^7$ cycles for both specimens, a uniform scatter of $T_s = 1{:}1.24$ (specimen A) and a uniform scatter of T_s = 1:1.30 (specimen B) and slopes of k = 7 - 8. For $N > N_k$ the slope of k* = 45 is derived by assuming a fatigue strength decrease of 5% per decade (Sonsino & Maddox, 2001). At the knee point, nominal stress amplitudes of $\sigma_{a,n}$ = 71 MPa for the butt joint and $\sigma_{a,n}$ = 72 MPa for the butt and fillet joint are obtained. To get more information between $10^6 < N < 10^7$, further test will be done.

The combined results of specimens A and B in a single SN-curve are shown in the hot spot stress and in the notch stress system in Figure 10.

The results also permitted an evaluation according to the concept of uniform SN-curve (Hobbacher, 1996) with an appointed knee point at $N_k = 10^7$ cycles for both specimens, a uniform scatter of T_s = 1:1.27 and slopes of k = 8. For $N > N_k$ the slope of k* = 45 is derived by assuming a fatigue strength decrease of 5% per decade (Sonsino, C.M. & Maddox, S.J. 2001). At the knee point, hot-spot stress amplitudes of $\sigma_{a,\ hs}$ = 208 MPa and notch stress amplitudes of $\sigma_{a,k}$ = 300 MPa for both connections are obtained. Concerning the notch stress concept, the results were obtained with a fictitious radius r_f = 1.0 mm.

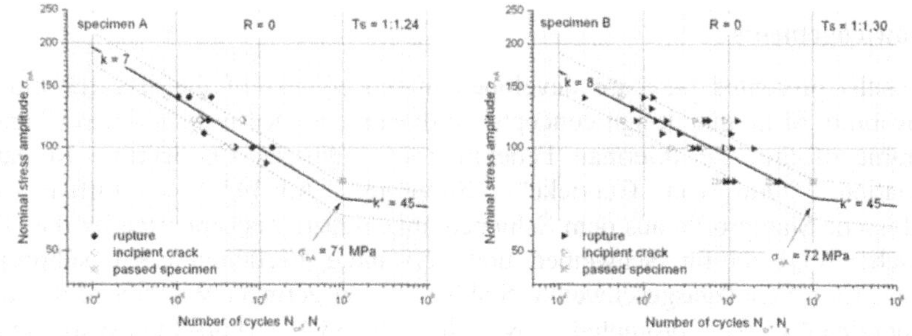

Figure 9. SN-curves of specimen A and B in the nominal stress system under pulsating loading

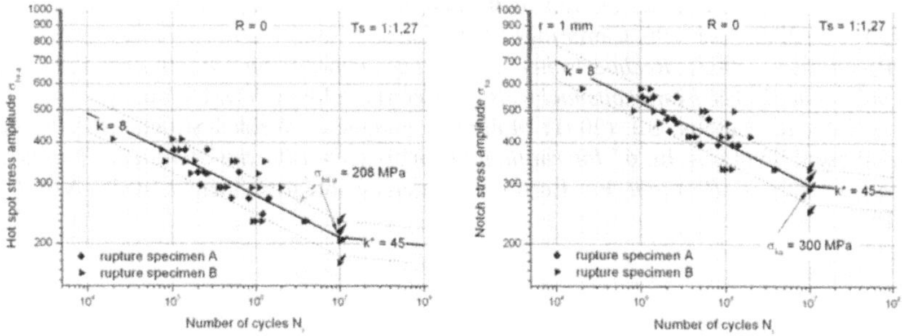

Figure 10. SN-curves of specimen A and B in the hot spot stress and notch stress system under pulsating loading

6 Conclusions

The investigation showed a good correlation between the calculated strains in the FE-models and the measured strains of the specimens with the help of strain gauges. Finally the calculated hot-spots are the areas of failure of the specimens in the experiments

When the strain gauge is located near the fatigue crack, the investigation showed that it is possible to detect the incipient crack. The ultrasonic-burst-phase-thermography is a further possibility to get information about the incipient crack.

The fatigue strength behaviour was determined with load-controlled test with constant amplitude and pulsating load of two different specimen types. It was possible to present the results in a single SN-curve in the nominal, hot-spot and notch stress system with a uniform scatter below $T_s = 1{:}1.50$ and slopes of $k = 8$ for $N < N_K$. Further investigations are the load-controlled test with constant amplitude and pulsating load with the described components (Figure 1). Afterwards a comparison between the fatigue strength of the components and the tested specimens will be possible.

Acknowledgements

The results presented here are developed in the context of the research cluster "applicability of fatigue design concepts for dynamic loaded weld structures" under the frame of the AiF (German Federation of Industrial Cooperative Research Associations "Otto von Guericke"). Subproject AiF 14519N ("Offene und geschlossene Stahlprofile aus dem Schienenfahrzeugbau") is supported by the DVS (Deutscher Verband für Schweißen und verwandte Verfahren e.V.), subproject AiF 14521N („Schutzgasgeschweißte Stahlstrukturen geringer Wanddicke aus dem Automobilbau") is promoted by the FOSTA (Forschungsvereinigung Stahlanwendung e. V). We thank the AiF for the financial support.

References

Hobbacher, A. (1996) *Fatigue Design of Welded Joints and Components.* IIW-Doc. XIII – 1539 –96 / XV – 845 – 96, Cambridge: Abington

FKM-Richtlinien (2002) *Rechnerischer Festigkeitsnachweis für Maschinenbauteile und Bruchmechanischer Festigkeitsnachweis*, Frankfurt am Main: VDMA-Verlag

Sonsino, C.M. & Maddox, S.J. (2001) Multiaxial Fatigue of Welded Structures – Problems and Present Solutions. In: *6th International Conference on Biaxial/Multiaxial Fatigue and Fracture, Lisboa, Portugal,* Ed. European Structural Integrity Society (ESIS) Vol. I, pp. 3-15.

Section 5
Frames

5.1 Large Displacement Analysis of Spatial Frames under Creep Regime

Domagoj Lanc, Goran Turkalj, Josip Brnic
Faculty of Engineering of University of Rijeka, HR-51000 Rijeka, Croatia, dlanc@riteh.hr

Abstract

The paper presents a beam finite element for creep analysis of spatial framed structures. Displacements and rotations are allowed to be large while strains are assumed to be small. The corresponding equilibrium equations are formulated in the framework of co-rotational description, using the virtual work principle. In contrast to conventional co-rotational formulation, linear on element level, in this paper an additional non-linear part of stiffness matrix is evaluated in order to model Wagner effect.

Keywords: *beam element, frames, large rotations, large displacements, creep*

1 Introduction

In the field of structural engineering beam columns and frames take a very important place, Alfutov (2000). Such structures display very complex structural behaviour and thus the development of advanced non-linear analysis tools has been a major activity of many structural engineering researchers in past years. An important consideration concerns the accurate prediction of their limit load-carrying capacity in the large displacement and large rotation regime, Galambos (1998). That imposes numerical modelling as adequate method because theoretical solutions are limited on cases of simple geometry.

Columns under sustained loads are generally unstable in the regime of creep. This means that loss of stability may occur during a period of exploitation of structure even for loads lower than critical buckling load, Bažant & Cedolin. (1991). Due to this reason stability is characterized by critical buckling time defined as load duration for which buckling deflections becomes infinitive. From that aspect any load can become long time critical buckling load while of course the larger load supposes the shorter time to collapse.
Firstly, geometrical non-linearities of an elastic, straight and prismatic beam member are treated. It is assumed that cross-section is not deformed in its own plane. Displacements and rotations are allowed to be large but strains are small. The shear strain in the middle surface can be neglected. Loading of a considering structure is static and conservative.

The large-displacement analysis is carried out through load deflection manner using incremental descriptions. It means that load-deformation path is divided into a number of sub-steps or increments and results in the form of a set of non-linear equilibrium equations of the structure, which should be solved using some incremental-iterative scheme. The corresponding equilibrium equations are formulated in the framework of co-rotational description, using the virtual work principle, Izzuddin & Elnashai (1993). In contrast to conventional co-rotational formulation, which is linear on element level and unable to model Wagner effect, an additional nonlinear part of stiffness matrix is evaluated and added to standard elastic stiffness.

On the basis of above-mentioned assumptions an own computer program BMCA is developed and its implementation is tested on several typical examples. Obtained solutions are for verification compared with theoretical results as well as with available finite element results of the other authors.

2 Beam kinematics

A right-handed Cartesian co-ordinate system (z, x, y) is chosen in such a way that axis z coincides with the beam axis passing through the centroid O of each cross-section, while the x- and y-axes are the principal inertial axes of the cross-section. Displacement measures of a cross-section are defined as

$$\begin{gathered} w_0 = w_0(z), \quad u_0 = u_0(z), \quad v_0 = v_0(z) \\ \varphi_z = \varphi_z(z), \quad \varphi_x = -\frac{\mathrm{d}v_0}{\mathrm{d}z} = \varphi_x(z), \quad \varphi_y = \frac{\mathrm{d}u_0}{\mathrm{d}z} = \varphi_y(z) \end{gathered} \tag{1}$$

where w_0, u_0 and v_0 are the rigid-body translations of the cross-section in the z-, x- and y-directions, respectively; φ_z, φ_x, and φ_y are the rigid-body rotation about the z-, x- and y-axis, respectively. All the measures are associated with the centroid of each cross-section.
If rotations are small in the local (co-rotational) coordinate system, the displacements of an arbitrary point on the cross-section defined by the position coordinates x and y and the warping function $\omega(x, y)$ can be expressed as

$$w(z,x,y) = w_0(z) - y\frac{\mathrm{d}v_0}{\mathrm{d}z}(z) - x\frac{\mathrm{d}u_0}{\mathrm{d}z}(z) - \omega(x,y)\frac{\mathrm{d}\varphi_z}{\mathrm{d}z}(z) \tag{2}$$

$$u(z,x,y) = u_0(z) - y\,\varphi_z(z) \tag{3}$$

$$v(z,x,y) = v_0(z) + x\,\varphi_z(z) \tag{4}$$

According to the St. Venant torsion theory, the term $\mathrm{d}\varphi_z/\mathrm{d}z$ in Eq.2. is equal to a constant.
The strain tensor contains only three components, i.e.

$$\begin{gathered} \varepsilon_z \approx \frac{\mathrm{d}w_0}{\mathrm{d}z} + y\frac{\mathrm{d}\varphi_x}{\mathrm{d}z^2} - x\frac{\mathrm{d}\varphi_y}{\mathrm{d}z^2} + \frac{x^2+y^2}{2}\left(\frac{\mathrm{d}\varphi_z}{\mathrm{d}z}\right)^2; \\ \gamma_{zx} - \left(y + \frac{\partial\omega}{\partial x}\right)\frac{\mathrm{d}\varphi_z}{\mathrm{d}z}; \quad \gamma_{zy} = \left(x - \frac{\partial\omega}{\partial y}\right)\frac{\mathrm{d}\varphi_z}{\mathrm{d}z} \end{gathered} \tag{5}$$

Since the co-rotational approach assumes small displacements in the local co-ordinate system, the first order approximation of the strain components is employed in the above equations and the only quadratic term occurs in Eq.8. as necessary for modelling Wagner's effect, Chen & Atsuta (1977).
Assuming the rigid in-plane deformations ($\sigma_x = \sigma_y = \tau_{xy} = 0$), then according to the engineering theories for bending and torsion, stress resultants can be defined as follows:

$$F_z = \int_A \sigma_z \, dA; \; F_x = \int_A \tau_{zx} \, dA; \; F_y = \int_A \tau_{zy} \, dA;$$
$$M_x = \int_A \sigma_z y \, dA; \; M_y = -\int_A \sigma_z x \, dA; \; M_z = \int_A \left(\tau_{zy} x - \tau_{zx} y\right) dA \tag{6}$$

where F_z represents an axial force, F_x and F_y are shear forces, M_x and M_y are bending moments with respect to x and y axes, respectively and M_z is a torsional moment. According to Hook's law: $\sigma_z = E e_{zz}$, $\tau_{zx} = G e_{zx}$, $\tau_{zy} = G e_{zy}$, the incremental force-displacement relations can be written as:

$$F_z = \int_A E \frac{dw_0}{dz} dA; \quad M_z = \int_A G \frac{d\varphi_z}{dz} \left[\left(x - \omega_y\right)^2 - \left(y + \omega_x\right)^2 \right] dA;$$
$$M_x = -\int_A E \frac{d^2 v_0}{dz^2} y^2 \, dA; \quad M_y = \int_A E \frac{d^2 u_0}{dz^2} x^2 dA \tag{7}$$

where ω_x and ω_y denotes partial derivation of warping function ω with respect to x and y.

3 Finite element formulation

Figure 1 shows a two-node beam finite element with six degrees of freedom, where (z, x, y) is the local Cartesian co-ordinate system, which in a co-rotational formulation continuously translates and rotates with the element as the deformation proceeds.

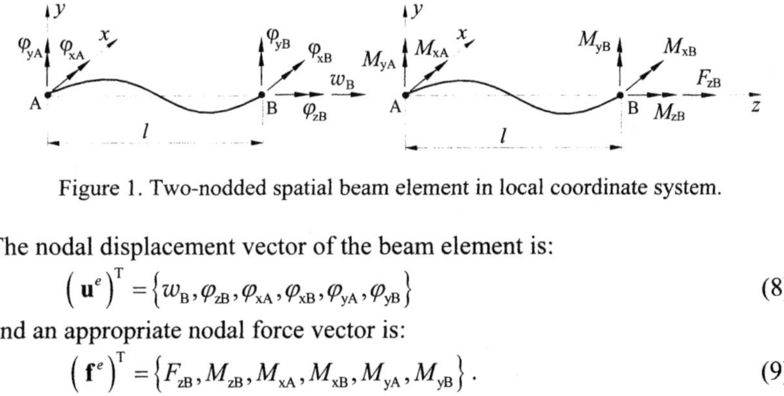

Figure 1. Two-nodded spatial beam element in local coordinate system.

The nodal displacement vector of the beam element is:

$$\left(\mathbf{u}^e\right)^{\mathrm{T}} = \left\{ w_{\mathrm{B}}, \varphi_{\mathrm{zB}}, \varphi_{\mathrm{xA}}, \varphi_{\mathrm{xB}}, \varphi_{\mathrm{yA}}, \varphi_{\mathrm{yB}} \right\} \tag{8}$$

and an appropriate nodal force vector is:

$$\left(\mathbf{f}^e\right)^{\mathrm{T}} = \left\{ F_{\mathrm{zB}}, M_{\mathrm{zB}}, M_{\mathrm{xA}}, M_{\mathrm{xB}}, M_{\mathrm{yA}}, M_{\mathrm{yB}} \right\}. \tag{9}$$

Incremental analysis supposes that a load-deflection path is subdivided into a number of steps or increments. This path is usually described using three configurations: the initial or undeformed configuration C_0; the last calculated equilibrium configuration C_1 and current unknown configuration C_2. Adopting co-rotational formulation, all system quantities should be referred to configuration C_2, Yang & Kuo (1994).

Applying the virtual work principle and neglecting the body forces, the equilibrium of a finite element can be expressed as, Turkalj et. al. (2004):

$$\delta \mathrm{U} = \delta \mathrm{W} \tag{10}$$

in which U is potential energy of internal forces, W is the virtual work of external forces, while δ is variation in.

After making the first variation of Eq.10., the following incremental equations can be obtain:

$$\delta \mathrm{W} = \left(\delta \mathbf{u}^e\right)^{\mathrm{T}} \mathbf{f}^e; \quad \delta \mathrm{U} = \left(\delta \mathbf{u}^e\right)^{\mathrm{T}} \mathbf{k}_{\mathrm{T}}^e \, \mathbf{u}^e \tag{11}$$

In Eq.11. $\mathbf{k}_{\mathrm{T}}^{\mathrm{e}}$ denotes the local tangent stiffness matrix of the e-th beam element, which can be evaluated acording to procedure explaned in Lanc et. al. (2007). Now the incremental equilibrium equation can be written in the following form:

$$\mathbf{k}_{\mathrm{T}}^{\mathrm{e}} \, \Delta \mathbf{u}^{\mathrm{e}} = \Delta \mathbf{f}^{\mathrm{e}} \tag{12}$$

The element global tangent stiffness matrix $\overline{\mathbf{k}}_{\mathrm{T}}^e$ can be obtained as follows:

$$\overline{\mathbf{k}}_{\mathrm{T}}^e = \mathbf{t}_1^e \mathbf{k}_{\mathrm{T}}^e \mathbf{t}_1^e + \mathbf{t}_2^e \mathbf{f}^e . \tag{13}$$

Matrices $\mathbf{t}_1^e$ and $\mathbf{t}_2^e$ are standard transformation matrices from local co-rotational to global coordinate system explained in Izzuddin (2001). Matrix $\mathbf{t}_1^e$ is of dimension 12×6 and contains first derivations of local with respect to global displacements while $\mathbf{t}_2^e$ is 12×12×6 matrix containing second derivations. Matrix $\mathbf{t}_2^e$ presents geometric stiffness contribution because it contains effects on global forces caused with change in geometry. Element force vector transformed from local to global coordinate system is:

$$\overline{\mathbf{f}}^e = \mathbf{t}_1^e \mathbf{f}^e . \tag{14}$$

After the standard assembling procedure, Turkalj & Brnić (2004), the overall incremental equilibrium equations can be obtained as:

$$\mathbf{K}_{\mathrm{T}} \mathbf{U} = \mathbf{P}, \quad \mathbf{K}_{\mathrm{T}} = \sum_{e} \overline{\mathbf{k}}_{\mathrm{T}}^e, \quad \mathbf{P} = {}^2\mathbf{P} - {}^1\mathbf{P}, \tag{15}$$

where $\mathbf{K}_{\mathrm{T}}$ is tangential stiffness matrix of a structure, while $\mathbf{U}$ and $\mathbf{P}$ are the incremental displacement vector and the incremental external loads of the structure. ${}^2\mathbf{P}$ and ${}^1\mathbf{P}$ are the vectors of external loads applied to the structure at C_2 and C_1 configurations, respectively.

4 Constitutive equations

Since the stress deviator tensor for the beam element is:

$$\mathbf{s}^{\mathrm{T}} = \left\{ s_{\mathrm{z}} \quad s_{\mathrm{zx}} \quad s_{\mathrm{zy}} \right\} = \left\{ 2\sigma_{\mathrm{z}}/3 \quad \tau_{\mathrm{zx}} \quad \tau_{\mathrm{zx}} \right\} \tag{16}$$

by assuming the isochoric-isothermal strains in the elastic-plastic-creep response, the constitutive equations for the beam element at configuration C_2 can be written in the following form, Kojić & Bathe (1987):

$${}^2s_{\mathrm{z}} = \frac{2E}{3}\left({}^2\varepsilon_{\mathrm{z}} - {}^2\varepsilon_{\mathrm{z}}^{\mathrm{c}}\right); \qquad {}^2\varepsilon_{\mathrm{z}}^{\mathrm{c}} = {}^1\varepsilon_{\mathrm{z}}^{\mathrm{c}} + \Delta\varepsilon_{\mathrm{z}}^{\mathrm{c}} \tag{17}$$

$${}^2s_{\mathrm{zx}} = G\left({}^2\gamma_{\mathrm{zx}} - {}^2\gamma_{\mathrm{zx}}^{\mathrm{c}}\right), \qquad {}^2\gamma_{\mathrm{zx}}^{\mathrm{c}} = {}^1\gamma_{\mathrm{zx}}^{\mathrm{c}} + \Delta\gamma_{\mathrm{zx}}^{\mathrm{c}} \tag{18}$$

$${}^2s_{\mathrm{zy}} = G\left({}^2\gamma_{\mathrm{zy}} - {}^2\gamma_{\mathrm{zy}}^{\mathrm{c}}\right), \qquad {}^2\gamma_{\mathrm{zy}}^{\mathrm{c}} = {}^1\gamma_{\mathrm{zy}}^{\mathrm{c}} + \Delta\gamma_{\mathrm{zy}}^{\mathrm{c}} \tag{19}$$

where ε_{ij}^{c} denotes the creep strain components, respectively.
The creep strain increment can be calculated as

$$\Delta\boldsymbol{\varepsilon}^{c} = {}^{1}k\,{}^{1}\mathbf{s} \tag{20}$$

where the factor ${}^{1}k$ is

$${}^{1}k = \frac{3}{2}\frac{{}^{1}\dot{\varepsilon}_{eq}^{c}}{{}^{1}\sigma_{eq}}\Delta t \tag{21}$$

with $\dot{\varepsilon}_{eq}^{c}$ as the effective creep strain rate. In the case of creep, configurations C_1 and C_2 are the time configurations, so a time increment Δt represents the real time passed during the element movement from configuration C_1 to configuration C_2, Bathe (1996).

$$\Delta t = {}^{2}t - {}^{1}t \tag{22}$$

The effective creep strain rate $\dot{\varepsilon}_{eq}^{c}$ occurring in Eq.21. can be obtained by using some of the creep laws. In the present study, the Norton power creep law

$$\dot{\varepsilon}_{eq}^{c} = K\sigma_{eq}^{n} \tag{23}$$

is adopted. In this, K and n are material constants determined from uniaxial tests at constant temperature conditions.

5 Examples

5.1 *Flexural creep buckling*

A cantilever with a 5×2 cm rectangular cross-section is axially loaded by a constant axial force F = 6000 N, Figure 2. The material moduli are E = 210 GPa, G = 80.77 GPa. The creep parameters are: $K = 0.58\cdot10^{-15}$, n = 3. The analysis is performed for the pure creep flexural buckling by using four beam elements, and a time increment of 50 h while 10 integration areas in the Y-direction are applied for cross sectional integration. A constant lateral perturbation force ΔF = 10 N is also added to initiate buckling. The same problem has been analyzed by Walczak et al. (1981), who applied eight 2-D finite elements, and whose results are given for comparison in Figure 3. together with those obtained by the program NASTRAN using 400 solid finite elements. As it can be seen, the critical time at which creep buckling occurs is reached approximately after $15\cdot10^{3}$ h.

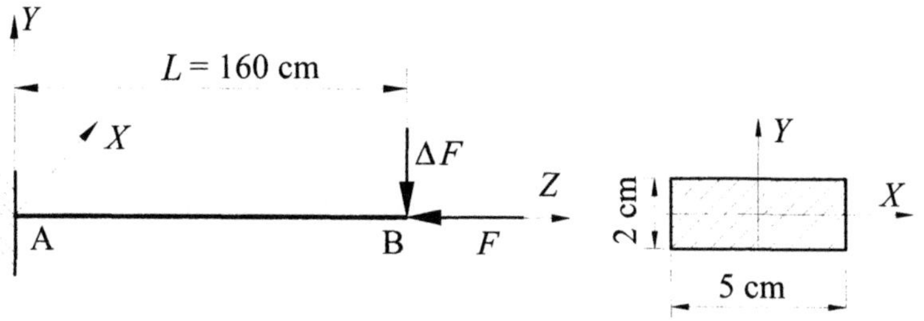

Figure 2. Axially loaded cantilever.

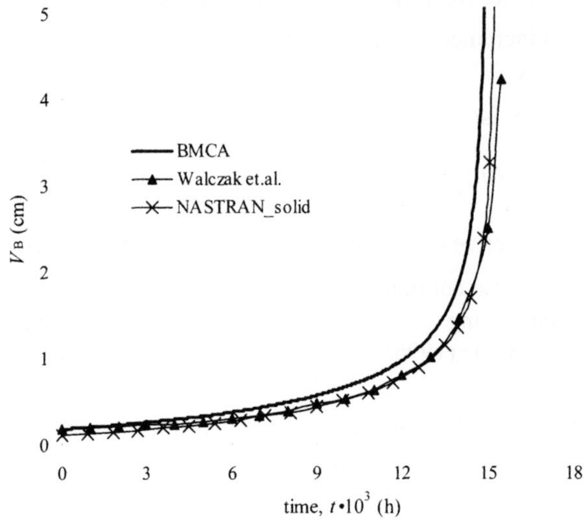

Figure 3. Creep flexural buckling of axially loaded cantilever ($F = 6000$ N).

5.2 *Torsional creep buckling*

The simulation is performed in the case of pure creep torsional buckling of an axially loaded cantilever with a square cross-section 0.1×0.1 m at a constant axial force $F = 550$ MN, Figure 4. The material moduli $E = 210$ GPa and $G = 80.77$ GPa. The creep parameters $K = 0.58 \cdot 10^{-15}$, $n = 3$ are adopted. The cantilever is modelled by 4 beam elements, while the cross section integration is performed by using a 20×20 integrational areas. To initiate the pure creep torsional buckling occurrence, a disturbing twisting moment $\Delta M = 0.001F$ is introduced. The elastic critical load is $F_{cr} = 673.92$ MN.

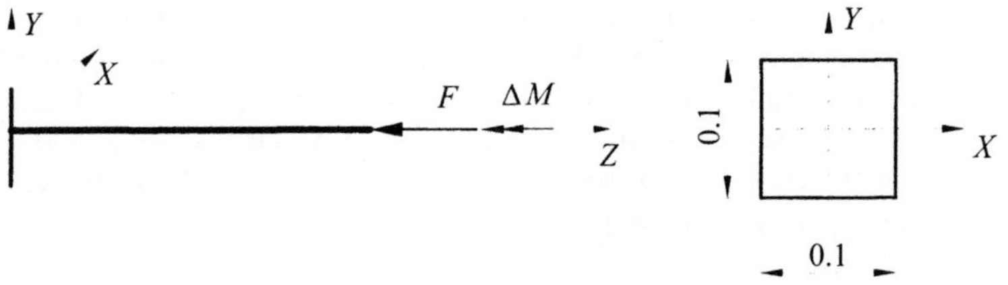

Figure 4. Cantilever from example 5.2.

Figure 5 shows the variation of twist angle vs. creep time using the two different time increments, i.e. $\Delta t = 10^4$ h and $\Delta t = 10^5$ h. As it can be seen, the creep buckling time $t_{cr} \approx 250 \cdot 10^4$ h. In this figure, the result obtained after neglecting Wagner's effect in equation (26) is also presented. As one can see, such a model is not capable to recognize the creep torsional instability.

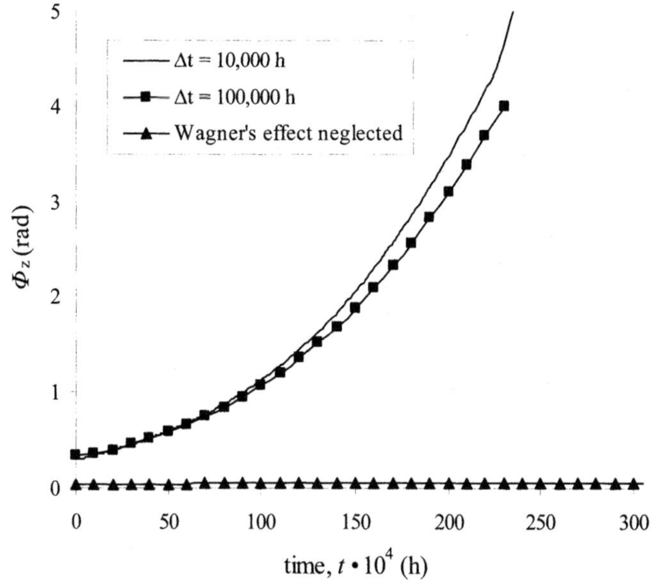

Figure 5. Creep torsional buckling of cantilever from example 5.2.

6 Conclusion

Co-rotational formulation for the non-linear analysis of beam columns in the regime of large displacements as well as creep material behaviour is proposed. The governing incremental equilibrium equations of a two-node space beam element are developed using the linearized virtual work principle and the non-linear cross-sectional displacement field accounting for Wagner effects. Applying the linear interpolation for axial and torsional displacement components and the cubic interpolations for flexural displacement components, the non-linear tangental stiffness matrix of beam element are obtained. Its importance is illustrated on pure torsional buckling example. Presented test examples suggest that developed numerical model is accurate tool for modelling beam structure non-linear behaviour.

Acknowledgement

The research presented in this paper was made possible by the financial support of the Ministry of Science, Education and Sports of the Republic of Croatia, under the project No. 069-0691736-1731.

References

Alfutov, N. A. (2000) *Stability of Elastic Structures*. Berlin: Springer-Verlag.

Bathe, K. J. (1996) *Finite Element Procedure.* London: Prentice-Hall.

Bažant, Z. P. & Cedolin, L. (1991) *Stability of structures: elastic, inelastic, fracture and damage theories*. Oxford: Oxford University Press.

Chen, W. F. & Atsuta, T. (1977) *Theory of Beam-Columns, Vol. 2.* New York: McGraw-Hill.

Galambos, T. V. (1998) *Guide to Stability Design Criteria for Metal Structures*. New York: John Wiley & Sons.

Izzuddin, B. A. & Elnashai, A. S. (1993) Eulerian formulation for large displacement analysis of space frames. *Journal of Engineering Mechanics* **119** 549-569

Izzuddin, B. A. (2001) Conceptual issues in geometrically nonlinear analysis of 3D framed structures. *Computer Methods in Applied Mechanics and Engineering* **191** 1029-1053

Kojić, M. & Bathe, K. J. (1987)T he Effective-Stress-Function algorithm for thermo-elasto-plasticity and creep. *International Journal for Numerical Methods in Engineering* **24** 1509-1532

Lanc, D., Turkalj, G. & Brnic, J. (2007) Finite-element model for creep buckling analysis of beam-type structures, *Communications in Numerical Methods in Engineering*, DOI: 10.1002/cnm.1004

Turkalj, G., Brnic, J. & Prpic Orsic, J. (2004) ESA formulation for large displacement analysis of framed structures with elastic-plasticity. *Computers & Structures* **82** 2001-2013

Turkalj, G. & Brnic, J. (2004) Nonlinear stability analysis of thin-walled frames using UL-ESA formulation. *International Journal of Structural Stability and Dynamics* **4** No.1, 45-67

Walczak, J. (1981) On creep buckling analysis of structures. *Computers & Structures* **13:** 683-689

Yang, Y. B. & Kuo, S.R. (1994) *Theory & Analysis of Nonlinear Framed Structures*. New York: Prentice Hall.

5.2 Plastic Limit and Shakedown Analysis of Elasto-plastic Steel Frames with Semi-rigid Connections

János Lógó[1], Sándor Kaliszky[1], Mohammed Hjiaj[2] and Majid Movahredi Rad[1]
[1]*Department of Structural Mechanics, Budapest University of Technology and Economics, H-1521, Budapest, HUNGARY, logo@ep-mech-me.bme.hu*
[2]*LGCGM, INSA de Rennes, 20 avenue des Buttes de Coësmes, 35043 Rennes cedex - France*

Abstract

The aim of this paper is to study the influence of the semi-rigid connections on the plastic behaviour of elasto-plastic steel frames subjected to dead load and quasi-static working loads. In the presented methods the static theorem of plastic limit analysis and the static theorem of shakedown analysis are applied and to control the plastic deformation bound on the complementary strain energy of residual forces is also used. The semi-rigid connections are represented by different elasto-plastic models (rigid, strong, medium, soft and pinned) which are incorporated in the elementary stiffness matrix of the beam elements. The presented methods are suitable for the construction of plastic limit and shakedown curves, respectively, which provide the plastic limit load and shakedown multipliers for different ratios of working loads and bound the safe domains against collapse and unrestricted plastic deformations, respectively. The numerical calculations show that the semi-rigid connection can influence significantly the plastic behaviour of steel structures.

Keywords: *semi-rigid connection, shakedown, limit analysis, mathematical programming*

1 Introduction

In simple calculations it is usually assumed that the connections between the beams and the columns of framed steel structures are either rigid or pinned. In reality, however, these connections are semi-rigid. Their complex behaviour has been investigated by experimental and analytical methods and has been described in several papers and standards [Eurocode 3 (1999), Iványi (2000)]. It was shown that the semi-rigid connections (SRC) might influence significantly the behaviour of structures. The aim of this paper is to study the influence of the SRC on the plastic behaviour of the elasto-plastic steel frames subjected to dead load and quasi-static working loads. In the methods to be presented the static theorem of limit analysis and the static theorem of shakedown analysis are applied and to control the plastic deformations bound on the complementary strain energy of residual forces are also used. The SRC are represented by different elasto-plastic models (rigid, strong, medium, soft and pinned), which are incorporated in the elementary stiffness matrix of the beam elements. The presented methods are suitable for the construction of plastic limit and shakedown curves, respectively, which provide the plastic limit load and shakedown multipliers for different ratios of working loads and bound the safe domains against collapse and unrestricted plastic deformations, respectively. The formulation of the problem yields to nonlinear mathematical programming, which is solved by the use of an iterative procedure. The investigation of the results of numerical examples shows, that the effect of the SRC can influence significantly the plastic behaviour of the steel frames. The parametric study is illustrated by the solution of an example.

In the paper the following notations are used:

$\mathbf{P}_d$: dead load ; $\mathbf{P}_1$, $\mathbf{P}_2$: Static working loads; $\mathbf{M}_h^e$, $\mathbf{M}_d^e$: Fictitious elastic moments calculated from the live and dead loads assuming that the structure is purely elastic; $\mathbf{Q}^r$, $\mathbf{M}^r$: residual internal forces and moments; $\mathbf{M}_d^p$, $\mathbf{M}_h^p$: plastic moments; $\overline{\mathbf{M}}^p$: limit moments of the SRC; W_0^h : allowable complementary strain energy of the residual forces; $\boldsymbol{\sigma}_y$, E : yield stress and Young's modulus; A_i, I_i, S_{0i} and ℓ_i : areas, moment of inertias of the cross-sections and length of the finite elements (*i=1,2,...,n*), respectively; $\overline{S}_j$: stiffness of the SRC; (*j=1,2,...,k*) is the number of SRC. They are subsets of (*i=1,2,...,n*), $\mathbf{F}, \mathbf{K}, \mathbf{G}, \mathbf{G}^*$: flexibility, stiffness, geometrical and equilibrium matrices, respectively.

Table 1. Loading combinations

H	*Multipliers*	*Loads*	*Limit multipliers*	
			Shakedown	*Plastic limit state*
1	$m_2 = 0$	$\mathbf{Q}_1 = \mathbf{P}_1$	m_{s1}	m_{p1}
2	$m_1 = 0$	$\mathbf{Q}_2 = \mathbf{P}_2$	m_{s2}	m_{p2}
3	$m_1 = 0.5m_2$	$\mathbf{Q}_3 = [0.5\mathbf{P}_1,\ (0.5\mathbf{P}_1 + \mathbf{P}_2),\ \mathbf{P}_2]$	m_{s3}	m_{p3}
4	$m_1 = m_2$	$\mathbf{Q}_4 = [\mathbf{P}_1,\ (\mathbf{P}_1 + \mathbf{P}_2),\ \mathbf{P}_2]$	m_{s4}	m_{p4}
5	$m_1 = 2m_2$	$\mathbf{Q}_5 = [2.0\mathbf{P}_1,\ (2.0\mathbf{P}_1 + \mathbf{P}_2),\ \mathbf{P}_2]$	m_{s5}	m_{p5}

2 Semi-rigid connections

The typical general behaviour of the SRC can be illustrated by a moment-rotation relationship shown in Figure 1. In this paper this relationship will be approximated by five different elasto-plastic models given in Figure 2. Here $\overline{M}^p$ is the plastic limit moment and $\overline{S}$ is the stiffness of the SRC, respectively. Their magnitudes can be assumed from the results of experiments. These models are incorporated in the elementary stiffness matrix of the beam elements.

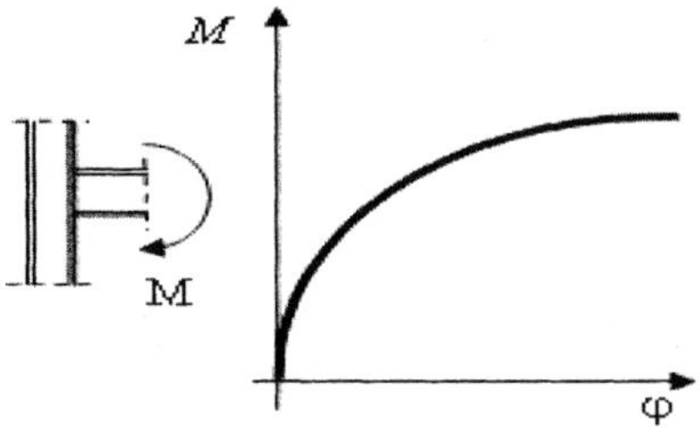

Figure 1. Real behaviour of the semi-rigid connection

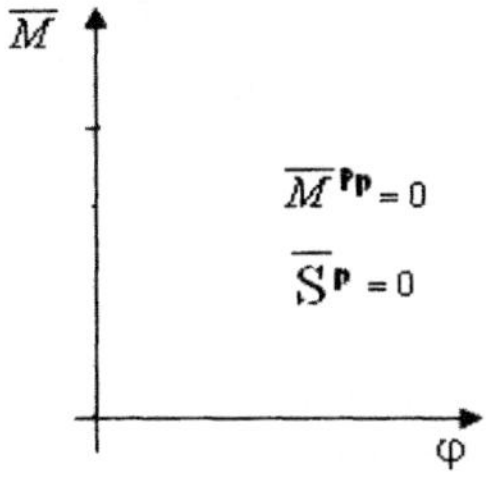

2.a, Pinned

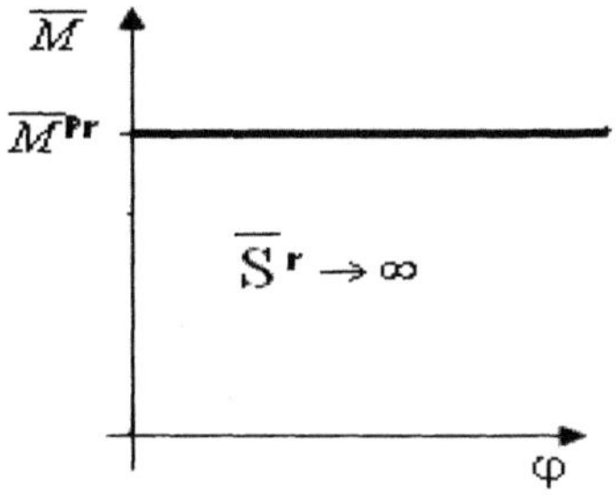

2.b, Rigid

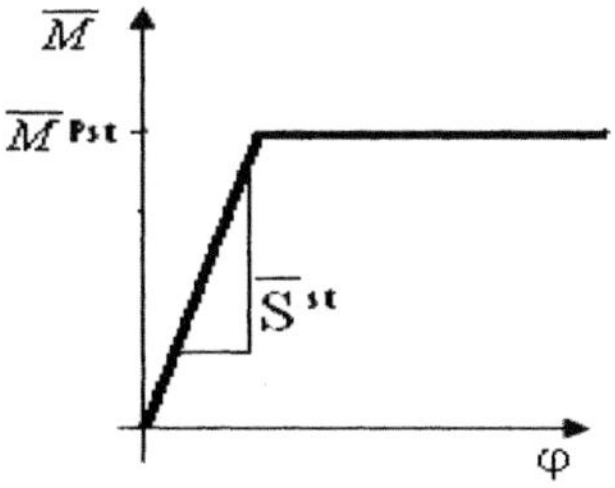

2.c, Strong

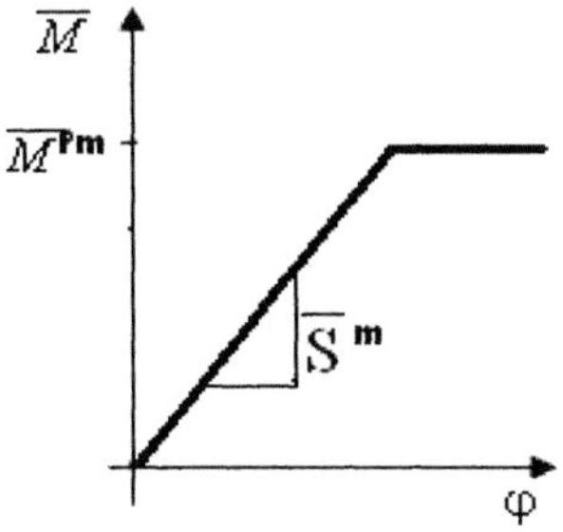

2.d, Medium

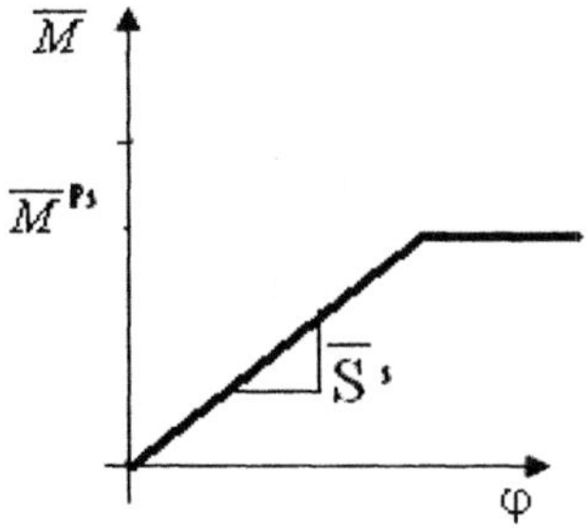

2.e, Soft

Figure 2.a-e Models of the semi-rigid connection

3 Loading: Plastic collapse and shakedown

The structure is subjected to a dead load $\mathbf{P}_d$ and two independent, static working loads $\mathbf{P}_1$ and $\mathbf{P}_2$ with multipliers $m_1 \geq 0$, $m_2 \geq 0$. In the analysis five loading cases (*h=1,2,...,5*) shown in Table 1 are taken into consideration. For each loading case a plastic limit load multiplier m_{ph} and a shakedown multiplier m_{sh} can be calculated. Making use of these multipliers two limit curves can be constructed in the m_1, m_2 planes (Figure 3). The structure does not collapse and shakes down, respectively, under the action of the loads $m_1\mathbf{P}_1$, $m_2\mathbf{P}_2$ if the points corresponding to the multipliers m_1, m_2 lies inside or on the plastic limit curve , respectively.

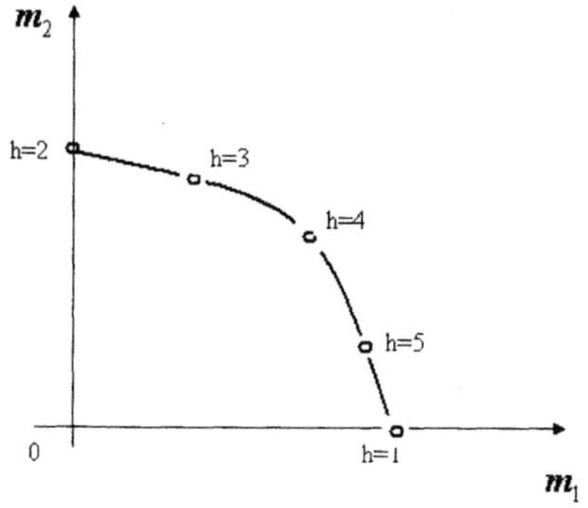

Figure 3. Limit curve and safe domain

4 Control of plastic deformations

At the application of the plastic analysis and design methods the control of the plastic behaviour of the structures is an important requirement. Since the shakedown analysis provides no information about the magnitude of the plastic deformations and residual displacements accumulated before the adaptation of the structure, therefore for their determination several bounding theorems and approximate methods have been proposed. Among others Kaliszky (1996), Kaliszky & Lógó (1995, 1997, 2002, 2006) suggested that the complementary strain energy of the residual forces could be considered an overall measure of the plastic performance of structures and the plastic deformations should be controlled by introducing a limit for magnitude of this energy:

$$\frac{1}{2}\sum_{i=1}^{n} \mathrm{Q}_i^r \mathbf{F}_i \mathrm{Q}_i^r \leq \mathrm{W}_{\mathrm{P0}} \tag{1}$$

This constraint can be expressed in terms of the residual moments M_{i1}^r and M_{i2}^r acting at the ends of the finite elements as follows:

$$\frac{1}{6\mathrm{E}}\sum_{i=1}^{n} \frac{\ell_i}{\mathrm{I}_i}\left[(\mathrm{M}_{i1}^r)^2 + (\mathrm{M}_{i1}^r)(\mathrm{M}_{i2}^r) + (\mathrm{M}_{i2}^r)^2\right] \leq \mathrm{W}_{\mathrm{P0}} \tag{2}$$

5 Limit analysis

The solution method based on the static theorem of limit analysis is formulated as below:

$$\text{Maximize} \quad m_{ph} \tag{4.a}$$

Subject to

$$\mathbf{G}^{*}\mathbf{M}_{d}^{p} + \mathbf{P}_{d} = \mathbf{0}; \tag{4.b}$$

$$\mathbf{G}^{*}\mathbf{M}_{h}^{p} + m_{ph}\mathbf{Q}_{h} = \mathbf{0}; \tag{4.c}$$

$$\mathbf{M}_{d}^{e} = \mathbf{F}^{-1}\mathbf{G}\mathbf{K}^{-1}\mathbf{P}_{d}; \quad \mathbf{M}_{h}^{e} = \mathbf{F}^{-1}\mathbf{G}\mathbf{K}^{-1}m_{ph}\mathbf{Q}_{h}; \tag{4.d}$$

$$-2\mathrm{S}_{0i}\sigma_{y} \le (\mathbf{M}_{di}^{p} + \max \mathbf{M}_{hi}^{p}) \le 2\mathrm{S}_{0i}\sigma_{y}, \ (i = 1, 2..., n); \tag{4.e}$$

$$-2\mathrm{S}_{0i}\sigma_{y} \le (\mathbf{M}_{di}^{p} + \min \mathbf{M}_{hi}^{p}) \le 2\mathrm{S}_{0i}\sigma_{y}, \ (i = 1, 2..., n); \tag{4.f}$$

$$-\overline{\mathbf{M}}_{j}^{p} \le (\mathbf{M}_{dj}^{p} + \max \mathbf{M}_{hj}^{p}) \le \overline{\mathbf{M}}_{j}^{p}, \ (j = 1, 2..., k); \tag{4.g}$$

$$-\overline{\mathbf{M}}_{j}^{p} \le (\mathbf{M}_{dj}^{p} + \min \mathbf{M}_{hj}^{p}) \le \overline{\mathbf{M}}_{j}^{p}, \ (j = 1, 2..., k); \tag{4.h}$$

$$\mathbf{M}_{hi}^{r} = \left[\left(\max \mathbf{M}_{hi}^{e} + \mathbf{M}_{di}^{e}\right)\right] - \left[\left(\max \mathbf{M}_{hi}^{p} + \mathbf{M}_{di}^{p}\right)\right], \ (i = 1, 2..., n); \tag{4.i}$$

$$\frac{1}{6\mathrm{E}}\sum_{i=1}^{n}\frac{\ell_{i}}{\mathrm{I}_{i}}\left[(\mathbf{M}_{hi1}^{r})^{2} + (\mathbf{M}_{hi1}^{r})(\mathbf{M}_{hi2}^{r}) + (\mathbf{M}_{hi2}^{r})^{2}\right] - \mathrm{W}_{0}^{h} \le 0. \tag{4.j}$$

This is a nonlinear mathematical programming problem which can be solved by any appropriate solution method (e.g. SPQL method). Selecting one of the semi-rigid connection models for each loading combination $\mathbf{Q}_h$; $(h = 1, 2, ..., 5)$ a plastic limit load multiplier m_{ph} can be determined and then the limit curve of the plastic limit state can be constructed.

6 Shakedown analysis

The solution method based on the static theorem of shakedown analysis is formulated as below:

$$\text{Maximize} \quad m_{sh} \tag{5.a}$$

Subject to

$$\mathbf{G}^{*}\mathbf{M}_{h}^{r} = \mathbf{0}; \tag{5.b}$$

$$\mathbf{M}_{d}^{e} = \mathbf{F}^{-1}\mathbf{G}\mathbf{K}^{-1}\mathbf{P}_{d}; \quad \mathbf{M}_{h}^{e} = \mathbf{F}^{-1}\mathbf{G}\mathbf{K}^{-1}m_{sh}\mathbf{Q}_{h}; \tag{5.c}$$

$$-2\mathrm{S}_{0i}\sigma_{y} \le (\mathbf{M}_{di}^{e} + \mathbf{M}_{hi}^{r} + \max \mathbf{M}_{hi}^{e}) \le 2\mathrm{S}_{0i}\sigma_{y}, (i = 1, 2..., n); \tag{5.d}$$

$$-2\mathrm{S}_{0i}\sigma_{y} \le (\mathbf{M}_{di}^{e} + \mathbf{M}_{hi}^{r} + \min \mathbf{M}_{hi}^{e}) \le 2\mathrm{S}_{0i}\sigma_{y}, \ (i = 1, 2..., n); \tag{5.e}$$

$$-\overline{\mathbf{M}}_{j}^{p} \le (\mathbf{M}_{dj}^{e} + \mathbf{M}_{hj}^{r} + \max \mathbf{M}_{hj}^{e}) \le \overline{\mathbf{M}}_{j}^{p}, \ (j = 1, 2..., k); \tag{5.f}$$

$$-\overline{\mathbf{M}}_{j}^{p} \le (\mathbf{M}_{dj}^{e} + \mathbf{M}_{hj}^{r} + \min \mathbf{M}_{hj}^{e}) \le \overline{\mathbf{M}}_{j}^{p}, \ (j = 1, 2..., k); \tag{5.h}$$

$$\frac{1}{6\mathrm{E}}\sum_{i=1}^{n}\frac{\ell_{i}}{\mathrm{I}_{i}}\left[(\mathbf{M}_{hi1}^{r})^{2} + (\mathbf{M}_{hi1}^{r})(\mathbf{M}_{hi2}^{r}) + (\mathbf{M}_{hi2}^{r})^{2}\right] - \mathrm{W}_{p0}^{h} \le 0. \tag{5.i}$$

This is a non-linear mathematical programming problem, which can be solved by any appropriate solution method (e.g. SPQL method). Selecting one of the semi-rigid connection models for each loading combination $Q_h;(h=1,2,...,5)$ a shakedown multiplier m_{sh} can be determined and then the limit curve of shakedown can be constructed.

7 Example

The application is illustrated by an example shown in Figure 4. At the joints 2 and 4 the portal frame has SRC. The working loads are $F_1 = 10\ kN$, $F_2 = 15\ kN$ and $F_d = 0$.

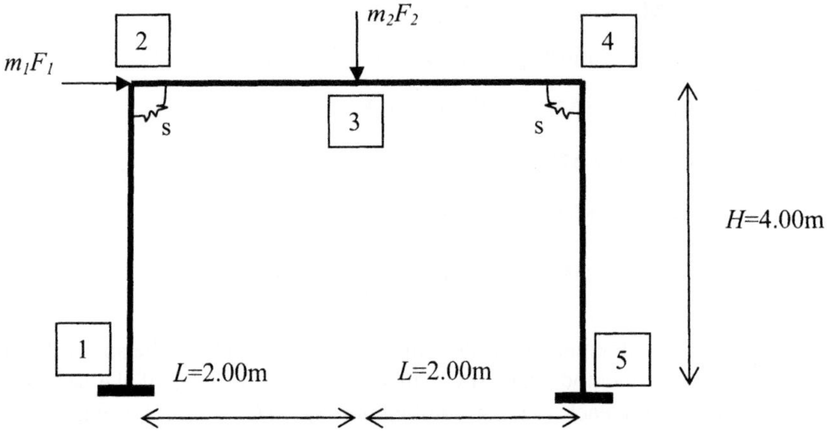

Figure 4. Portal frame as test problem

The yield stress and the Young's modulus are $\sigma_y = 21\ kN/cm^2$ and $E = 2.07 \cdot 10^6\ kN/cm^2$. The cross-sectional data of the beam are: $A_B = 28.5\ cm^2$, $I_B = 1943\ cm^4$, $S_{B0} = 130.0\ cm^3$, while for the columns $A_c = 39.0\ cm^2$, $I_c = 3891.6\ cm^4$, $S_{c0} = 210.0\ cm^3$.

The results of the solution are presented in Figure 5. The thin lines represent the solution of the shakedown while the thick ones belong to the limit analysis. As it is seen the stiffnesses of the SRC influence significantly the plastic behaviour of the frame. In the trivial case when the joints 2 and 4 have pinned connections the frame is statically determinate and the forces $m_1\mathbf{P}_1$ and $m_2\mathbf{P}_2$ have no interactions.

8 Conclusions

In the paper the semi-rigid behaviour is described by appropriate models and to control the plastic behaviour of the structure bound on the complementary strain energy of the residual forces is applied. Limit curves are presented for the plastic limit load and shakedown multipliers. The numerical analysis shows that the stiffness of the semi-rigid connections can influence significantly the magnitude of the plastic limit load and shakedown multipliers and in some cases the results are very sensitive on the stiffness of the semi-rigid connections The presented investigation drowns the attention to the importance of the problem but further investigations are necessary to make more general statements.

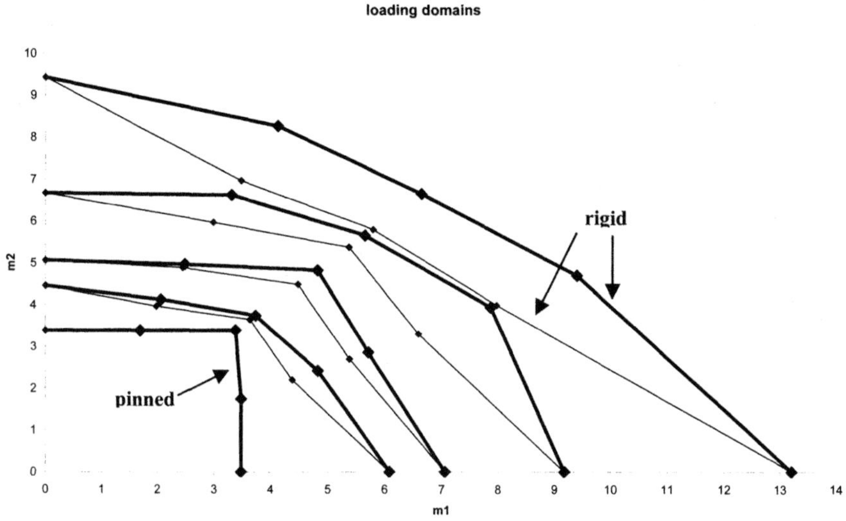

Figure 5. Maximum loading domains for limit analysis and shakedown

Acknowledgements

The present study was supported by The Hungarian National Scientific and Research Foundation (OTKA) (grant K 62555) and by Ministry of Education (TET grant F-44/05)

References

Eurocode 3 (1999) Design of Steel Structures, *Commission of the European Communities.*

Iványi, M. (2000) Semi-rigid Connections in Steel Frames, Part I. In *Semi-Rigid Connections in Structural Steelwork,* ed. by Iványi, M. & Baniotopoulos, C.C., CISM Courses and Lectures No. 419. Springer-Wien New York, 1-101.

Kaliszky, S. & Lógó, J. (1995) Elasto-Plastic Analysis and Optimal Design with Limited Plastic Deformations and Displacements, In: *Structural and Multidisciplinary Optimization, ed. by N. Olhoff, G.I.N. Rozvany, Pergamon Press,* 465-470.

Kaliszky, S. (1996) Elasto-plastic Analysis with Limited Plastic Deformations and Displacements, *Journal of Mechanics of Structures and Machines,* **24** No. 1, 39-50.

Kaliszky, S. & Lógó, J. (1997) Optimal Plastic Limit and Shakedown Design of Bar Structures with Constraints on Plastic Deformation, *Engineering Structures*, **19** No. 1, 19-27.

Kaliszky, S. & Lógó, J. (2002) Layout and Shape Optimization of Elastoplastic Disks with Bounds on Deformation and Displacement, *Journal of Mechanics of Structures and Machines*, **30** No.2, 177-191.

Lógó, J., Kaliszky, S. & Hjiaj, M. (2006) A Parametric Survey of the Influence of the Semi-rigid Connections on the Shakedown of Elasto-plastic Frames, *Periodica Polytechnica Civil Engineering*, **50** No.2,139 -147.

5.3 Welded Triangular Haunch for the Seismic Improvement of Steel Beam-to-column Connections

Marco Valente
Dipartimento di Ingegneria Strutturale, Politecnico di Milano, Milano, Italia, valente@stru.polimi.it

Abstract

The effectiveness of using welded haunch for the seismic improvement of steel welded beam-to-column connections was studied by numerical analyses. Finite element models of steel subassemblages tested previously at the Politecnico di Milano were developed with the addition of a triangular haunch beneath the beam bottom flange and comparative analyses were carried out. The haunch was aimed to protect the potentially vulnerable beam-to-column groove welded joint by relocating the plastic hinge away from the face of the column. Numerical results showed that the use of welded haunch significantly reduced strain demands at the beam flange weld and a strain concentration was observed at the haunch tip. The effects of a few local details, such as double haunch, beam web vertical stiffeners, continuity plates and doubler plates reinforcing the column panel zone, were investigated in the haunch models.

Keywords: *welded beam-to-column connections, haunch, finite element model*

1 Introduction

Experimental results from cyclic testing of full-scale specimens carried out in the past years within the SAC Steel Project in the USA demonstrated that welding a triangular haunch beneath the beam bottom flange could improve significantly the seismic performance of steel moment connections. To gain more insight into the behaviour of welded haunch connections, bare steel beam-to-column subassemblages tested previously at the Politecnico di Milano were reinforced with welded haunch and analyzed using the general-purpose finite element analysis program ABAQUS, (Valente & Castellani 2002). The main purpose of this work was to study the stress and strain distribution in beam-to-column connections reinforced with welded haunch and to compare the results to the case of unreinforced connections. The effectiveness of using welded haunch for the seismic improvement of steel beam-to-column connections was evaluated. Fig. 1 shows the details of a typical welded beam-to-column connection with welded haunch.

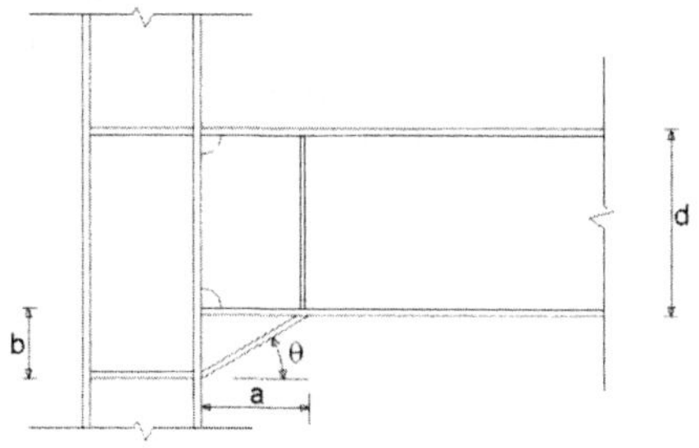

Figure 1. Typical scheme of the haunch connection

2 Finite element models

Steel welded beam-to-column subassemblages tested at the Politecnico di Milano were modelled and different finite element models including welded triangular haunch were generated from the unreinforced model. High stress and strain concentration regions that could lead to potential failure were identified.

The models were one-sided steel beam-to-column subassemblages representative of exterior beam-to-column connections consisting of a column (HEB300) with one beam (IPE450) welded to the column flange. Finer meshes were arranged close to the joint area, as shown in Fig. 2. The continuity plates, the weld access hole, the doubler plates were included in the models. The connection was reinforced by triangular haunch at the bottom beam flange. Fig. 2 shows different details of the models of the beam-to-column subassemblages analyzed in this study in order to evaluate the effects of local details on the performance of the joints.
The web and the flange of the haunch were supposed welded to the column flange and the beam flange. The haunch sizes were dimensioned according to the prescriptions of the Eurocode 3 and the AISC Code. It is suggested that the length a and angle θ of the haunch, Fig. 1, be taken as:
$a = (0.5 \text{ to } 0.6)d$
$\theta = 30^{\circ} \pm 5^{\circ}$
The haunch web and flange thickness were taken equal, respectively, to the beam web and flange. The haunch length was 270 mm and the haunch height was 156 mm. A pair of beam web stiffeners was provided at the end of the haunch. In the models the beam and column flange and the haunch were joined by constraining the nodes at common locations to have identical displacements. The haunch detail was designed to relocate the plastic hinge beyond the edge of the haunch and away from the face of the column. Material properties used for the models were taken from tensile coupon tests. The plasticity model used in the analyses was based on a von Mises yield surface and an associated flow rule. The plastic hardening was defined using a non-linear kinematic hardening law. In the experimental apparatus the column was in the horizontal position while the beam was vertical. In the numerical model two lines of nodes at each end of the column were restrained against translation to replicate approximately the support conditions used for the laboratory tests. Prescribed displacements of equal amount up to 100 mm but opposite in sign were imposed to the tip of the beam. Analyses were controlled by the interstory drift angle, which was defined as the displacement imposed at the tip of the beam divided by the beam span to the centreline of the column. The analysis was able to simulate both local buckling and lateral–torsional buckling in the beam. The base model was validated in terms of hysteresis loops and peak values of the force by the results from experimental tests performed at the Politecnico di Milano within the European Steelquake Project, (Valente & Castellani 2002).

2.1 *Stress and strain indices*

Two stress and strain indices were computed from the ABAQUS results to assess the effectiveness of the welded haunch in enhancing the seismic performance of steel beam-to-column connections.

Pressure Index: This index is defined as the ratio of the hydrostatic stress, σ_m, to the yield stress, σ_y, of the material:

$$\text{PRESS Index} = \frac{\sigma_m}{\sigma_y}$$

The hydrostatic stress, σ_m, is defined as follows:

$$\sigma_m = -\frac{1}{3}trace(\sigma_{ii})$$

where σ_{ii} are the Cauchy stress components (i ranges from 1 to 3). The Pressure Index and the hydrostatic stress are negative in value under tensile forces. Large principal stresses are usually accompanied by high hydrostatic stresses. High principal stresses can result in a large stress intensity factor at crack tips, which will increase the potential for brittle facture.

PEEQ Index: This index is defined as the ratio between the equivalent plastic strain, PEEQ, and the yield strain, ε_y:

$$\text{PEEQ Index} = \frac{\text{PEEQ}}{\varepsilon_y}$$

where the equivalent plastic strain is defined as:

$$PEEQ = \sqrt{\frac{2}{3}\varepsilon_{ij}^{p}\varepsilon_{ij}^{p}}$$

where ε_{ij}^{p} is the plastic strain components in the i and j directions. The PEEQ Index can be used as a measure of plastic strain demand.

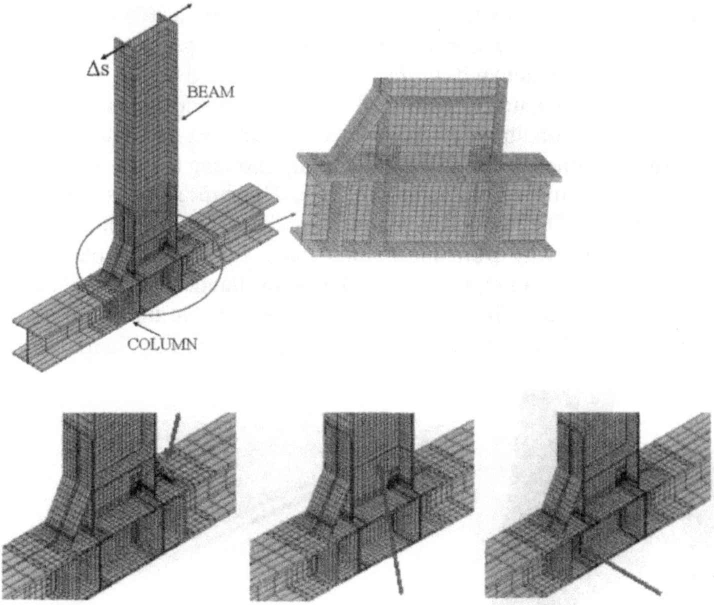

Figure 2. Finite element models of the beam-to-column subassemblage and different details of the haunch connection: double haunch, beam web stiffener, doubler plates and continuity plates.

3 Effects of welded haunch

Monotonic and cyclic non-linear static analyses were performed on the models reinforced with welded triangular haunch. A comparison with the unreinforced (standard) model was carried out. The haunch model provided a slight increase in initial elastic stiffness and strength as compared with the standard model, Fig. 3.

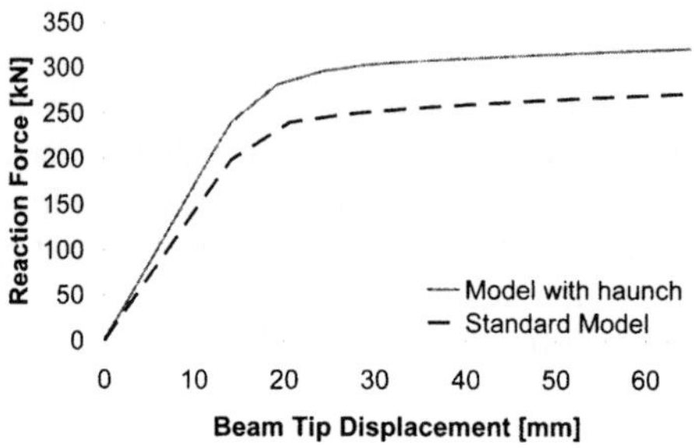

Figure 3. Reaction force-beam tip displacement for negative bending (haunch in compression)

The presence of the welded haunch changed the beam shear force transfer mechanism. Conventional beam theory cannot provide a reliable prediction of stress distributions in the haunch connection. Numerical results showed that the majority of the beam shear was transferred to the column through the haunch flange rather than through either the beam web or the beam flange welds. The reduction of the beam shear in the beam flange was one of the contributory factors for the improved performance of the haunch model. Providing sufficient axial stiffness and strength to the haunch flange, the force demand in the bottom flange weld was significantly reduced and the force demand in the top flange weld could be reduced to a reasonable level. The haunch web had a minor effect on the flexural stress distribution in the beam, but it was needed to stabilize the haunch flange. The welded triangular haunch featured a strut action, allowing for the beam shear to be transferred to the column via the haunch flange. The strut action of the haunch was identified in Fig. 4. The triangular haunch model effectively moved the plastic hinging of the beam away from the column face, Fig. 4. High plastic strain and stress concentration was found to occur in the beam flange at the intersection of the haunch toe with the beam. The first yielding was detected at the compression flange near the haunch toe followed by tensile yielding of the beam top flange and web.

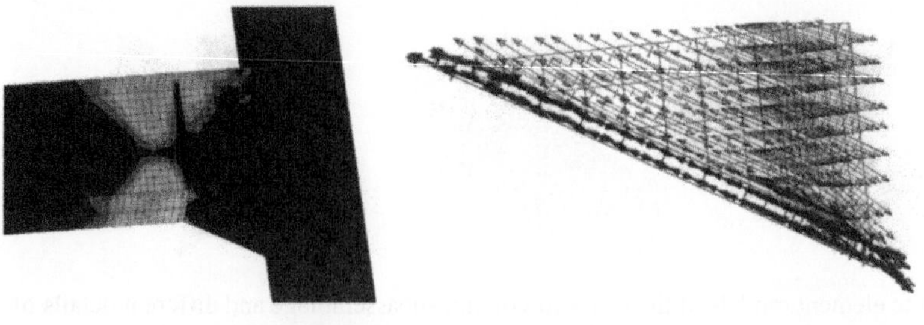

Figure 4. Plastic strain demand at the end of the haunch and principal stress distribution in the haunch

The distribution of the PEEQ Index along the beam flange width at the column interface is shown in Fig. 5 for the two models. Numerical results confirmed the effectiveness of the welded haunch in order to reduce the plastic strain concentration at the tensile beam flange, respect to the unreinforced connection. The standard model had a maximum PEEQ Index close to 55; with the addition of the haunch, the maximum PEEQ Index was reduced to approximately 10.
Fig. 5 shows the comparison between the two models of the PRESS Index across the beam flange width at the groove weld location. With the introduction of the welded haunch, significant reduction, from 0.8 to approximately 0.6, in the PRESS Index was registered at the column interface.

A considerable decrease of the values of the PEEQ Index near the beam weld access hole is shown in Fig. 6. Previous experimental and numerical studies (Valente & Castellani, 2002) demonstrated that the presence of the weld access hole could cause stress and strain concentration and improved geometry details of the hole were investigated. In the haunch connection the presence and the geometry of the weld access hole become less important because the value of the stress and strain indices are low in this region.

The peak value of the PEEQ Index in the haunch model was smaller than the peak value registered in the unreinforced connection. In particular the values of the PEEQ Index decrease significantly near the weld respect to the unreinforced model.
The critical sites for the two models were different: in the haunch model the maximum value of the PEEQ Index was registered at the edge of the haunch in the beam flange, while in the unreinforced model the maximum value was at the root of the weld access hole. With the addition of the haunch, the heat affected zone and the weld access hole region were less sensitive to low-cycle failures.

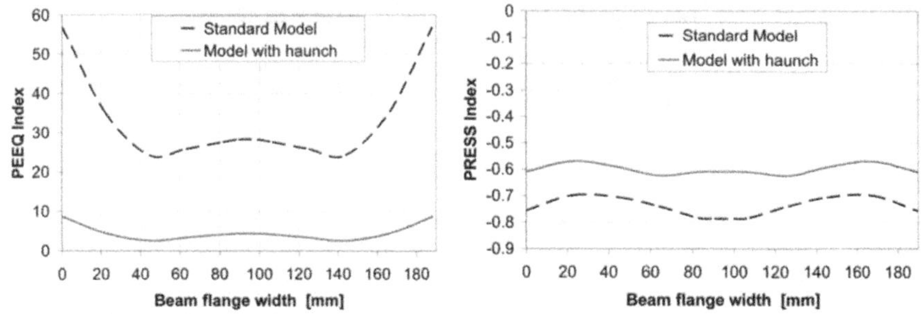

Figure 5. PEEQ and PRESS Index distribution along the beam flange width at the column interface

4 Effects of local details

Parametric analyses were carried out and different models were generated in order to evaluate the influence of different local details on the performance of beam-to-column connections.

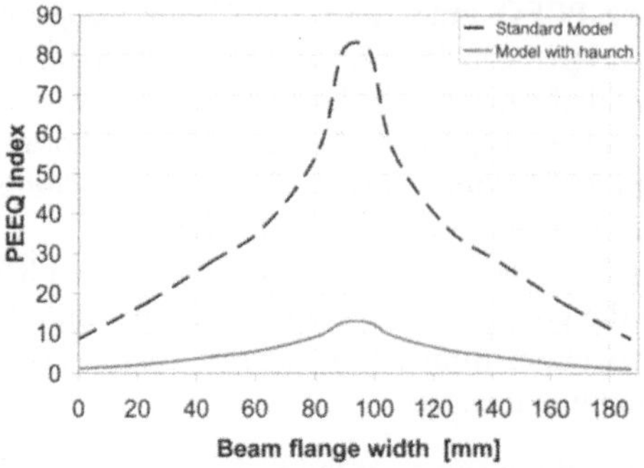

Figure 6. PEEQ Index distribution along the beam flange width near the weld access hole region

4.1 *Doubler plates*

In the haunch model the doubler plates reinforcing the panel zone were deleted to investigate the effect of the panel zone flexibility on the local response near the welds joining the beam flange to the column flange. In the model without doubler plates an increase of the PEEQ and PRESS Index was observed at the column interface, Fig. 7, while a decrease of the PEEQ Index at the edge of the haunch was registered, Fig. 8. The contribution of panel zone deformation to the drift of the model reduced the plastic strain demands on the beam at the edge of the haunch.

An increase of the shear stress along the haunch depth at the column interface was registered. In the absence of flange and web local buckling, the critical section of the model without doubler plate was the beam flange at the column interface.
Moreover, numerical analyses showed that, when a haunch was added beneath the beam, the panel zone was enlarged and a dual panel zone was formed.

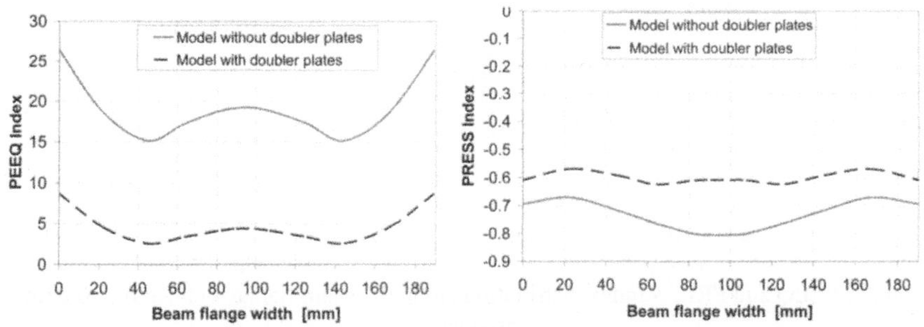

Figure 7. PEEQ and PRESS Index distribution along the beam flange width at the column interface

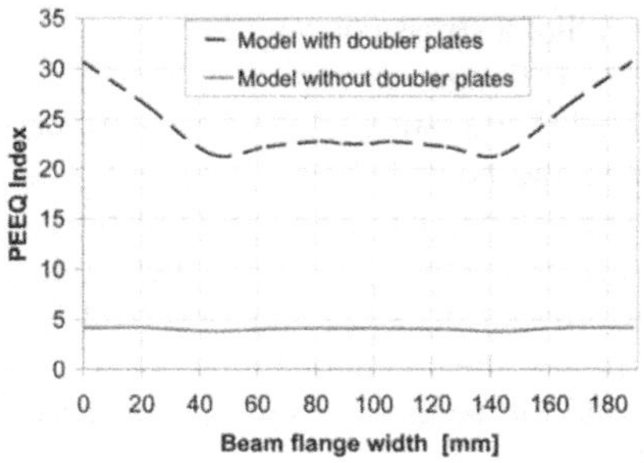

Figure 8. PEEQ Index distribution along the beam flange width at the end of the haunch

4.2 *Double haunch*

In this study the haunch was used for the bottom flange only or for both the top and bottom flange. The addition of a welded haunch at the beam top flanges was considered as a method for reinforcing the existing beam top flange groove weld. With top and bottom haunches, the PEEQ and Press Index decreased significantly at the column interface, Fig. 9, and a slight decrease was observed at the haunch end, Fig.10. The largest stress and strain demands did not occur at the beam flange weld.

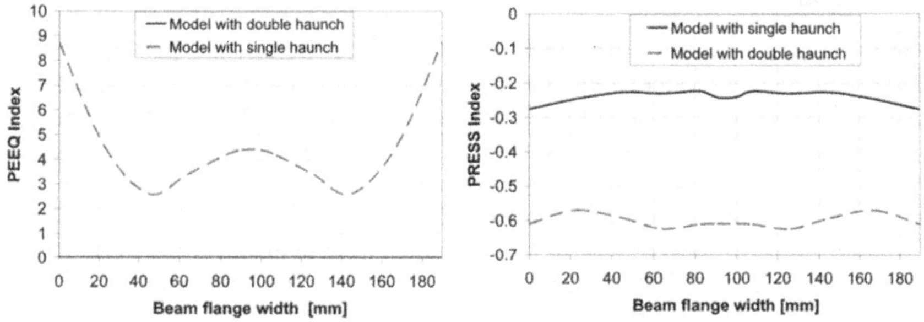

Figure 9. PEEQ and PRESS Index distribution along the beam flange width at the column interface and at the end of the haunch for the two numerical models

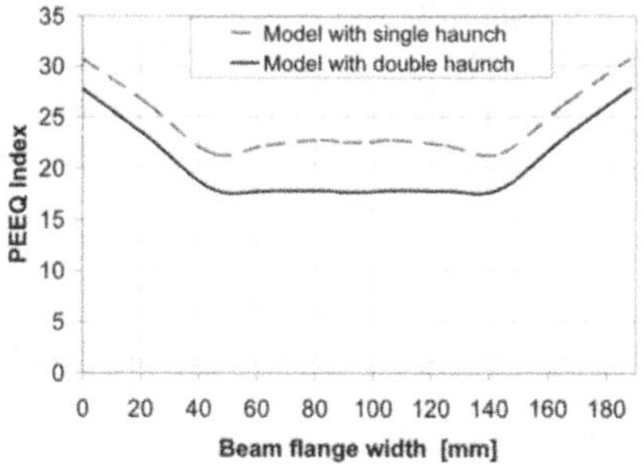

Figure 10. PEEQ Index distribution along the beam flange width at the end of the haunch

4.3 *Beam web vertical stiffeners*

Numerical results showed that a stress concentration was localized at the haunch toe section, where the haunch flange exerted a concentrated force on the beam. Therefore, it was suggested that a pair of transverse stiffeners be added to the beam web at the location where the haunch flange intersects the beam to distribute the vertical forces into the beam web. The analyses confirmed that the haunch toe can be strengthened effectively by means of web stiffeners extended to the full depth of the beam. No lateral distortional buckling was observed in the models. The use of full-depth stiffeners increases the likelihood that local buckling of the beam top flange would occur outside the haunch region, not next to the column face, at the location of groove weld.

4.4 *Continuity plates*

The effects of the presence or absence of the continuity plates in the column were investigated. Numerical results confirmed the use of a pair of continuity plates at the beam top flange level to reduce the stress concentration in the groove weld. Moreover, a pair of continuity plates should be added at the location where the haunch flange intersects the column. Numerical results indicated that continuity plates were not needed at the beam bottom flange level because flexural stress level was low.

5 Conclusions

The effectiveness of the welded haunch in enhancing the seismic performance of steel beam-to-column connections was demonstrated by numerical analyses. Finite element models were developed to investigate the stress and strain distribution near the connection region. The presence of a welded haunch changed the beam shear force transfer mechanism. The welded triangular haunch featured a strut action, allowing for the beam shear to be transferred to the column via the haunch flange. The welded haunch was effective in reducing the strain demand in the beam flange preventing weld failure. The triangular haunch model effectively moved the plastic hinging of the beam away from the region near the weld. High plastic strain and stress concentration was found to occur in the beam flange at the intersection of the haunch toe with the beam. Haunch toe was strengthened effectively by means of web stiffener to the full depth of the beam. Continuity plates had to be added at the location where the haunch flange intersects the column, while were not needed at the beam bottom flange level because of low flexural stress in that area. The double haunch was effective in reducing the plastic strain demand at the beam flange. The plastic deformation demand on the beam at the edge of the haunch was reduced due to yielding of the panel zone.

References

HKS. ABAQUS user's manual. Version 6.5. Providence (RI): Hibbit, Karlsson & Sorensen, Inc.; 2004.

Valente M., Castellani A. (2002) Cyclic behaviour and weld access hole detailing of steel welded beam-to-column joints. In: *Proc. of 3rd International Eurosteel Conference, Coimbra, Portugal.*

Valente M. (2008) Composite welded beam-to-column subassemblages reinforced with triangular and straight haunch. In: *Proc. Int. Conf. DFE 2008, Miskolc, Hungary.* Horwood Publishers, Proceedings pp. 253-260.

5.4 Composite Welded Beam-to-column Subassemblages Reinforced with Triangular and Straight Haunch

Marco Valente
Dipartimento di Ingegneria Strutturale, Politecnico di Milano, Milano, Italia, valente@stru.polimi.it

Abstract

The influence of the concrete slab on the cyclic performance of beam-to-column connections reinforced with welded haunch at the beam bottom flange was investigated. Finite element models of beam-to-column subassemblages with and without concrete slab were developed and analyzed under imposed cyclic displacements to investigate the stress and strain distribution in the connection region. The effectiveness of using welded triangular haunch technique for seismic improvement of composite beam-to-column connections was evaluated. A comparison between triangular and straight haunch was carried out and the soundness of a proposed scheme aiming to minimize stress concentration at the straight haunch tip was studied.

Keywords: *composite welded beam-to-column connections, concrete slab, triangular haunch, straight haunch, finite element model*

1 Introduction

Field observations on steel frames during recent earthquakes and results from experimental tests showed that composite beam-to-column connections subjected to seismic action may suffer premature failure near the weld at the beam bottom flange. Extensive research was conducted to propose methods to improve and strengthen composite joints. One scheme consists of the addition of a haunch at the bottom of the beam; it requires no modification to the beam top flange, minimizing the need to remove or alter the floor slab. In this study the effects of the concrete slab on the cyclic performance of beam-to-column connections reinforced with welded haunch were examined through a series of numerical analyses. The effectiveness of using welded haunch technique for seismic improvement of composite beam-to-column connections was investigated. Triangular and straight haunch models were compared and a solution aiming to minimize stress concentration at the straight haunch tip was studied.

2 Finite element models

Numerical models were used to investigate local behaviour and local ductility demand near the weld in the lower beam flange of composite beam-to-column connections. Three-dimensional finite element models created using Abaqus code were employed for the numerical analyses. The models represented exterior connections between an IPE450 beam and a HEB300 column attached by welding.
Fig. 1 shows the numerical models of the bare steel and composite subassemblages, called respectively M and MS. With the exception of the addition of the concrete slab in the composite model, the two models were identical. Other two numerical models, called MH and MHS, were created introducing a welded haunch at the beam bottom flange. The haunch was designed by following the design procedure

according to the AISC seismic provisions and the Eurocode 3 specification. Following the capacity design concept, the aim of haunch strengthening was to effectively move the plastic hinging of the beam outside the haunch region such that the haunch region remained essentially elastic. The model named M was validated by the experimental tests performed at the Politecnico di Milano within the European Steelquake Project, (Valente & Castellani, 2002). Based on this satisfactory correlation study, the finite-element models were then used to compute the stress and strain distributions of the welded haunch connection. Four basic models representing bare steel and composite subassemblages were analyzed, Fig. 1. Eight-node solid elements were used for the bare steel connection and the concrete slab. The beam web and flange were directly connected to the column flange in the model. Multi-linear isotropic hardening model was used for the steel beam and the steel column based on material test results.

A concrete slab was modelled using the damaged plasticity model, which provides a general capability for the analysis of concrete structures, based on a damaged plasticity algorithm. The expression proposed by Carreira & Chu (1985) to describe the uniaxial compressive stress-strain curve of concrete was used:

$$f_c = \frac{f_c'\beta\left(\frac{\varepsilon}{\varepsilon_c'}\right)}{\beta - 1 + \left(\frac{\varepsilon}{\varepsilon_c'}\right)^{\beta}} \qquad \varepsilon \le \varepsilon_u$$

where ε_c' is the strain corresponding to the maximum stress f_c' and

$$\beta = \left(\frac{f_c'}{32.4}\right)^3 + 1.55$$

(all strength units in MPa). The strain at which maximum compressive stress occurs is taken as 0.0022, while the stress at which the concrete crushes is usually about 0.0035. The tension behaviour of concrete was modelled by using a fracture energy cracking criterion. Reinforcing bars were modelled and embedded in the concrete slab mesh. The shear connection was modelled in a discrete manner using non-linear connector elements. Monotonously increasing and cyclic displacements up to 100 mm were applied at the tip end of the beam, reproducing the test setup and the displacement history of the experimental tests carried out on the bare models.

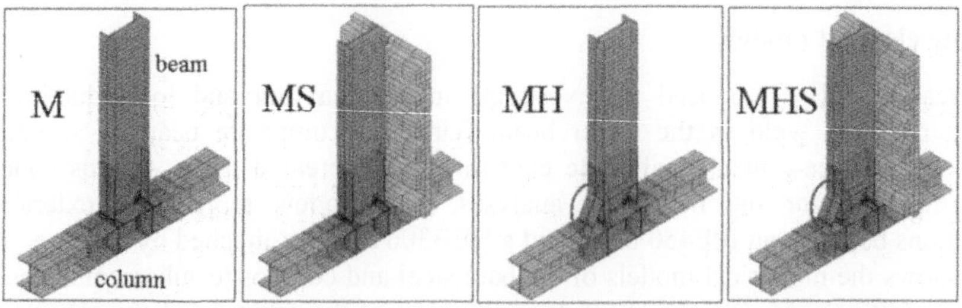

Figure 1. Finite element models of the beam-to-column subassemblages

Stress and strain indices, (Valente 2008), were computed at the critical locations, near the weld region at the beam-to-column interface.

3 Effects of concrete slab on the model reinforced with welded haunch

Numerical analyses included matched pairs of models reinforced with welded haunch, one bare steel (MH) and one with composite slab (MHS). Results that emphasize the influence of the concrete slab on connection behaviour are presented and discussed.

As expected, the model with composite slab (MHS) exhibited an increase of initial elastic stiffness and of positive moment capacity (slab in compression) as compared with the bare steel model (MH). The larger attained positive moments in the composite haunch models reflects the effects of composite action. Negative moment capacity was slightly increased by the composite slab, suggesting a potentially beneficial effect of the slab. The additional strength may be due to some tensile capacity contributed by the slab and the stabilizing effect of the slab on the beam. The slab delayed beam local and lateral torsional buckling, resulting in less severe strength degradation. Therefore, the delayed buckling in the composite models may permit the development of larger moments.

For positive moment (slab in compression), the addition of the composite slab shifted the neutral axis nearer the top of the sections, with respect to the bare steel sections. As the slab is much more effective in compression than in tension, the neutral axis locations for negative moment (slab in tension) were similar to those of the bare steel sections, despite the different negative moments achieved by the two models.

With the addition of a composite slab, the haunch models showed better performance. Beneficial effects of a composite slab included reducing the top flange stresses and delaying local and lateral torsional buckling of the beam. The strength of the models deteriorated more gradually and later due to reduced local and lateral buckling of the beams.

The yielding of the models during the analysis showed bottom flange yielding at the end of the haunch and top flange yielding was concentrated at a location much closer to the column face. However, composite model top flange strains were significantly reduced compared to the bare steel models. The presence of a composite slab did not increase considerably the bottom flange strains, which were similar between bare steel and composite models. Fig. 2 shows that the peak value of the PEEQ Index at the bottom flange near the weld at the column interface was not significantly affected. Only a slight increase of the peak value of the PEEQ Index at the haunch end along the beam bottom flange in the model MHS was observed. The maximum value of the PEEQ Index was registered for both the models at the haunch end, showing the effectiveness of the welded haunch to protect the weld region near the column face. Plastic strains were more uniform across the flange width at the critical section than at the column face in the composite model. Numerical results showed that plastic strains were decreased at the top flange when a composite slab was added, although strains at the bottom flange remained about similar. It is often assumed that the presence of a slab increases strain demand at the bottom flange. The increase in moment may be offset by the increased section depth, resulting in little change to bottom flange strains.

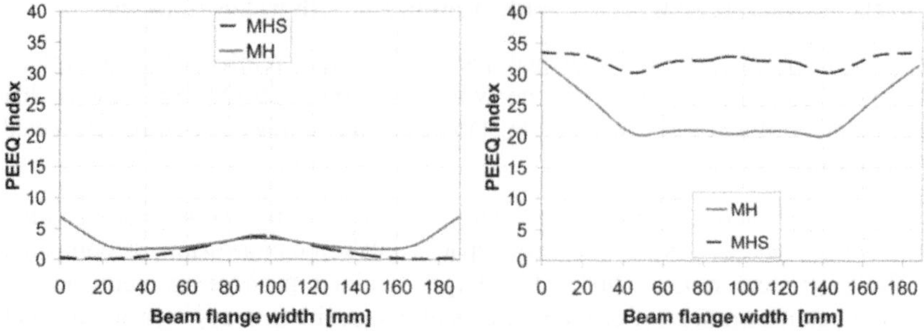

Figure 2. PEEQ Index along the beam bottom flange at the column interface and at the haunch tip

4 Buckling behaviour

To simulate the strength degradation of the models due to local buckling and lateral-torsional buckling, the modified Riks algorithm was used in the analyses so that the post-buckling behaviour could be predicted. The developed models were able to capture the beam flange and web local buckling at large beam story drift. In fact, as analyses proceeded to larger deformation levels, the beams of the models experienced varying degrees of local flange, local web and lateral torsional buckling. A significant effect of the slab on local instabilities could be seen. In bare steel models, local and lateral buckling of the beam occurred; both flanges of the beams experienced local buckling. Such behaviour was still evident, but at a much smaller scale, in the composite models. This different behaviour became more remarkable at large beam story drift. The top flange buckling was controlled by the slab and was significantly less severe than in the bare steel models, leading to reduced likelihood of top flange weld failures in the composite joints. If brittle fractures of welds were prevented, another source of weld fracture may be due to low cycle fatigue from high amplitude distortions. The presence of a slab appeared to control this behaviour very well at the top flange. The use of welded haunch reduced stresses on the bottom beam flange welds and the slab restraint to lateral torsional and local buckling could prevent low cycle fatigue failures at the top flange. The presence of a concrete slab altered the beam buckling mode. Lateral-torsional buckling was prevented due to the bracing effect of the slab, but flange local buckling still could be developed under negative bending. Therefore, strength degradation due to buckling was less severe in positive bending than in negative bending.

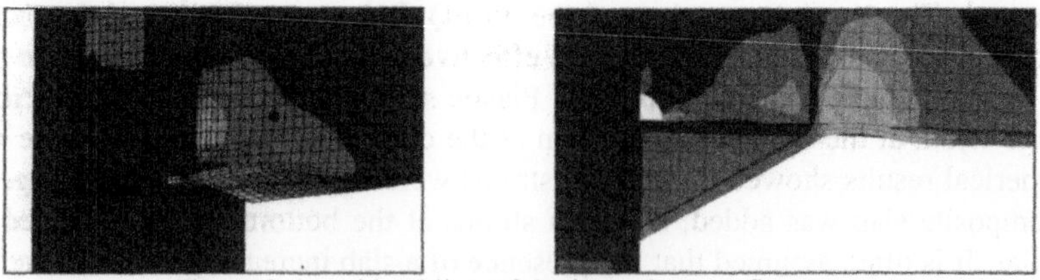

Figure 3. Local buckling of the beam bottom flange

The beam bottom flange buckling occurred near the column interface in the bare model and only at the end of the haunch in the reinforced composite model, Fig. 3.

5 Effects of haunch on the model with concrete slab

The effectiveness of the welded haunch on composite beam-to-column subassemblages was evaluated comparing the numerical results derived from analyses carried out on the models MS and MHS. In the model MHS the plastic strain concentration shifted from the beam flange weld region to the haunch end, Fig. 4. A significant decrease of the PEEQ Index at the beam bottom flange in the model MHS was registered, Fig. 5. Moreover, the maximum values of the PEEQ Index observed at the haunch end in the model MHS were smaller than those registered at the column interface in the model MS. A significant decrease of the PRESS Index was observed in the model MHS at the column interface along the beam flange width, Fig. 5. The PEEQ and PRESS Index decreased considerably near the weld access hole, where the peak values were achieved in the model MS. The flange local buckling amplitudes were smaller in negative bending (beam bottom flange in compression) in the model MHS due to the restraint provided by the haunch. When a composite floor slab is present, the use of a welded bottom haunch appears to provide significantly better structural performance than the unreinforced model.

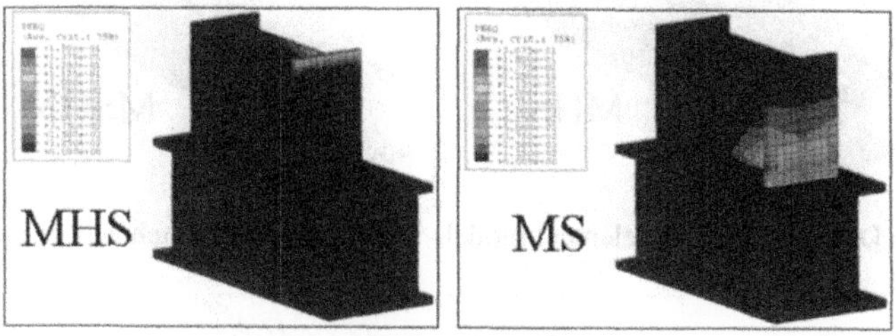

Figure 4. Contour plot of the PEEQ at the beam bottom flange

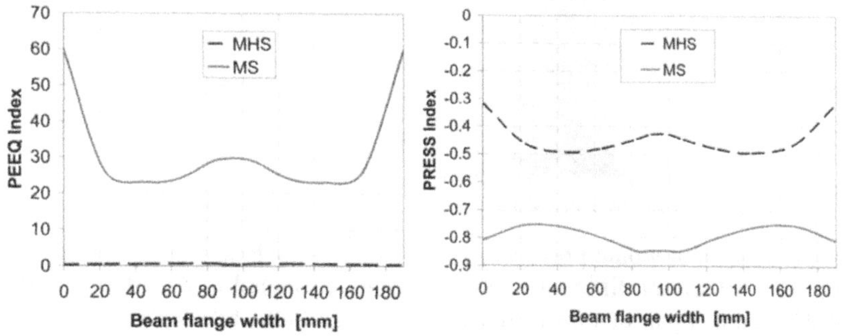

Figure 5. PEEQ and PRESS Index distribution along the beam bottom flange at the column interface

6 Composite model with straight haunch

In the triangular haunch connection, the complete joint penetration groove weld at both ends of the haunch flange with an inclined angle requires a significant amount of overhead welding. To minimize construction cost, the use of a straight haunch with one free end was proposed within the SAC Steel Project. In Fig. 6 the finite element model with straight haunch welded beneath the bottom flange, named MSHS, is shown. Also in this case, numerical analyses showed that the welded haunch changed the force transfer mechanism that could not be predicted reliably by the beam theory. An inclined strip in the web of the straight haunch acted as a strut, Fig 7. Both high shear and normal stresses exist at the interface between the beam and haunch. High stress concentration was also evident at the upper corner. The majority of the normal force at the beam–haunch interface was concentrated near the haunch tip. Fig. 8 compares the contour plot of PEEQ in the two models, MHS and MSHS, indicating a strain concentration near the haunch web - beam flange interface in the model MSHS. A more uniform distribution of the PEEQ Index was observed in the model MHS respect to the model MSHS, Fig. 9. Moreover a peak value of PRESS Index was observed at the middle of the beam flange near the haunch tip in the model MSHS, Fig. 9. These results are confirmed by experimental tests, showing weld fractures initiating from the haunch tip, where stress concentration was the highest.

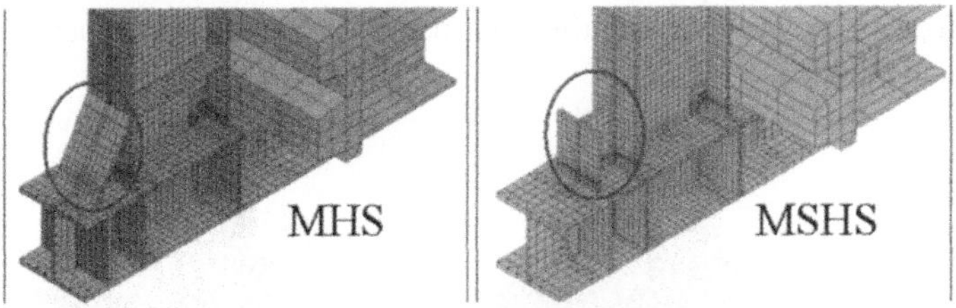

Figure 6. Detail of the finite element models with triangular haunch and straight haunch

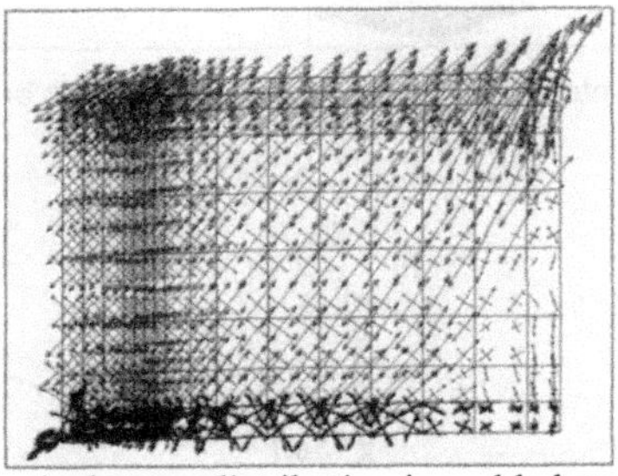

Figure 7. Principal stress distribution in welded straight haunch

Different strategies that would prevent cracking at the haunch tip were proposed and developed. Fig. 10 shows different haunch tip details adopted in this study for the model MSHS. First, a pair of web stiffeners were added to the full depth of the beam in order to strengthen the haunch tip and to distribute the vertical forces into the beam web. Then, a pair of beam web stiffeners was fully extended from the beam flange to the haunch flange.

A larger degree of restraint and redundancy at the haunch tip was obtained in the model. Fig. 11 shows a significant decrease of the peak value of the PEEQ Index along the beam flange width at the haunch tip respect to the model without web stiffeners. Numerical results indicated that this local detail may successfully prevent failure at the haunch tip.

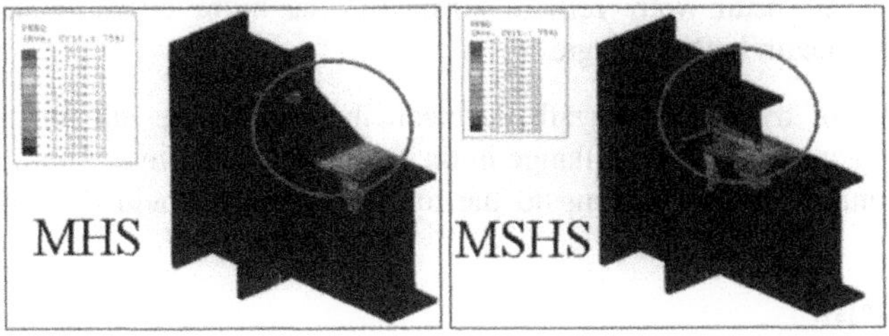

Figure 8. Contour plot of the PEEQ near the haunch tip at the beam bottom flange

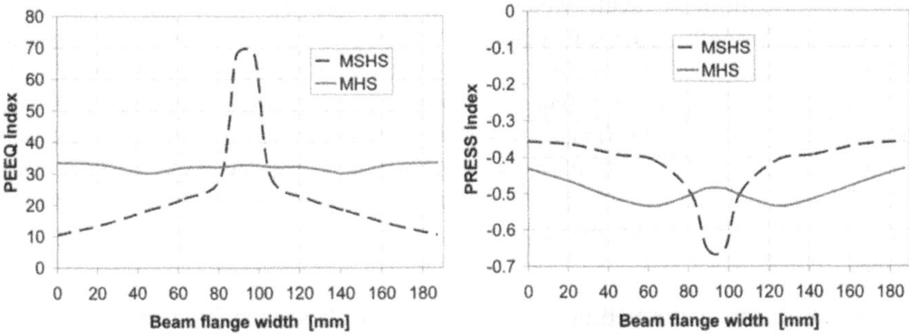

Figure 9. PEEQ and PRESS Index distribution along the beam flange width at the haunch tip

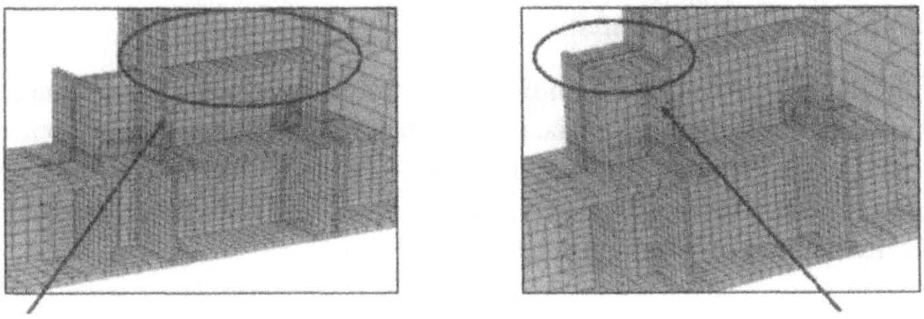

Figure 10. Straight haunch model with different haunch tip details.

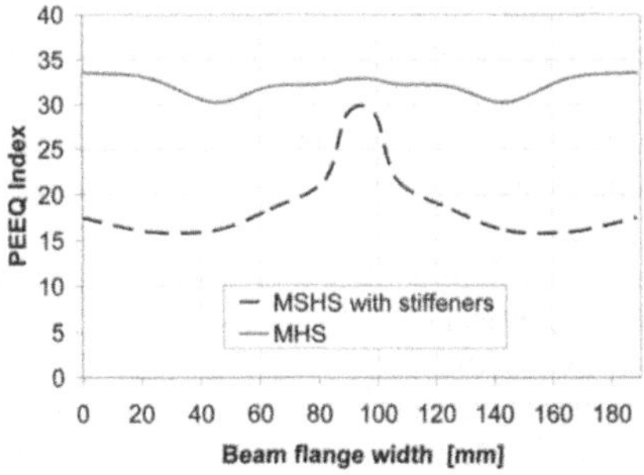

Figure 11. Effect of haunch tip details on PEEQ Index distribution along the beam flange width at the haunch tip

As in the model with triangular haunch, the straight haunch was effective in providing restraint to the beam bottom flange, decreasing the lateral torsional and flange buckling amplitudes in proportion to the degree of the haunch tip restraint.

The model with beam web vertical stiffeners was more effective in providing restraint to the beam bottom flange.

The presence of a haunch beneath the beam bottom flange did not provide any bracing effects to the beam top flange in the bare model. However, in the composite model, continuous lateral bracing to the top flange was provided by the concrete slab.

7 Conclusions

The influence of a composite slab on the performance of beam-to-column connections reinforced with welded haunch at the beam bottom flange was investigated through numerical analyses. Numerical results showed that the slab significantly reduced the severity of local and lateral buckling of the beam. The reduced severity of beam top flange buckling decreased strain demands at the top beam flange weld. Bottom flange strains were about similar between bare steel and composite models. Plastic strain at the bottom flange near the weld at the column interface was not significantly affected. The reduction in the severity of top flange buckling was important in preventing top flange weld fracture in the composite haunch models.

Numerical results confirmed the effectiveness of welded haunch when a composite floor slab was present. The plastic strain concentration shifted from the beam flange weld region to the haunch end and a notable decrease of the PEEQ Index at the beam bottom flange was registered.

The straight haunch proved to be an effective tool to reduce the stress concentration near the weld at the column interface. However, a stress and strain concentration was observed at the haunch tip. A pair of beam web stiffeners was fully extended from the beam web to the haunch web and a significant decrease of the strain demand at the haunch tip was registered.

References

Carreira D.J. & Chu K.H. (1985) Stress-strain relationship for plain concrete in compression. *ACI Journal* **82** No. 6, 797-804.

HKS. ABAQUS user's manual. Version 6.5. Providence (RI): Hibbit, Karlsson & Sorensen, Inc.; 2004.

Valente M. & Castellani A. (2002) Cyclic behaviour and weld access hole detailing of steel welded beam-to-column joints. In: *Proc. of 3rd International Eurosteel Conference, Coimbra, Portugal.*

Valente M. & Castiglioni C. A. (2003) Effects of concrete slabs on the behaviour of steel beam-to-column joints: experimental and numerical study. In: *Proc. Int. Conf. on Steel Structures in seismic areas, Napoli, Italy.*

Valente M. (2008) Welded triangular haunch for the seismic improvement of steel beam-to-column connections. In: *Proc. Int. Conf. DFE 2008, Miskolc, Hungary.* Horwood Publishers, Proceedings pp. 261-268.

5.5 Calculation of the Stiffness and Resistance of Minor Axis and 3D Connections

Katalin Vértes[1], Miklós Iványi[2]
[1]Budapest University of Technology and Economics, Department of Structural Engineering, vertes@vbt.bme.hu, [2]University of Pécs, Pollack Mihály Faculty of Engineering

Abstract

Minor axis and 3D connections often appear in steel structures; yet the behaviour and the calculation process of these structural connections still hides much uncertainty. There are three ways to determine the main characteristics (stiffness, strength, rotation capacity) of a connection: analytical method, numerical method and experimental method. The analytical solution of the Eurocode 3 is called the component method. The main problem that arises when calculating minor axis or 3D connections is that a new component appears which is not defined in the Eurocode: this is the column web in bending. This study introduces a calculation method for the determination of the main characteristics of flush end-plate minor axis and 3D connections through the extension of the component method of the Eurocode 3.

Keywords: *minor axis, 3D connection, component method*

1 Introduction

When calculating a steel framework, the connections are considered as ideally pinned or fixed, although the real behaviour of the joint is different: according to experiments joints have semi-rigid behaviour, which means that it can rotate but a bending moment develops in it as well. The main characteristics of a joint are its strength, stiffness and rotation capacity. The Eurocode 3 (2004) provides the component method for the analysis of steel connections. According to this calculation method, the real characteristics of the joint can be taken into account during the design. Through this modern analysis a more economical structure can be designed.

The component method considers the joint as a set of basic component, where each component possesses its own strength and stiffness. During the calculation process first the basic components have to be identified, then their characteristics have to be calculated and finally these have to be put together in a statically appropriate way.
The component method of the Eurocode provides a detailed description of the calculation of so-called major axis connections (Figure 1.).

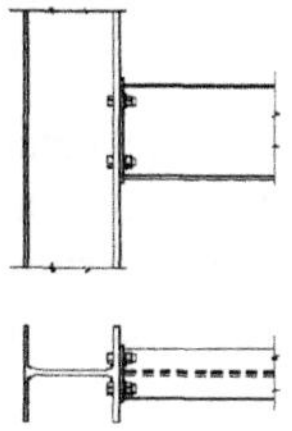

Figure 1. Major axis connection

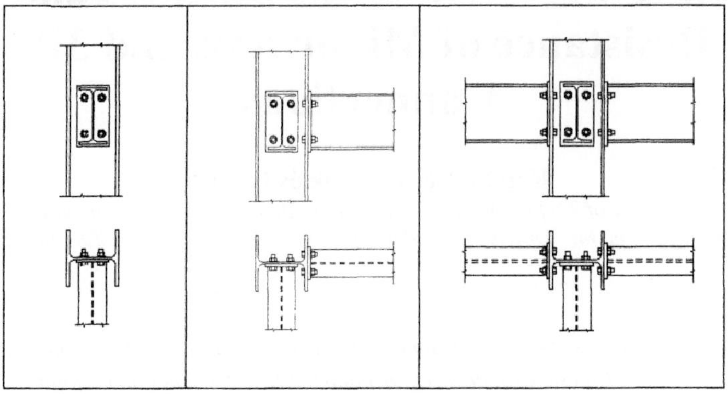

a.) Minor axis joint b.) Corner (3D) joint c.) Full (3D) joint

Figure 2. Minor axis and 3D joints

In the case of minor axis connections the beam is connected directly to the web of the column, so it is bent around its minor axis. In the case of 3D connections, beams are also connected to the column from major and minor directions (Figure 2.). The characteristics of these types of joints cannot be calculated according to the component method. This is because a new component appears in these connections which are not defined in the Eurocode. This component is the **column web in bending**. In order to be able to determine the characteristics of these connections this new component has to be defined and calculated. This study provides a method of which help the characteristic of minor axis and 3D connections can be determined (Vértes, Iványi, 2006).

2 Determination of the stiffness

2.1 *Stiffness of the minor axis connection*

The basic components of a minor axis direction are the following:

- End-plate in bending
- Bolts in tension
- Column web in bending

The web of the column is directly bent through the connecting beam. This bending moment is transmitted to the column web by the end-plate and the tensioned bolts. Firstly, this bending is distributed to tension and compression according to Figure 3. The consideration of the compression zone is different from the Literature (Gomes et al. 1994, Gomes et al. 1996, Neves et al. 1996, Steenhuis et al. 1998). For detailed description see (Vértes 2006).

Figure 3 shows the compression and the tension zone of the column web. The tension zone is a plate with hinged supports at the sides which is loaded by the concentrated force of the bolts. The compression zone is a plate loaded by uniformly distributed load at the area of c. b according to Figure 3.

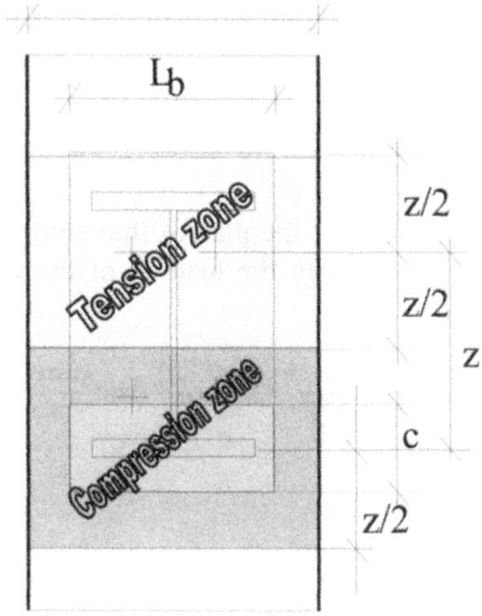

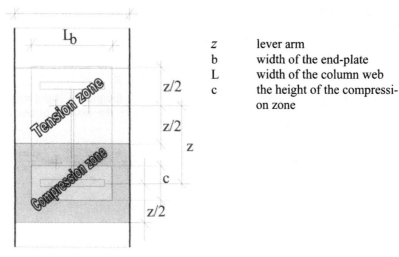

z	lever arm
b	width of the end-plate
L	width of the column web
c	the height of the compressi-on zone

Figure 3. The consideration of the tension and compression zone of the column web

So the deflection of the plates (compression zone and tension zone) can be calculated according to the solution of Szilárd (1974).
In the case of the tension zone it is the following (for one bolt, Figure 4.):

$$w_1(x,y) = \frac{4P}{\pi^4 abD}\sum_{m=1}^{\infty}\sum_{n=1}^{\infty}\frac{\sin(m\pi\xi/a)\sin(n\pi\eta/b)}{\left[(m^2/a^2)+(n^2/b^2)\right]^2}\sin\frac{m\pi x}{a}\sin\frac{n\pi y}{b},$$

$(\mathrm{m,n} = 1,2,3,\ldots)$, (1)

with,

$$D = \frac{Et_{wc}^{\ 3}}{12(1-\nu^2)},$$

where, E, the Young's modulus
t_{wc}, the thickness of the column web
ν, Poisson's ratio

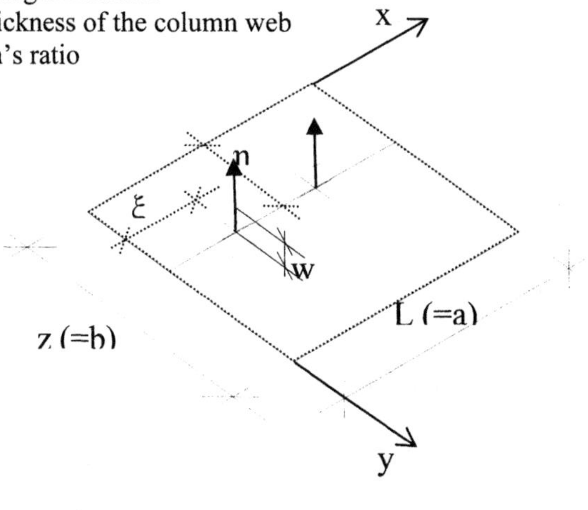

Figure 4. Deflection of the tension zone

The stiffness of the tension zone considering P = 1 kN unit load is the following:

$$k_{cwt} = \frac{1}{Ew}, \qquad (2)$$

where, k_{cwt}, the stiffness of the tensioned column web
w, the deflection of the plate under unit load

In the case of the compression zone the deflection of the plate of the compression zone can be determined as in the previous case. Only the loading of the plate is different (Figure 5.). The deflection is the following:

$$w(x,y) = \frac{16p_0}{\pi^6 D}\sum_{m=1}^{\infty}\sum_{n=1}^{\infty}\frac{\sin(m\pi\xi/a)\sin(n\pi\eta/b)\sin(m\pi c/2a)\sin(n\pi d/2b)}{mn\left[(m^2/a^2)+(n^2/b^2)\right]^2}\cdot\sin\frac{m\pi x}{a}\cdot\sin\frac{n\pi y}{b} \qquad (3)$$

The stiffness (kcwc) is the same as in the previous case.

$$k_{cwc} = \frac{1}{Ew}, \qquad (4)$$

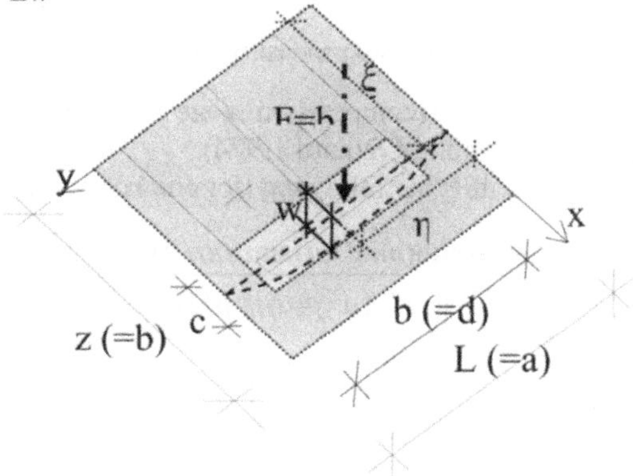

Figure 5. Deflection of the compression zone

After the calculation of the component stiffnesses the stiffness of the connection:

$$S_{j,ini} = \frac{M}{\Phi} = \frac{Ez^2}{\sum\frac{1}{k_i}} = \frac{Ez^2}{\frac{1}{k_5}+\frac{1}{k_{10}}+\frac{1}{k_{cwt}}+\frac{1}{k_{cwc}}}, \qquad (5)$$

where, k_5, is the stiffness of the end-plate in bending according to EC 3 1-8
k_{10}, is the stiffness of the bolt sin tension according to EC 3 1-8

2.2 *Stiffness of 3D connections*

In the case of 3D connections the effect of the major axis connection has to be considered. This is taken into account by the factor ψ (see Vértes 2006). The stiffness of a 3D connection can be calculated as follows:

$$S_{j,ini,3D} = \frac{M}{\Phi} = \frac{Ez^2}{\sum \frac{1}{k_i}} = \frac{Ez^2}{\frac{1}{k_5} + \frac{1}{k_{10}} + \frac{1}{k_{cwt}} + \frac{1}{k_{cvc}} - \frac{\psi}{\nu k_3}}, \tag{6}$$

where, k_3 the stiffness of the column web in the case of major axis tension
ν Poisson's ratio
$\psi = 0{,}5$ in the case of corner joint,
$\psi = 1{,}0$ in the case of full 3D joint.

3 Determination of the resistance

3.1 *Resistance of minor axis connections*

The resistances of the components of the minor axis connection are:

- The column web in bending and punching,
- Bolts in tension,
- End-plate in bending,
- Beam web in tension,
- Beam web and flange in compression.

The above written components are defined in the Eurocode 3, except the bending of the column web. The failure of the column web can be global and local failure. According to the yield theory of the plates the resistance of the column web has been determined for global and local failure mode as well.

Resistance in the case of local failure

The considered yield mechanism is shown in Figure 6 and 6. The yield moment is:

$$m_{pl} = 0{,}25 \cdot t_{wc}^2 \cdot f_y, \tag{7}$$

where, f_y, the yield limit of the column web
t_{wc}, the thickness of the column web

The resistance belonging to local failure is:

$$F_{local,1} = f_y \cdot t_{wc}^2 \cdot \frac{4 \cdot \sqrt{L}}{\sqrt{L - b_0}} \tag{8}$$

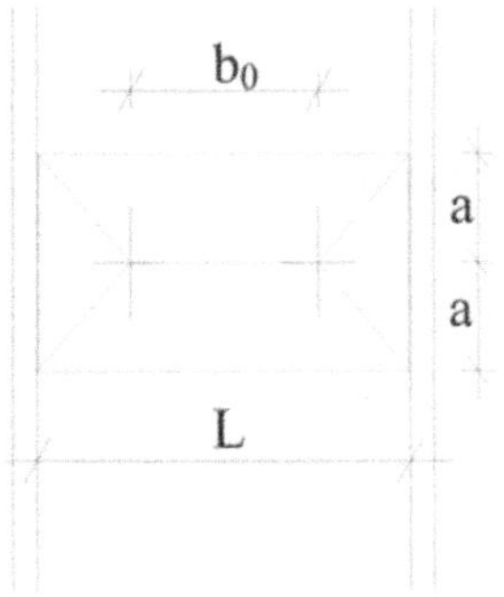

Figure 6. Considered local mechanism

Figure 7. Local mechanism according to numerical calculation

If the bolts are near to the flanges, the previous solution is not correct, so an another yield mechanism is considered for these cases (Figure 8)

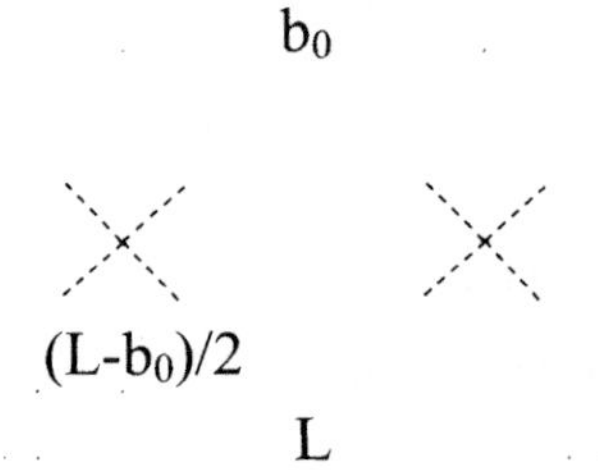

Figure 8. Local yield mechanism if the bolts are near to the flanges

The resistance in this case:

$$F_{local,2} = 2 \cdot t_{wc}^2 \cdot \pi \cdot f_y \tag{9}$$

So the resistance in the case of local failure:

$$F_{local,\,Rd} = \min \begin{cases} F_{local1} \\ \\ F_{local2} \end{cases} \tag{10}$$

Resistance in the case of global failure

The considered yield mechanism is shown in Figure 9 and 10.

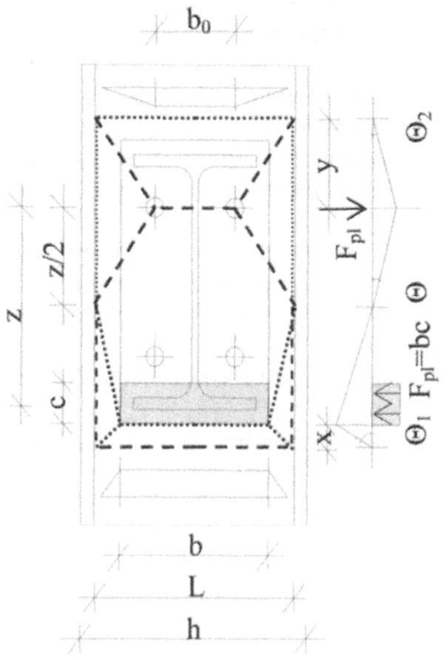

··········· Positive yield line
- - - - Negative yield line

Figure 10. Global mechanism

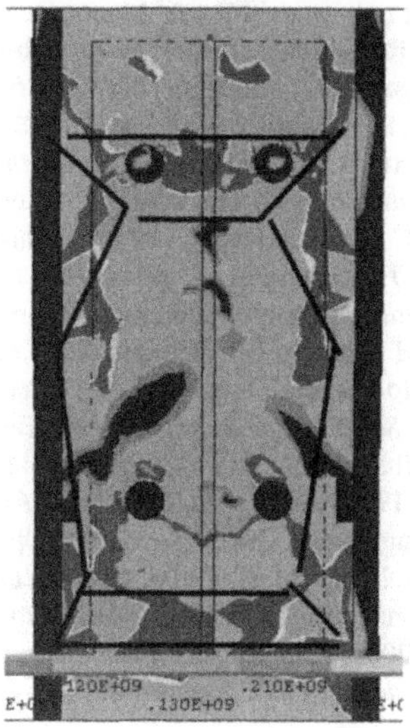

Figure 11. Global mechanism

The resistance in case of global failure is:

$$F_{Global,Rd} = 0,5 \cdot t_{wc}^{2} \cdot f_{y} \cdot \left(\frac{2\sqrt{L}}{\sqrt{L-b_{0}}} + \frac{2\sqrt{L}}{\sqrt{L-b}} + \frac{L}{z} + \frac{z}{L-b_{0}} + \frac{z}{L-b} \right) \tag{11}$$

3.2 *Resistance of 3D connections*

The resistance of 3D connection can be determined by using a ρ reduction factor for the yield limit. For detailed description see Vértes (2006).

4 Summary

This study provides an analytical calculation method for the determination of the stiffness and resistance of minor axis and 3D bolted end-plate connections. The presented calculation method is based on the component method of the Eurocode 3.

Acknowledgement

The financial support for this research given OTKA grants No. T 048814 is gratefully acknowledged.

References

MSZ ENV 1993 *Design of steel structures, Part 1-8* (2002) Design of joints.

Gomes,F.C.T., Jaspart,J-P. & Maquoi,R. (1994) Behaviour of Minor-Axis Joints and 3D Joints, *Proceedings of the Second State of the Art Workshop COST C1*, Ed. Wald, F., Czech Technical University, Prague, 26-28. October 1994, pp. 111-120.

Gomes,F.C.T., Jaspart,J-P. & Maquoi,R. (1996) Moment Capacity of Beam-to-Column Minor-Axis Joints, *Proceedings of the IABSE International Colloquium: Semi-Rigid Structural Connections*, Turkey, September 1996, pp. 319-326.

Neves,L.F.C., Gomes,F.T.C. (1996) Semi-Rigid Behaviour of Beam-to-Column Minor-Axis Joints, *Proceedings of the IABSE International Colloquium: Semi-Rigid Structural Connections*, Turkey, September 1996, pp. 319-326.

Steenhuis,M., Jaspart,J-P., Gomes,F.C.T. & Leino,T. (1998) Application of the Component Method to Steel Joints, COST C1 *Proceedings of the International Conference, Control of the Semi-Rigid Behaviour of Civil Engineering Structural Connections*, Ed. By Maquoi, R., Liege, 17-19. September 1998, pp. 125-143.

Szilárd,R. (1974) *Theory and Analysis of Plates, Classical and Numerical Methods*, Prentice-Hall, Englewood Cliffs, New Jersey, pp.60-78.

Vértes,K. & Iványi,M. (2006) Analytical calculation method of minor axis and 3d flush end-plate connections, *Építés és Építészettudomány* **34** No. 3-4. 293-307.

Vértes,K. (2006) *Analysis of minor axis and 3D bolted end-plate connections and eye-bars of chain bridges*, PhD Dissertation.

Section 6
Hollow sections

6.1 A New Yield Line Theory Based Design Approach for Ultimate Capacity of Welded RHS X-joints

Timo Björk, Gary Marquis
Lappeenranta University of Technology, Lappeenranta, Finland
timo.bjork@lut.fi gary.marquis@lut.fi

Abstract

The load capacity equations for structural hollow section joints given in design guidance documents are frequently based on the yield line theory. These predictions are then confirmed or adjusted based on experimental findings. The calculated capacities are valid only for a certain failure pattern. For welded rectangular hollow sections (RHS) X- and T-joints, the failure can take place in the brace member, in the chord member or partly in both members. In this study the last mechanism is taken as general failure mode which covers other mechanisms. Theoretical load carrying capacity for this new mechanism is presented and it is compared with some experimental test results. The derived mechanism is applied to define optimal joint geometries for X- and T-joints.

Keywords: *RHS sections, design of joints, optimization, yield line theory, ultimate capacity*

1 Introduction

The rectangular hollow sections (RHS) are widely used structural components due to their good strength properties and structural appearance. The joints are typically the weakest location of a structure fabricated from RHS. The often competing demands of load-carrying capacity and fabrication costs influence the joint design. Thus, simple X- and T-joints are frequently used connections and for many applications the width of the brace member is chosen to be less than width of the chord (β < 1.0). The typical failure mechanisms for X- and T-joint are presented in Eurocode 3 (EN1993, 2005). If joint failure occurs in the brace member, and the design load-carrying capacity is fulfilled, the deformation capacity of the structure is considered maximal. However, this failure mode generally requires relatively thick and, therefore, uneconomical chord members. If the brace member has thick walls and β is low, failure may occur according to the yield pattern illustrated in Fig. 1a or failure may take place so that both joint members yield as presented in Fig. 1b. In this paper the failure in Fig. 1b is termed the general failure mode of T- and X-joints, and is not included in current design code (EN1993,2005). Some investigations concerning the general failure mechanism are seen in references (Davies 1981, Wardenier 1982, Morita et al. 1988, Packer et al. 1989, Partanen & Björk 1993, Osman & Korol 1993. Björk 2005). Both the chord flange yield pattern and brace yielding are special cases of the general failure mechanism. In this study the load carrying capacity and the limits for this mechanism will be presented and optimal joint geometry can be defined.

2 General yield line pattern

In the general failure mode of X- and T-joints the plastic hinges can move to the inside of the brace member as shown in Fig. 2. The required dimensions for the yield pattern can be defined by applying yield line analysis. As an example, the simple T-joint geometry with brace angle of 90° is analyzed and the plastic work done by each plastic hinge and membrane zone is given in the Table 1.

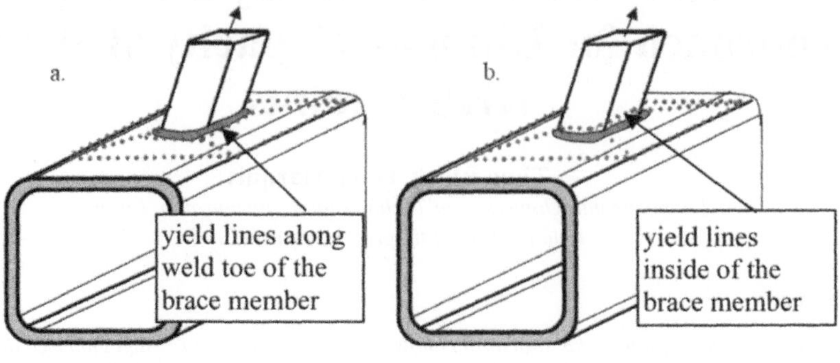

Figure 1. Conventional yield pattern a) and b) a general failure mode

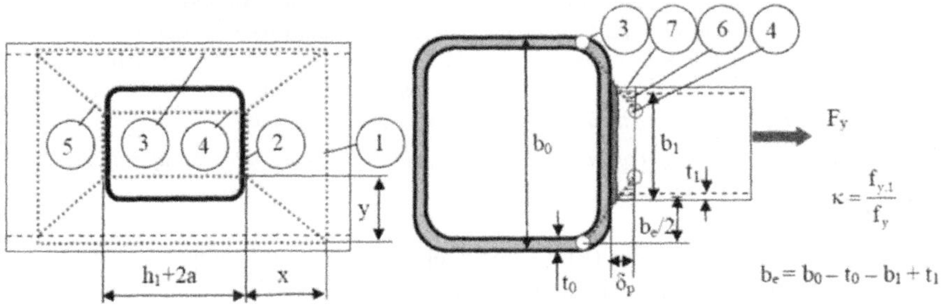

Figure 2. Yield line pattern for general failure mechanism of a T-joint

Table 1. Parameters for plastic hinges and plastic work for the general mechanism

#	Number of zones	Length of yield zone	Force unit length	Plastic deformation	Plastic work $W_{in,i}$
1	2	$b_0 - t_0$	$\frac{t_0^2}{4} f_y$	$\frac{\delta_p}{x}$	$\frac{t_0^2}{2} f_y \frac{b_0 - t_0}{x} \delta_p$
2	2	$b_0 - t_0 - 2y$	$\frac{t_0^2}{4} f_y$	$\frac{\delta_p}{x}$	$\frac{t_0^2}{2} f_y \frac{b_0 - t_0 - 2y}{x} \delta_p$
3	2	$h_1 + 2a + 2x$	$\frac{t_0^2}{4} f_y$	$\frac{\delta_p}{y}$	$\frac{t_0^2}{2} f_y \frac{h_1 + 2a + 2x}{y} \delta_p$
4	2	$h_1 + 2a$	$\frac{t_0^2}{4} f_y$	$\frac{\delta_p}{y}$	$\frac{t_0^2}{2} f_y \frac{h_1 + 2a}{y} \delta_p$
5	4	$\sqrt{x^2 + y^2}$	$\frac{t_0^2}{4} f_y$	$\sqrt{x^2 + y^2} \frac{\delta_p}{xy}$	$t_0^2 f_y \left[\frac{x}{y} + \frac{y}{x} \right] \delta_p$
6	4	$\frac{2y - b_e}{2}$	$t_1 \kappa f_y$	$\frac{2y - b_e}{2y} \frac{\delta_p}{2}$	$\frac{t_1 \kappa f_y (2y - b_e)^2}{2y} \delta_p$
7	2	$h_1 - 2t_1$	$t_1 \kappa f_y$	$\frac{2y - b_e}{2y} \delta_p$	$\frac{t_1 \kappa f_y (h_1 - 2t_1)(2y - b_e)}{y} \delta_p$

The internal plastic energy of the mechanism is

$$W_{in} = f_y \left\{ t_0^2 \left[\frac{b_0 - t_0}{x} + \frac{h_1 + 2a + 2x}{y} \right] + \kappa t_1 \frac{(2y - b_e)(2y - b_e + 2h_1 - 4t_1)}{2y} \right\} \delta_p \tag{1}$$

By equating the internal and external, the yield capacity for the general mechanism can be computed

$$F_{G,y} = f_y \left\{ t_0^2 \left[\frac{b_0 - t_0}{x} + \frac{h_1 + 2a + 2x}{y} \right] + \kappa t_1 \frac{(2y - b_e)(2y - b_e + 2h_1 - 4t_1)}{2y} \right\} \tag{2}$$

The failure mechanism has two independent variable x and y, which can be found by the principle of minimum energy

$$\frac{d\Pi}{dx} = \frac{d(W_{in} - W_{out})}{dx} = 0 \Rightarrow x = \sqrt{\frac{b_0 - t_0}{2} y} \tag{3}$$

$$\frac{d\Pi}{dy} = \frac{d(W_{in} - W_{out})}{dy} = 0 \Rightarrow y = \sqrt{\frac{(h_1 + 2a + 2x)t_0^2 + b_e(0.5b_e - h_1 + 2t_1)\kappa t_1}{2\kappa t_1}} \tag{4}$$

Equations (3) and (4) yield a bi-quadratic equation which, when solved, gives

$$x = \frac{b_0 - t_0}{4} \left[C_4 + \sqrt{\frac{4t_0^2}{C_4 (b_0 - t_0)\kappa t_1} - C_4^2} \right] \qquad y = \frac{b_0 - t_0}{8} \left[C_4 + \sqrt{\frac{4t_0^2}{C_4 (b_0 - t_0)\kappa t_1} - C_4^2} \right]^2 \tag{5}$$

where $C_1 = \dfrac{2t_0^4}{\kappa^2 t_1^2 (b_0 - t_o)^2}$, $C_2 = \dfrac{8}{3} \left[\dfrac{(h_1 + 2a)t_0^2 + b_e \kappa t_1 (0.5b_e - h_1 + 2t_1)}{(b_0 - t_0)^2 \kappa t_1} \right]$,

$C_3 = \sqrt{C_1^2 + C_2^3}$ and $C_4 = \sqrt{\sqrt[3]{C_1 + C_3} - \sqrt[3]{C_3 - C_1}}$

The capacity, $F_{G,y}$, of the general failure mechanism can be calculated from equation (2). If the distance between two parallel yield lines is less than two times the plate thickness, a shear stress reduction will be needed. For greater distances the shear reduction effect is less than 8 % from the plastic moment capacity of the plate (Björk 1988). If the straight hinge mechanism in Fig. 2 is replaced with a fan mechanism, the capacity of the joint is

$$F_{G,y} = f_y \left\{ t_0^2 \left[\frac{b_0 - t_0 - 2y}{x} + \frac{h_1 + 2a}{y} + \pi \right] + \kappa t_1 \frac{(2y - b_e)(2y - b_e + 2h_1 - 4t_1)}{2y} \right\} \tag{6}$$

The capacity of the brace member, when it is fully effective, can be evaluated by the following formula

$$F_{B,y} = \kappa f_y t_1 \left[2b_1 + 2h_1 - 16t_1 + 3\pi t_1 \right] \tag{7}$$

If the joint capacity is dived by the tension capacity of brace member, the effectiveness factor χ for the joint can be defined for the general failure mechanism as

$$\chi_G = \frac{1}{\kappa A_b}\left\{t_0^2\left[\frac{b_0 - t_0 - 2y}{x} + \frac{h_1 + 2a}{y} + \pi\right] + \kappa t_1 \frac{(2y - b_e)(2y - b_e + 2h_1 - 4t_1)}{2y}\right\} \tag{8}$$

where A_b is the brace member area.

The effectiveness factor of the general mechanism must be compared to the effectiveness factor χ_F obtained by the conventional flange mechanism model, see Fig. 1a.

$$\chi_F = \frac{t_0^2}{\kappa A_b}\left[\frac{4}{\sqrt{1-\beta}} + \frac{2h_1 + 4a}{(b_0 - t_0)(1-\beta)}\right] \tag{9}$$

When the $\beta \rightarrow 1.0$, the punching shear mechanism can be critical and the effectiveness factor χ_S becomes

$$\chi_S = \frac{t_0}{\sqrt{3}\kappa A_b}\left[2(h_1 + 2a) + \frac{20 b_1 t_0}{b_0}\right] \tag{10}$$

In the case the $\beta = 1.0$ the yielding due to membrane stress can take place in the web of the chord member. In this case the matching plastic hinges develop in the chord flange. The effectiveness factor of the joint in this case is

$$\chi_W = \frac{2t_0}{\kappa A_b}\left[\sqrt{b_0 t_0} + (h_1 + 2a)\right] \tag{11}$$

In this study the brace member is loaded in tension. All the formulas above are valid also for compression load as long as local buckling failure can be avoided by choosing a sufficiently small flat face width-to-thickness ratio according to the following formula

$$\frac{b_1 - 3t_1}{t_1} \leq \frac{0.673\pi\sqrt{E}}{\sqrt{3\kappa f_y(1-\nu^2)}} \tag{12}$$

The computed effectiveness factors χ for alternative failure mechanisms in X-joints with typical geometries are illustrated in Fig. 3.

3 Experimental tests

Capacities of 15 T-type joints were measured experimentally in order to verify the validation of the theoretical model. The test specimen is seen in Fig. 4. After welding the specimens were stress-relieved for 30 minutes at 650 °C. Specimen dimensions, measured after heat treatment, are seen in the Table 2. Material properties are given in Table 3.

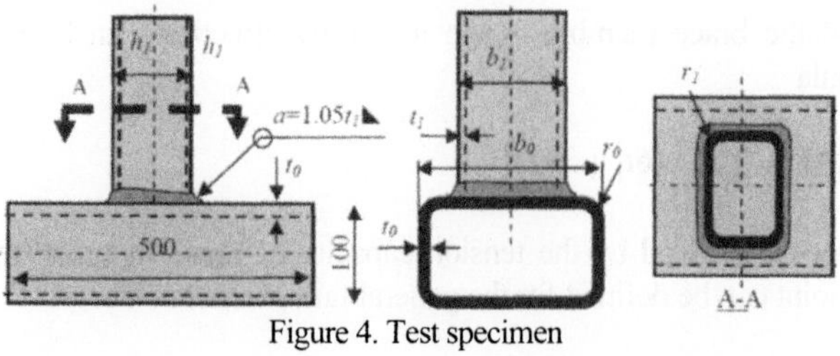

Figure 4. Test specimen

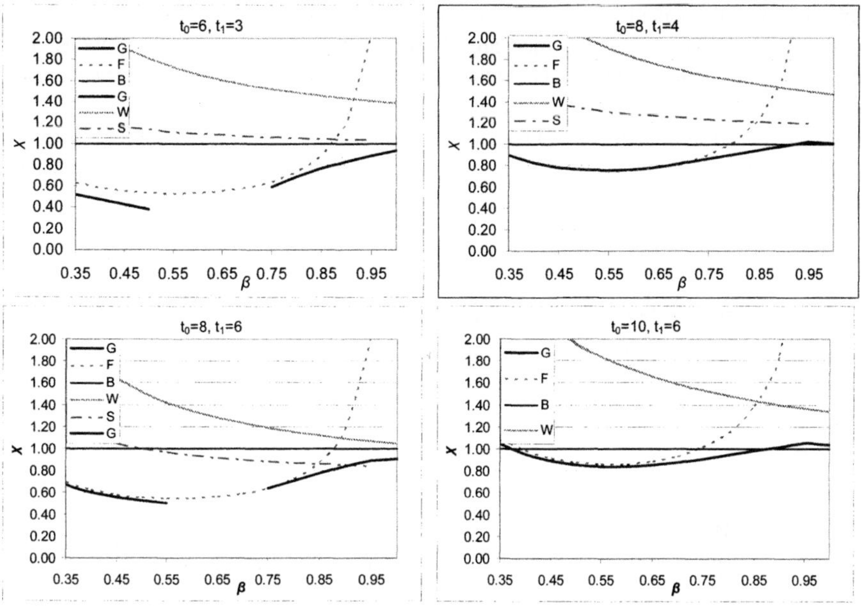

Figure 3. Validation of different failure mechanisms ($b_0 = 100$, $b_i = h_1$). Symbols in these figures denote to the following failure mechanisms: G -general, B - brace, F - flange, W - web and S - punching shear mechanism.

Table 2. Geometry of the tests specimens (refer to Fig. 4)

ID	b_0	t_0	r_0	b_1	h_1	t_1	r_1	a	β	h_1/b_1
1	199.1	7.85	18	39.8	40.3	1.99	6	4	0.19	1.00
2	199.1	7.84	18	60.1	39.7	1.97	5	4	0.30	0.66
3	199.0	7.84	18	99.8	79.4	1.98	5	5	0.50	0.80
4	199.1	7.72	16	120.6	80.8	1.99	6	5	0.61	0.67
5	199.3	7.72	16	118.7	39.2	3.88	9	4.5	0.60	0.33
6	199.3	7.69	15	159.2	80.4	3.90	8	5	0.80	0.51
7	199.2	7.80	18	181.0	79.8	3.90	9	5	0.91	0.44
8	199.2	7.86	18	159.4	122.0	3.90	8	6.5	0.80	0.77
9	199.2	7.69	16	180.5	122.2	3.90	8	5	0.91	0.68
10	199.0	7.76	18	159.6	159.3	3.96	8	6	0.80	1.00
11	199.2	7.77	18	200.0	160.8	3.89	8	8	1.00	0.80
12	199.0	7.80	18	158.8	81.0	5.99	11	5.5	0.80	0.51
13	198.8	7.80	18	178.7	79.5	6.30	11	6	0.90	0.44
14	199.2	7.84	18	179.7	119.7	6.02	12	5.5	0.90	0.67
15	198.9	7.81	18	200.0	118.3	6.00	11	8.5	1.00	0.59

Table 3. Mechanical properties of test material

ID	brace		chord	
	f_y	f_b	f_y	f_b
1, 2, 4	351	471		
3	255	334	288	422
5, 6, 7, 8, 9, 10, 11	319	444		
12, 13, 14, 15	324	465		

Table 4 shows the calculated theoretical load capacities of the test specimens based on measured dimensions, measured mechanical properties and potential failure modes. The bold values for each specimen indicate the critical, i.e., minimum load, yield mechanism. Column $F_{G,y}$ is computed using Eq. (6) and $F_{B,y}$ is computed using Eq. (7), while $F_{F,y}$, $F_{W,y}$ and $F_{S,y}$ are based on effectiveness factors from Eqs. (9 - 11). These critical yield loads are reproduced in the column marked $F_{cal,y}$. In the next column, $F_{cal,u}$, critical joint capacities based on the ultimate tensile strength of the brace and chord members are presented.

Table 4. Computed load capacities based on different failure mechanisms.

ID	$F_{G,y}$	$F_{F,y}$	$F_{B,y}$	$F_{W,y}$	$F_{S,y}$	x	y	$F_{cal,y}$	$F_{cal,u}$
1	**76**	112	204	415	180	87	79	76	111
2	**83**	120	254	407	196	83	73	83	119
3	126	**119**	345	596	348	85	76	119	113
4	**156**	186	542	596	368	78	64	156	224
5	**119**	145	1394	402	253	69	50	119	174
6	272	**261**	2201	592	404	69	49	261	363
7	**355**	463	2406	594	430	69	50	355	505
8	362	**322**	2606	818	540	69	50	322	448
9	**453**	530	2813	772	531	69	49	453	642
10	425	**353**	3060	968	626	66	46	353	491
11	**647**	-	3360	1003	688	71	53	647	912
12	313	**273**	5117	608	416	62	41	273	392
13	**435**	443	6108	606	435	62	40	435	630
14	573	**541**	6566	787	542	63	41	541	776
15	**721**	-	6965	817	581	63	42	721	1040

The experimental load carrying capacities referring to the yield strength $F_{test,y}$ were measured for each specimen. The yield line pattern from experimental test was defined by measuring the out of plane displacements of the chord flange. The displacements were measured during the test at several load levels along the centreline of the chord member and in the transverse direction along the brace weld. From the plotted displacements the locations of the plastic hinges, i.e., x and y in Fig. 2, could be estimated.

In Table 5, comparisons between the calculated and experimentally determined capacities are given. The measured distances x and y for the general yield pattern and the failure mode observed for the joint are also given.

4 Discussion

The X- and T-joints prepared by fillet welds are more economical than joints prepared by bevel welds. However, this necessitates that the brace member is narrower than the chord member ($\beta < 1$) so that the fillet welds can located on the flat section of the chord flange as illustrated in Fig. 4. On the other hand, it can be seen from the effectiveness curves in Fig. 3 that the brace member should be as wide as possible compared to width of the chord, i.e., β near unity. Typically the optimal β-value which can be reached by using fillet welds is only about 0.8…0.85. If the joints are also loaded in compression a square section is optimal for brace members and the ratio b_1/t_1 should be limited so as to be within limits of cross section class 3. The slenderness limit from Eq. 12 is $b_1/t_1 \approx 30$ depending on the steel grade used ($b_1/t_1 = 34$ for steel grade 355). Applying those strength and fabrication limitations one can define the optimal brace member for a certain chord member.

Table 5. Comparison between tested and calculated values.

ID	*Failure mode*	x	y	$F_{test,y}$	$F_{test,u}$	$\frac{F_{ytest,y}}{F_{calc,y}}$	$\frac{F_{ytest,u}}{F_{calc,y}}$	$\frac{F_{ytest,u}}{F_{calc,u}}$	$\frac{x_{test}}{x_{cal}}$	$\frac{y_{test}}{y_{cal}}$
1	B	60	86	66	138	0.87	1.82	1.24	0.69	1.09
2	G	65	76	74	177	0.90	2.14	1.49	0.78	1.04
3	G	70	66	105	185	0.88	1.55	1.64	0.82	0.87
4	G	68	56	142	260	0.91	1.66	1.16	0.87	0.88
5	G	62	43	110	263	0.93	2.21	1.51	0.90	0.86
6	G	56	30	242	395	0.93	1.51	1.09	0.81	0.61
7	G	67	39	347	475	0.98	1.34	0.94	0.97	0.78
8	G	62	27	300	400	0.93	1.24	0.89	0.90	0.54
9	G	63	36	460	625	1.02	1.38	0.97	0.95	0.74
10	G	59	34	390	735	1.10	2.08	1.50	0.89	0.74
11	G	70	50	600	946	0.93	1.46	1.04	0.99	0.94
12	G	62	25	275	450	1.01	1.65	1.15	1.00	0.61
13	G	68	26	395	590	0.91	1.36	0.94	1.10	0.65
14	G	65	30	525	800	0.97	1.48	1.03	1.03	0.73
15	G	70	38	740	1158	1.03	1.61	1.11	1.11	0.90

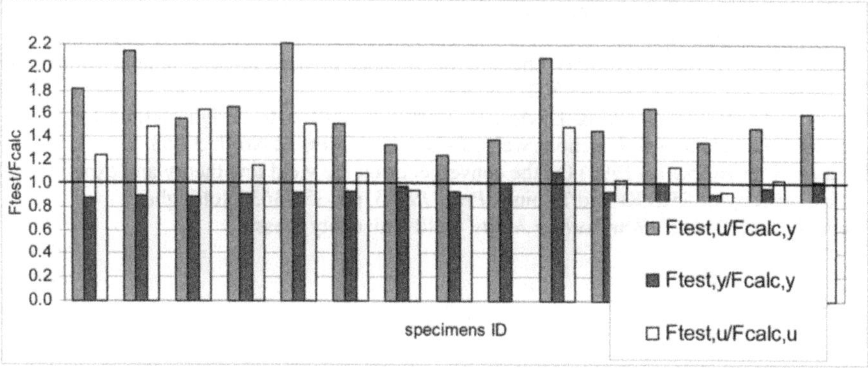

Figure 5. Comparison between experimental and theoretical capacities

In Fig. 5 the comparison between theoretical and test results is shown graphically. It can be seen that the estimated and computed failure load are in best agreement is yield, rather than ultimate strength is used for comparison. The theoretical capacities are close to the experimental value and the largest observed difference is less than 13 %. Also, the calculated yield line pattern, based on x and y, matched quite well with experimental pattern. From Tables 4 and 5 it can be seen that the general mechanism is the critical failure mode for most joints. This was the same for both experiments and computed yield loads. Thus it seems difficult to avoid brace member yielding. Thus, the general mechanism should be considered whenever economical and safe T- and X-joints are designed.

5 Conclusions

A new failure model, the general yield mechanism, has been developed and verified experimentally. This mechanism is based on yield line theory for T- and X-joints. The following conclusions can be made:

- The general yield mechanism considers all competing yield failure mechanisms for tension loaded X- and Y-joints
- The computed load capacities calculated by the general yield mechanism were in good agreement with experimentally measured capacities for 15 alternate T-joints.
- The yield line pattern for the general yield mechanism predicts the critical failure mode and minimum load capacity for many T- and X- joints relevant in design.
- The general yield mechanism can be used in the design of joints to avoid low effectiveness or poor deformation capacity.

References

Björk, T. (1988) Applications of the yield line theory. *Steel structures I*, RIL 167, pp. 188-200. Association of Finnish Civil Engineers. Helsinki, (in Finnish).

Björk, T. (2005) *Ductility and ultimate strength of cold-formed rectangular hollow section joints at subzero temperatures*, Lappeenranta University of Technology, PhD thesis.

Davies, G. (1981) The effective width of brace crosswalls for RR cross joints in tension. *Stevin report 6-81-7*, Delft University of Technology.

EN 1993-1-8, Eurocode 3 (2005) Design of steel structures. Part 1-8: Design of joints

Morita, K., Yamamoto, N. & Ebato, K.(1988) Analysis of the strength of unstiffened beam flange to RHS column connections based on combined yield line model, *Proc. ISTS3*, pp. 164-171. Lappeenranta.

Osman, A. & Korol, M. (1993) Strength of T-joints in back-to-back double chord HSS trusses. *Proc. ISTS5*, pp. 495-502. London.

Packer, J. A., Morris, G. A. & Davies, G. (1989) A limit state design method for welded tension connections to I-sections webs, *J Cons. Steel Res,* **12** No. 1. 33-53.

Partanen, T. & Björk, T. (1993) On the convergence on the yield line theory and experimental test capacity of RHS K- and T-joints, *Proc. ISTS5*, pp. 353-363. Nottingham.

Wardenier, J. (1982) *Hollow Section Joints*, Delft University Press.

6.2 Non-linear Behaviour of Concrete-filled Welded Steel Box-section Arches

Mark A Bradford
The University of New South Wales, UNSW Sydney, NSW, Australia,
m.bradford@unsw.edu.au

Abstract

This paper presents a non-linear analysis of pinned circular composite arches made by filling a welded steel box-section with concrete, once the steel arch has been set in place. In this application, the composite member is unpropped, with the steel box-section supporting the self weight of the steel and concrete infill, and any further imposed loading being supported by the section compositely. It is widely recognised that quantifying the behaviour of arches, which are shallow requires recourse to non-linear formulations for the strain-displacement relationship, and such a formulation is used in the current paper. The analysis considers the steel box-section, then the composite section with an initial curvature induced by the self-weight loading and following the derivation of the equations of equilibrium the use of the non-linear method is demonstrated with an illustrative example.

Keywords: *Arches, box-section, composite, non-linear, welded.*

1 Introduction

Arches resist external actions by the development of primarily compressive internal actions, and because of this they are especially suited to brittle materials such as concrete and stone. Steel box-section arches that are hollow during erection, but which are then filled with concrete, provide an efficient design for an arch bridge structure, which following curing of the concrete ensures that the concrete-filled box-section behaves compositely (Oehlers & Bradford, 1995). The efficiency of this construction sequence is reliant on the ability to deploy a lightweight steel box section that may be fabricated by welding, which acts as integral and permanent shuttering for the concrete. Because the concrete is best placed by pumping, it is advantageous if the arch is shallow to eliminate excessive pressure heads as well as segregation of aggregate. However, it is well-known that shallow arches experience geometric non-linearity (Pi *et al.*, 2002; Bradford *et al.*, 2007) and it has been shown that the accepted 'classical' theory used in Timoshenko & Gere (1961), Vlasov (1961) and elsewhere fails to predict the response of a shallow arch accurately, and may lead to overestimates of in-plane buckling loads. Potential also exists for possible "creep buckling" of a shallow arch (Wang *et al.* 2006).

This paper considers the non-linear analysis and behaviour of pin-ended welded steel box section arches that are filled with concrete so as to produce an unpropped composite member, for which the self weight is resisted by the steel box section and then the imposed loading by the composite member, with full-interaction between the steel and concrete being assumed. The arch is taken as having a circular profile, and being sufficiently shallow that the loading acts radially, i.e. normal to the initially curved axis of the arch. The use of the closed form solution is illustrated with a numerical example.

2 Strain-displacement relationship and deformations for steel box-section

Figure 1 shows the elevation and cross-section of the box-section arch. Initially, the steel section alone is subjected to a loading q_0 distributed uniformly around its axis, and when composite the steel and concrete are subjected to a short-term additional loading q_i. Considering the steel box-section only under wet concrete loading, the strain-displacement formulation derived by Pi *et al.* (2002) can be written as

$$\varepsilon_0 = \varepsilon_{0m} - y v_0'' , \qquad (1)$$

in which the membrane strain is

$$\varepsilon_{0m} = w_0' - v_0 \kappa_0 + \tfrac{1}{2} v_0'^2 \qquad (2)$$

and for which w_0 is the axial displacement in the tangential s direction (Figure 1), v_0 is the displacement in the y direction orthogonal to the arch axis, $\kappa_0 = 1/R$ is the constant initial curvature of the arch and $(\ )' \equiv \mathrm{d}(\ )/\mathrm{d}s$.

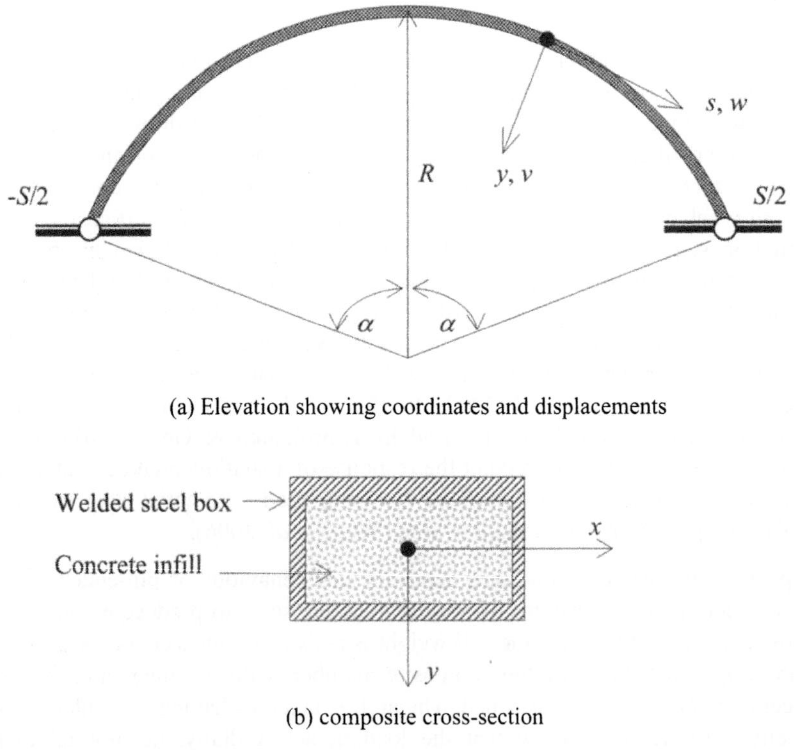

Figure 1. Arch profile, axes, deformations and cross-section.

Under the action of the load q_0, the total potential can be written in a format so as to use variational calculus as

$$\Pi_0 = \int_{-S/2}^{S/2} F_0(w_0', v_0, v_0', v_0'')\,\mathrm{d}s\,, \tag{3}$$

for which

$$F_0 = \tfrac{1}{2}\left[(EA)_0\varepsilon_{0m}^2 + (EI)_0 v_0''^2\right] - q_0 v_0\,. \tag{4}$$

Using the Euler-Lagrange equations in the axial direction,

$$-\left(\partial F_0/\partial w_0'\right)' = 0 \tag{5}$$

which leads to $\varepsilon_{0m}' = 0$ and so the membrane strain is constant and can be written as

$$\varepsilon_{0m} = -\frac{N_0}{(EA)_0}\,, \tag{6}$$

in which N_0 is the axial force in the steel section, and $(EA)_0$ and $(EI)_0$ are its axial and flexural rigidities respectively. Using the Euler-Lagrange equations in the transverse or radial direction, viz.

$$\partial F_0/\partial v_0 - \left(\partial F_0/\partial v_0'\right)' + \left(\partial F_0/\partial v_0''\right)'' = 0\,, \tag{7}$$

produces

$$(EI)_0 v_0^{iv} + N_0 v_0'' = q_0 - N_0/R\,. \tag{8}$$

The solution of Eq. 8 which satisfies the boundary conditions

$$v_0 = v_0'' \text{ at } s = \pm S/2 \tag{9}$$

is

$$v_0 = \frac{\omega_0}{\mu_0^2 R}\left[\cos\theta_0 \sec\alpha_0 - 1 + \tfrac{1}{2}\left(\theta_0^2 - \alpha_0^2\right)\right] \tag{10}$$

in which

$$\theta_0 = \mu_0 s \quad \text{and} \quad \alpha_0 = \mu_0 S/2\,, \tag{11}$$

μ_0 is a load parameter obtained from

$$\mu_0^2 = \frac{N_0}{(EI)_0} \tag{12}$$

and

$$\omega_0 = q_0 R / N_0 - 1 \tag{13}$$

is a dimensionless parameter that is a measure of the departure of the linear solution for the axial force q_0R from its actual value N_0.

3 Non-linear equilibrium equation for steel box-section

The non-linear equation of equilibrium for the steel box-section under the distributed loading q_0 can be obtained by the reasoning that the constant membrane strain in Eq. 6 is equal to its value in Eq. 2, when averaged mathematically over the domain $s \in [-S/2, S/2]$. Hence

$$-\frac{N_0}{(EA)_0} = \frac{1}{S} \int_{-S/2}^{S/2} \left(w_0' - \frac{v_0}{R} + \frac{v_0'^2}{2} \right) \mathrm{d}s \, . \tag{14}$$

Noting that $\int w'_0 \mathrm{d}s = w_0| = w_0(S/2) - w_0(-S/2) = 0$, using Eq. 10 and its derivatives in Eq. 14 leads to the transcendental formulation for the non-linear equation of equilibrium as

$$A_0 \omega_0^2 + B_0 \omega_0 + C_0 = 0 \tag{15}$$

in which

$$A_0 = \frac{1}{4\alpha_0^2} \left(5 - \frac{5 \tan \alpha_0}{\alpha_0} + \tan^2 \alpha_0 \right) + \frac{1}{6}; \;\; B_0 = \frac{1}{\alpha_0^2} \left(1 - \frac{\tan \alpha_0}{\alpha_0} \right) + \frac{1}{3}; \;\; C_0 = \frac{\alpha_0^2}{\lambda_0^2} \tag{16}$$

where

$$\lambda_0 = \frac{S^2}{4 r_0 R} \tag{17}$$

is a slenderness parameter where $r_0 = \sqrt{(I_0/A_0)}$ is the radius of gyration of the steel box-section.

4 Strain-displacement relationship and deformations for composite section

Under the self-weight loading for the unpropped steel box-section, Eq. 15 represents an equilibrium configuration, and a new equilibrium configuration is sought after the additional imposed load q_i is applied. This load induces counterpart additional deformations v_i and w_i to v_0 and w_0 respectively, so that the total displacements are

$$v = v_0 + v_i \quad \text{and} \quad w = w_0 + w_i \, . \tag{18}$$

In a Lagrangian description of the displacements from the initially circular arch configuration, the strain-displacement relationship (Eqs. 1 and 2) is

$$\varepsilon = w' - v/R + \tfrac{1}{2} v'^2 \, , \tag{19}$$

and so the additional strain caused by the loading q_i is obtained by subtracting Eq. 1 from Eq. 19 to give

$$\varepsilon_i = \varepsilon_{im} - y v_i'' , \tag{20}$$

for which the membrane strain is

$$\varepsilon_{im} = w_i' - v_i/R + \tfrac{1}{2} v_i'^2 - v_0' v_i' . \tag{21}$$

The increase in total potential can be written similarly to Eqs. 3 and 4 as

$$\Pi_i = \int_{-S/2}^{S/2} F_i\left(w_i', v_i, v_i', v_i''\right) \mathrm{d}s \tag{22}$$

for which

$$F_i = \tfrac{1}{2}\left[(EA)_i \varepsilon_{im}^2 + (EI)_i v_i''^2\right] - q_i v_i \tag{23}$$

and which leads to

$$\varepsilon_{im} = \text{const} = -\frac{N_i}{(EA)_i} \tag{24}$$

in the axial direction, and to

$$(EI)_i v_i^{iv} + N_i v_i'' = q_i - N_i\left(1/R + v_0''\right) \tag{25}$$

in the radial direction, in which N_i is the axial force induced by the loading q_i and $(EA)_i$ and $(EI)_i$ are the axial and flexural stiffnesses of the composite section respectively, in deference to their counterpart values for the steel box-section only. It is worth noting that Eq. 25 is the counterpart of Eq. 8 for the transverse displacements v_i when measured from a deformed configuration v_0 and for which the curvature is $(1/R + v''_0)$.

The solution of Eq. 25, which satisfies the boundary conditions that $v_i = v''_i$ at $s = \pm S/2$ is

$$v_i = \left(\frac{\omega_i + \omega_0}{\mu_i^2 R}\right)\left[\sec\alpha_i \cos\theta_i - 1 + \tfrac{1}{2}\left(\theta_i^2 - \alpha_i^2\right)\right] + \frac{\omega_0\left(\mu_0^2 \sec\alpha_i \cos\theta_i - \mu_i^2 \sec\alpha_0 \cos\theta_0\right)}{\mu_0^2 R\left(\mu_o^2 - \mu_i^2\right)} \tag{26}$$

in which $\mu_i = \sqrt{(N_i/(EI)_i)}$, $\theta_i = \mu_i s$, $\alpha_i = \mu_i S/2$ and $\omega_i = q_i R/N - 1$.

5 Non-linear equilibrium equation for composite section

The same procedure for determining the equations of equilibrium for the composite section using the constant membrane strain can be used as was done for the steel box-section, so that

$$-\frac{N_i}{(EA)_i}=\frac{1}{S}\int_{-S/2}^{S/2}\left(w_i'-\frac{v_i}{R}+\frac{v_i'^2}{2}-v_i'v_0'\right)\mathrm{d}s\,. \tag{27}$$

This produces the non-linear equations of equilibrium in the quadratic transcendental form

$$A_i\omega_i^2+B_i\omega_i+C_i=0 \tag{28}$$

in which

$$A_i=\frac{1}{4\alpha_i^2}\left(5-\frac{5\tan\alpha_i}{\alpha_i}+\tan^2\alpha_i\right)+\frac{1}{6}, \tag{29}$$

$$B_i=\frac{1}{\alpha_i^2}\left(1-\frac{\tan\alpha_i}{\alpha_i}\right)+\frac{1}{3}+2A_i\omega_0, \tag{30}$$

$$C_i=\frac{\alpha_i^2}{\lambda_i^2}+A_i\omega_0^2+B_i\omega_0, \tag{31}$$

and where $\lambda_i = S^2/(4r_iR)$ in which $r_i = \sqrt{(I_i/A_i)}$ is the radius of gyration for the composite cross-section.

6 Illustration

In order to illustrate the analytical procedure developed in the previous sections, a steel welded box-section arch whose cross-section is 2000 mm wide and 1000 mm deep, with top and bottom flange thicknesses of 60 mm and web thicknesses of 30 mm is considered. The radius of the arch is 150 m and the arc length of the arch rib is 60 m so that the included angle is 0.4 radians (the span or chord length is L = 59.601 m and the rise is f = 2.990 m $\approx L/20$). For this steel section, $A_0 = 300 \times 10^3$ mm^2, $I_0 = 65.07 \times 10^9$ mm^4, r_0 = 465.7 mm, λ_0 = 12.88 and under the assumed self weight, q_0 = 73 kN/m with E_0 being taken as 200 kN/mm^2.

Initially, under the wet concrete loading q_0 = 73 kN/m, Eqs. 15 and 16 may be solved iteratively to produce

$$\alpha_0 = 0.8615,\ \ \omega_0 = 0.0314,\ \ N_0 = 10.617 \times 10^6\ \text{N},$$

and from Eq. 10, the deflection at the crown ($\theta_0 = 0$) is $v_0(0)$ = 41.7 mm $\approx f/70$. Using Eq. 6, the membrane strain is $\varepsilon_{0m} = 177 \times 10^{-6}$ and by differentiating Eq. 10, the maximum curvature at the crown is $v''_0(0) = -110.4 \times 10^{-9}$ mm^{-1}, leading to extreme fibre strains of 232.2×10^{-6} and 121.8×10^{-6}, and corresponding stresses of 46.44 N/mm^2 and 24.23 N/mm^2, which are well below the yield stresses of steel. These stresses are shown in Figure 2.

Under the additional action of a superimposed load q_i = 100 kN/m that acts on the composite cross-section (and for which the elastic modulus of the concrete core is taken as 25 kN/mm^2), using transformed area theory (Hall 1984) with a modular ratio of n = 25/200 = 1/8, $(EA)_i = 102.7 \times 10^9$ N, $(EI)_i = 15.77 \times 10^{15}$ Nmm^2, r_i =

391.8 mm and λ_i = 15.314. For this, Eqs. 28 to 31 can be solved iteratively to produce

$$\alpha_i = 0.9563, \ \omega_i = -0.0106, \ N_i = 16.024 \times 10^6 \text{ N},$$

and from Eq. 26, the additional deflection at the crown is $v_i(0)$ = 37.8 mm ≈ f/80. The additional membrane strain is ε_{im} = 156 × 10^{-6} and the additional curvature at the crown is $v''_i(0)$ = -101.9 × 10^{-9} mm^{-1}, leading to extreme fibre strains of 207.0 × 10^{-6} and 101.1 × 10^{-6}, which produce additional steel stresses of 41.4 N/mm^2 and 21.0 N/mm^2 and concrete stresses of 5.17 N/mm^2 and 2.63 N/mm^2. These stresses are shown in Figure 2.

It should be noted that under "classical" theories for linear behaviour, the compression in the arch is qR; the non-vanishing of ω_0 and ω_i indicate that linear theory is inaccurate. In addition, under linear theory for circular arches the bending moment is zero, and the bending stresses in Figure 2 show that flexural effects are significant for the arch in question.

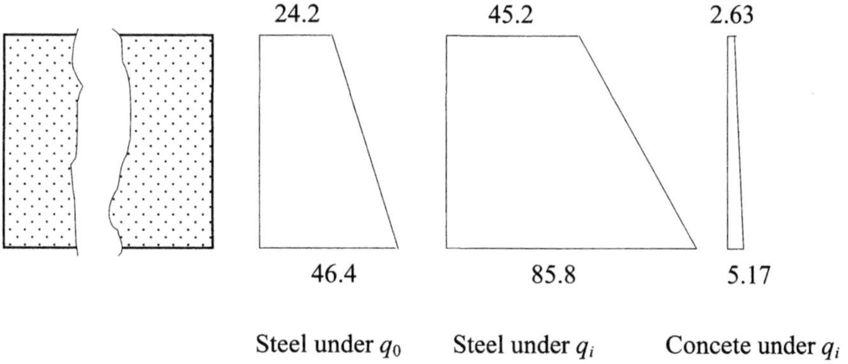

Figure 2. Stresses in composite box-section (N/mm^2).

7 Conclusions

This paper has presented an analysis of a shallow welded steel box-section arch in an analytic formulation that includes the necessary second order strain-displacement relationship required for non-linear elastic analysis. The motivation for the study is the use of box-section arches in bridges, for which a pre-fabricated welded hollow box section is deployed in the initial stages of construction, and then filled with concrete. Construction in this form is unpropped, with the steel section supporting the self weight of the concrete infill and the weight of the steel box itself, and the composite section then supporting additional imposed loading. Because of the limitations of the rise of the arch owing constraints imposed by concrete placement, the arch is necessarily shallow and so non-linear geometric formulations must be used.

A formulation was presented first in which the steel box-section deflected under self weight loading from a constant initial curvature of $1/R$ (R is the radius of the arch), and the equations of equilibrium were formulated in a transcendental form which can be solved simply by iteration. The formulation was then repeated for a composite cross-section, starting from an initial deformation established by the wet loading phase, and similarly the equations of equilibrium were established in transcendental form. The technique was then illustrated with an example of an arch with a span to rise ratio of approximately 20, and it was shown that the axial compression developed in the arch differed from the "classical" value and that bending effects were significant; these bending effects being omitted from classic formulations in which the internal moment is zero. The non-linear effects are significant, cannot be ignored in design, and may have ramifications on the propensity of a shallow arch to buckle either under instantaneous loading or in the long-term by the phenomenon of "creep buckling".

Acknowledgement

The work in this paper was supported by the Australian Research Council through a Federation Fellowship and a Discovery Project awarded to the author.

References

Bradford, M.A. (2006) Buckling of circular steel arches subjected to fire loading. *Welding in the World* **50**, 394-399.

Bradford, M.A., Wang, T., Pi, Y.-L. & Gilbert, R.I. (2007) In-plane stability of parabolic arches with horizontal spring supports. I: Theory. *Journal of Structural Engineering*, ASCE **133** No. 8, 1130-1137.

Hall, A.S. (1984) *An Introduction to the Mechanics of Solids*. Sydney: John Wiley.

Oehlers, D.J. & Bradford, M.A. (1995) *Composite Steel and Concrete Structural Members: Fundamental Behaviour*. Oxford: Pergamon Press.

Pi, Y.-L., Bradford, M.A. & Uy, B. (2002) In-plane stability of arches. *International Journal of Solids and Structures* **39** 105-125.

Timoshenko, S.P. & Gere, J.M. (1961) *Theory of Elastic Stability*. New York: McGraw Hill.

Vlasov, V.Z. (1961) *Thin-Walled Elastic Beams*. Jerusalem: Israel Program for Scientific Translation.

Wang, T., Bradford, M.A. & Gilbert, R.I. (2006) Creep buckling of shallow parabolic concrete arches. *Journal of Structural Engineering*, ASCE **132** No. 10, 1641-1649.

6.3 Numerical Simulation of Buckling of Metal Tubes

Róbert Dúl, Károly Jármai
University of Miskolc, H-3515 Miskolc, Hungary,
robert.dul@cfdengineering.hu, altjar@uni-miskolc.hu

Abstract

This article deals with the deformation specialities of aluminium circular tubes. These tubes could be used to absorb energy of collisions. For example in a test rig created to measure and test seat belts a batch of tubes slow down the test rig and simulate the deformation of a car chassis. Circular tubes with the same cross section and tube length parameters made of three different aluminium magnesium alloys were analysed by finite element method to determine the effect of increasing yield strength values on energy absorbed.

Keywords: *aluminium, buckling, finite element method*

1 Introduction

In case of metal structures built for the automotive industry it is a must to design reliable and cheap parts and assemblies. Besides there is also a strong need to have lighter and safer car bodies. Saving weight and increasing safety by designing more rigid car bodies could be tough challenges, but even those two requirements can be built into one structure by using light metals like aluminium.

Passenger safety can be increased in the case of a car built with an aluminium space frame – like the one that can be seen on Figure 1 – for example with energy absorber elements. Those elements absorb collision energy by deformation. Their form, cross section can be different, some elements can even have buckling initiators to guide the deformation. Such a buckling initiator can be seen on Figure 2. This absorbing element is an aluminium rectangular hollow section with 50mm flange width and 100mm web height.

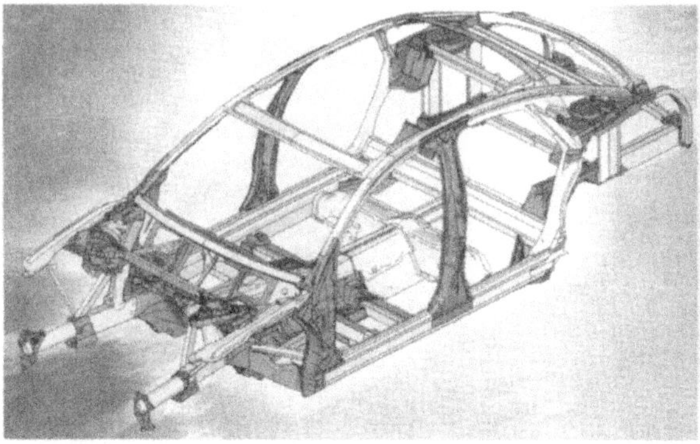

Figure 1.
Aluminium space frame

Figure 2.
Energy absorbing element
with buckling initiators

To determine the dimensions of such aluminium profiles and to choose the best suitable alloy for the purpose a finite element analysis is required.

This article deals with the deformation specialities of an aluminium circular tube that can be used as an energy absorber at the front of a car space frame. Finite element method is used to analyze the deformation of this axially compressed aluminium tube to figure out how the geometry and material properties affect the way of deformation.

2 Tube geometry

To find the initial values of the tube cross section parameters for the FEA we relied on the numerous references written on axially compressed tubes. Andrews et al. (1983) determined the deformation modes of axially compressed circular aluminium tubes based on quasi-static experiments. In their article a table was published that contained the different deformation modes in the function of *L/D* (tube length / inner diameter of the tube) and *t/D* (thickness / inner diameter of the tube).

Seven different deformation modes were derived from the experiments: a) concertina, b) diamond, c) Euler-type, d) concertina and 2 and/or 3-lobe diamond, e) concertina and axisymmetric crushing, f) 2-lobe diamond, g) crushing plus tilting of tube axis.

From the seven types only two of them are considered important in the frame of energy absorption: a) concertina shown on Figure 3.a and b) diamond shown on Figure 3.b. Andrews et al. stated that concertina mode absorbs the highest amount of energy per unit length of tube, so in our finite element analysis we would like to receive only this type of deformation.

The table Andrews et al. published contained the seven modes as a map (see Figure 4), where one can find the predictable tube deformations based on *L/D* and *t/D* ratios.

Figure 3a. Concertina type of deformation

Figure 3b. 3-lobe diamond type of deformation

As the table of Fig. 4. shows the concertina deformation mode can approximately be restricted to the area of $0.5 < L/D < 5$ and $0{,}02 < t/D < 0.1$ where D is the inner diameter of the tube, t is the thickness of the tube, L is the initial length of the tube.

Abramovitz et al. (1997) made detailed experiments with quasi-static and dynamic loads. Referencing their work and based on numerical simulations N.K. Gupta (2000) states in his article that D/t should be between 70 and 90 to receive a concertina mode for deformation. Above that area diamond deformation mode can be predicted.

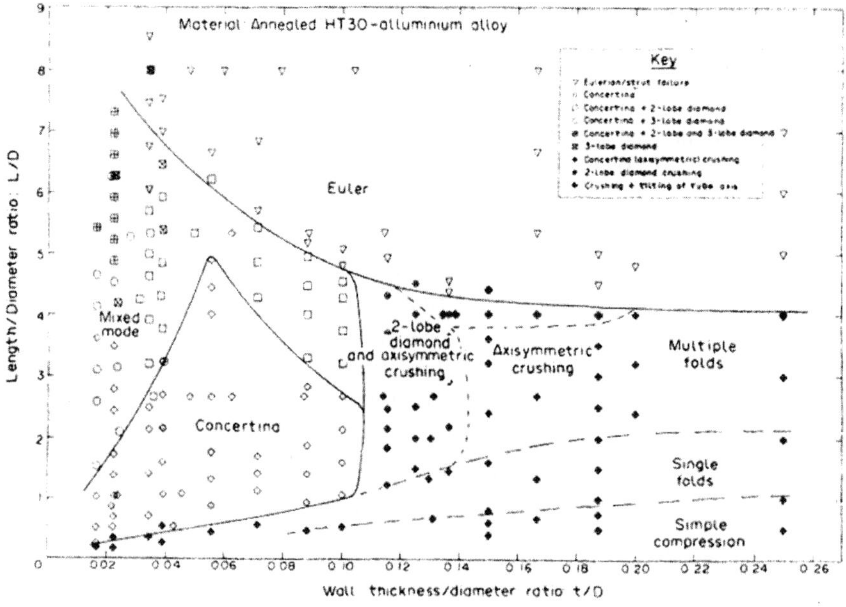

Figure 4. Deformation modes of aluminium tubes created by Andrews et al.

To set the dimensions of the tube cross section the catalogue of standard sections of ALCOA Extrusions was used. The values were chosen from the "middle" of the section table in order to preserve the possibility of both decreasing and increasing the value of the parameters. Supposing that the initial length of the tube is L=250mm, the initial values of the cross section parameters are shown in Table 1.

Table 1. Cross section parameters for finite element analysis

D' [mm]	t [mm]	L/D	t/D
90	4	3.048	0.048

where D' is the outer diameter of the tube.

3 Aluminium alloys

The aluminium – magnesium alloys are widely used in automotive, marine applications since they can be manufactured easily in cold state and they are weldable. With heat treatment their mechanical properties can be significantly improved. From those alloys three were selected for our purposes.

Table 2. Properties of aluminium alloys used for the analysis

Alloy	*State*	$R_{p0.2}$ [MPa]	R_m [Mpa]	A_5 [%]	*E* [MPa]	ν
6060	T1	65	130	15	69500	0.33
6082	T4	110	205	14	70000	0.33
6060	T5	150	190	10	69500	0.33

The finite element analysis had to be made with the same geometry for all aluminium alloys in order to determine the best alloy suitable for energy absorber elements.

4 Analysis

For the quasi-static analysis MSC.PATRAN pre- and post processor and the explicit finite element solver MSC.Dytran were used.

The middle surface of the 250mm high tube was modelled and 50 nodes both on the circumference and length of the tube were defined. Quadratic shell elements were created based on the nodes. To compress the tube two rigid surfaces were also defined at both ends of the tube. One of the rigid surfaces was fixed, the other (we can refer to it as "ram") was moved downwards (in –z direction) 125mm parallel with the axis of the tube. At the fixed end the translational degrees of freedom of the nodes were fixed, the rotation was allowed for both ends.

With defining such a loading condition it is possible to explore the elastic and plastic strain energies while compressing the tube. The software is capable of calculating both of those energies so the energy absorption of such an element can be analysed.

Aluminium was defined as an isotropic material with two constitutive models:

1. Elastic: defined with elastic modulus and Poisson's ratio.
2. Plastic: with elastic-plastic type, with isortopic hardening rule and von Mises yield criterion. The strain rate method was set to piecewise linear.

The elements are expected to touch both the rigid end surfaces and each other so contact was also incorporated into the finite element model.

5 Results

The results of the simulations are in good correlation with the results of the experiments and theoretical investigations that can be found in the references.

The special barrelling that is described by Singace A.A. (1999) can be experienced near the two rigid end plates as it is shown on Figure 5a. Figure 5b and 5c. show the deformed shape of the tube that were reached later in the simulation.

MSC.Dytran is capable of listing several simulation parameters and calculated results. The diagram below shows the energy of distortion expressed in Joule. It can be seen that an alloy with higher yield strength requires higher energy levels to reach the same level of distortion. It can be said that it is recommended to use an alloy with higher yield strength for the same tube length because more energy can be absorbed on unit length.

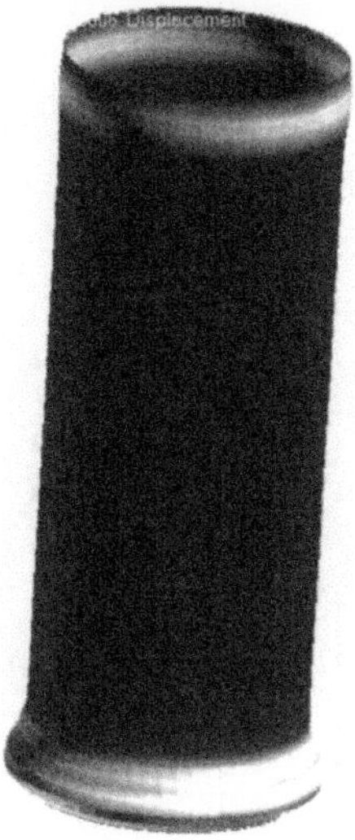

Figure 5.a Barrelling at both ends

Figure 5.b Development of bends in concertina mode

Figure 5.c Development of bends later in the analysis

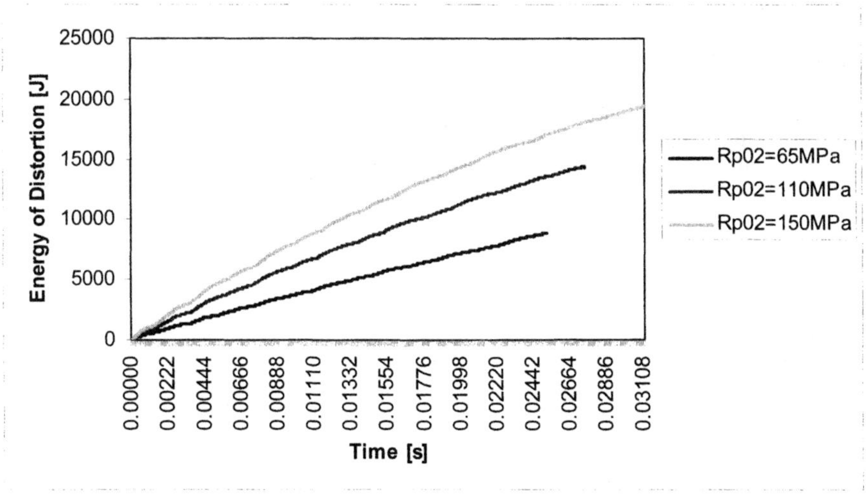

Figure 6. Values of energy of distortion

On Figure 7 the deceleration values calculated on the moving ram can be seen.

From the graphs of decelerations we can derive that the higher the yield strength is the bigger the deceleration of the ram will be. The periodic shape of the graphs came from the periodic way of deformation: when the material collapses and one ring of the deformed tube reaches the ring before, the deceleration values rise and drop the same way.

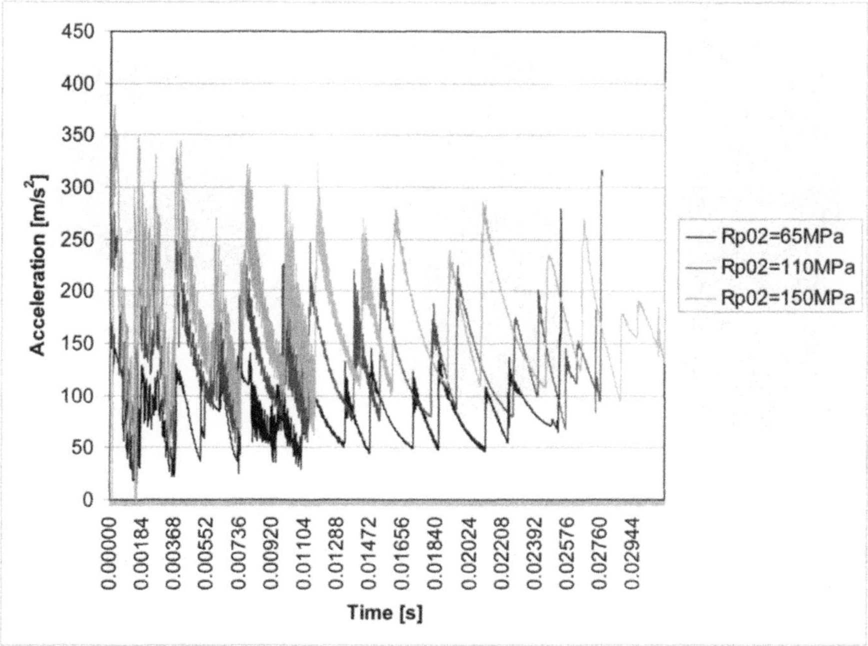

Figure 7. Decelerations of the ram

6 Conclusions

With the help of quasi-static numerical simulations the deformations of aluminium cylindrical tubes were examined. It was shown based on analyses with different yield strengths that the tube made of an alloy with higher yield strength can absorb more energy per unit length than a tube with smaller yield strength value. Therefore, it is recommended to use high-strength aluminium for the shock absorber elements of vehicle frames.

7 References

Andrews,K.R.F., England,G.L. & Ghani,E. (1983) Classification of the axial collapse of cylindrical tubes under quasi-static loading. *Int. Journal of Mechanical Sciences*. **25** No 9-10, 687-696.

Abramovitz, W. & Jones, N. (1997) Transition from initial global bending to progressive buckling of tubes loaded statically and dynamically. *Int. J. Impact Engng.* **19** Nos. 5-6, 415-437.

Gupta N.K. (2000) On non-Axisymmetric collapse of thin tubes. Department of Applied Mechanics, Indian Institute of Technology, Delhi, Hauz Khas, New Delhi – 110016 India. *Internet.*

Singace, A.A. (1999) Axial crushing analysis of tubes deforming in the multi-lobe mode. *Int. Journal of Mechanical Sciences*. **41** 865-890.

6.4 Effect of Longitudinal Weld on Residual Stresses and Strength of Stainless Steel Hollow Sections

Michal Jandera, Josef Machacek
Czech Technical University in Prague, Thakurova 7, 166 29 Praha, Czech Republic, michal.jandera@fsv.cvut.cz, machacek@fsv.cvut.cz

Abstract

Current research at CTU in Prague on presence and influence of residual stresses in austenitic 1.4301 stainless steel hollow sections is listed and discussed. The investigation is focused on cold rolled and longitudinally welded rectangular or square hollow sections (RHS/SHS). Results of experimental investigation of residual stresses induced by cold-forming and welding process are described in detail. Both surface and through thickness residual stresses have been measured by X-ray diffraction method. Subsequently eleven stub column tests were carried out and used to calibrate FEM model. The numerical model originates from ABAQUS software and includes non-linear stress-strain diagram and enhanced strength corner properties to obtain relevant credibility of the solution. Parametrical study investigating influence of both membrane and bending residual stresses on local and global buckling are based on Cruise (2007) residual stress model. Finally, summary of the results and some recommendations for inclusion of the residual stresses in such profiles into design are given.

Keywords: *stainless steel, hollow section, residual stress, stub column test*

1 Introduction

In previous years, the need of alloyed steels in structures, namely high strength and stainless steels has significantly increased. Stainless steels are becoming used as load carrying construction material because of superior corrosion resistance and attractive surface finish together with good strength to weight ratio. These steels, despite a higher material cost, may lead to lower-cost constructions provided all life cycle of the structure is taken into account.

The grade used in this study is 1.4301 austenitic steel, which is probably the most common stainless steel grade used for load carrying structures. The whole research concerns the typical stainless steel element, a cold-rolled box section. In previous decades intensive research was performed describing significant behaviour differences and dissimilarities of stainless steels in comparison with common carbon ones, such as non-linear and asymmetric stress-strain diagram, anisotropy of stainless steels, increase of yield proof and collapse strengths due to cold-working process and influence of initial imperfections, etc., see Euro-Inox (2006). This paper aims to show an importance of residual stresses due to welding and forming in rectangular hollow sections analysed in parametrical study.

2 Residual stresses measurement

To find complete residual stress pattern both along and trough-thickness of the hollow sections the decision of measuring residual stresses by combination of two methods, i.e. X-ray diffraction method and sectioning method was adopted.

Sectioning method is usually very reliable method which with strain-gauges glued on both surfaces of specimen allows both membrane and bending stresses to be monitored. The results of the sectioning method are usually rather coarse and the through-thickness distribution of stresses by this method is not exactly predictable.

The X-ray diffraction method is based on measuring of crystalline planes distance change giving elastic strain, which is described by Brag's law (e.g. in Jandera & Machacek (2007)). The result represents stress at specimen surface up to a depth of 5-10 μm. For through-thickness stress pattern measurement separate layers are electrolytically removed and new measurements follow.

2.1 *X-ray diffraction method*

Results of the measurements on the outer surface of a quarter of rectangular hollow specimen are presented (Figure 1) for both longitudinal and transversal directions with respect to specimen axis. Most of the section surface is compressed what subsequently can positively affect stress-corrosion-cracking resistance of cold-rolled sections. The magnitude of the transversal stresses in general is doubled in comparison with the longitudinal one.

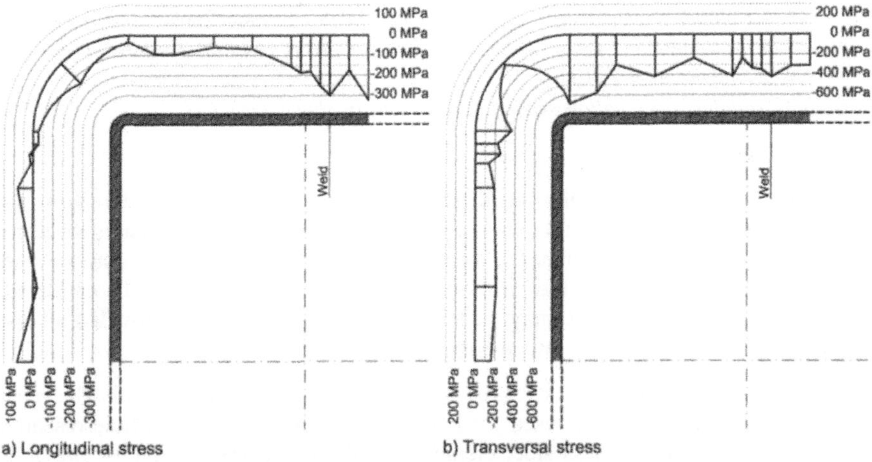

Figure 1. Outside surface residual stresses of RHS 100x80x2 mm relating to a) longitudinal and b) transversal direction of specimen axis.

The through-thickness gradients were measured in ten positions and within half of a web-thickness, but only two of them, in the weld area, gave successful results (Figure 2). Other measurements did not show a good diffraction pattern due to large grains of austenitic structure under the surface of material, even despite of 10 mm oscillation of irradiating X-ray beam. More information is given in Jandera & Machacek (2007).

The two trough-thickness measurements proved tensile residual stresses along the half-thickness of the specimen, excluding thin compressive surface layer induced probably by contact with forming machine. Longitudinal stress pattern seems to form plastic block-like distribution what supports idealisation of Cruise's (2007) bending stress distribution. To verify this idealisation more measurements are under progress. From the residual stress distributions in these two positions is also evident that the tensile membrane stress blocks usually used to interpret influence of weld in common carbon steel cannot be used to describe residual stresses due to high-frequency welding of these thin-walled cold formed stainless steel sections.

2.2 *Sectioning method*

Destructive sectioning method for residual stress detection was also used in this investigation (for SHS 100x100x3 and 120x120x4 mm) and the evaluation of the results is under progress. The results should enable to compare and verify the X-ray diffraction method results and supplement Cruise's (2007) measurements to receive general residual stress pattern. The width of each stripe in the sectioning of the specimen was 20 mm, which together with 10x5 mm measuring grid for strain-gauges placed on both sides of the stripe led to high accuracy of results. However, the strong gradient of stresses in a section is not possible to measure and have to be provided from X-ray diffraction or other suitable method measurements.

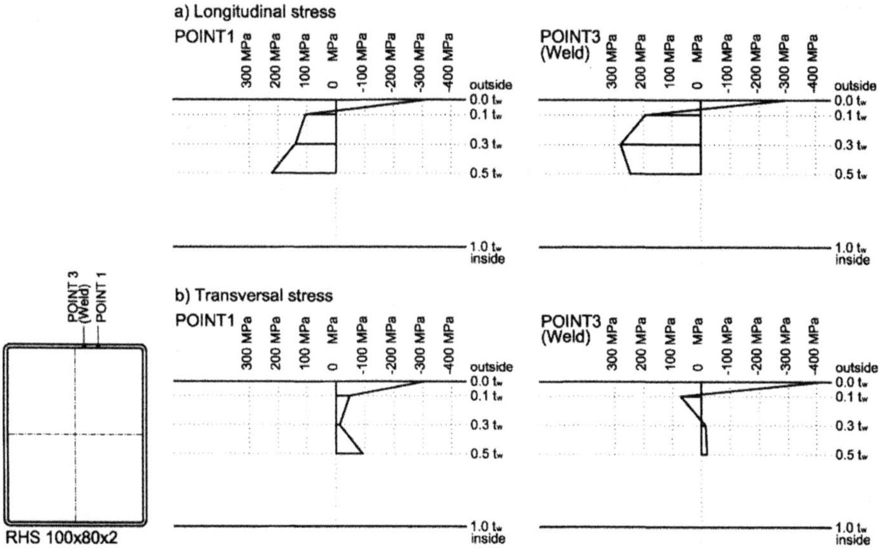

Figure 2. Two half through-thickness residual stress measurements in weld area with respect to a) longitudinal and b) transversal direction of specimen axis for RHS 100x80x2 mm.

3 SHS stub column tests

Eleven SHS stub columns were tested (one of them with annealed residual stresses). Web of the specimens ranged from 60 to 100 mm with thickness 2, 3 and 4 mm. The length of each column was equal to three width of the specimen, which is expected to provide stable column without global buckling.

The tests were accomplished with help of force-controlled hydraulic jack with strain stabilisation for several load levels to get dynamic response free results. The values received from the testing are presented in Tab. 1 and Figure 3. The end shortenings of tested specimens were recorded by three inductive gauges and the strain in corners by strain-gauges glued in each corner of relevant section.

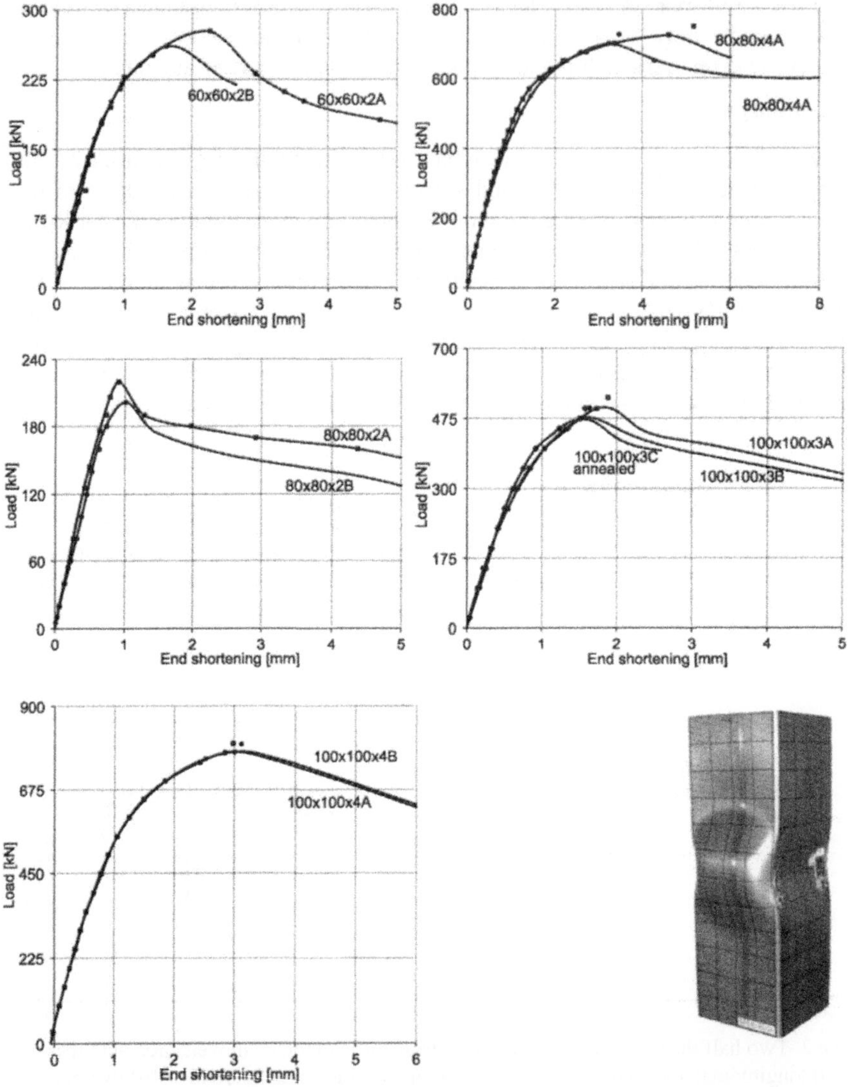

Figure 3. Load-shortening stub column relationships and photo of tested specimen.

4 FE model and parametrical study of residual stress influence

The influence of residual stresses was studied numerically on geometrically and materially non-linear model including imperfections (GMNIA) using ABAQUS (2004) software package. Authors verified the numerical model on tests described above and in addition also on tests performed by Gardner (2002). The following parametrical study on influence of residual stresses was carried out based on

Cruise's (2007) bending residual stress model for upper 5% fractile of stress magnitudes. Both membrane and bending parts of residual stresses observed and measured for SHS 150x150x4 mm were used.

Table 1. Stub column test results.

Specimen	Length [mm]	Depth [mm]	Width [mm]	Thickness [mm]	Area [mm^2]	Ultimate load F_u [kN]	End shortening at F_u [mm]
SHS 60x60x2A	180	60.06	60.14	2.22	528.5	274	2.32
SHS 60x60x2B	180	60.07	60.10	2.11	505.6	260	1.61
SHS 80x80x2A	240	79.86	79.92	1.86	597.9	222	0.90
SHS 80x80x2B	240	79.76	80.04	1.82	585.1	201.5	1.01
SHS 80x80x4A	240	80.28	80.41	3.88	1284.8	750	5.18
SHS 80x80x4B	240	80.17	80.42	3.80	1259.8	725	3.47
SHS 100x100x3A	300	99.91	100.00	2.71	1105.1	576	1.89
SHS 100x100x3B	300	99.91	100.10	3.00	1218.6	550	1.64
SHS 100x100x3C*	300	99.98	100.11	3.08	1249.0	548	-
SHS 100x100x4A	300	99.86	99.92	3.69	1497.7	801	2.99
SHS 100x100x4B	300	99.86	99.92	3.69	1497.7	798	3.14

* Annealing process could cause decrease of a yield and proof strength of cold-worked material

For modelling of thin-walled cross sections the second order thin-shell element S9R5 with reduced integration and five degrees of freedom at each node was used. These elements are suitable for modelling of very thin plates with thickness-to-span ratio lower than 1/15, where the transverse shear strains are assumed to vanish. This is fulfilled for all the parametrical study and justified by numerical verification. S9R5 does not lead to hour-glassing because of the second order character of the element, which may occur at elements with reduced integration and the element in general is computational-timesaving. The choice followed testing of several shell elements and mesh-sensitivity analysis.

4.1 *Influence of residual stresses on global buckling*

The influence of bending and membrane residual stresses is presented in Figure 4. Positive influence of residual stresses on load capacity for non-dimensional slenderness up to 1.4 is evident following negative one beyond this value. The increase of load strength in the watched region reaches -2 to 10 % and is caused by non-linearity of stress-strain diagram where bending residual stresses increase the tangential modulus of elasticity for important part of the relationship.

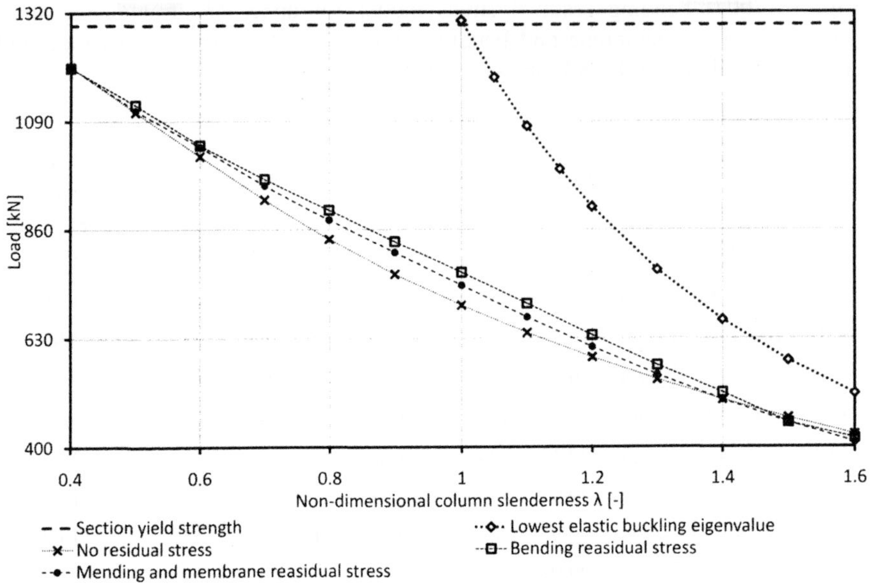

Figure 4. Parametrical study of residual stress influence on load capacity of initially deflected columns subjected to global buckling. $\lambda=\sqrt{(F_{0.2}/F_{crit})}$, where F_{crit} is the elastic buckling strength of column subjected to global buckling and $F_{0.2}$ the yield strength capacity of the section.

This is illustrated in Figure 5a) where the maximal possible value of bending residual stresses with box-like plastic bending stress distribution was employed. A strong decrease of relationship containing the residual stresses (dashed line) is evident. Consequently, the tangential modulus of elasticity plotted in Figure 5 b) shows significant decrease up to $\varepsilon = 0.1$ % and increase beyond this value. These higher strains ($\varepsilon \approx 0.15$ %) are important for the most sensitive non-dimensional web slenderness range ($\lambda \approx 0.8 \div 1.2$) and therefore, cause increase of the load capacity.

On the contrary, when higher column slenderness ($\lambda > 1.5$) is considered the lower strains are reached at ultimate strength. Therefore the tangential modulus for material where bending stresses were included is lower than tangential modulus of material without residual stresses (Figure 5b) and residual stresses can lead to reduction of ultimate strength. Nevertheless, the higher web slenderness is practically residual stress non-sensitive and the decrease due to residual stresses is up to 2 % of the load capacity.

4.2 *Influence of residual stresses on local buckling*

The similar conclusions result from another parametrical study concerning local buckling shown in Figure 6. The residual stresses increase the load capacity of slender compressed section up to 5.3 % according to web slenderness. The influence of membrane residual stresses seems to be insignificant in comparison with the influence of the bending part of the residual stresses.

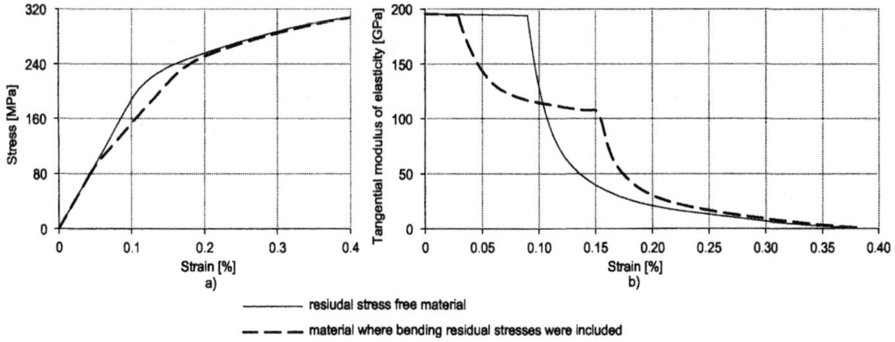

Figure 5. An example of influence of bending residual stress on a) stress-strain diagram and b) tangential modulus of elasticity.

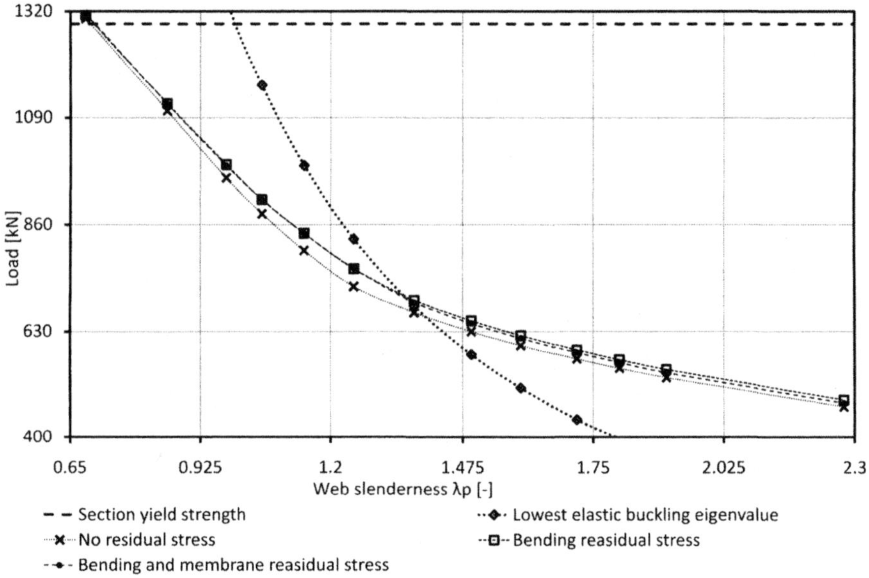

Figure 6. Parametrical study of residual stress influence on load capacity of slender hollow sections subjected to local web-buckling. $\lambda_p=\sqrt{(F_{0.2}/F_{p.crit})}$, where $F_{p.crit}$ is the elastic buckling strength of short column subjected to local buckling and $F_{0.2}$ the yield strength capacity of the section.

5 Conclusions

Investigation of hollow structural elements made of austenitic steels with special attention to introduction and influence of residual stresses was described.

X-ray diffraction method was used for detection of residual stresses due to cold forming and welding of SHS/RHS elements. The measured residual stresses at thin layer of outer surface of the specimen were in compression nearly in all positions of measured surface. The trough-thickness measurements displayed that the tension area in vicinity of the weld is much larger than commonly regarded in mild carbon steel. Tension residual stresses inside the thickness of the web are partly compensated with surface compression stresses. Outside of the weld area the surface residual stresses are moderate and in web without weld even change in sign.

Experimental results of large series of SHS stainless steel stub columns under compression are presented including the full load-shortening relationships.

Finally, numerical modelling and GMNIA of the columns is described. Paradoxically, the residual stresses in these sections improve the behaviour and increase the load capacity. Parametrical study of residual stresses influence showed increase of load capacity by 0 to 5.3 % for section subjected to local buckling and -2 to 10 % for columns subjected to global buckling according to web and column slenderness, respectively, due to effect of these stresses.

Acknowledgement

The financial support of the Czech Ministry of Education (Grant MSM 6840770001) is gratefully acknowledged.

References

ABAQUS (2004) *Analysis User's Manual*, Volumes I.-VI., Version 6.5, Hibbitt, Karlsson & Sorensen, Inc. Pawtucket, USA.

Cruise, R. B. (2007) The influence of production route on the response of structural stainless steel members, *PhD thesis, Imperial College London, UK*, 428 p.

Euro-Inox (2006) *Design manual for structural stainless steel*, Brussels, Belgium, 199 p.

Gardner, L. (2002) A new approach to structural stainless steel design, *PhD thesis, Imperial College London, UK*, 299 p.

Jandera, M. & Machacek, J. (2007) Residual stresses and strength of hollow stainless steel sections. In: *Proc. 9th International Conference Modern Building Materials, Structures and Techniques, Vilnius, Lithuania*, pp. 262-263 + CD

6.5 Optimum Design of Trussed Columns

Ferenc Orbán
University of Pécs, H-7624 Pécs, Hungary, e-mail:orb@witch.pmmf.hu

Abstract

Trussed columns for overhead lines for power transmission were designed about 40 years ago. The aim of this study is to elaborate the optimum design for these columns with different cross sections of bars. The weight was the objective function. Different constraints were taken into account eg. global buckling constraints for compressed members, displacement constraints for the column and frequency analysis. Mathematical problem-solving software was used to calculate the optimum values.

Keywords: *truss structure, optimum design, FEM analysis*

1 Introduction

Columns for overhead lines for power transmission are often constructed with rectangular trussed steel section. Earlier the cross – section of struts were designed only by angle profile. The goal of this study is to investigate how economic these columns are, however, we take only the mass into account. Further examinations were carried out for these structures when we used CHS (circular hollow section).
Supporting columns are proportioned to different loads and to their corresponding combinations.
These are:

- Dead load,
- Wind load,
- Uneven wiring internal forces.

The load carrying capacity of the column can be characterized by the so called forces reduced to the peak. For optimum design two loads were chosen: wind load and force reduced to the peak. Both of them are constrained by buckling capacity of the bars. While carrying out an optimum design the displacement and eigenvalue frequency were taken into account.
The general arrangement of trussed columns for overhead lines for power transmission can be seen in Figure 1.

Our main goal is to determine dimensions of parts of the columns. Design of head structures is not part of this study as it has several electric and constructional limitations.

2 Optimum design of trussed columns for buckling

The internal bar forces of truss structure are calculated according to pin joint analysis.

The chord member force:

$$S_c = \frac{N}{4} + \frac{F_r l}{2 \cdot a} \tag{1}$$

The dimensioning condition relates to buckling capacity of struts.
Check for global buckling according to Eurocode 3 (2002):

$$\frac{S_c}{A} \leq \chi \cdot \frac{f_y}{\gamma_1} \quad ; \quad \gamma_1 = 1{,}1 \tag{2}$$

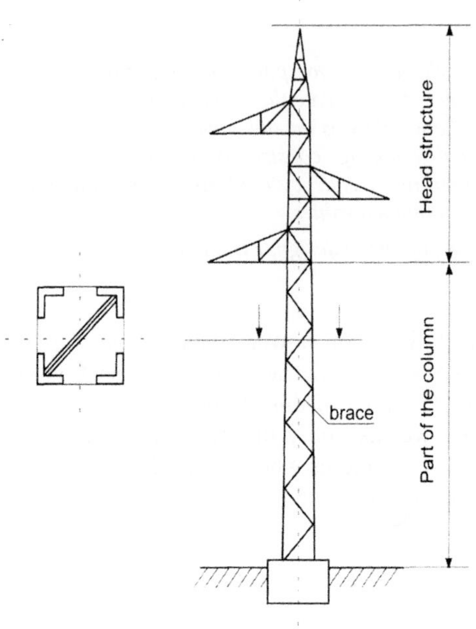

Figure 1. General arrangement of trussed columns.

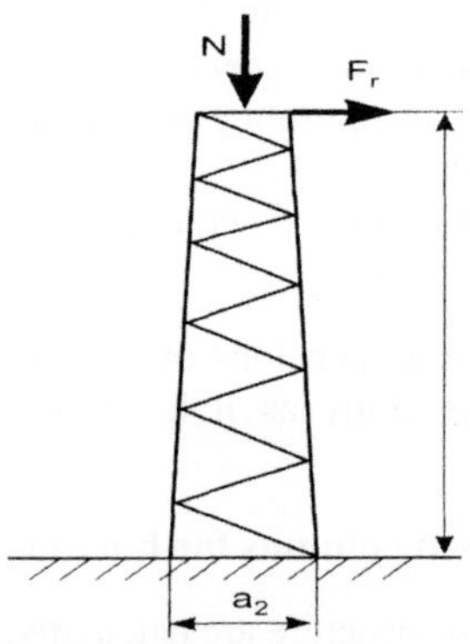

Figure 2. Dimensions and loads of the trussed columns.

χ can be determined in function of slenderness.

$$\lambda = \frac{KL}{r \cdot \lambda_E} \tag{3}$$

L is the length of the compression member,
r is the radius of gyration, K is effective length factor. For brace members the slenderness is limited.

$$r \geq \frac{K \cdot L}{\lambda_{MAX}} \quad ; \quad \lambda_{MAX} = 150 \tag{4}$$

The effective length factor of the chords is 0.76.
At the brace members $K = 1$

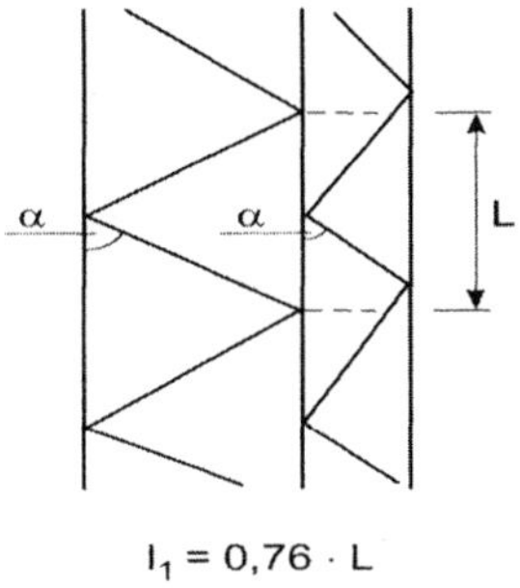

Figure 3. Effective length of chords.

In case of angle profile the cross-section is:

$$A = 2 \cdot b \cdot t \tag{5}$$

b is the side length of the equal leg angle profile.
The radius of gyration is

$$r = \frac{b}{5{,}17} \tag{6}$$

We make an approach for the flexural buckling factor

$$\chi = 1{,}122 - 0 \cdot 57476\bar{\lambda} \quad if: \quad 0{,}2 \leq \bar{\lambda} \leq 1 \tag{7}$$

At CHS of the cross section:

$$A = D\pi \cdot t \tag{8}$$

D – middle diameter.
The radius of gyration is

$$r = \frac{D}{\sqrt{8}} \tag{9}$$

The flexural buckling factor is

$$\chi = 1{,}00658 + 0{,}018107\bar{\lambda} - 0{,}3566\bar{\lambda}^2 \tag{10}$$

3 Optimum design for the part of column

The type of the column under investigation is V20 – 2500, means that in this case the column is 20 m high and the force reduced to the peak is 25 kN (2500 kp).

Forces reduction to the peak is based on the requirement of the moment of all the forces should be equal with the moment reduced to the foundation (Figure 5).

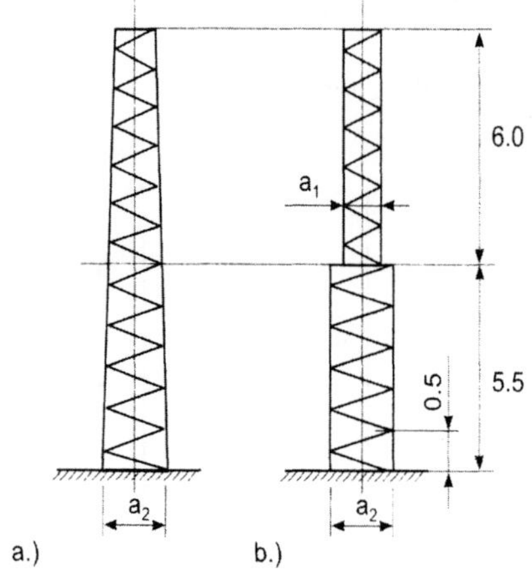

Figure 4. Arrangement of columns.

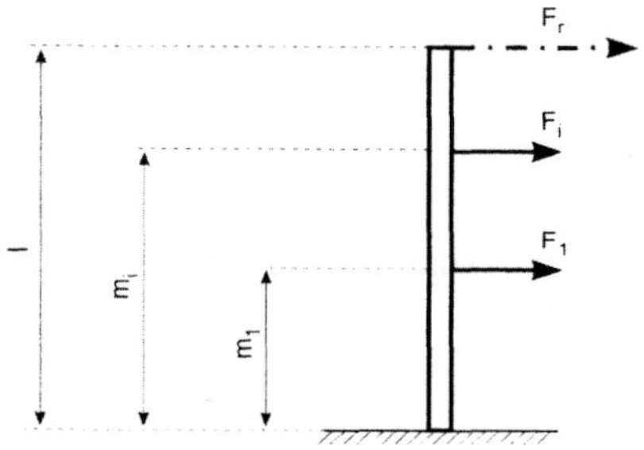

Figure 5. Explanation of reduced force

$$F_r = \frac{\sum F_i \cdot m_i}{l} \tag{11}$$

The optimization was carried out when the load is peak force and the cross section is angle profile. (Table 1 and Table 2.) *m* is mass of the column with standardized values.

Table 1. Optimized values for upper part of the column.

a_1 *mm*	S_c *kN*	*Chord member*	l_r *mm brace*	*brace section*	*m/kg*
800	228	L65x65x11	943	L35x35x4	147
900	202,7	L75x75x8	1029	L40x40x4	**134**
1000	182,5	L75x75x8	1118	L45x45x5	140,8
1100	166	L75x75x7	1208	L45x45x5	144
1200	152	L70x70x7	1300	L50x50x5	141

Table 2. Optimized values for bottom part of columns

a_2 *mm*	S_c *kN*	*Chord member*	l_r*mm brace*	*brace section*	*m kg*
1000	502,5	L80x80x10	1118	L40x40x4	160,6
1100	228,4	L75x75x8	1208	L45x45x5	167
1200	209,3	L75x75x8	1300	L45x45x5	**154,6**
1300	193,2	L75x75x8	1393	L50x50x5	157,6
1400	179,4	L75x75x8	1480	L55x55x5	167,3

Optimization was carried out when cross-section is CHS.

Table 3. Optimized values for upper part of columns.

a_1 *mm*	*S kN*	*D cm*	l_r*mm brace*	*brace section*	*m kg*
1000	182,5	6,98	1118	∅35x3	115,7
1100	166	6,39	1208	∅35x3	112,4
1200	152	5,9	1300	∅35x3	107,4
1300	140,3	5,49	1393	∅35x3	**105,8**
1400	130,3	5,14	1480	∅35x3	106,7

Table 4. Optimized values for bottom part of columns.

a_2 *mm*	S_c*kN*	*D cm*	l_r *mm brace*	*brace section*	*m kg*
1300	193,25	7,36	1393	∅35x3	118,7
1400	179,4	6,874	1486	∅35x3	115,7
1500	167,5	6,449	1581	∅35x3	112,8
1600	157	6,079	1676	∅35x3	**109,9**
1700	147,8	5,75	1772	∅35x3	115

Standardized values for the upper part of column is ∅60x4 and the bottom part is ∅65x4. The dimensions of the brace members are ∅35x3.

4 Optimum design of the column for wind load

The load case when the direction of the wind is perpendicular to the wire was examined. F_{wo} is value of the wind load for wire, when the distance between columns is 150 m. F_{w1} is the wind load acting on the head structure.
The wind load to column depends on the wind pressure the surface and the form of the cross-section. The form factor of the angle profile is 1,4 and CHS is 0,7 .

Table 5. Optimized values for angle profile (*optimum is marked by bold letters*).

a_2	*b cm*	*m kg*	*Chord section*	*m kg*
700	6,66	181,2	L70x70x6	194,6
800	6,0	176,1	L60x60x6	**177,08**
900	5,64	174,9	L60x60x6	181,9
1000	5,3	184,2	L55x55x6	179,5

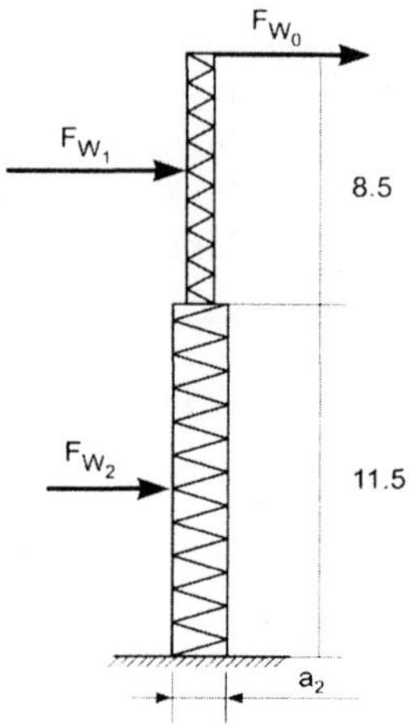

Figure 6. Wind forces to the column.

Table 6. Optimized values for CHS

A_2	*D cm*	*m kg*	*Chord section*	*M kg*
800	4,23	124,1	∅48x4	151,2
900	3,848	120,7	∅42x4	**142,3**
1000	3,55	124,6	∅40x4	145,0
1100	3,31	121,2	∅38x4	142,9

5 Optimum design for displacement

The displacement of truss structure can be calculated with the formula

$$w = \sum \frac{S_i \cdot s_i \cdot l_i}{A_i \cdot E} \tag{12}$$

Where: S_i , internal force, s_i internal force when acting force is 1 N, l_i, length of bars, A_i cross-section of bars.
At cantilever:

$$w_{adm} = \frac{l}{150} \tag{13}$$

Figure 4.b shows a simplification of the Figure 4.a structure to ease the calculation of the mass. According to Figure 4.b we got the following condition:

$$w = \frac{14.79}{A_1 \cdot a_1^2} + \frac{39.36}{A_2 \cdot a_2^2} + 9{,}65 \leq 44\ mm \tag{14}$$

A_1 , A_2 , are the cross-section of the chord members.
The minimum cross-section area was determined from buckling condition.

Table 7. Optimized values for angle profile.

a_1 *mm*	A_1 *mm*^{2}	a_2 *mm*	A_2 *mm*^{2}	*m kg*
900	1000	1600	1000	298,5
1000	1000	1400	1026	**295,55**
1100	1000	1300	1052,5	296,73

The standardized section for optimized value is L75x75x8

Table 8. Optimized values for CHS

a_1 *mm*	A_1 *mm*^{2}	a_2 *mm*	A_2 *mm*^{2}	*m kg*
1000	800	1800	800	225,2
1100	800	1600	806	**223,3**
1200	800	1500	813	224,3

The standardized section for optimized value is ∅70x4.

6 Design for natural frequency

In general, the minimum eigenvalue frequency is determined.

$$f_1 \geq f_{adm} \tag{15}$$

The frequency examination was carried out by FEM. We make comparisons between values, FEM calculations and Eq 16. See Table 9.

$$\alpha_1 = \frac{3{,}52}{L^2}\sqrt{\frac{EI}{m_0}} \quad ; \quad f_1 = \frac{\alpha_1}{2\pi} \tag{16}$$

The trussed column was assumed to be a beam.

$$I = A \cdot a_1^2 \tag{17}$$

m_0, mass of a 1 m long column.

$$m_0 = \left(4 \cdot A_{chord} + 8 \cdot \sqrt{a_1^2 + 0{,}5^2}\, A_{brace}\right)\rho \tag{18}$$

The column length was 20 m.
According to FEM and Eq. (16) calculations, the eigenvalue frequency is increasing, if a_1 dimension is bigger but the chords cross–section area has less influence on the value of frequency.

Table 9. Eigenfrequency values for column.

a_2 *mm*	*Chord section*	f_1 H_z *COSMOS*	*m kg*	f_1 *theoretical*
1500	L70x70x7	3,33	1543	3,36
	L75x75x7	3,52	1587	3,43
	L90x90x9	3,79	1926	3,86
1800	L70x70x7	3,74	1714	3,9
	L75x75x7	3,8	1759	3,91
	L90x90x9	4,1	2098	4,44

7 Conclusion

According to optimum values we got smaller mass for CHS both for peak force and wind load, when the constraint is the buckling of strut. For displacement restriction the CHS gave smaller mass.

The standards determine a minimum eigenvalue frequency value. It is possible to fulfil the frequency condition, if we increase the distance of legs then we have to look for a cross section area value to get prescribed frequency.

References

Eurocode 3 (2002) Design of steel structures. Part 1-1: *General structural rules.* Brussel, CEN.

Farkas, J. & Jármai, K. (1997) *Analysis and optimum design of metal structures* Balkema, Rotterdam-Brookfield.

Farkas J. & Jármai, K. (2003) *Economic design of metal structures.* Rotterdam, Millpress.

Section 7

Plated structures

7.1 Elastic Flexural-torsional Buckling of Web-tapered Fabricated Cantilevers

Mark A Bradford
The University of New South Wales, UNSW Sydney, NSW, Australia,
m.bradford@unsw.edu.au

Abstract

This paper develops a simple, finite element-based method for predicting the elastic flexural-torsional buckling loads of steel I-section members that contain tapered webs, which are fabricated by welding. The formulation is bifurcative, and is based on a beam or line element with eight assemblable buckling degrees of freedom. Some previously reported studies have been shown to fail to account for the component of the flange moments during buckling in the longitudinal direction, which vanishes for a prismatic member, and whose omission leads to erroneous results. The formulation herein accounts for the inclined flange moments. The results of the buckling analysis are useful, insofar as the Eurocode 3 does not present prescriptive equations for the design of tapered members and directs the user to the results of research studies of the type developed herein. Results are given for tapered cantilever beams.

Keywords: *Buckling, cantilever, elastic, non-prismatic, web-tapered.*

1 Introduction

Contemporary welding practice allows for the easy fabrication of tapered steel members, and this tapering can be extremely beneficial if the member is tapered in such a way as to optimize material usage, in particular, by tapering the member with respect to the variation of the bending moment along its length. One such example is a cantilever loaded transversely by a tip load or by a uniformly distributed load, for which the bending moment varies from zero at the tip to a maximum at the cantilever root, and so an I-section member whose web is tapered allows for this variation of bending moment to be utilised optimally. A web-tapered member may be fabricated by diagonally cutting the web of a hot-rolled member, rotating it, and welding the web, or by welding two flange plates to a trapezoidal tapered web plate.

Tapering members in such a way is not new, and was reported in 1952 by Amirikian, and by many others since. Research on the flexural-torsional buckling of tapered steel members has also been quite extensive, with Kitipornchai & Trahair (1972) providing a detailed review of work prior to 1971. In general, the problem cannot be solved with solutions in closed form, and so numerical formulations have been developed; some of the earlier contributions being reported by Lee *et al.* (1972), Nethercot (1973), Morrell & Lee (1974), Taylor *et al.* (1974), Horne & Morris (1977), Brown (1981), Nakane (1984), Wekezer (1985), Bradford & Cuk (1988) and Chan (1990). One of the fundamental issues in the buckling of I-sections whose webs are tapered is that of bimoments; in a member that is tapered the flange moments have components in the longitudinal direction that do not exist in a prismatic member and as a result, some of the early formulations have been shown to be erroneous (Ronagh & Bradford, 1994) because this term is not included. In particular, the use of prismatic members for modelling buckling leads to convergence to the wrong result. More general formulations for the buckling of arbitrary members which are not restricted to prismatic geometries have been reported by Ronagh & Bradford (1999), Ronagh *et al.* (2000a,b) and Pi & Bradford (2001a,b).

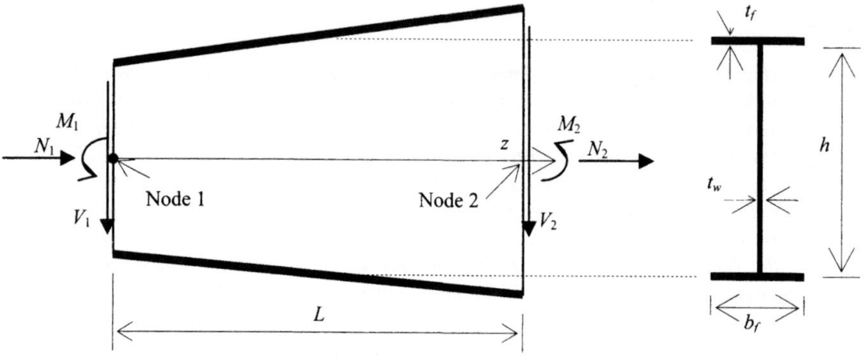

(a) Elevation of element (b) Cross-section at $z \in [0, L]$

Figure 1. Tapered web finite element.

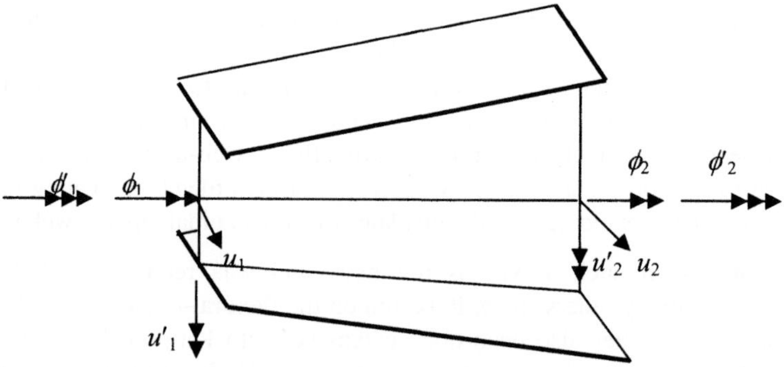

Figure 2. Assemblable buckling degrees of freedom.

Several design codes of practice provide guidance for the strength design of tapered members incorporating flexural-torsional buckling, although some disquiet has been reported on the methodology for generally designing non-prismatic members in these codes (Bradford *et al.*, 2002). The Australian AS4100 (SA, 1998) provides both prescriptive equations or allows for the use of elastic buckling solutions in the process of "design by buckling analysis" (Trahair *et al.*, 2007); the Eurocode 3 (BSI, 2005) also allows for this process but it does not provide any prescriptive guidance and mentions in Clause 6.3.4(3) that finite element analysis may be used. In response to the need for solutions for the elastic buckling of tapered beams designed according to EC3, this paper outlines a simple bifurcation-type analysis that may be used to produce buckling solutions to aid the design of tapered I-beams fabricated by welding, and that can lead to a comprehensive design formulation being developed for EC3.

2 Finite element formulation

2.1 *Deformations*

The web-tapered element is shown in Figure 1; the z axis is located at the mid-height of the web and the element is subjected to end actions λN_i, λM_i, λV_i (i = 1 denotes node 1 at z = 0, i = 2 denotes node 2 at z = L) and N is the axial force, M is the bending moment and V is the shear force. The load factor λ, by which the stress resultants in the member are scaled, reaches its bifurcation value λ_{cr} at elastic flexural-torsional buckling so that $\lambda_{cr}N_i$, $\lambda_{cr}M_i$, $\lambda_{cr}V_i$ (i = 1, 2) are the applied end actions at buckling. The buckled configuration of the rigid cross-section (distortional buckling is not included in the formulation) is defined by the lateral deformation u and twist ϕ from an initial unbuckled configuration defined by $u = \phi = 0$ using cubic interpolation functions lengthwise, so that

$$\mathbf{u} = \begin{Bmatrix} u \\ \phi \end{Bmatrix} = \mathbf{M}\boldsymbol{\alpha} = \begin{bmatrix} L & L\xi & L\xi^2 & L\xi^3 & 0 & 0 & 0 & 0 \\ 0 & 0 & 0 & 0 & 1 & \xi & \xi^2 & \xi^3 \end{bmatrix} \begin{Bmatrix} \alpha_1 \\ \alpha_2 \\ \vdots \\ \alpha_8 \end{Bmatrix}, \tag{1}$$

where $\xi = z/L$ and $\alpha_1, \alpha_2, .., \alpha_8$ are kernel buckling freedoms. By defining the vector of assemblable freedoms shown in Figure 2 as

$$\mathbf{q} = \{u_1 \quad u_2 \quad u_1' \quad u_2' \quad \phi_1 \quad \phi_2 \quad \phi_1' \quad \phi_2'\}^T \tag{2}$$

where $(\ )' \equiv d(\ )/dz$, the buckling deformations can be written in terms of the assemblable freedoms from

$$\mathbf{u} = \mathbf{M}\cdot\mathbf{C}\cdot\mathbf{q} = \mathbf{M}\cdot\begin{bmatrix} \mathbf{C}_u & \mathbf{0} \\ \mathbf{0} & \mathbf{C}_\phi \end{bmatrix}\cdot\mathbf{q}, \tag{3}$$

in which

$$\mathbf{C}_\phi = L\mathbf{C}_u = \begin{bmatrix} 1 & 0 & 0 & 0 \\ 0 & 0 & L & 0 \\ -3 & 3 & -2L & -L \\ 2 & -2 & L & L \end{bmatrix}. \tag{4}$$

2.2 *Stiffness matrix*

The element can be considered to comprise of two flanges experiencing flexural and membrane actions, and a web experiencing flexural actions, during buckling. The resistance of the flange to membrane actions is provided by its minor axis second moment of area I_f and to torsional actions by its torsion constant J_f; the web resists flexural action by bending through its minor axis second moment of area I_w and torsion by its torsion constant J_w. The increase in strain energy during buckling is

$$U = \tfrac{1}{2}\int_0^L \left[EI_f u_T''^2 + EI_f u_B''^2 + EI_w u_w''^2 + \left(2GJ_f + GJ_w\right)\phi'^2 \right] \mathrm{d}z \,, \tag{5}$$

where the flange displacements are

$$u_T = u + (h/2)\phi; \quad u_B = u - (h/2)\phi \,. \tag{6}$$

Substituting Eq. 6 into Eq. 5 and noting that $I_w << I_f$ produces

$$U = \tfrac{1}{2}\int_0^L \left[2EI_f u''^2 + 2h'^2 EI_f \phi'^2 + \frac{h^2 EI_f}{2}\phi''^2 + 2hh'EI_f \phi'\phi'' + GJ\,\phi'^2 \right] \mathrm{d}z \tag{7}$$

where $J = 2J_f + J_w$ and E is the elastic modulus and G the shear modulus. Appropriate differentiation of the terms in Eq. 1 allows Eq. 7 to be written as

$$U = \tfrac{1}{2}\mathbf{q}^T \mathbf{k}_e \mathbf{q} \,, \tag{8}$$

where $\mathbf{k}_e$ is the element stiffness matrix.

2.3 *Stability matrix*

The element stability matrix can be obtained by considering the work done during flexural-torsional buckling. During buckling, each longitudinal fibre (thought of as a fibre of spaghetti of area δA) deflects laterally and becomes inclined to its original position. The web mid-height o, which is the location of the shear centre, displaces laterally during buckling by u to a position o_1, and a general point P in the cross-section displaces laterally u to P_1, and then rotates through an angle ϕ about o_1 to a position P_2. The coordinates of P are (x, y) and those of P_2 are $(x + u - y\phi, y + x\phi)$, so that the buckling displacements of P are $(u^*, v^*) = (u - y\phi, x\phi)$. The angles of inclination of the fibre δA through P2 are

$$\theta_y = \mathrm{d}u^*/\mathrm{d}z = u' - y\phi'; \quad \theta_x = \mathrm{d}v^*/\mathrm{d}z = x\phi' \tag{9}$$

while when the fibre of length δz rotates through an angle θ, it shortens by

$$\delta\Delta = \delta z - \delta z\cos\theta \approx \tfrac{1}{2}\theta^2 \delta z \,. \tag{10}$$

The work done by the axial force λN can be found by calculating the work done by the axial stresses on the spaghetti fibre of area δA, so that

$$\delta V_N = \left(\frac{\lambda N}{A}\right) \cdot \delta A\left(\tfrac{1}{2}\theta_x^2 + \tfrac{1}{2}\theta_y^2\right)\mathrm{d}z \,, \tag{11}$$

and integrating this over the whole member, and making use of centroidal axis properties produces

$$V_N = \tfrac{1}{2}\int_0^L \lambda N\left(u'^2 + r_0^2\phi'^2\right)\mathrm{d}z \tag{12}$$

where $r_0 = \sqrt{(I_x + I_y)/A}$ is the polar radius of gyration.

The applied major axis moment λM_x causes the section to twist during buckling, and when coupled with shears causes an additional deflection v_B in the plane of the web. The moment produces an axial stress given by

$$\lambda\sigma_b = -\lambda M_x y / I_x \tag{13}$$

and the work done by this stress on the fibre is

$$\delta V_M = -\left(\frac{\lambda M_x y}{I_x}\right)\cdot\delta A\left(\tfrac{1}{2}\theta_x^2 + \tfrac{1}{2}\theta_y^2\right), \tag{14}$$

and which when integrated over the entire element produces

$$V_M = \int_0^L \lambda M_x u'\phi'\,\mathrm{d}z\,. \tag{15}$$

The work done by the moments and shears when the line of action of these through o displaces v_B is given by

$$V_B = \int_0^L \lambda\left(V_1 v_{B1} + V_2 v_{B2} - M_1 v'_{B1} - M_2 v'_{B2}\right)\mathrm{d}z \tag{16}$$

which can be integrated by parts using $V_y = \mathrm{d}M_x/\mathrm{d}z$ to produce

$$V_B = -\int_0^L \lambda M_x v_B''^2\,\mathrm{d}z\,. \tag{17}$$

It can be shown (Trahair, 1993) that the buckling curvature $v''_B \approx \phi u''$, so that

$$V_B = -\int_0^L \lambda M_x \phi u''\,\mathrm{d}z\,. \tag{18}$$

The effect of any off-axis vertical loading in doing work during buckling is represented by

$$V_W = -\tfrac{1}{2}\sum_m \lambda W_m \bar{a}_m \phi_m^2\,, \tag{19}$$

where λW_m is the applied load at section m which twists ϕ_m during buckling, and $\bar{a}_m$ is the height of application of the load above o; this term is included by manipulation of the structure stability matrix.

The total work done by the axial forces, shear forces and bending moment on an element is therefore given by

$$V = \tfrac{1}{2}\int_0^L \lambda\left\{Nu'^2 + r_0^2\phi'^2 - 2M_x\phi u''\right\}\mathrm{d}z\,, \tag{20}$$

which can be written after appropriate differentiation of u and ϕ as

$$V = \tfrac{1}{2}\mathbf{q}^T\lambda\mathbf{g}_e\mathbf{q} \tag{21}$$

where $\mathbf{g}_e$ is the element stability matrix.

2.4 *Buckling solution*

Standard assembly algorithms allow the total change in potential Π to be assembled from the elements as

$$\Pi = \tfrac{1}{2}\mathbf{Q}^T\left(\mathbf{k} - \lambda\mathbf{g}\right)\mathbf{Q}\,, \tag{22}$$

where **k** and **g** are the structural stiffness and stability matrices respectively, and **Q** is the vector of assemblable structural buckling freedoms. Using familiar methodologies, minimising the total potential so that $\delta\Pi = 0$ leads to the eigenproblem

$$\left(\mathbf{k} - \lambda\mathbf{g}\right)\mathbf{Q} = \mathbf{0} \tag{23}$$

which can be solved to produce the buckling load factor λ and the eigenvector **Q** that describes the buckled shape.

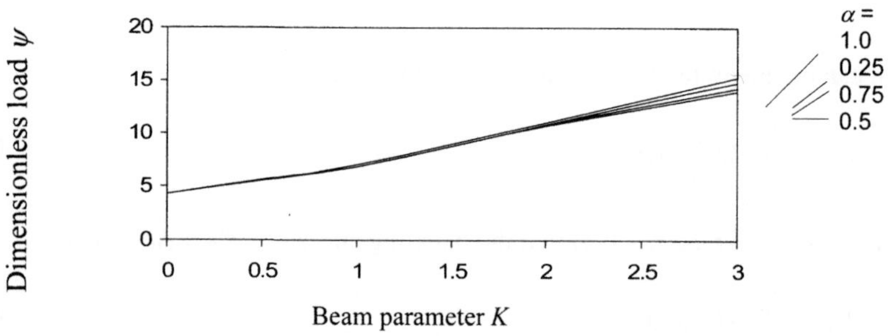

Figure 3. Buckling load for tip-loaded cantilever.

3 Numerical solutions

The finite element method has been programmed to determine the elastic buckling load of web-tapered cantilevers. Kitipornchai & Trahair (1972) derived the governing differential equations for this problem; their formulations indicates that the solutions depends on the independent parameters

$$\psi = \frac{W\ell^2}{\sqrt{\left(EI_y GJ\right)_0}} \quad \text{and} \quad K = \frac{\pi}{\ell}\sqrt{\left(\frac{EI_\omega}{GJ}\right)_0} \tag{24}$$

in which ℓ is the member length, the subscript 0 denotes the rigidities being determined at the largest section, I_ω is the warping section constant and K is known as the beam parameter (Trahair, 1993, Trahair *et al.*, 2007). Figure 3 shows a plot of the dimensionless buckling load (Eq. 24) against a taper parameter α, which is the ratio of the web depth at the smallest section to that at the largest section and for which the loading is applied centroidally. It can be seen that in this case the effects of tapering are not great; this is attributable to the resistance against flexural-torsional buckling being derived from the minor axis second moment of area of the flanges and the torsional rigidity of the members (I_w has been ignored in the derivation as small compared with I_f), and so the effect of web tapering is only to reduce the torsional rigidity GJ_f. Although not reported herein, the effects of web tapering on the buckling of a cantilever with the point load on the top flange are more profound than for the shear centre loading in Figure 3.

4 Conclusions

This paper has presented a simple line-type finite element method of analysis for the elastic flexural-torsional buckling of I-section beams with tapered webs; these beams being typically fabricated by welding. The buckling formulation for these members is complicated by the presence of inclined flange moments that do not exist for prismatic members. Nevertheless, the formulation herein using simple cubic interpolation functions within an energy-based framework of buckling analysis, together with simple assemblable buckling freedoms, provides an attractive bifurcation buckling approach. The results are valuable and much needed, because EC3 does not provide any prescriptive equations for buckling of tapered members, and its use requires recourse to alternative solutions, such as finite element results.

Acknowledgement

The work in this paper was supported by the Australian Research Council through a Federation Fellowship and a Discovery Project awarded to the author.

References

Amirikian A. (1952) Wedge-beam framing. *Transactions*, ASCE **117** 596-652.

Bradford, M.A. & Cuk, P.E. (1988) Lateral buckling of tapered monosymmetric I-beams. *Journal of Structural Engineering*, ASCE **114** No. 5, 977-996.

Bradford, M.A., Woolcock, S.T. & Kitipornchai, S. (2002) Lateral buckling design of crane runway beams. *Second International Conference on Structural Stability and Dynamics*, Singapore.

British Standards Institution (2005) *Eurocode 3: Design of steel structures – Part 1-1: General rules and rules for buildings*. London: BSI.

Brown, T.G. (1981) Lateral torsional buckling of tapered I-beams. *Journal of the Structural Division*, ASCE **107** No. ST4, 689-697.

Chan, S.L. (1990) Buckling analysis of structures composed of tapered members. *Journal of Structural Engineering*, ASCE **116**, No. 7, 1893-1906.

Horne, M.R. & Morris, L.J. (1977) The design against lateral stability of haunched members restrained at intervals along the tension flange. *Proceedings of Second International Colloquium on Stability*, Washington DC, 618-625.

Kitipornchai, S. & Trahair, N.S. (1972) Elastic stability of tapered I-beams. *Journal of the Structural Division*, ASCE **98**, No. ST3, 713-728.

Lee, G.C., Morrell, M.L. & Ketter, R.L. (1972) Design of tapered members. *Bulletin No 173*, Welding Research Council, USA.

Morrell, M.L. & Lee, G.C. (1974). Allowable stress for web tapered beams. *Bulletin No. 192*, Welding Research Council, USA.

Nakane, K. (1984) The design for instability of non-uniform beams. *9th Australasian Conference of the Mechanics of Structures and Materials*, Sydney, 18-22.

Nethercot, D.A. (1973) Lateral buckling of tapered beams. *Publications*, IABSE **33**, No II, 173-192.

Pi, Y.-L. & Bradford, M.A. (2001) Effects of approximations in analyses of beams of open, thin-walled cross-section. Part 1: Flexural-torsional stability. *International Journal for Numerical Methods in Engineering* **51**, No. 7, 757-772.

Pi, Y.-L. & Bradford, M.A. (2001) Effects of approximations in analyses of beams of open, thin-walled cross-section. Part 2: 3-D Nonlinear behaviour. *International Journal for Numerical Methods in Engineering* **51**, No. 7, 773-790.

Ronagh, H.R. & Bradford, M.A. (1994) Some notes on finite element buckling formulations for beams. *Computers and Structures* **52**, No. 6, 1119-1126.

Ronagh, H.R. & Bradford, M.A. (1999) Nonlinear analysis of thin-walled members of open cross-section. *International Journal for Numerical Methods in Engineering* **46**, No. 3, 535-552.

Ronagh, H.R., Bradford, M.A. & Attard, M.M. (2000a) Nonlinear analysis of thin-walled members of variable cross-section. Part 1: Theory. *Computers and Structures* **77**, No. 3, 285-299.

Ronagh, H.R., Bradford, M.A. & Attard, M.M. (2000a) Nonlinear analysis of thin-walled members of variable cross-section. Part 2: Application. *Computers and Structures* **77**, No. 3, 301-313.

Taylor, J.C., Dwight, J.B. & Nethercot, D.A. (1974) Buckling of beams and struts: proposals for a new British code. *Proceedings of Conference on Metal Structures and the Practising Engineer,* Melbourne, Australia, 37-41.

Trahair, N.S. (1993) *Flexural-Torsional Buckling of Structures.* London: E&FN Spon.

Trahair, N.S., Bradford, M.A., Nethercot, D.A. & Gardner, L. (2007) *The Behaviour and Design of Steel Structures to EC3.* 4th edn. London: Taylor & Francis.

Wekezer, J.W. (1985) Instability of thin-walled bars. *Journal of Engineering Mechanics*, ASCE **111**, No. 7, 923-935.

7.2 Plate Girders Fabricated by Single Sided Fillet Weld: Imperfections, Tests, Resistances

László Dunai[1], Gábor Jakab[1]
[1]*Budapest University of Technology and Economics,*
Department of Structural Engineering,
Bertalan Lajos u. 2., H-1111 Budapest, Hungary, gjakab@vbt.bme.hu

Abstract

An extended experimental and numerical research program is completed on plate girders fabricated by automatic single sided fillet welds. The imperfections of the test specimens are measured and characterized. Load-bearing tests are carried out to study stability behaviour modes: local and shear buckling, web crippling, lateral torsional buckling. An advanced finite element model is developed and virtual testing program is done on the imperfection sensitivity. The results are compared, evaluated and concluded in the light of the Eurocode design resistances.

Keywords: *welded structure, plate girder, imperfection, experiment, load-bearing capacity*

1 Introduction

The subject of the research is the experimental and numerical investigation of the effect of welding technology on the imperfections and load-bearing capacity of welded plate girders.

In the first phase of the research the plate components are investigated in the factory during the welding process; the temperature distribution is measured using a radiation (infrared) thermometer. Two types of measurements are carried out: with the camera fixed or moving relative to the welding arc. These measurements are being used as base data for the finite element (FE) modelling of the welding process, Nezo et al. (2006). In the next phase of the research a series of laboratory tests is carried out on 14 specimens. Previous to the testing the out-of-plane geometrical imperfections of the members are measured using a device developed at BME, and then the load-bearing tests are completed in the Structural Laboratory of the Department of Structural Engineering. The experimental programme is designed to check different behaviour modes: plate buckling due to bending and/or shear, distortional buckling, lateral-torsional buckling, web crippling etc. in order to compare the results to the pertinent design resistances calculated according to the application rules of Eurocode 3 (EC3). As a part of the experimental study the residual stresses – mechanical imperfections – are also measured using the hole-drilling method, Dunai et al. (2005).

A global shell FE model is developed to help the design of tests. This model is also used to investigate the effect of different imperfection shapes, by geometric and material non-linear imperfect analyses.

In this paper the focus is on the laboratory tests and on the effect of geometrical imperfections on the load-bearing capacity.

Test specimens

Preliminary calculations according to EC3 are completed to determine the basic sizes of the specimens; cross-section dimensions and length. The test specimens are fabricated with stiffeners only at the supports. The final test arrangements, in particular the position of the load(s), the supports and the stiffeners are designed using the shell FE model; the additional longitudinal and transversal stiffeners are welded in the workshop of the Laboratory. An overview on the test specimens is given in Table 1. A detailed description of the preliminary measurements and the measurement system used during the load-bearing tests can be found in Jakab et al. (2005).

2 Imperfections

2.1 *Geometrical imperfections*

The imperfections of the web are measured by using a half-automatic device developed for this purpose. The shape of the web is measured with a transducer in transverse direction applying the flange edges as bases. The position of the transducer and the measured value are registered to obtain the shape of the web. The range of the transducer measuring perpendicular to the web was +/– 15 mm, with a resolution of 0.01 mm. The measured data is recorded with a sample rate of 100 Hz using a HBM SPIDER 8 device connected to a computer. The imperfection measurements are performed twice; before and after welding the additional stiffeners on the specimens. In longitudinal direction the distance of the measured cross-sections ranged from 50 to 100 mm. The integrity of the measured data is checked after measuring each cross-section using a Matlab program developed for this purpose. Based on the measured data a contour surface is defined using splines to interpolate between the measured points in the two orthogonal directions, as illustrated in Figure 1.

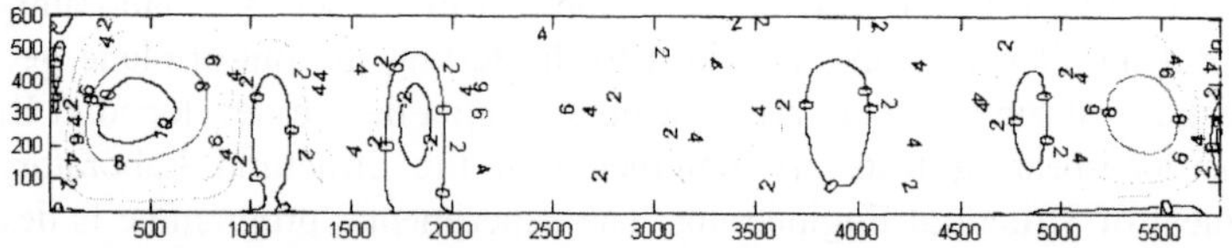

Figure 1. Contour surface of the web based on processed data; specimen L9, [mm]

The flange imperfections and the leaning of the specimens are measured in the same cross-sections after placing the specimen in the loading frame using tape-measure and plummet, providing data with an accuracy of ~0.2 mm.

A summary of the measured web imperfections can be found in Table 1. The values are observed amplitudes of the imperfections of the specimens. According to the measured data it is clear that the imperfections of the web are usually greater at the end of the girder where the welding starts, smaller in the middle of the specimen, and again bigger at the other end, but smaller than those at the beginning. The detailed analysis of the measured imperfections is under process, here the measured minima, maxima and mean values for each specimen are presented.

Table 1. Characteristics of specimens, [mm]

Specimen	Flange	Web	Weld	Length	Web Imperfections		
					Max	Min	Mean
L1	150x6	250x4	Single	5970	2,34	0,51	1,55
L2	150x6	250x4	Single	5970	3,83	0,74	2,48
L3	150x6	600x4	Single	5970	-	-	-
L4	150x6	600x4	Single	5970	8,59	1,08	4,34
L5	200x10	600x4	Single	5970	6,94	1,07	3,22
L6	200x10	600x4	Single	5970	6,94	0,91	3,16
L7	200x10	250-600x4	Single	5970	6,98	0,7	1,98
L8	200x10	250-600x4	Single	5970	2,96	0,66	1,34
L9	150x6	600x4	Double	5970	11,26	1,26	5,04
L10	200x10	600x4-8	Single	5970	-	-	-
L11	200x10	600x4	Single	5970	6,46	0,74	2,88
S1	200x10	600x4	Single	2970	7,26	1,26	3,77
S2	150x6	600x4	Single	2970	8,06	2,96	5,38
S3	150x6	250x4	Single	2970	4,48	1,75	3,62

2.2 *Mechanical imperfections*

A hole-drilling method using uniaxial strain gauges is used to determine the residual stress level in one of the flanges (6 locations) and in half of the web (2 or 3 locations) of the girders. A hole is drilled directly next to each gauge at both ends of it. The current strain values at each gauge of the cross-section are recorded both before and after a hole is drilled in order to capture any interaction among the measurement locations, if any. Both the longitudinal and the transverse residual stresses are measured. The placement of the gauges and the execution of the drilling are very prone to inaccuracies. The effect of these is not known yet; they are under investigation using FE models and the high resolution photographs made during the measurement.

Figure 2 shows the distribution of stresses in longitudinal direction on specimens with automatic single-sided (Middle, Single (I60x15S)) and double-sided fillet weld (Middle, Double (I60x15D)) in the middle cross-section of the specimen. End, Double (I60x15MV) refers to the end zone of the specimen where the welding is made manually.

The diagrams show that the use of single-sided welding results in significantly smaller residual stresses, comparing to the double-sided case; however, these are asymmetric whilst the double-sided fillet weld results in larger stresses but with a symmetric distribution.

The measured maximum and minimum residual stress values are summarized in Table 2. The results are used in the verification of the FE simulation of the welding process (Dunai et al. 2005).

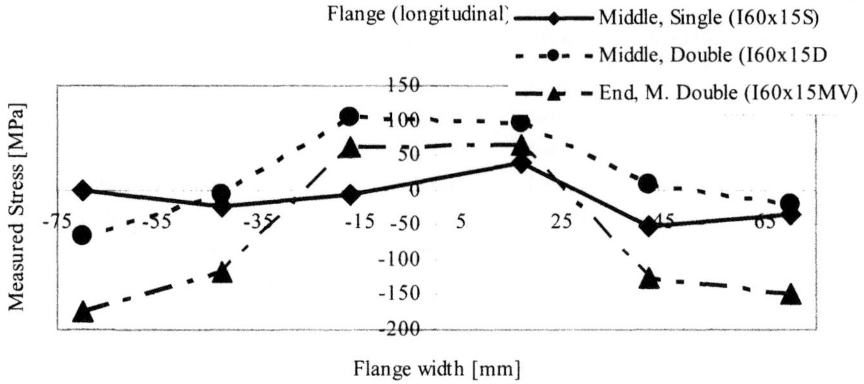

Figure 2. Stress distribution in the flange (longitudinal measurements)

Table 2. Measured residual stresses

Flange	Web	Weld	Length	Cross-section	Gauge orientation	Measured Stress [MPa]	
						Max	Min
150x6	250x4	Single	2970	Centre	Long.	124.671	-77.374
150x6	600x4	Single	2970	Centre	Long.	38.646	-51.335
150x6	600x4	Single	2970	End	Long.	-20.765	-153.264
150x6	600x4	Single	2970	Centre	Trans.	-1.401	-35.350
150x6	600x4	Single	2970	End	Trans.	23.237	-162.081
150x6	600x4	Single	2970	End	Long.	65.426	-173.946
150x6	600x4	Double	2970	Centre	Long.	105.719	-66.826
200x10	600x4	Single	2970	Centre	Long.	23.731	-46.391

3 Experimental results

In this chapter a short overview of the completed tests is given: test arrangements and results are presented. The load-bearing tests are performed in 3- and 4-point-bending arrangements, as shown in Figures 3 and 4. The span of the specimens is basically 5700 mm, with the exception of tests T10 and T11 where 4350 mm span is used. For the short specimens a span of 1800 mm is applied.

For the long specimens "fork" supports are applied and the short specimen are supported by hinged end supports. No lateral supports are applied to the upper (compression) flanges but the loading system provided it by friction.

The load was applied using hydraulic jacks; the force was distributed on the full width of the flange by a thick steel plate. In the measurement system the strains are observed by strain gauges and the horizontal and vertical displacements are measured by transducers.

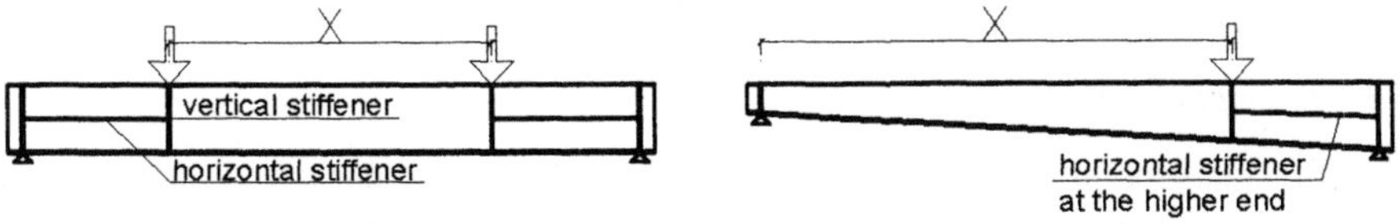

Figure 3. Long specimen in 4-point-bending and in 3-point-bending

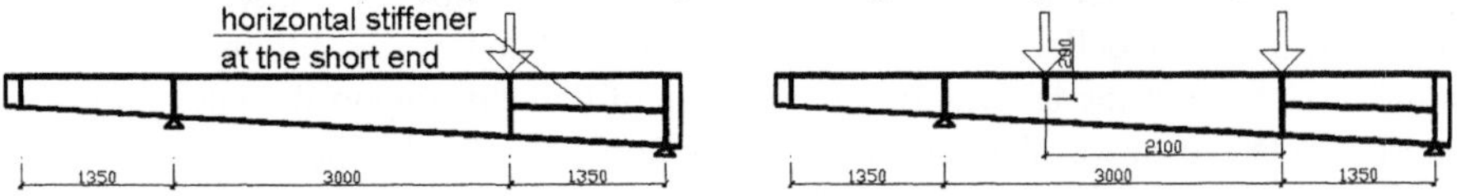

Figure 4. Unique arrangements for test T10 (left) and test T11 (right)

Further details on the test arrangements, failure modes and measured load-bearing capacities with Eurocode-based resistances calculated for the pertinent failure modes are presented in Table 3. Note, that the listed failure modes are in most tests not "pure" modes but the dominant ones of several interacting phenomena, for instance the local effects of the concentrated force in the load-drive-in areas. The measured and EC3 calculated ultimate load show – in some cases – significant differences; more details on this are presented in Jakab et al. (2005).

Table 3. Test setups, failure modes and ultimate load results

Test No.	Specimen	Arrangement		Failure mode	Ultimate Load [kN]	
		Type	X		Test	EC3
T1	L5	4point	1000	web buckling	248	204
T2	L3	3point	2350	flange buckling	174	115
T3	L4	4point	2000	web crippling	116	106
T4	L6	4point	3000	-	-	269
T5	L6	3point	2850	web crippling	-	115
T6	L1	4point	3000	LT buckling	74	40
T7	L2	4point	3000	flange buckling	-	-
T8	L2	4point	3000	LT buckling	78	40
T9	L7	4point	3000	local buckling	261	-
T10	L7	3point	4350	-	-	-
T11	L7	4point	2100	dist. LT buckling	221	-
T12	L9	4point	3000	dist. LT buckling	176	96
T13	L11	3point	4350	dist. LT buckling	540	454
T14	L10	3point	4350	dist. LT buckling	641	513
T15	L8	3point	4350	comb. LT buckling	472	-
T16	S1	3point	600	shear buckling	686	289
T17	S2	3point	600	LT buckling	581	-
T18	S3	3point	600	shear and LT buckling	-	-

4 Finite element model

4.1 *Description of the model*

The FE analyses are completed by Ansys (2002) general purpose finite element software package. The 4-node, 6-DOF SHELL43 element with capabilities of plasticity, large deflection and large strain is used to follow the target phenomena. The plates are modelled in their mid-plane. The boundary conditions are defined similar to those in the laboratory tests; in case of the end support the relevant nodes are restricted to move; the partial lateral support provided by the hydraulic jack is replaced with a restricted degree of freedom in the transversal direction. The load is applied as concentrated forces on the nodes at the intersection of the upper flange and the vertical stiffener. The applied material model is elastic–linearly plastic, with the von Mises yield criterion and kinematic hardening; the material properties are derived from tensile tests. The average element edge size is defined to be 50 mm based on a convergence study. A Matlab-based pre- and postprocessor program is developed to automatically create the FE model and display the results.

4.2 *Imperfections*

Measured, EC3 proposed, eigenshape-based and sinusoidal imperfections are applied in the FE model. In each case with the same maximum imperfection amplitude is added to the initially perfect plates to study the imperfection sensitivity of the girders.

The measured imperfections are added to the model after preprocessing the raw measurement data, without any change in the amplitude.

The finite element analysis-based design methodology is detailed in the EC3 Part 1.5 Annex C. Two alternatives are proposed by the standard: either to apply an out-of-plane imperfection to the web, twist to the flanges/longitudinal stiffeners and global bow-shaped imperfection to the model, or to use eigenshapes of the model as imperfect shape. For the amplitude of the geometrical imperfections equivalent values are defined, which means these include other effects as well (e.g. residual stresses) and therefore these may be larger than those measured on the specimens.

The sinusoidal shape imperfections are similar to those proposed by EC3 but in this case the number of the half-waves in the flanges is defined in accordance with the waves of the web in the longitudinal direction.

No separate mechanical imperfections are considered in the model, since these are included in the equivalent geometrical imperfections.

4.3 *Magnitude of geometrical imperfections*

The maximum imperfection amplitudes of the assumed different imperfections are adjusted according to proposal of EC3 1.5, as follows: flange imperfection: 1/50 slope; web imperfection: min(a/200, b/200), where a and b are the two sides of the investigated web panel. In the standard cases these results in a deviation from the perfect shape by 3 mm in case of the specimen with the bigger cross-section height.

5 Analyses and results

In this chapter the results of three selected tests and their simulations are presented and discussed. The measured load-bearing capacities are compared with the values obtained from the numerical simulations assuming different imperfect shapes.

As example, Figure 5 shows the obtained force-vertical displacement diagrams for the test T6 (failure: lateral-torsional buckling). Figure 6 shows the different shapes from the simulations with different imperfect shapes. The obtained load-bearing capacities for the three tests are summarized in Table 4, with buckling ($F_{b,Rd}$) and elastic ($F_{c,Rd,el}$) and plastic ($F_{c,Rd,pl}$) cross-section resistances calculated by EC3.

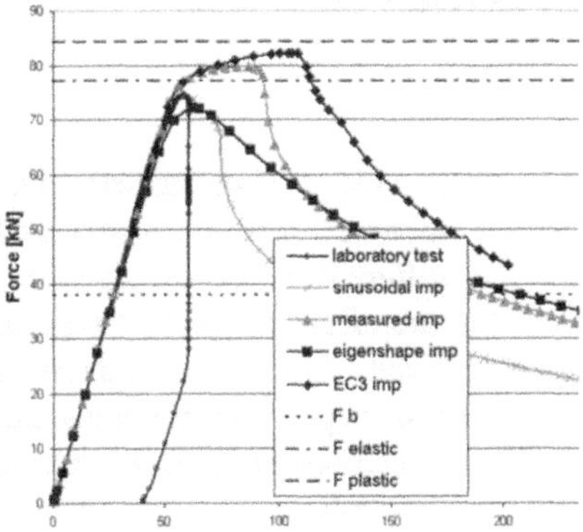

Figure 5. Force-vertical displacement diagrams for T6

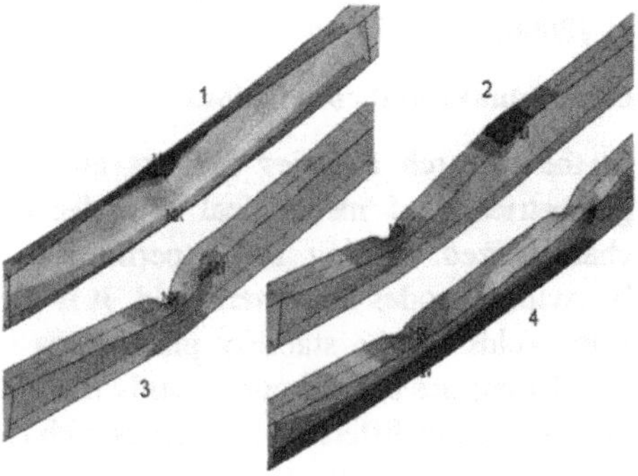

Figure 6. Failure shapes of T6 – initial imperfections: sinusoidal (1), measured (2), buckling shape (3), EC3 (4)

Table 4. Results of tests and simulations

Test		Ultimate load/test	Ultimate load/simulations Imperfection applied				EC3 resistances		
			Measured	Sin	Eigen	EC3	$F_{b,Rd}$	$F_{c,Rd,el}$	$F_{c,Rd,pl}$
T1	kN	248	237	224	235	235	223	239	262
	%	100	96	90	95	95	90	96	106
T3	kN	117	81	85	84	83	-	167	196
	%	100	69	72	72	71	-	143	168
T6	kN	75	80	75	72	82	38	77	84
	%	100	107	100	96	110	51	103	112

In all simulations the numerical model accurately followed the test in the linear range: the initial rigidity of the model is independent from the applied imperfection. However, in the non-linear range of the force-displacement diagrams – near the failure – deviations are experienced between the test and simulations. Differences can be seen in the load-bearing capacity, in the non-linear part of the force-displacement diagram and in the shape of the failed models.

The accuracy of the model depens on (i) the type of failure mode (ii) the applied imperfect shape. Global failure modes can be simulated quite accurately using this model if the imperfect shape applied reflects the failed shape of the member. On the other hand, failure modes influenced by local effects cannot be modelled with the desired accuracy using this model.

Comparison of the load-bearing capacities obtained assuming different imperfections shows that the recommendations of EC3 do not always provide the most accurate ultimate load, in some cases the ultimate load obtained is on the unsafe side. The main reason of this may be that the imperfections to be taken into account defined by the standard are based on rather simple rules and therefore can not reflect all possible – complex – failure modes.

A more detailed discussion of the FE model and its results can be found in Jakab et al. (2006).

6 Summary and conclusions

In the research a series of tests are carried out on welded plate girders. The geometrical and mechanical imperfections of the members are measured and characterized. During the experimental study the different types of nonlinear behaviour modes are determined. It is observed that the effect of the single-sided fillet welds on the stability phenomena is not significant. A series of numerical simulations are carried out to study the effect of different shaped imperfections and to compare the different approaches with reality. The comparison showed that taking into account geometrical imperfections in the FE simulation-based design approach is inevitable, but the shape of the imperfections applied can strongly influence the result of the analysis.

Acknowledgement

The numerical part of the presented research is conducted under the financial support of the OTKA T049305 project, and the experimental study is supported by the Lindab Buildings Ltd, Hungary.

References

ANSYS, Release 9.0A1 (2002)

Dunai, L., Jakab, G., Nezo, J. & Topping, B.H.V. (2005). Experiments on welded plate girders: fabrication, imperfection and behaviour. In: *Proc. of the first Int. Conf. on Advances in Experimental Struct. Eng. 2005* (Vol 1), Nagoya, Japan pp. 51-58.

EC3-1-5. 2005. *prEN1993-1-5:2005 Eurocode 3: Design of steel structures. Part 1-5: Plated structural elements.*

Jakab, G., Dunai, L. & Macdonald, R.T. (2005). Behaviour of plate girders with single-sided fillet welds. *Proc. of the Eurosteel Conf. 2005 on Steel and Composite Struct. (Vol B)*, Maastricht, The Netherlands pp. 2.2-2 – 2.2-9.

Jakab, G., Szabo, G. & Dunai, L. (2006). Imperfection sensitivity of welded beams: experiment and simulation. *Proc. of Steel: A new and traditional material for building*, Poiana, Romania pp. 173-181.

Nezo, J., Topping, B.H.V. & Dunai, L. (2002). Virtual fabrication of steel welded plate girders. In: *Proc. of the 6th Int. Conf. on Computational Structures Technology*, Prague, Czech Republic, pp. 241-242 (in CD-ROM 14 pages).

7.3 Determination of Elastic Constants of Panels with Structural Orthotropy

Vladimír Ivančo[1], Peter Heinze[2]

[1]*Technical University of Košice, Letná 9, 042 00 Košice, Slovakia, vladimir.ivanco@tuke.sk,*
[2]*University of Technology, Business and Design, Wismar, Germany*

Abstract

The paper deals with determination of basic stiffness parameters of rectangular panels with structural orthotropy. The reason for determining these parameters is to substitute structural parts created as stiffened panels by equivalent orthotropic plates for static and dynamic computations. Main purpose of the substitution is reduction of degrees of freedom in finite element analyses of complex structures.

Keywords: orthotropy, elastic constants, FEA

1 Introduction

Laser welded structural steel sandwich panels; see Figure 1, that are now available on the market offer strong and lightweight materials for heavy industries. They usually enable 20 to 50% weight reduction over standard steel panels, 90 to 95% weight reduction compared to traditional structural concrete panel systems, two thirds reduction in structural volume and weight over traditional building methods and improved energy absorption and damage tolerance over solid steel plates. Another advantage is rapid installation and finishing compared to welded plate steel. These panels are typically used for trucks and trailers, rail cars, bridge decks and transportation infrastructure, constructions and shipbuilding and offshore structures.

There exists many analytical solutions of shape orthotropic structural members. Analytical solution was presented by Bareš & Machan (1962) and many other authors (Romanoff et al. 2007). Though these analytical methods can be applied for sandwich panels, their common disadvantage is complexity and elaborateness of calculations. This mostly precludes utilization of analytical methods. In engineering practice, sandwich panels are usually considered as orthotropic plates. The main reason for it, is the simplification of computation in structural analyses. A substitution of a panel with structural orthotropy by equivalent orthotropic plate needs determination of elastic constants describing stiffness of orthotropic plates. There exists many experimental methods for determination of these constant, most of them were originally developed for determination of elastic constants of orthotropic materials like plywood (e.g. Hearmon) and composites and corrugated boards (Luo et al. 1995). Experimental methods are usually based on measurement of static deflections or natural frequencies of specimens. Common practice is to determine bending and torsional stiffnesses by formulas (Farkas & Jármai 1997).

Substitution of sandwich panels by equivalent orthotropic plates can be important in FEM analyses of complex structures. Proper finite element representation of a panel requires sufficiently fine finite element mesh. This usually leads to large number of nodes and elements and too large FE model of the whole structure. On other hand, substitution of the panel by the orthotropic plate may significantly reduce the FE model size with subsequent reduction of cost of analyses.

The paper presented deals with substitution of structurally orthotropic panels by orthotropic plates for purposes of simplifications of linear FEM analyses. Elastic material constants necessary for definition of shell finite elements based on Mindlin's theory of thick plates are determined by FEM simulations of various tests of sandwich panels.

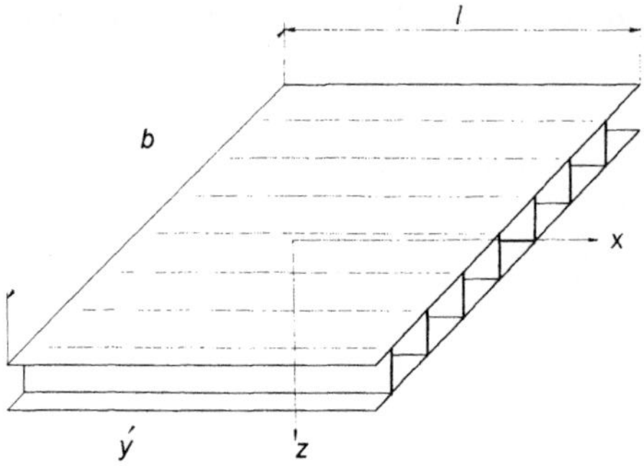

Figure 1. Orthotropic sandwich panel.

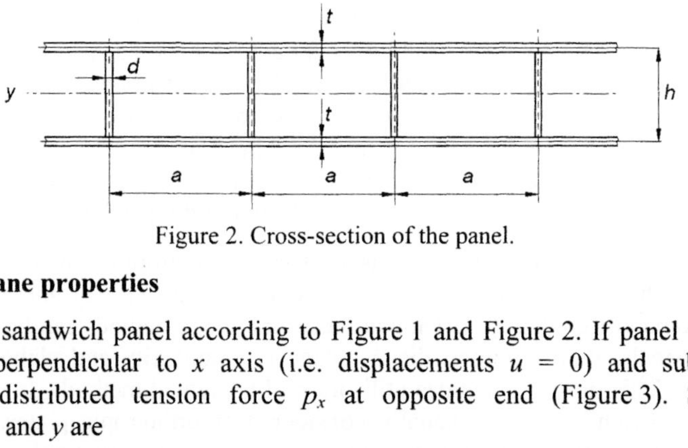

Figure 2. Cross-section of the panel.

2 Membrane properties

Consider a sandwich panel according to Figure 1 and Figure 2. If panel is fixed at one edge perpendicular to x axis (i.e. displacements $u = 0$) and subjected to uniformly distributed tension force p_x at opposite end (Figure 3). Strains in directions x and y are

$$\varepsilon_x = \frac{p_x}{E} \qquad \text{and} \qquad \varepsilon_y = -\nu \frac{p_x}{E} \tag{1}$$

where E and ν are modulus of elasticity and Poisson's ratio respectively. Strains in substitute orthotropic plate are

$$\varepsilon_x = \frac{\sigma_x}{E_x} = \frac{p_x A_x}{E_x b h_0} \qquad \varepsilon_y = -\frac{\nu_{xy} p_x A_x}{E_x b h_0} \tag{2}$$

where h_0 is thickness of orthotropic plate and A_x is area of the panel cross section. Strains of panel and plate must be equal, hence by comparison of Eq. 1 and Eq. 2 one can obtain

$$E_x h_0 = E\frac{A_x}{b} = S_x \qquad \text{and} \qquad \nu_{xy} = \nu \tag{3}$$

The same approach, i.e. to consider tension in y direction where one of edges perpendicular to x axis is fixed and the second loaded by uniformly distributed force cannot be used as due to effect of stiffeners, strains are not constant.

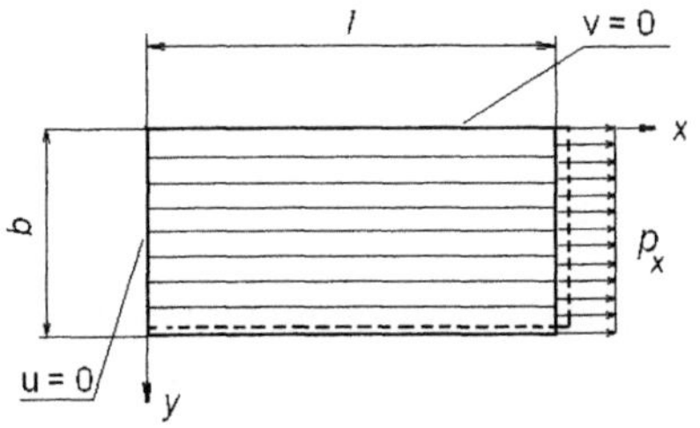

Figure 3. Simulation of tension tests.

Uniform strain fields can be obtained e.g. if boundary conditions are represented by prescribed displacements u and v: $u = 0$ for $x = 0$ and $x = l$, $v = 0$ for $y = 0$ and $v = \overline{v}$ for $y = l$. Stresses in the panel then are

$$\overline{\sigma}_x = \nu \overline{\sigma}_y \qquad \overline{\sigma}_y = \frac{E}{1-\nu^2} \frac{\overline{v}}{b} \tag{4}$$

and reactions R_x and R_y on edges $x = 0$ and $y = 0$ are

$$R_x = 2bt\overline{\sigma}_x \qquad R_y = 2lt\overline{\sigma}_y \tag{5}$$

Strains in orthotropic plate are (Altenbach et al. 2004)

$$\varepsilon_x = \frac{\sigma_x}{E_x} - \nu_{xy} \frac{\sigma_y}{E_x} \qquad \varepsilon_y = \frac{\sigma_y}{E_y} - \nu_{yx} \frac{\sigma_x}{E_y} \tag{6}$$

In the case considered, stresses in the plate can be expressed as

$$\sigma_x = \frac{R_x}{bh_0} \quad \text{and} \quad \sigma_y = \frac{R_y}{lh_0} \tag{7}$$

From conditions that strains of the orthotropic plate are equal to strains of the panel, i.e. $\varepsilon_x = 0$ and $\varepsilon_y = v/l$, we obtain

$$\frac{E_y h_0}{1 - \nu_{xy}\nu_{yx}} = 2\frac{bt}{l}\frac{E}{1-\nu^2} \tag{8}$$

Eq. 3 and Eq. 8 together with condition of symmetry

$$\frac{\nu_{yx}}{E_y} = \frac{\nu_{xy}}{E_x} \tag{9}$$

enable determination of numeric values of $E_x h_0 = S_x$, $E_y h_0 = S_y$ and Poisson's ratios ν_{xy} and ν_{yx} of the orthotropic plate.

3 Bending and torsion properties

It is known (Lekhnitskiy 1957) that in a case when bending moments m_x, m_y and torsional moment m_{xy} are constant, deflection of specially orthotropic plate (edges of quadrilateral plate are parallel with orthotropic axes) is represented by biquadratic polynomial

$$w(x, y) = a_0 + a_1 x + a_2 x^2 + a_3 y + a_4 y^2 + a_5 xy \tag{10}$$

If origin of coordinates is selected that $w = 0$ and $\partial w/\partial x = 0$ for $x = 0$ and $\partial w/\partial x = 0$ for y = 0, constants a_0, a_1 and a_3 are zero. In the case of pure bending and twisting, strains can be expressed as (Hearmon (1961))

$$\varepsilon_x = -\frac{\partial^2 w}{\partial x^2} z = -2a_2 z \qquad \varepsilon_y = -\frac{\partial^2 w}{\partial y^2} z = -2a_4 z \tag{11}$$

$$\gamma_{xy} = -2\frac{\partial^2 w}{\partial x \partial y} z - 2a_5 z \tag{12}$$

Strains due to pure bending and torsion of orthotropic plate are (Lekhnitskiy (1957))

$$\varepsilon_x = \left(\frac{12m_x}{E_x h_0^3} - \nu_{xy}\frac{12m_y}{E_x h_0^3}\right) z \qquad \varepsilon_y = \left(\frac{12m_y}{E_y h_0^3} - \nu_{yx}\frac{12m_x}{E_y h_0^3}\right) z \tag{13}$$

$$\gamma_{xy} = \frac{\tau_{xy}}{G_{xy}} = \frac{12m_{xy}}{G_{xy} h_0^3} z \tag{14}$$

Constants a_2, a_4, a_5 can be expressed in terms of bending and torsion moments as

$$a_2 = -\left(\frac{6m_x}{E_x h_0^3} - \nu_{xy}\frac{6m_y}{E_x h_0^3}\right)^2 \qquad a_4 = -\left(\frac{6m_y}{E_y h_0^3} - \nu_{yx}\frac{6m_x}{E_y h_0^3}\right) \tag{15}$$

$$a_5 = -\frac{6m_{xy}}{G_{xy} h_0^3} \tag{16}$$

Values of elastic constants E_x, E_y and ν_{yx} can be determined from deflections of the panel computed by FEM by following manner:

3.1 *Pure bending,* $m_x \neq 0,\ m_y = m_{xy} = 0\ m_y$

From computed displacements $w(x,0)$ along line $y = 0$ can be determined

$$a_{2,x} = \frac{w(x,0)}{x^2} \quad \text{and} \quad E_x h_0^3 = -\frac{6m_x}{a_{2,x}} \tag{17}$$

and from displacements $w(0,y)$ along line $x = 0$ can be determined

$$a_{4,x} = \frac{w(0,y)}{y^2} \quad \text{and} \quad \frac{E_y h_0^3}{\nu_{yx}} = \frac{E_x h_0^3}{\nu_{xy}} = \frac{6m_x}{a_{4,x}} \tag{18}$$

From comparison of second Eq. 17 and Eq. 18 it follows that

$$\nu_{yx} = -\frac{a_{4,x}}{a_{2,x}} \tag{19}$$

3.2 *Pure bending,* $m_x = m_{xy} = 0,\quad m_y \neq 0$

From displacements $w(x,0)$ along line $y = 0$ it is possible to calculate

$$a_{2,y} = \frac{w(x,0)}{x^2} \quad \text{and} \quad \frac{E_x h_0^3}{\nu_{xy}} = \frac{6m_y}{a_{2,y}} \tag{20}$$

and from displacements $w(0,y)$ along line $x = 0$ it can be determined

$$a_{4,y} = \frac{w(0,y)}{y^2} \qquad E_x h_0^3 = -\frac{6m_y}{a_{4,y}} \tag{21}$$

3.3 *Torsion,* $m_{xy} \neq 0, \quad m_x = m_y = 0$

In this case, constants a_2 and a_4 are zero and

$$a_5 = \frac{w(x,y)}{xy} \qquad G_{xy} = -\frac{6m_{xv}}{a_5} \tag{22}$$

In all simulations mentioned, only one quarter of the panel need to be modeled. As an example, in Figure 4 a setup of the computational model loaded by m_y.is shown. On edges $x = 0$ and $y = 0$ symmetrical boundary conditions are prescribed (displacements u and rotations φ_y and φ_z about y and z directions are zero for $x = 0$ and zero v, φ_x and φ_z are for $y = 0$). Bending moment at edges $y = b$ is $m_y = -p_y th$, where h is distance between middle surfaces of both plates of thickness t. Setup of computational model for m_x is analogical. In a case of torsion, boundary conditions in planes of symmetry are anti-symmetrical. Torsional moment can be modeled by opposite tractions along edges of both plates ($p_y = \pm p$ for $x = l$ and $p_x = \pm p$ for $y = b$).

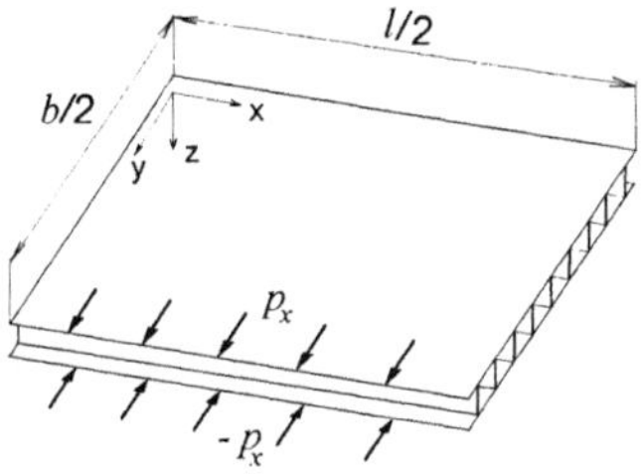

Figure 4. Simulation of pure bending.

3.4 *Determination of elastic constants*

Generally, for any value of thickness h_0 of the plate, e.g. $h_0 = h$, moduli of elasticity E_x, E_y and Poisson ratio ν_{yx} in tension differ from values determined for bending. From condition that moduli E_x for tension and bending are equal we obtain

$$h_{0,x} = \sqrt{T_x / S_x} \tag{23}$$

that gives different modules E_y for tension and bending. Equal values of these moduli can be achieved if thickness is selected as

$$h_{0,y} = \sqrt{T_y / S_y} \tag{24}$$

but in this case are different moduli E_x for tension and bending. Differences can be minimized by proper selection of h_0 from interval between $h_{0,x}$ and $h_{0,y}$.
Density of the orthotropic plate that is necessary for solution of dynamic problems can be then calculated from condition of equal mass of the plate and the panel

$$\rho_0 = \frac{A_x}{h_0 l} \tag{25}$$

where ρ is density of material of the panel.

4 Effect of transverse shear

To consider effect of shearing forces, shearing modules G_{xz} and G_{yz} have to be defined as material properties of finite element formulated according to Mindlin's theory. They can be determined by comparing deflections of the panel and substituted orthotropic plate loaded by transverse load. Boundary conditions, the same for the panel and the plate, should be defined in such a way that deflections of the plate do not depend on one of the moduli G_{xz} and G_{yz}. Possible boundary conditions for identification of G_{yz} are in Figure 5, symmetrical boundary conditions are at all edges and deflections $w = 0$ are prescribed at the edge $y = b/2$. Under these boundary conditions deflections are constant along direction of x axis and do not depend on value of G_{xz}.

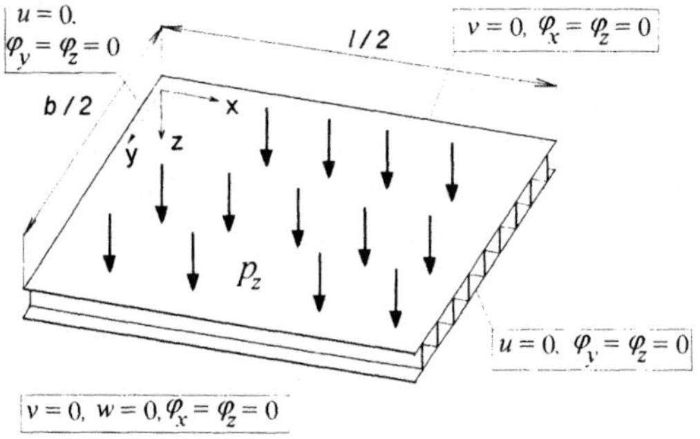

Figure 5. Boundary conditions for identification of G_{yz}.

Deflections of the plate can be considered as composed of two parts

$$w = a + \frac{b}{G_{yz}} \tag{26}$$

where both a and b are constant for fixed values of p_z, E_x, E_y, ν_{xy}, ν_{yx} and G_{xy}. Value of G_{yz} can be then determined from deflections computed for two different test values of G_{yz}. Analogical way can be used for determination of G_{xz}.

5 Example

A steel sandwich panel is considered with dimensions $a = 120$ mm, $b = 3120$ mm, $h = 42{,}5$ mm, $d = 4$ mm, $t = 2{,}5$ mm, $l = 3120$ mm and material constants $E = 2{,}1 \cdot 10^5$ MPa, $\nu = 0{,}3$ and $\rho = 7\,850$ kgm^{-3}. From Eq. 13, Eq. 18 and Eq. 10 we obtain $S_x = E_x h_0 = 1{,}3475 \cdot 10^6$ Nmm^{-1}, $S_y = E_y h_0 = 1{,}0713 \cdot 10^6$ Nmm^{-1}, $\nu_{xy} = \nu = 0{,}3$ and $\nu_{xy} = 0{,}2385$.

Detail of the FE model of a quarter of the panel is in Figure 6. Comparatively fine FE mesh with average element size about 7 mm consists of 120 666 elements and has 118 770 nodes. This represents, in dependence of boundary conditions, about 706 000 equations in FEA. Hence, importance of reduction problem size by substitution of the panel by an orthotropic plate is obvious.

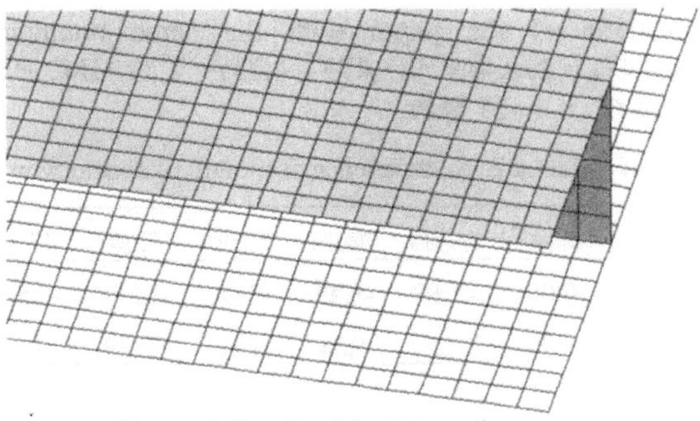

Figure 6. Detail of the FE mesh.

Values computed by finite element analyses are following:
$T_x = E_x h_0^3 =$ 6,2355·10^9 Nmm, $T_y = E_y h_0^3 =$ 5,7416·10^9 Nmm, ν_{xy} = 0,3005 and ν_{xy} = 0,2767. These values can be considered as good representation of pure bending properties of the panel because deflections w computed by FEM satisfy Eq. 10 very well. This is documented in Figure 7 where the dependence of deflections $w(x, 0)$ along line $y = 0$ on x^2 for m_x = - 10625 N and $m_y = m_{xy} = 0$ is displayed. From linear regression it follows that deflections are proportional to the square of coordinate x with and reliability of determination of $E_x h_0^3$ is 1,0000. The same, i.e. very good compliance with Eq. 10 was achieved for $w(0, y)$. It can be shown that results for pure bending by m_y and for torsion give the same very good satisfaction of this equation.
Values of the orthotropic plate thickness can be selected from interval

$$h_{0,x} = \sqrt{T_x / S_x} = 68{,}3 \text{ mm} \qquad \text{and} \qquad h_{0,y} = \sqrt{T_y / S_y} = 73{,}21 \text{ mm}.$$

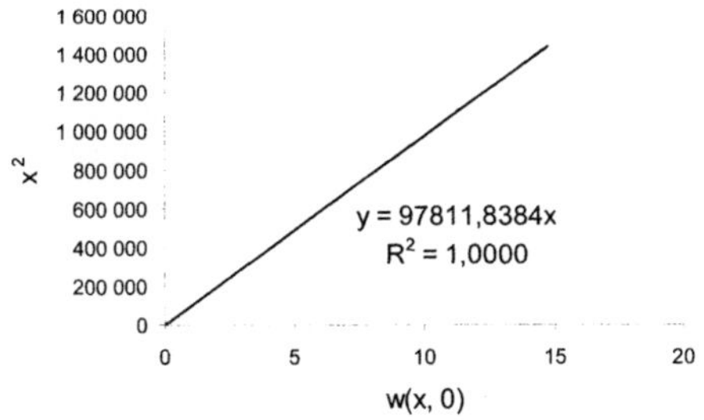

Figure 7. Displacements $w(x, 0)$ versus x^2 for $m_x \neq 0$ and $m_y = m_{xy} = 0$.

Lower value gives modulus E_x for tension equal to E_x for bending while upper value gives equal moduli E_x. Differences between moduli can be minimized by proper selection of h_0 as it is shown in Table 1. For h_0 = 70 mm is $E_x = T_x / h_0^3$ = 18 179 MPa, $E_y = T_y / h_0^3$ = 16 739 MPa and G_{xy} = 6 395 MPa. Remaining values of and G_{xz} can be determined by comparison of deflection of panel and orthotropic plate with boundary conditions according to Figure 5. Determination of G_{yz} is illustrated in Figure 8. Constants from Eq. 26 are determined by linear regression of values w corresponding to different values of G_{yz} of the plate.

Maximum deflection of the panel with the same boundary conditions was 31,479 mm, then $G_{yz} = 208{,}489 / (31{,}479 - 4{,}73) = 7{,}79$ MPa. By the same method can be determined $G_{xz} = 1\,137$ MPa.

Table 1. Influence of value of h_0 on difference of moduli of elasticity in tension and bending.

h_0 (mm)	68,03	69,00	**70,00**	71,00	72,00	73,00	73,21
ΔE_x (%)	0,00	-2,89	**-5,89**	-8,94	-12,03	-15,16	-15,82
ΔE_y (%)	15,82	12,57	**9,38**	6,32	3,39	0,57	0,00

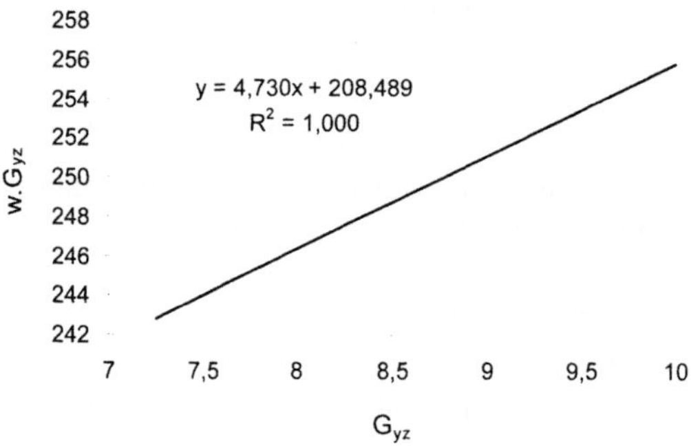

Figure 8. Determination of G_{yz}

6 Conclusions

Method of determination of material elastic constants for substitution of sandwich panels by orthotropic plates was presented. Equivalence of displacements can be achieved only for bending with small differences in membrane behaviour.

Acknowledgement

Support of the Scientific Grant Agency of the Ministry of Education of Slovak Republic and Slovak Academy of Sciences under grant VEGA 1/4160/07 is greatly appreciated.

References

Altenbach, H., Altenbach, J. & Kissing, W. (2004) *Mechanics of composite structural elements*, Berlin Heidelberg: Springer-Verlag.

Bareš, R. & Machan, P. (1962) *Exact calculus of shape orthotropic plates*, Czechoslovak Academy of Sciences, Prague

Farkas, J. & Jármai, K. (1997) *Analysis and optimum design of metal structures*, Rotterdam: A. A. Balkema.

Farkas, J. & Jármai, K. (2003) *Economic Design of Metal Structures*, Rotterdam: Millpress.

Hearmon, R. F. S. (1961) *An Introduction to Applied Anisotropic Elasticity*. The Clarendon Press, Oxford.

Lekhnitskiy, S. G. (1957) *Anisotropic Plates*, Moscow

Luo, S., Suhling, J. C. & Laufenberg, T. L. (1995) Bending and twisting test for measurement of the stiffnesses of corrugated board. *Mechanics of Cellulosic Materials* ASME AMD-Vol. 209 MD-Vol. 60, 91-109.

Romanoff, J., Varsta, P. & Klanac, A. (2007) Stress Analysis of Homogenized Web-Core Sandwich Beams. *Composite Structures* **79** No. 3, 411-422.

7.4 The Elastic-plastic Local Stability and Load-carrying Capacity of Compressed Thin-walled Steel Members

Pavol Juhás[1], Mohamad Al Ali[1], Zuzana Kokorudová[1]
[1]*Technical University in Kosice, 042 00, Kosice, Slovak Republic,*
pavol.juhas@tuke.sk, mohamad.alali@tuke.sk, zuzana.kokorudova@centrum.sk

Abstract

The paper presents fundamental information about realized experimental-theoretical research to determinate the load-carrying capacities for thin-walled compressed steel members with quasi-homogenous and hybrid cross-sections. The webs of such members are stressed in the elastic-plastic region. The research joins on previous research of the first author of the paper (Juhás, P. 2006). The aim of this research is to investigate and analyze the elastic-plastic post-critical behaviour of thin web and its interaction with flanges. The experimental program, test members and their geometrical parameters and material properties are evident from Tables 1, 2 and Figures 1 and 2. The test arrangement and failures of the test members are illustrated on Figures 3, 4 and 5. Some partial results are presented in the paper, too.

Keywords: *post-critical behaviour, elastic-plastic carrying capacity.*

1 Introduction

The continued effort for economic design of steel structures lead to decrease their weight by shape and material optimization, which lead to use the high-stress steels combined with standard structural steels. The efficiency of high-stress steels using and their combination with standard structural steels is evident in case of bending members – beams mostly subjected to bending loading. From the complex optimization analyses follow, that the high-stress steels combined with standard structural steels may be also advantage in case of members subjected to compression loading, mainly in case of thin-walled members. This paper presents basic information about ongoing experimental/theoretical research of elastic-plastic bearing capacities of thin-walled compression steel members with quasi-homogenous and combined cross-sections. The experimental part of this research is just finished. This research program is resulting on previous research of first author of this paper. This research is distinctively oriented on the investigation and analyses of post-critical behaviour and interaction of slender/ultra-slender web with flanges in the process of their transformation and failure (Juhás et. al. 1993, 2000, 2007), (Kriváček 2007).

2 The research experimental program and tested members

The experimental research includes the testing of 24 welded compression steel models/members having quasi-homogenous and combined I cross-sections with different dimensions, advisable elected to show, in decisive extent, the elastic-plastic post-critical effect of slender webs and their interaction with flanges in process of their strain, transformation and failure. Table 1 presents the total research program, designed geometrical dimensions and designed materials for flanges and webs of several testing models/members groups. Scheme of testing models/members is illustrated in figure 1. Basic geometrical and material characteristics of several designed testing models/members are presented in table 2.

Table 1. The total research program, geometrical dimensions and materials of flanges and webs of testing models

Testing models/members			Geometrical dimensions [mm]						Steel	
M.G.	C.G.	Marking	L	h	b	t_f	d	t_w	flanges	webs
A	1	AS11, AS12, AS13	250	112	60	6	100	2	S235	S235
	2	AS21, AS22, AS23	500	312	90		200			
	3	AS31, AS32, AS33	750	312	120		300			
	4	AS41, AS42, AS43	1000	412	150		400			
B	1	BS11, BS12, BS13	250	112	60	6	100	2	S355	S235
	2	BS21, BS22, BS23	500	212	90		200			
	3	BS31, BS32, BS33	750	312	120		300			
	4	BS41, BS42, BS43	1000	412	150		400			

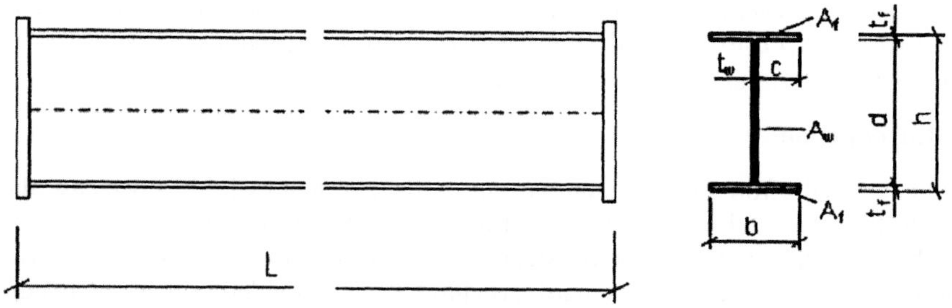

Figure 1. Scheme of testing models/members

Table 2. Basic geometrical and material characteristics of designed testing models/members

Members		Geometrical characteristics					Material characteristics
M. G.	C.G.	$\lambda_y = L / i_y$	$\lambda_z = L / i_z$	$\beta_f = c / t_f$	$\beta_w = d / t_w$	$\gamma = A_w / A$	$m_y = f_{yf} / f_{yw}$
A	1	5,03	16,02	4,83	50,0	0,217	1,000
	2	5,01	20,80	7,33	100,0	0,270	
	3	4,99	23,02	9,83	150,0	0,294	
	4	4,98	24,28	12,33	200,0	0,308	
B	1	5,03	16,02	4,83	50,0	0,217	1,511
	2	5,01	20,80	7,33	100,0	0,270	
	3	4,99	23,02	9,83	150,0	0,294	
	4	4,98	24,28	12,33	200,0	0,308	

All of testing models/members are divided to 2 material groups (M.G.: A, B) and 4 cross-sectional groups (C.G.: 1, 2, 3 and 4). The materials group A is created by members with homogenous cross-section made from steel S235 and group B is created by members with combined cross-section made from steels S355 (flanges) and S235 (webs). The several cross-sectional groups have different dimensions, but first of all have different web slenderness β_w. It is apparently, that the members are thin-walled at the compression loading. At the same time, according to local stability aspects, the flange dimensions are designed to be compact (slenderness β_f), when subjected to elastic loading. At last, according to global stability aspects and dimensions of several cross-sectional groups, the lengths of members L are designed to be quasi-compact (slenderness λ_y, λ_z, $\lambda_z > \lambda_y$). The ratio γ give an evident characteristic of the economic efficiency of designed cross-sections – table 2.

The flanges of all members were made out from 2 sheets, 6 mm thick (steel S235 and S355) and the webs from 2 sheets, 2 mm thick (steel S235). Three material specimens were taken form each of used sheet to make normative shaped testing bars. The testing bars underwent a tension tests to find out the strain-stress diagrams and required material characteristics. Characteristic strain-stress diagrams are illustrated in figure 2, where the values of averaged determined yield stress f_y and ultimate tensile strength f_u are presented. Mentioned yield stresses f_y and ultimate tensile strength f_u was assigned to the relevant flanges and webs of several members.

For consistent evaluation and analyses of experimental knowledge and results, it is also necessary to know the real geometrical dimensions of the testing models/members. Therefore and before testing at the beginning, the detailed dimension measuring of all members was done. Dimensions of cross-sections: height h, width b, thicknesses of flanges t_f and webs t_w was measured on the top, middle and bottom of each member. Averaged values of measured dimensions are considered as real.

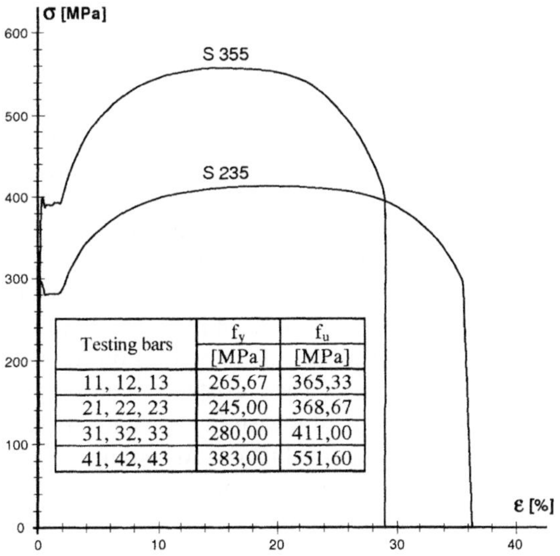

Testing bars	f_y [MPa]	f_u [MPa]
11, 12, 13	265,67	365,33
21, 22, 23	245,00	368,67
31, 32, 33	280,00	411,00
41, 42, 43	383,00	551,60

Figure 2. Characteristic strain-stress diagrams and determined material characteristics

Figure 2 show the good quality of testing members' material characteristics. Determined yield stresses of flanges and webs f_{yf} and f_{yw} are higher than the normative values. Also in the case of materials group A members with designed homogenous cross-sections, the determined flanges yield stresses values were higher than the webs yield stresses, $f_{yf} > f_{yw}$. This means that they are of combined material (m_y = 1,054, event. 1,143). In the case of materials group B it is categorical to go about members with material combined cross-sections (m_y = 1,442, event. 1,563). In the case of all testing bars a good material ductility was found, $A_5 > 29$ %.

3 Methodology and test content

The tests have to bring out detailed investigation about transformation, failure and gross bearing capacity of the above-mentioned members, in consideration of several designed geometrical and material parameters.

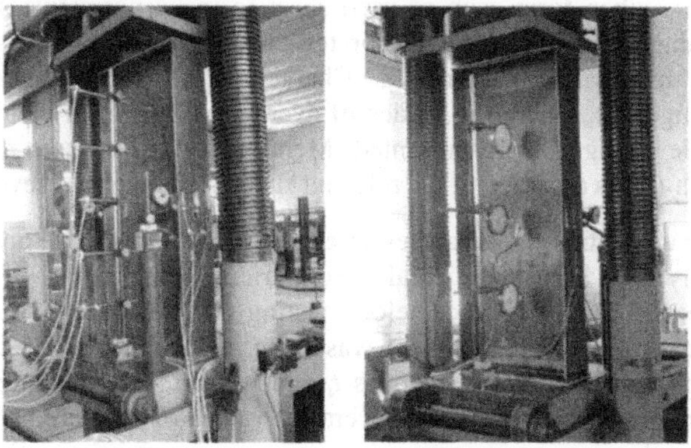

Figure 3. General layout of the test, measurement of strains ε, deflections of the web w and buckling of member BS41

In accordance with the research target, the emphasis is imposed on elastic-plastic post-critical behaviour of slender webs and their interaction with flanges. In context with that, the initial shape deflections of members, mainly the initial buckling of slender webs are significant for the experimental results valuation and connected theoretical analyses. Therefore and before the beginning of the testing the initial buckling of all members' webs are finding out on previously drawn raster by means of inductive sensors. The tests of members in compression are realized by means of press at the Bearing structures laboratory. During consecutive programmed overloading of member, the strains ε in the middle cross-section were measured. Measurement was realized in 12 places, double-faced on the web in 6 places and also on the flanges in 6 places. The resistance tensiometers were used to measurement the strains ε by means of measuring apparatus Hottinger Balwin UPM 60 connected to computer for direct evaluation. According to member's length, the deflections of web w are measured in 3 or more places elected in the characteristic positions. The web deflections w were measured using inductive sensors connected to computer and also using mechanic gauges. In the case of members with ultra-slender webs (AS41 ~ AS43 and BS41 ~ BS43), the global buckling was also investigated in the middle cross-section on the edges of flanges. The member's global buckling v was measured by means of mechanic gauges. General layout of the test, measurement of strains ε, lateral deflections of the web w and global buckling v are illustrated in figure 3.

The members during the test were consecutively overloaded and released. The member overloading was regulated close to its behaviour, measured values of strains ε and deflections of the web w. The test continue up to total failure, defined by the beginning of consecutive, continuous increasing of strains ε and deflections of the web w. Figure 4 illustrate the total failure of members AS22 a BS22 by local buckling of webs and flanges.

Figure 4. The total failure of members AS22 a BS22 by local buckling of webs and flanges

4 Partial theoretical and experimental results

According to procedures and formulae of the first author, the theoretical ultimate loading of all testing members were calculated as following:

N_{el} ultimate elastic loading definite by attaining the web yield stress f_{yw} of cross-section,

N_{pl} ultimate plastic loading definite by attaining the flanges yield stress f_{yf} of cross-section,

$N_{ul,el}$ ultimate elastic post-critical loading definite by attaining the yield stress f_{yw} in the outer fibers of cross-sectional web,

$N_{ul,ep}$ ultimate elastic-plastic post-critical loading definite by attaining the ultimate strain $\varepsilon_u = \varepsilon_{yf}$ in the outer fibers of cross-sectional web,

$N_{u,y}$, $N_{u,z}$ ultimate buckling loading of member according to axis y and z considering the elastic- plastic postcritical behaviour of the web.

Table 3 presents the relevant values of several members' ultimate loadings. All of members' ultimate loadings were calculated according to real – measured dimensions and determined yield stresses of their flanges and webs.

In general, ultimate $N_{u,z}$ buckling loadings have the smallest values. However, these values are very close to the ultimate elastic-plastic post-critical loading values – $N_{ul,el}$. When the real boundary conditions of members are considered in accordance with the research target, the elastic-plastic post-critical interaction between thin webs and flanges may appear in conclusive rate.
In this time, all of 24 members were tested. All of tested members failed by local failure of flanges in consequence of webs' local deflection in multiple waves with

different shapes. Conclusive buckling of web and flanges was mainly concentrated in the ending areas of members – obviously because of concentrated loading transfer, Figures 4 and 5.

Table 3. Theoretical and experimental bearing capacities [kN]

Member	N_{el}	N_{pl}	$N_{ul,el}$	$N_{ul,ep}$	$N_{u,z}$	$N_{u,y}$	$N_{u,exp}$	$N_{u,exp}/N_{u,z}$
AS11	249,9	260,2	249,9	260,2	260,2	260,2	278,0	1,068
AS12	251,3	261,8	251,3	261,8	261,8	261,8	275,0	1,050
AS13	249,9	260,3	249,9	260,3	260,3	260,3	280,0	1,076
AS21	377,6	416,7	322,8	360,9	356,1	360,9	357,0	1,003
AS22	377,1	415,6	321,9	359,2	354,3	359,2	373,0	1,053
AS23	376,0	414,4	320,9	358,2	353,3	358,2	359,0	1,016
AS31	540,2	560,3	424,5	444,8	435,6	444,8	447,0	1,026
AS32	542,8	563,0	426,8	447,2	438,1	447,2	432,0	0,986
AS33	549,7	570,0	432,2	452,7	443,4	452,7	442,0	0,997
AS41	651,5	713,8	491,0	552,0	539,1	552,0	490,0	0,909
AS42	653,9	716,6	493,9	555,4	542,0	555,4	512,5	0,946
AS43	654,6	717,4	494,6	556,2	542,8	556,2	480,0	0,884
BS11	253,6	339,8	253,6	339,8	335,5	339,8	357,0	1,064
BS12	261,6	350,6	261,6	350,6	346,5	350,6	363,0	1,048
BS13	257,0	345,1	257,0	345,1	341,0	345,1	357,0	1,047
BS21	376,4	527,8	321,2	469,2	454,9	469,2	466,0	1,024
BS22	376,0	527,3	320,9	468,7	454,7	468,7	492,0	1,082
BS23	370,1	517,9	314,9	459,2	445,7	459,2	482,0	1,081
BS31	544,6	711,3	429,2	597,0	569,9	592,8	565,0	0,991
BS32	549,0	716,6	432,6	595,0	574,3	597,2	577,5	1,006
BS33	546,1	713,5	430,7	691,8	572,2	595,0	562,5	0,983
BS41	643,0	888,2	486,7	691,8	659,5	691,8	657,5	0,997
BS42	654,7	902,1	494,4	709,2	676,1	709,2	590,0	0,873
BS43	650,9	896,2	490,5	695,7	663,0	695,7	625,0	0,943

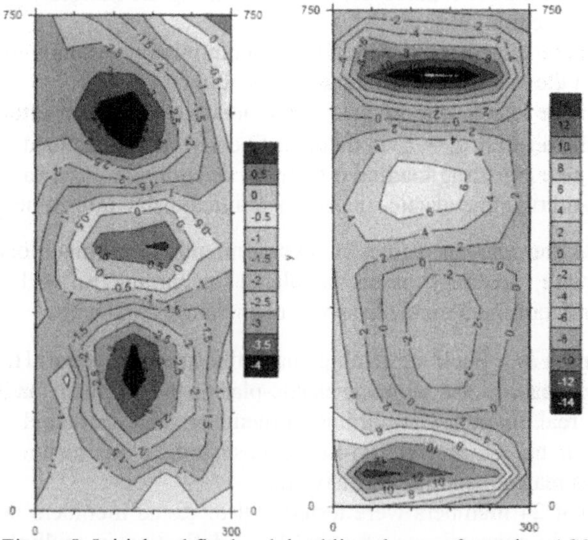

Figure 5. Initial and final web buckling shapes of member AS31

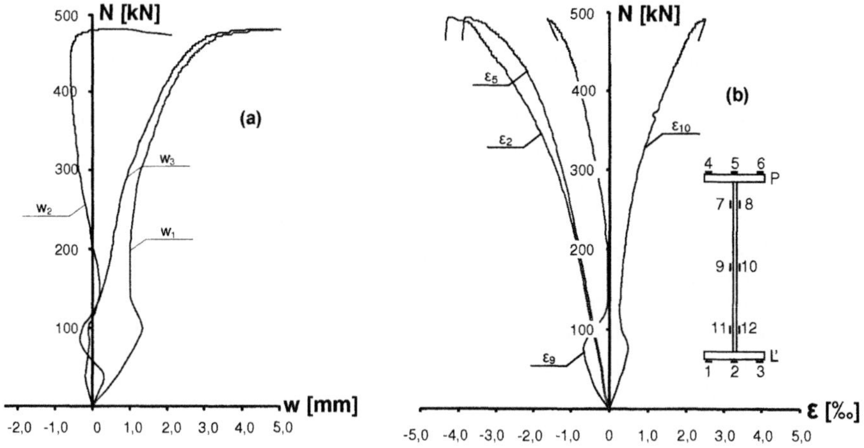

Figure 6. The web deflections w in the middle and quarters of member BS 23 (a), the web and flanges strains ε of member BS 22 (b)

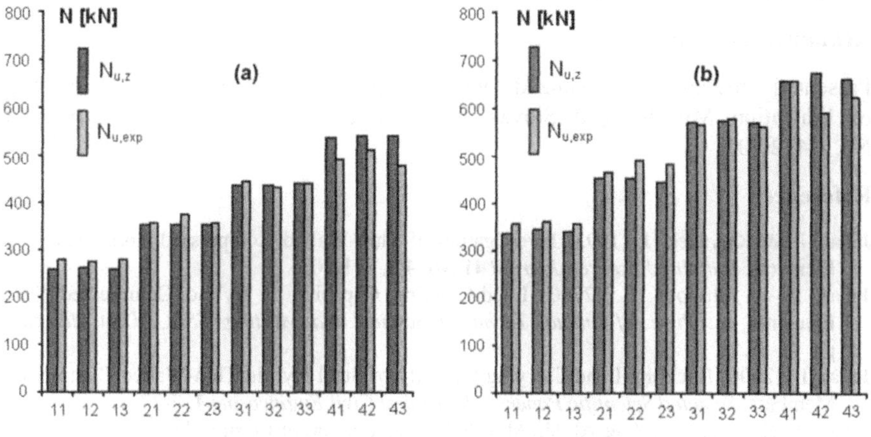

Figure 7. The comparing of theoretical and experimental capacities of members Material group A (a), material group B (b)

The interaction between web and flanges is evidently manifested here. Determined ultimate experimental loadings, eventually bearing capacities $N_{u,exp}$ are also illustrated in table 3. Very good consonance can by found from the preliminary realized comparison between determined experimental bearing capacities $N_{u,exp}$ and theoretical ultimate loadings $N_{u,z}$, $N_{ul,ep}$.

In the case of members with ultra-slender webs: $\beta = 200$, (AS41 ~ AS43 and BS41 ~ BS43) more significant differences was registered. Some of obtained experimental results and relations $N - w$ and $N - \varepsilon$ are illustrated in figures 5, 6 and 7.

5 Conclusion

- The results of presented research affirm and expand the knowledge of previous research about the elastic-plastic behaviour and bearing capacities of thin-walled compression members with quasi-homogenous and combined cross-sections.
- Elastic-plastic post-critical bearing capacity of compression members' thin webs is on a large scale dependent on their initial buckling shape and also dependent on their forming shape during the loading process.
- Post-critical bearing capacity of compression members' thin webs increases by increasing the number of buckling waves during the elastic-plastic and plastic stage of strain.
- Thin webs with slenderness $\beta \leq 150$ (A), event. $\beta \leq 100$ (B) prove a sufficient support of compression members' flanges. Theoretical local bearing capacities $N_{ul,ep}$ and total bearing capacities $N_{u,z}$ are in a good consonance with the obtained experimental bearing capacities $N_{u,exp}$.
- In the case of members with webs' slenderness $\beta = 200$, the influence of non-sufficient support of compression flanges by ultra thin web was also manifested here. This effect was significant near the members' ending which can be caused by local transfer of loading to the flanges and web. In the case of these members, theoretical local bearing capacities $N_{ul,ep}$ and total bearing capacities $N_{u,z}$ are a bit less than the experimental bearing capacities $N_{u,exp}$.

Acknowledgement

Presented research is realized with support of Science Grant Agency of Education Ministry and Slovak Academy of Science in frame of project No. 1/4220/07.

Reference

Juhás, P. & Kriváček, J. (1993) Investigation of Thin-Walled Compressed Combined Steel Elements. *Building Research Journal* **41** No. 4/2, 871-907.

Juhás, P. & Juhásová, E. (2000) Load-Carrying Capacity of Hybrid Compressed Steel Elements. In: *Proc. of Annual Technical Session and Meeting, SSRC, USA-Memphis,* pp. 75-88.

Juhás, P. (2006) Buckling Load-Carrying Capacity of Steel Hybrid Thin-Walled Compressed Members. *Selected Scientific Papers – Journal of Civil Engineering* **1**, 7-27.

Juhás, P., Kokoruďová, Z. & Al Ali, M. (2007) Investigation of Elastic-Plastic Load-Carrying Capacity of Thin-Walled Compressed Steel Elements. In: *Proceedings of the 8th Scientific Conference the Technical University Civil Engineering Faculty in Košice, Košice,* pp. 99-106.

Juhás, P., Al Ali, M. & Kokoruďová, Z. (2007) Experimental Investigation of Elastic-Plastic Carrying Capacity of Thin-Walled Compressed Steel Members. In: *Proceedings of Czech – Slovak Conference: Experiment'07, Brno, Czech Republic,* pp. 139-146.

Kriváček, J. (2007) The knowledge from Experiment – Basis for Numerical Simulation of Thin-Walled Steel Structures' behaviour. In: *Proceedings of Czech – Slovak Conference: Experiment'07, Brno, Czech Republic,* pp. 229-234.

7.5 A User-friendly Design of the Webs of Steel Bridges Subjected to Many Times Repeated Loading

Miroslav Škaloud and Marie Zörnerová
Institute of Theoretical and Applied Mechanics, AS CR, v.v.i.
skaloud@itam.cas.cz, zornerova@itam.cas.cz

Abstract

In the first part of the paper, two ways of steel bridge construction are discussed, viz. (i) thin-walled construction and (ii) economic-fabrication one. Then it is reported about new stages of the Prague research, both theoretical and experimental, on the fatigue limit state of steel plate girders whose webs breathe under repeated loading. Based on analysis of the new results and conclusions, simple formulae are established such as to give (i) the maximum web slenderness or (ii) the maximum load for which the impact of the complex problem of web breathing can entirely be disregarded in design. This approach can substantially simplify the analysis of steel girders subjected to repeated loading.

Keywords: *Thin-walled construction, Economic-fabrication construction, Plated structures, Post-buckled behaviour, Repeated loading.*

1 Thin-walled construction

One of the most promising trends in our striving to save steel is to use thin-walled structures, i.e. structural systems made of slender (usually plate) elements. Of course, it can be argued that such elements are liable to buckle so that the limit state of the system is substantially reduced by stability phenomena. The situation is however remedied by, and the promising idea of thin-walled construction can be materialized thanks to, the miracle of post-buckled behaviour, in the light of which a thin-walled plated system behaves like a (so called) super-smart structure, i.e. like one which is able not only to diagnose its own situation, but also to generate a means of powerful defence. In the case of plates and plated systems, this self-defence is brought about by the stabilising effect of membrane stresses, which come into play when plate deflections are of the order of plate thickness, and thereafter very considerably slow down any further deflection growth and therefore also substantially increase the load-bearing capacity of the system. The (so-called) critical load, provided by linear buckling theory, has no practical meaning for the limit state of the structure because, thanks to the post-buckled behaviour, the ultimate strength of the system is usually very significantly (and in the case of slender webs subjected to combined shear and bending even several times) higher than the linear-buckling-theory critical load.

That is why great attention has been internationally paid to research on the post-buckled behaviour and ultimate strength of slender webs, flanges and other plate elements, the Czech research always striving to play a useful role in these activities.

For example, the authors of this paper and their co-workers spent about three decades in investigating the post-critical reserve of strength and ultimate load behaviour of steel plate girders, box girders, thin-walled columns etc.

In past decades much attention was paid to the derivation of an ultimate load approach to the design of the webs of steel plate and box girders. It was found, using both theoretical and experimental investigations, that the (so-called) critical load,

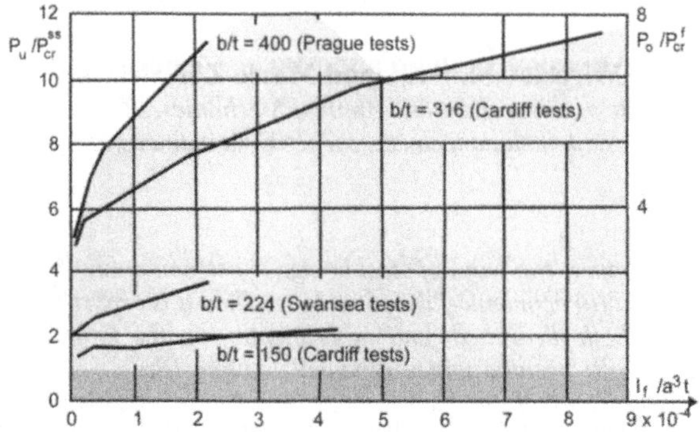

Figure 1: Swansea, Cardiff and Prague shear girder test results

determined via linear buckling analysis, was in no relation to the actual ultimate strength of the web, since in a great majority of cases there has always existed a very substantial post-critical reserve of strength. This is demonstrated in Figure 1, where the results of tests on shear girders carried out by (i) K.C. Rockey and one of the authors in Swansea and Cardiff and (ii) by the authors of this paper in Prague are plotted. They are plotted in terms of the flange rigidity parameter $I_f / a^3 t$, because it was found that flange stiffness very substantially affected the ultimate load of the test girders. On the left-hand vertical axis, the ultimate load is related to the critical load of a web simply supported on all boundaries, while on the right-hand vertical axis the critical load of a clamped web is used in the denominator of the ratio ultimate strength/critical load. But it can be seen in the figure that, whether we consider the former or the latter value of the critical load, the post-buckled reserve of strength is always very great.

2 Economic-fabrication construction

There are many situations where an application of thin-walled construction is very advantageous. For example, in steel bridgework with larger spans.

However, there are also numerous cases where another approach is more economical. This results from the fact that the price of a steel structure is not only given by the price of the steel used, but is also considerably affected by the cost of the fabrication of the structure. In many situations, it is the latter aspect that prevails; then it does not matter much that the structure in question is by a few tonnes heavier if this is compensated (and frequently outweighed) by substantially reduced fabrication costs.

In bridge construction this occurs with small and medium-span bridges, where a simple structural system, composed of a simple welded I-beam (without any longitudinal ribs and with as few transverse stiffeners as possible) made composite with a concrete slab is becoming very popular. The first author of this paper is grateful to (i) Prof. Joël Raoul, who showed him numerous examples of bridges of this kind during their common journey from Paris to Normandy, and (ii) J. Kunrt and D. Riedl (respectively Director and Deputy Director of Division 7 of the METROSTAV, Prague, plc., which fabricates steel bridges) for supporting him in the above way of thinking.

But the economic-fabrication approach is not only connected with the structural system chosen, but also with the way it is fabricated. In this respect an important question ought to be asked, viz. should we straighten the plate elements (for example, the webs of steel plate and box girders) of which the structural system is composed, which, as a result of the fabrication of the system (in particular as a result of the welding procedures used), always exhibit an initial curvature?

3 To straighten or not to straighten the webs of steel bridge girders

It is practically impossible to fabricate welded steel plated structures without their plate elements exhibiting initial curvatures. Therefore, it is understandable that various standards require that this initial "dishing" be kept under control via prescribed tolerances, and that in the case of need the magnitude of the initial curvature be reduced by straightening, usually heat straightening.

But is this really indispensable and desirable?

This problem was dealt with some time ago by the Task Group "Tolerances in Steel Plated Structures", sponsored by the IABSC, chaired by Prof. Ch. Massonnet (1980), with one of the authors being a member of the Task Group. The observations made during and the conclusions drawn from, the activities of the Task Group are very much of interest even now.

For example, the following definition of an optimum set of tolerances as that which minimizes the total cost of a definite structure, required to sustain definite sets of loads and to respect the stress, displacement, stability, corrosion, etc.

It is true that discussions within the Task Group exhibited some differences of opinion. For example, in one of the introductory sections of the final report of the Task Group, you can read that (i) in Belgium it is generally forbidden to apply any type of plate straightening, (ii) in the USA they are in favour of so doing, and (iii) in Germany it is a golden-mean position because in this country there seems to be in practice an unofficial mutual agreement between all parties concerned that (a) geometrical imperfections within generally accepted limits should not be made a reason for straightening procedures, (b) where heat straightening is unavoidable and seemingly advantageous, it should always be restricted to rare occasions only and that such procedures should be used very carefully and only be done by skilled experts with the unanimous approval of all parties concerned.

However, the above differencies notwithstanding, the Task Group tended to the opinion that heat straightening in steel plated structures is not desirable.

The authors of this paper endorse the above stand point, the reasons being twofold:

(i) The procedure of straightening is rather costly (not only directly but also due to blocking some space in the steel fabricator, which can be used for other operations) and therefore, would not be compatible with the aforesaid economic-fabrication strategy.

(ii) While it is understandable that the straightening of webs should frequently be employed for aesthetical and psychological reasons, it is not certain that the actual stability behaviour of a girder with a straightened web is better than that of the original girder. To straighten the web of a girder by heat treatment only means that one initial imperfection (initial web curvature) is replaced by another initial imperfection (additional residual stresses induced by the heat treatment applied). The aforementioned Task Group also turned attention to another important fact; namely, while in the case of compressed plates (e.g. the compression flanges of steel box girder bridges) the influence of an initial curvature on load-carrying capacity can be significant (in ordinary cases even 20%), with webs subjected to combined shear and bending the same effect is (of course, when the initial "dishing" is not too large – and it should not be when the plated structure is fabricated in a good enough steel fabricator) much less important (a few % only).

This conclusion was also confirmed by the results of sensitivity analysis carried out by one of the writers, Z. Kala, et al. (2004), see Fig. 2, where it can be seen that other factors (for example, web and flange dimensions and material properties) have a much more significant effect on the ultimate limit state of the girder than initial web curvature.

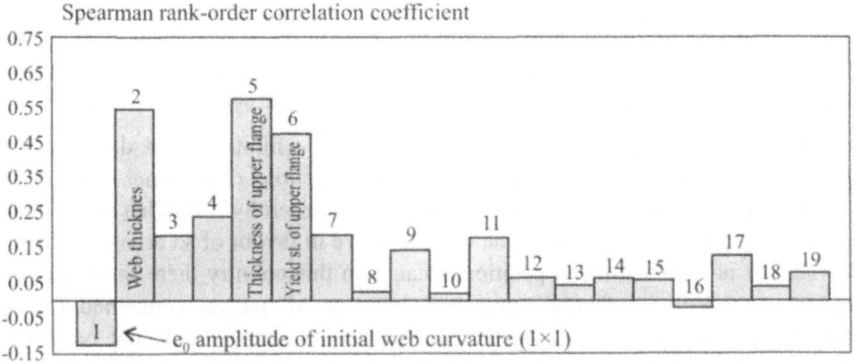

Figure 2. Sensitivity analysis of the load-carrying capacity

As said above, the authors take the view that the use of heat straightening in steel plated structures need not be advantageous and hence is not desirable. This statement, however, holds true only for the products of highly-qualified steel fabricators, such as to guarantee a high enough standard of workmanship, i.e. steel fabricators being well enough equipped and having a staff with high expertise and experience. Only in such cases can it be expected that the standard of workmanship achieved will be sufficiently high and resulting initial imperfections reasonably small.

For the products of other plants a check of imperfections is indispensable.

Even for the high-standard steel fabricators the International Task Group "Tolerances in Steel Plated Structures" recommended the following:

(i) Efficient design in balancing welds about neutral axis,

(ii) Avoidance of excessive use of weld metal (this applying to both the number and size of welds),

(iii) Ensure that fit up is as perfect as can be achieved,

(iv) Use appropriate welding procedures and sequences, noting how and to what extent the work distorts as welding proceeds. Be aware of the fact that generally automatic and semi-automatic welding yields better results than manual welding.

4 User-friendly design of the webs of economic-fabrication steel bridge girders

When the structural system is simple, it is understandably desired that the design method applied be also as simple and user-friendly as possible. This can be achieved by establishing design procedures such as to be able to disregard some (or all) secondary detrimental effects. The main objective of this paper is to find out conditions for one of the detrimental effects, viz. that of the complex problem of web breathing, being able to be neglected. Such webs are usually called "non-breathing webs" even though it would be more accurate to call them "no-breathing-effect webs" or "quasi non-breathing webs", since such webs can slightly breathe under repeated loading but the impact of such moderate breathing can entirely be disregarded.

The problem of web breathing and the related cumulative damage process being rather complex, experiments have always proved to be a most reliable way to investigate them, and the authors have been following it for more than 10 years now. To date they have tested 215 girders and thereby gained a lot of experience with experiments on breathing structural systems and plenty of information about various aspects of the breathing phenomenon. The large number of tests proved to be indispensable in view of the large scatter, which is characteristic of all breathing (and also of most other kinds of fatigue) experiments.

One of the new stages of the authors' experimental investigation was used so as to give background for a user-friendly design method for plate girders with quasi-non-breathing webs.

As this juncture it is useful to say a few words about the character of the test girders used.

Like most girders tested now by the writers at the Institute of Theoretical and Applied Mechanics in Prague, they are fairly large, having a web 1000mm deep, so that their character is not far from that of ordinary girders. One of the experimental girders in the testing position is shown in Fig. 3.

All the girders were fabricated in the steel fabricator of Division 7 of the Company METROSTAV plc., (which is one of the best steel fabricators in the Czech Republic) using the same technological procedures as are applied there in the fabrication of ordinary steel bridges. It is important to note that, compatibly with the recommendations presented in the previous section 3, in the fabrication of the test girders no attempt was made to diminish (by heat treatment) the initial curvature of the web generated during the process of girder fabrication.

So, the girders tested by the authors had webs whose initial curvature was as large as that resulting from the very fabrication of the girder, without any straightening used. Then the amplitudes w_0 of the web initial "dishing" were in the interval 2-11mm, i.e. $d/500$-$d/91$, d being the depth of the web. The average value w_0^{av} = 7.07mm, i.e. d/141.5. Hence it can be expected that the formulae established hereafter will hold for ordinary girders having the same order of initial irregularities.

Figure 3. One of the Prague girders in the testing position

An inspection of Fig. 3 reveals that the webs of the writers′ test girders were under the action of combined shear and bending and therefore, the formulae obtained can be used for a user-friendly design of webs under the same loading conditions. These represent the most frequently encountered loading in the case of the webs of steel bridge plate and box girders.

Let us now proceed from the general to the specific.

A detailed analysis of the experimental results obtained shows that so-called non-breathing webs can be obtained by limiting either their slenderness or load acting on them.

For each approach, two criteria are going to be applied:
(i) that of the initiation of the first fatigue crack,
(ii) that of the fatigue failure of the girder

4.1 *Limitation of web slenderness*

The maximum depth-to-thickness ratio λ^i_{lim} such that no fatigue crack can initiate under a number N of loading cycles:

$$\lambda^i_{\text{lim}} = 111.77\sqrt{k_\tau}\,\frac{1}{\sqrt[12.255]{N}}\sqrt{\frac{235}{f_y}} \tag{1a}$$

The maximum depth-to-thickness ratio λ^f_{lim} such that even if a fatigue crack (or cracks) initiate, all of them stabilize so that, under a given number N of loading cycles, no fatigue failure of the whole girder occurs:

$$\lambda^f_{\text{lim}} = 161.45\sqrt{k_\tau}\,\frac{1}{\sqrt[9.577]{N}}\sqrt{\frac{235}{f_y}} \tag{1b}$$

In the above two formulae k_τ = the buckling coefficient for a web in shear, in terms of its boundary conditions and side ratio, f_y = the yield strength (in Newton/mm^2) of the web material, and N = the number of loading cycles to which the web panel is expected to be exposed.

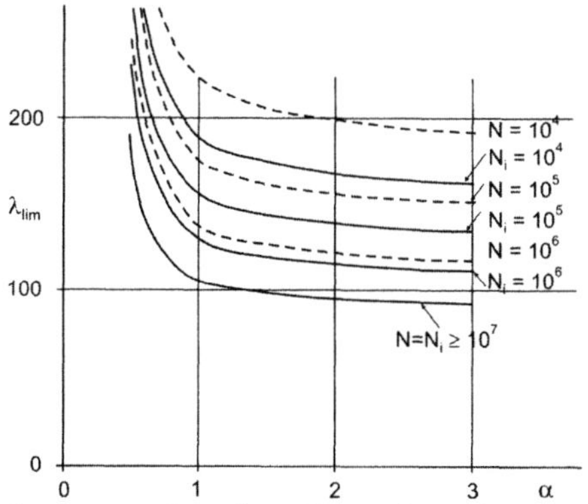

Figure 4. The limiting slenderness of non-breathing webs subjected to predominantly shear

Formulae 1a,b hold for $N \geq 10^4$. For lesser numbers of load cycles the effect of web breathing on the limiting slenderness can be disregarded. On the other side for $N > 10^7$ the limiting slenderness attains its lower bound $\lambda_{lim,0}$, so that for all $N > 10^7$, $\lambda_{lim,0} = \lambda_{lim}\,(N = 10^7)$. The formula is valid for webs subjected to predominantly shear, i.e. for web panels under the action of (i) shear τ or (ii) combined shear τ and bending σ

when $\sigma/\tau \leq 1$. When $\sigma/\tau > 1$, the values of the limiting depth-to-thickness ratio will be larger.

As said above, the buckling coefficient k_τ is a function of the boundary conditions of the web. For example, for a web panel clamped into girder flanges and simply supported on vertical stiffeners:

$$k_\tau = \frac{5.34}{\alpha^2} + \frac{2.31}{\alpha} - 3.44 + 8.39\alpha \qquad \text{for } \alpha \leq 1 \tag{2a}$$

$$k_\tau = 8.98 + \frac{5.61}{\alpha^2} - \frac{1.99}{\alpha^3} \qquad \text{for } \alpha \geq 1 \tag{2b}$$

α being the side ratio of the web panel.

The numerical impact of formulae 1a,b and 2a,b can be seen in Fig. 4, which is plotted for $f_y = 235$ MPa and for the above boundary conditions.

The full lines are there related to formulae 1a, the dashed ones to formulae 1b.

4.2 *Limitation of load acting on web*

Another possibility consists in limiting not the web slenderness, but the load acting on the web, and namely (it seems at the first sight) so that the maximum value of the cycling loads never exceeds the critical loading of the corresponding "ideal" web without initial imperfections.

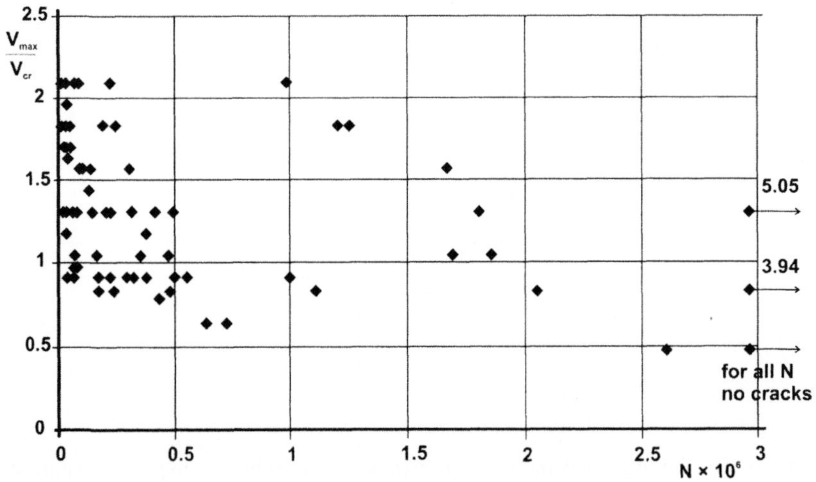

Figure 5. The initiation of the first fatigue crack in terms of (i) the ratio V_{max} / V_{cr} and (ii) the number N of the loading cycles

Then the maximum average shear stress such that no fatigue crack can initiate under a number N of loading cycles:

$$\tau_{max} = (-0.5 \cdot 10^{-6} N + 0.9)\tau_{cr} \qquad \text{for} \quad 10^4 \leq N \leq 6 \cdot 10^5 \tag{3a}$$

$$\tau_{max} = 0.6\tau_{cr} \qquad \text{for} \qquad N \geq 6 \cdot 10^5 \tag{3b}$$

The maximum average shear stress is such, that even if a fatigue crack (or cracks) initiate, all of them stabilize so that, under a given number N of loading cycles, no fatigue failure of the whole girder occurs:

$$\tau_{max} = (-10^{-7}N + 0.8)\tau_{cr} \qquad \text{for} \quad 10^4 \leq N \leq 2 \cdot 10^6 \tag{4a}$$

$$\tau_{max} = 0.6\tau_{cr} \qquad \text{for} \qquad N \geq 2 \cdot 10^6 \tag{4b}$$

Of course, the critical value τ_{cr} is calculated for the corresponding boundary conditions of the web.

It was said above that the substantial reduction of the maximum load τ_{max} to a value under the critical value τ_{cr} was due to the effect of initial imperfections. At this juncture it is therefore, pertinent to pose the question of what is the relationship between the amplitude w_0 of the initial web curvature and the quantity τ_{max}.

It was again an analysis of the authors′ experimental results that made it possible to arrive at the following equation:

$$\frac{w_0}{d} = -0.0069\frac{\tau_{max}}{\tau_{cr}} + 0.0107 \tag{5}$$

Having however, this relationship at our disposal, we can ask the very important question of what is the maximum allowable amplitude $w_{0,all}$ of the initial web "dishing" such that $\tau_{max} = \tau_{cr}$. From equation (5) it then follows that for the authors′ test girders $w_{0,all} = 3.8$ mm i.e.

$$w_{0,all} = \frac{d}{263} \tag{6}$$

If a little (by 14% only, since Eq (5) has been established based practically on the lower bound of experimental results) enlarging the safety of the formulae established above, with the view to allow for a larger scatter of the unavoidable imperfections of the webs of ordinary bridge plated girders (without any straightening procedures being applied) than that encountered in the authors′ tests, it can be stated that if w_0 is kept (for any value of N) under $d/300$, the maximum load τ_{max} can go as high as τ_{cr}.

4.3 *Other design criteria*

The fulfilment of the above formulae only means that the impact of web breathing can be completely disregarded in design. Of course, the design is then governed by other phenomena and criteria.

5 Conclusions

Based on more than 215 tests conducted on steel plate girders, where web panels were subjected to many times repeated predominantly shear, the authors establish user-friendly formulae for (i) maximum web slenderness, (ii) maximum load acting on the web, so that the effect of web "breathing" can entirely be disregarded in the design of economic-fabrication steel bridge girders.

Acknowledgement

The authors express their gratitude to (i) the Czech Science Foundation for the financial support of their research carried out within the projects 103/06/0064, 103/05/2003 and 103/05/2059, (ii) the Grant Agency of the Czech Academy of Sciences for the support of the project IAA200710603, and (iii) the ITAM AS CR, v.v.i. for the support within the project AVOZ 20710524.

References

Kala Z. & Kala J. & Škaloud M. & Teplý B. (2004) Sensitivity analysis of the effect of initial imperfections on the stress state in the crack-prone areas of breathing webs. In: *Thin-Walled Structures – Advances in Research, Design and Manufacturing Technology, Proc. of the Fourth Int. Conf. on Thin-Walled Structures,* Loughborough, GB, pp. 499 -506.

Massonnet Ch. & et al. (1980) Tolerances in steel plated structures. *IABSE Survey* – 14/80

7.6 Numerical Model for Buckling Analysis of Flexibly Connected Beam-type Structures

Goran Turkalj, Josip Brnic, Domagoj Lanc

Department of Engineering Mechanics, Faculty of Engineering, University of Rijeka, Vukovarska 58, HR-51000 Rijeka, Croatia, e-mail: goran.turkalj@riteh.hr

Abstract

This work presents a one-dimensional numerical formulation for buckling analysis of beam-type structures accounting for connection flexibility and warping deformation effects. Equilibrium equations of a 14-degree of freedom beam element are derived by applying linearized virtual work principle. The influence of connection behaviour is introduced in the numerical model by transforming stiffness matrices of a conventional beam element. For this purpose a special transformation matrix is derived.

Keywords: *buckling analysis, beam-type structures, space beam element, flexible connections*

1 Introduction

Load-carrying structures composed of thin-walled beam-type structural components of different configurations and cross-sectional shapes are extensively used in engineering practice, both in stand-alone forms and as stiffeners for plate- or shell-like structures. Unfortunately, such weight-optimized structural components, especially those with open profiles, are commonly susceptive to buckling failure, Gambhir (2004), López et al. (2007). Buckling analysis of such structures is frequently carried out under the assumption of an idealised structural connections behaviour, in which fully rigid of ideally pinned connections are assumed to occur, Kim et al. (1996). In contrary, real connections may exhibit the significantly different behaviour under exploitation, having characteristics somewhere in-between those ideal cases, Tan (2001), Urbonas & Daniūnas (2006). Such a flexible behaviour is a result of complex interaction between various components of the connection construction itself, Chen & Lui (2000). Therefore, conventional numerical analysis procedures must be broadened by incorporating real connection characteristics and improving the accuracy of structural analysis.

In this work a finite element simulation, based upon space beam element, of buckling behaviour of thin-walled beam-type structures comprising flexible connections is presented. The basic assumptions used in the analysis are: beam members are prismatic and straight; the cross-section is not deformed in its own plane but is subjected to warping in the longitudinal direction; displacements are large but strains are small; material is homogeneous, isotropic and linear-elastic; internal moments are represented as the resultants of stresses calculated by engineering theories: Euler-Bernoulli-Navier for bending and Timoshenko-Vlasov for torsion; external load is static and conservative.

Flexible connections are allowed to occur at finite element nodes by modifying elastic and geometric stiffness matrices of a conventional beam element and for this a special transformation matrix is derived. The effects of partially restrained warping are also considered. Buckling analysis is performed in an eigenvalue manner, neglecting pre-buckling deformations and assuming an ideal structural geometry and loading conditions.

2 Basic considerations

The deformation of an initially straight prismatic beam with a thin-walled cross-section of a wall of thickness t is considered. For the sake of simplicity, it is assumed that the shear centre and centroid of the cross-section coincide. A right-handed Cartesian coordinate system (z, x, y) is chosen in such a way that z-axis coincides with the beam axis passing through the centroid O of each cross-section, while the x- and y-axes are the principal inertial axes of the cross-section. Incremental displacement measures of a cross-section are defined as:

$$w_o(z),\ u_o(z),\ v_o(z),\ \varphi_z(z),\ \varphi_x(z) = -v',\ \varphi_y(z) = u_o',\ \theta(z) = -\varphi_z' \tag{1}$$

where w_o, u_o and v_o are the rigid-body translations of the cross-section associated with the centroid in the z-, x- and y-directions, respectively; φ_z, φ_x and φ_y are the rigid-body rotations about the z-, x- and y-axes, respectively; θ is a parameter defining the warping of the cross-section. The superscript 'prime' indicates the derivative with respect to z.

If rotations are small, the incremental displacement field of a thin-walled cross-section contains only the first-order displacement terms:

$$w = w_o - y\,v_o' - x u_o' - \omega\varphi_z',\quad u = u_o - y\,\varphi_z,\quad v = v_o + x\,\varphi_z \tag{2}$$

in which w, u and v are the linear or first-order displacement increments of an arbitrary point on the cross-section defined by the position coordinates x and y and the warping function $\omega(x, y)$.

If the assumption of small rotations is invalid, i.e. if the large rotation effects are to be taken into account, then the total displacement increments can be written as, Turkalj et al. (2003):

$$w + \tilde{w},\quad u + \tilde{u},\quad v + \tilde{v} \tag{3}$$

where

$$\begin{aligned} \tilde{w} &= 0.5\left(x\,\varphi_z\,\varphi_x + y\,\varphi_z\,\varphi_y\right) \\ \tilde{u} &= 0.5\left[-x\left(\varphi_z^2 + \varphi_y^2\right) + y\,\varphi_x\,\varphi_y\right] \\ \tilde{v} &= 0.5\left[x\,\varphi_x\,\varphi_y - y\left(\varphi_z^2 + \varphi_x^2\right)\right] \end{aligned} \tag{4}$$

represent the additional or second-order displacement terms due to large rotations. The corresponding Green-Lagrange incremental strain tensor can be written as:

$$\varepsilon_{ij} = e_{ij} + \eta_{ij} + \tilde{e}_{ij} \tag{5}$$

where

$$2e_{ij} = u_{i,j} + u_{j,i},\quad 2\eta_{ij} = u_{k,i}\, u_{k,j}\ ,\quad 2\tilde{e}_{ij} = \tilde{u}_{i,j} + \tilde{u}_{j,i} \tag{6}$$

Einstein's summation convention is assumed for each tensorial term with repeated dummy index, i.e. whenever the same index appears twice in a term, the index is to be given all possible values (z, x and y) and results added together. Furthermore, according to the geometrical hypothesis of the cross-sectional in-plane rigidity, the strain components due to the adopted assumption of in-plane cross-sectional rigidity, the strain tensor components ε_{xx}, ε_{yy} and ε_{xy} in Eq.5 should be equal to zero.

The beam stress resultant can be defined using seven components: F_z = axial force acting at the centroid, F_x and F_y = shear forces acting at the centroid in the x- and y-directions, respectively; M_z = torque with respect to the centroid consisting of two parts T_{sv} and T_ω representing the Saint-Venant or uniform torque and the warping or non-uniform torque, respectively; M_x and M_y = bending moments with respect to the x- and y-axes, respectively; M_ω = bimoment. Assuming Hook's law is valid, the linearized force-displacement relations can be written as:

$$\begin{gathered} F_z = E\,A\,w_o',\ M_z = T_{sv} + T_\omega,\ T_{sv} = G\,I_t\,\varphi_z',\ T_\omega = M_\omega',\ M_\omega = -E\,I_\omega\,\varphi_z'', \\ M_x = -E\,I_x\,v_o'',\ M_y = E\,I_y\,u_o'',\ \bar{K} = F_z\,\alpha_z + M_x\,\alpha_x + M_y\,\alpha_y + M_\omega\,\alpha_\omega \end{gathered} \tag{7}$$

where A is the cross-sectional area; I_x and I_y are, respectively, the principal moments of inertia about the x- and y-axes; I_t is the Saint-Venant torsion constant, I_ω is the warping moment of inertia, while $\bar{K}$ is the Wagner coefficient with α_z, α_x, α_y and α_ω representing the correspondig cross-section parameters, Chen & Atsuta (1977). Using the linearized virtual work principle, the equilibrium equations of the buckled beam element can be written as, Turkalj & Brnić (2004):

$$\delta U_E + \delta U_G = \delta W \tag{8}$$

where the left-hand side is the internal virtual work composed of the virtual elastic strain energy

$$\delta U_E = \int_V C_{ijkl}\, e_{kl}\, \delta e_{ij}\, dV \tag{9}$$

and the virtual geometric potential due to initial stresses and surface tractions, i.e.

$$\delta U_G = \int_V {}^0S_{ij}\, \delta\eta_{ij}\, dV + \int_V {}^0S_{ij}\, \delta\tilde{e}_{ij}\, dV - \int_{A_\sigma} {}^0t_i\, \delta\tilde{u}_i\, dA_\sigma \tag{10}$$

while the right-hand side is the virtual work of external forces

$$\delta W = \int_{A_\sigma} t_i\, \delta u_i\, dA_\sigma \tag{11}$$

In the above three equations C_{ijkl}, S_{ij} and t_i are, respectively, the incremental constitutive tensor, the second Piola-Kirchhoff stress tensor and surface tractions. The superscript '0' denotes the initial quantity values (prior the occurrence of buckling), no superscript the incremental quantity, while symbol 'δ' the virtual quantities. It should be noted that due to the strain terms $\tilde{e}_{ij}$ the geometric potential of all the internal moments involved in Eq.10 corresponds to that of the semi-tangential moment, ensuring the joint moment equilibrium conditions of adjacent non-collinear elements.

3 Beam element

Figure 1 shows a conventional 14 degrees-of-freedom beam-type finite element, with (z, x, y) denoting the local coordinate system of the element.

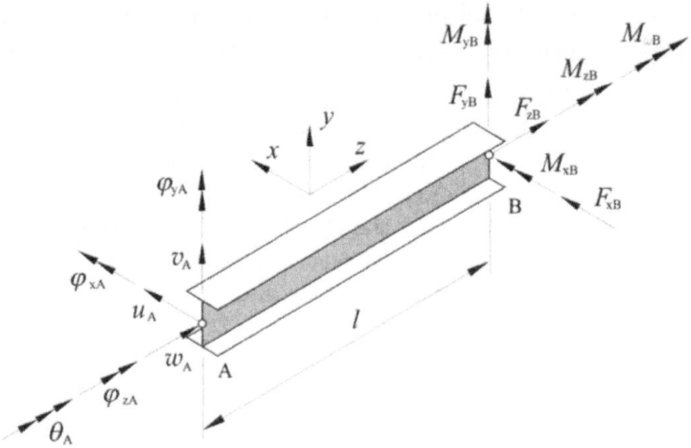

Figure 1. Thin-walled beam element: nodal displacements and nodal forces.

The corresponding nodal displacement and force vectors are:

$$\left(\mathbf{u}^{e}\right)^{T} = \left\{w_A, u_A, v_A, \varphi_{zA}, \varphi_{xA}, \varphi_{yA}, w_B, u_B, v_B, \varphi_{zB}, \varphi_{xB}, \varphi_{yB}, \theta_A, \theta_B\right\} \quad (12)$$

$$\left(\mathbf{f}^{e}\right)^{T} = \left\{F_{zA}, F_{xA}, F_{yA}, M_{zA}, M_{xA}, M_{yA}, F_{zB}, F_{xB}, F_{yB}, M_{zB}, M_{xB}, M_{yB}, M_{\omega A}, M_{\omega B}\right\} \quad (13)$$

where the right superscript 'e' denotes the e-th finite element. By adopting a linear interpolation for w_o and a cubic interpolation for u_o, v_o and φ_z, and by relating the beam stress resultants at the z-section to those at the element nodes, from Eqs 9-11 one can derive the following:

$$\delta U_E = \left(\delta \mathbf{u}^{e}\right)^{T} \mathbf{k}_E^{e}\, \mathbf{u}^{e},\; \delta U_G = \left(\delta \mathbf{u}^{e}\right)^{T} \mathbf{k}_G^{e}\, \mathbf{u}^{e},\; \delta W = \left(\delta \mathbf{u}^{e}\right)^{T} \left(\mathbf{f}^{e} + \mathbf{f}_{ekv}^{e}\right) \quad (14)$$

where $\mathbf{k}_E^{e}$ and $\mathbf{k}_G^{e}$ are, respectively, the elastic and geometric stiffness matrices of the thin-walled beam element from Figure 1, while $\mathbf{f}^{e}$ contains nodal forces applied to the beam element by other elements of the structure. The entries contained in $\mathbf{k}_{\mathrm{E}}^{\mathrm{e}}$ and $\mathbf{k}_{\mathrm{G}}^{\mathrm{e}}$ can be found in Turkalj & Brnic (2004).

By substituting Eq.14 into Eq.8, it follows:

$$\left(\mathbf{k}_{\mathrm{E}}^{\mathrm{e}}+\mathbf{k}_{\mathrm{G}}^{\mathrm{e}}\right)\mathbf{u}^{\mathrm{e}}=\mathbf{f}^{\mathrm{e}} \tag{15}$$

4 Hybrid element

The effects of connection flexibility are modelled by introducing the translational and rotational springs, each of stiffness S_{i}, at the element nodes. Figure 2 shows such a hybrid or semi-rigid (SR) beam element for the modelling of planar frames.

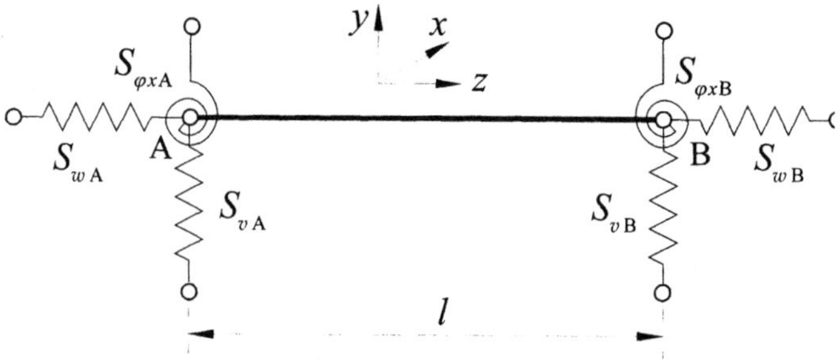

Figure 2. Hybrid or semi-rigid (SR) beam element.

For the buckled SR element, Eq.15 can be rewritten as:

$$\left(\mathbf{k}_{\mathrm{E,\,SR}}^{\mathrm{e}}+\mathbf{k}_{\mathrm{G,\,SR}}^{\mathrm{e}}\right)\mathbf{u}_{\mathrm{SR}}^{\mathrm{e}}=\mathbf{f}^{\mathrm{e}} \tag{16}$$

The nodal force vector $\mathbf{f}^{\mathrm{e}}$ can be defined with the respect to displacement terms $\mathbf{u}_{\mathrm{r}}^{\mathrm{e}}$ due to the relative nodal (connection) deformation as:

$$\mathbf{f}^{\mathrm{e}}=\mathbf{S}^{\mathrm{e}}\,\mathbf{u}_{\mathrm{r}}^{\mathrm{e}} \tag{17}$$

where matrix $\mathbf{S}^{\mathrm{e}}$ contains stiffness parameters pertaining to nodal degrees-of freedom. i.e.

$$\begin{aligned}
\mathbf{S}^{\mathrm{e}} &= \left\lceil \mathbf{S}_{\mathrm{A}}^{\mathrm{e}} \quad \mathbf{S}_{\mathrm{B}}^{\mathrm{e}} \quad \mathbf{S}_{\omega}^{\mathrm{e}} \right\rfloor \\
\mathbf{S}_{\mathrm{i}}^{\mathrm{e}} &= \left\lceil S_{w\mathrm{i}} \quad S_{u\mathrm{i}} \quad S_{v\mathrm{i}} \quad S_{\varphi z\mathrm{i}} \quad S_{\varphi x\mathrm{i}} \quad S_{\varphi y\mathrm{i}} \right\rfloor, \quad (\mathrm{i}=\mathrm{A,\,B}) \\
\mathbf{S}_{\omega}^{\mathrm{e}} &= \left\lceil S_{\theta\mathrm{A}} \quad S_{\theta\mathrm{B}} \right\rfloor
\end{aligned} \tag{18}$$

If the initial stability analysis is to be carried out, $\mathbf{S}^{\mathrm{e}}$ contains the initial connection stiffness parameters. The nodal displacement vector of the conventional element can be defined as a difference of the nodal displacement vector pertaining to the SR beam element, occurring in Eq.16, and the relative connection deformation vector from Eq.17, i.e.

$$\mathbf{u}^{e} = \mathbf{u}^{e}_{SR} - \mathbf{u}^{e}_{r} = \mathbf{u}^{e}_{SR} - \left(\mathbf{S}^{e}\right)^{-1}\mathbf{f}^{e} \tag{19}$$

By substituting Eq.19 into Eq.15, it follows:

$$\left(\mathbf{T}^{e}_{SR}\right)^{-1}\left(\mathbf{k}^{e}_{E} + \mathbf{k}^{e}_{G}\right)\mathbf{u}^{e}_{SR} = \mathbf{f}^{e} \tag{20}$$

where $\mathbf{T}^{e}_{SR}$ represents the transformation matrix

$$\mathbf{T}^{e}_{SR} = \mathbf{I} + \left(\mathbf{k}^{e}_{E} + \mathbf{k}^{e}_{G}\right)\left(\mathbf{S}^{e}\right)^{-1} \tag{21}$$

which transforms a conventional beam element into the SR (hybrid) one, while **I** denotes the 14×14 identity matrix. After equating Eq.16 and Eq.20, the stiffness matrices and the equivalent load vector of the SR element can be rewritten as:

$$\mathbf{k}^{e}_{E,\,SR} = \left(\mathbf{T}^{e}_{SR}\right)^{-1}\mathbf{k}^{e}_{E},\ \mathbf{k}^{e}_{G,\,SR} = \left(\mathbf{T}^{e}_{SR}\right)^{-1}\mathbf{k}^{e}_{G} \tag{22}$$

5 Numerical example

The FE procedure presented above is implemented in a computer program entitled EIGEN V.2, which has the capabilities of performing the structural stability analysis in an eigenvalue manner. Such an approach assumes ideal structural and loading conditions but allows us to determine the instability load of the structure in a direct manner without calculating the exact magnitude of deformations. The lowest eigenvalue obtained is recognized as the critical or buckling load and the corresponding eigenvector as the shape of buckling.

Figure 3 shows a space frame loaded by the vertical forces, each of value F, at the corners. Each member has an I-type cross-section corresponding to W 12×53 (A = 100.645 cm^2, I_x = 17689.836 cm^4, I_y = 3987.497 cm^4, I_t = 65.765 cm^4, I_ω = 848573.338 cm^6) of length L =365.76 cm. The material moduli are E = 200 GPa and G = 80 GPa. The buckling analysis of the frame is performed using three mesh configurations consisting of one, two and four elements per member. A warping restraint factor $0 \le a_F \le 1$, introduced by Yang & McGure (1984), is used to represent the flange warping conditions at the frame corners. In this, $a_F = 0$ means the girder flanges are free to warp, while $a_F = 1$ represents the fully restrained warping conditions. Three gorder-to-column connection conditions are considered: $S_{\varphi z} = 10^{12}$, $S_{\varphi x} = 10^{12}$, $S_{\varphi y} = 10^{12}$ and $S_\theta = 10^{16}$ (a_F = 1) for case A (rigid case); $S_{\varphi z}$ = $3GI_t/L$, $S_{\varphi x} = 5EI_x/L$, $S_{\varphi y} = 5EI_y/L$ and $S_\theta = 0.3822EI_\omega/L$ (a_F = 0.5) for case B; $S_{\varphi z}$ = GI_t/L, $S_{\varphi x} = EI_x/L$, $S_{\varphi y} = EI_y/L$ and $S_\theta \approx 0$ ($a_F = 0$) for case C.

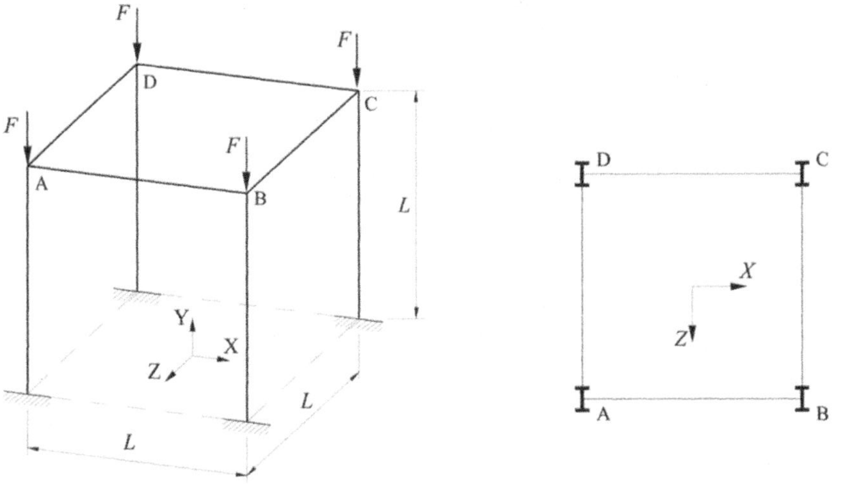

Figure 3. Portal space frame.

Obtained results for the critical buckling load normalized with respect to the Euler buckling load ($\pi^2 EI_y/L^2$) are given in Table 1. The results are compared with those reported by Blandford (1994), obtained by using four beam element per frame member and a special beam-to-column element for modelling the semi-rigid connections behaviour.

Table 1. Critical buckling load ratios

Case	Elem./memb.	EIGEN V.2	Blandford
	1	0.937	
A	2	0.933	0.927
	4	0.927	
	1	0.862	
B	2	0.854	0.854
	4	0.854	
	1	0.664	
C	2	0.659	0.659
	4	0.659	

6 Conclusion

The beam element capable of modelling thin-walled frames with flexible connections and various warping deformation conditions was developed by transforming stiffness matrices of the conventional thin-walled beam element. For this purpose, a special transformation matrix was introduced into the numerical procedure. The numerical simulation of frame instability behaviour was approached in the eigenvalue manner by the computer program, originally developed by the authors. The topic of our future research activities is to extend the proposed numerical algorithm to non-linear stability problems for which the entire connection characteristic and elastic-plastic material behaviour should be taken into account.

Acknowledgement

The research presented in this paper was made possible by the financial support of the Ministry of Science, Education and Sports of the Republic of Croatia, under the project No. 069-0691736-1731.

References

Blandford, G. E. (1994) Stability analysis of flexibly connected thin-walled space frames, *Computers & Structures* **53** 839-847.

Chen, W. F. & Atsuta, T. (1977) *Theory of Beam-Columns*, Vol. 2, New York: McGraw-Hill.

Chen, W. F. & Lui, E. M. (2000) *Practical Analysis for Semi-Rigid Frame Design*, Singapore: World Scientific Publishing Company.

Farkas, J. & Jármai, K. (2003) *Economic Design of Metal Structures*, Rotterdam: Millpress.

Gambhir, M. L. (2004) *Stability Analysis and Design of Structures*, Berlin: Springer-Verlag.

Kim, M. Y., Chang, S. P. & Kim S.B. (1996) Spatial stability analysis of thin walled space frames, *International Journal of Numerical Methods in Engineering* **39** 499-525.

López, A., Puente, I. & Serna, M. A. (2007) Direct evaluation of the buckling loads of semi-rigidly jointed single-layer latticed domes under symmetric loading, *Engineering Structures* **29** 101-109.

Tan, S. H. (2001) Channel frames with semi-rigid joints, *Computers & Structures* **79** 715-725.

Turkalj, G., Brnic, J. & Prpic-Orsic, J. (2003) Large rotation analysis of elastic thin-walled beam-type structures using ESA approach, *Computers & Structures* **81** 1851-1864.

Turkalj, G. & Brnić, J. (2004) Nonlinear stability analysis of thin-walled frames using UL-ESA formulation, *International Journal of Structural Stability and Dynamics* **4** 45-67.

Urbonas, K. & Daniūnas, A. (2006) Behaviour of semi-rigid steel beam-to-beam joints under bending and axial forces, *Journal of Constructional Steel Research*, **62** 1244-1249.

Yang, Y. B. & McGuire, W. (1984) A procedure for analyzing space frames with partial warping restraint, *International Journal of Numerical Methods in Engineering* **20** 1377-1398.

Section 8

Residual stresses and distortion

8.1 Repair by Heating/pressing for Locally Buckled Steel Members

Mikihito Hirohata and You-Chul Kim
Joining and Welding Research Institute, Osaka University,Osaka, Japan, e-mail: kimyc@jwri.osaka-u.ac.jp

Abstract

In order to investigate applicability and usability of correction by heating/pressing for steel members damaged by compressive loads, a series of experiments was carried out for cruciform and box columns. Steel members locally buckled by compressive loads could be corrected by heating/pressing without cracks in the welds. It was found that the maximum compressive load of the repaired columns was the same as that of the virgin ones in spite of the residual imperfection due to incomplete correction.

Keywords: *repair, correction by heating/pressing, buckling, ultimate strength, residual imperfection*

1 Introduction

When large structures, like bridges, are damaged by fire, traffic accident or earthquake, they should be quickly repaired to ensure the traffic of emergency and transportation of aid goods. Sometimes, local buckling damages of steel members, can be rapidly repaired on site with correction by heating/pressing (Editorial Committee for the Report on the Hanshin-Awaji Earthquake Disaster 1999). The repair of damaged steel members, correction by heating/pressing is one of the useful and important procedures, which are advantageous regarding the cost and construction time because it can be performed on site and it is no need of new members for repair. The effect of correction by heating/pressing on mechanical behaviour of steel members is still not elucidated clearly. It is necessary to confirm safety and reliability of members corrected by heating/pressing.

A series of experiments is carried out to investigate the effect of correction by heating/pressing on mechanical behaviour of steel members. Compressive experiments of virgin specimens are conducted and local buckling is generated. Buckling deformation is corrected by heating/pressing below A_1 transformation temperature. After that, compressive experiments are carried out again. By comparing the results of each compressive experiment, applicability and usability of correction by heating/pressing for steel members damaged by compressive loads is investigated.

2 Compressive Experiment in Virgin Situation

2.1 *Test Specimens*

Specimens are a cruciform column and a box column. Figure 1 shows the sizes of the specimens. The material is SM490Y. The thickness is 9mm in the case of the cruciform column and that is 6 mm in the case of the box column. Thick steel plates (thickness is 12mm) are welded at the top and bottom of the columns so that compressive loads act on the columns uniformly.

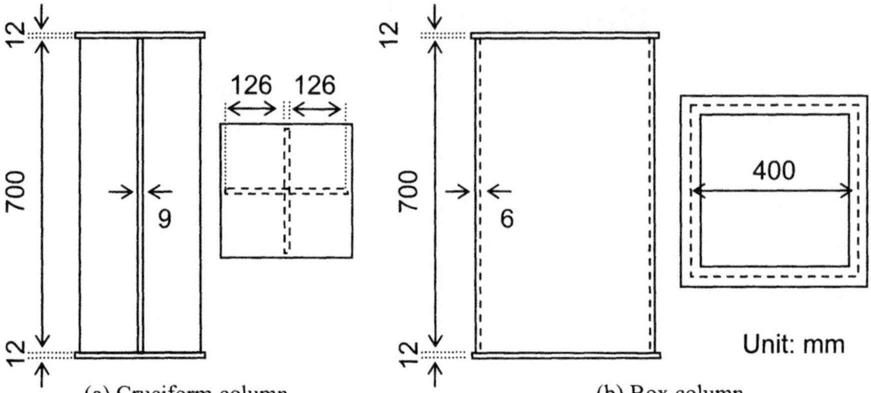

(a) Cruciform column. (b) Box column.

Figure 1. Test specimen.

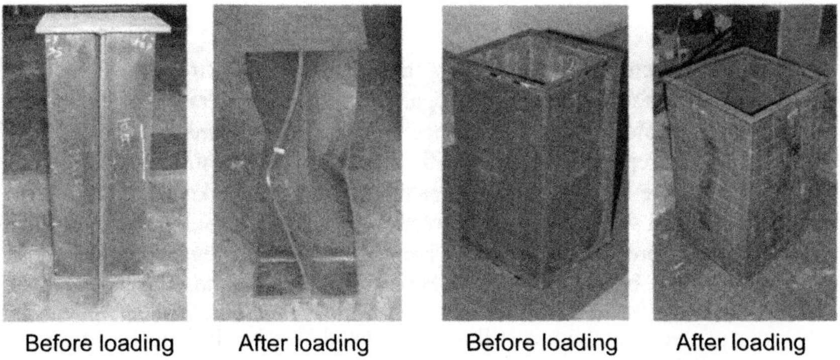

(a) Cruciform column. (b) Box column.

Figure 2. Buckling mode in the virgin situation.

Initial deflections in the panels are below 1.5mm toward the out-of-plane direction in each panel.

2.2 *Compressive Experiment*

Compressive experiments are carried out for the specimens.

Monotonic compressive loads are gradually increased and out-of-plane deformation in the panel is increased around the ultimate situation.

Figure 2 shows the buckling mode under compressive loads. In the case of the cruciform column, large out-of-plane deformation appears at the center in each projection panel. In the case of the box column, buckling mode is symmetric with respect to the center in the column's axial direction and the shape is one cycle of a sine curve.

No crack is observed in the welds after the experiment.

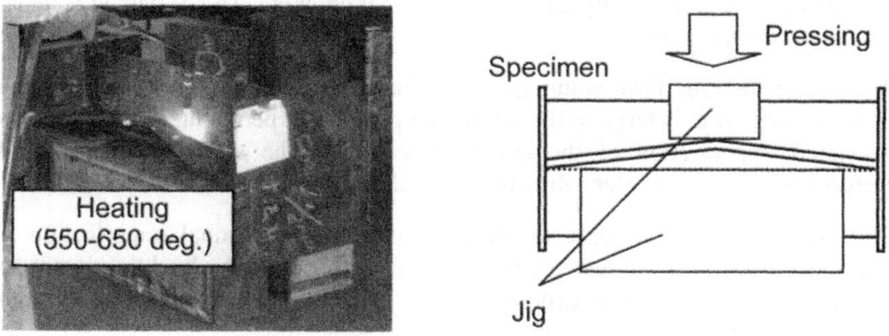

(a) Cruciform column.

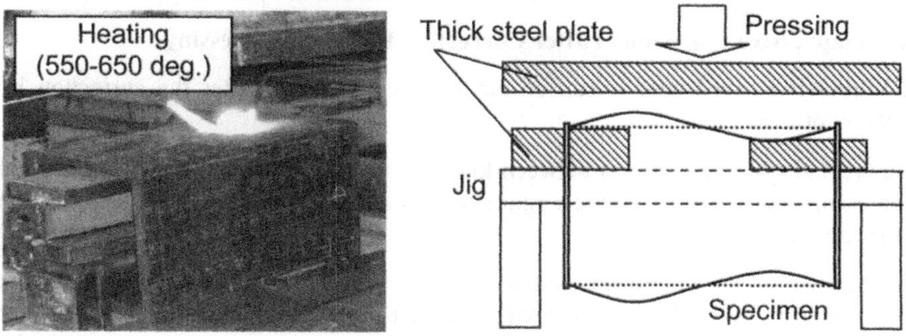

(b) Box column.

Figure 3. Correction by heating/pressing.

3 Correction by Heating/Pressing

Local buckling parts of the specimens are corrected by heating/pressing.

Figure 3 shows the view of the correction by heating/pressing. Local buckling parts are heated by using a gas burner. Heating temperature is 550-650 degrees centigrade, which should be kept below A_1 transformation temperature (about 700 degrees centigrade) in order not to change the mechanical properties of the materials, Japan Road Association (2002). The buckling parts are corrected by using a press machine without dismantling the specimen, which is supposed an actual work on site.

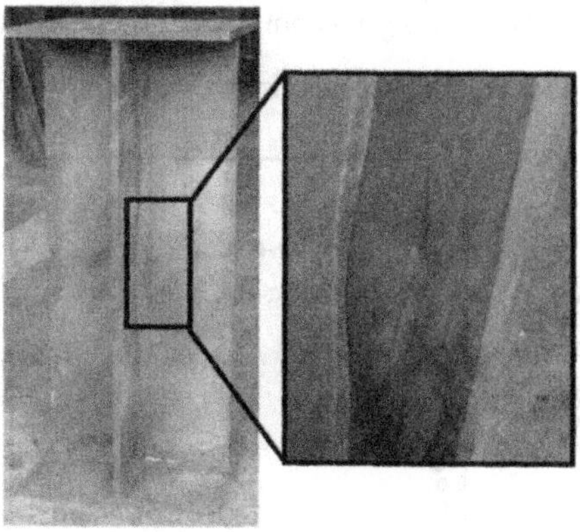

Figure 4. Residual imperfection (Cruciform column).

It is impossible that the buckling parts are corrected completely because there is a possibility of occurrence of cracks in the welds by excessive correction. Therefore, some imperfection is inevitably left so as to prevent cracking. This is named residual imperfection in this study.

In the case of the cruciform column, the residual imperfection is large because the buckling deformation is large in the virgin compressive experiment. Figure 4 shows the residual imperfection of the cruciform column. The value of the residual imperfection in the out-of-plane direction is 10-20mm.

On the other hand, in the case of the box column, the residual imperfection is considerably controlled because the out-of-plane deformation by buckling is not so large in the virgin compressive experiment. The value of the residual imperfection in the out-of-plane direction is 2-3mm.

In any case, no crack appears in the welds with correction by heating/pressing.

4 Compressive Experiment after Correction by Heating/Pressing

Compressive experiment is carried out for the specimens after the correction by heating/pressing.

Figure 5 shows the relations between load and vertical displacement. Open symbol represents the behaviour in the virgin situation and solid symbol does that after the correction by heating/pressing.

In the case of the cruciform column after the correction by heating/pressing, the stiffness in the axial direction is lower than that in the virgin situation. On the other hand, in the case of the box column after the correction by heating/pressing, the stiffness in the axial direction is almost the same as that in the virgin situation.

Both the cruciform and box columns after the correction by heating/pressing, ultimate strength is not lower compared with that in the virgin situation.

No crack is observed in the welds during the experiment.

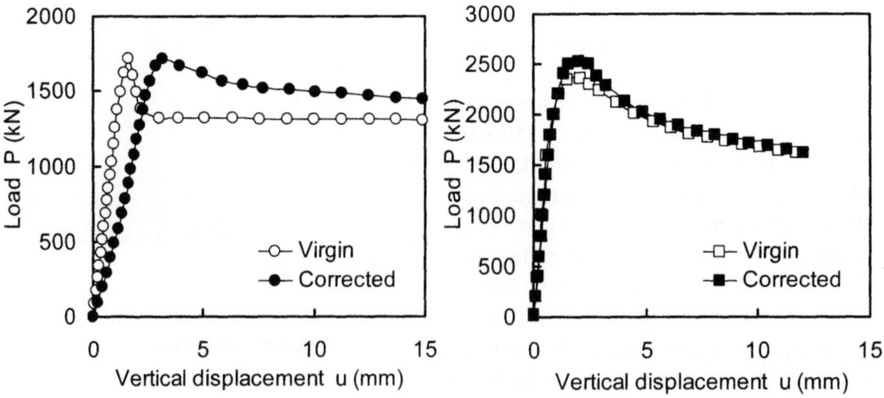

(a) Cruciform column. (b) Box column.

Figure 5. Relation between load and vertical displacement.

Figure 6 shows the buckling mode of the specimens after the correction by heating/pressing. The buckling mode is different from that in the virgin situation.

In the case of the virgin cruciform column, the out-of-plane deformation is largest at the center of the panel. After the correction by heating/pressing, the out-of-plane deformation is largest not at the center but at the upper side of the panel (refer to Figure 2(a) and Figure 6 (a)).

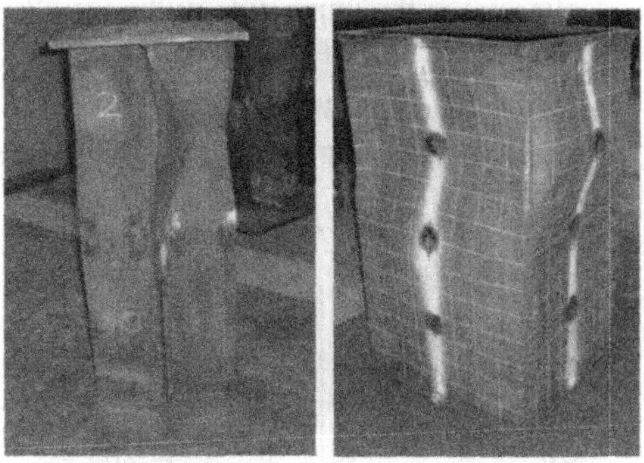

(a) Cluciform column. (b) Box column.

Figure 6. Buckling mode (after the correction).

In the case of the virgin box column, buckling mode is symmetric with respect to the center in the column's axial direction and one cycle of a sine curve. After the correction by heating/pressing, it is not symmetric. The out-of-plane deformation at one peak point is large (30mm), that at another peak point is small (10mm) (refer to Figure 2(b) and Figure 6 (b)).

5 Considerations

5.1 *Effect of correction by heating/pressing on mechanical behaviour of steel members*

In the compressive experiment for the specimens after the correction by heating/pressing, ultimate strength is not lower than that in the virgin situation and no crack occurs during the experiment. This result indicates that mechanical properties of the materials are not deteriorated with correction by heating/pressing, Japan Welding Society (2003).

In the case of the cruciform column after the correction by heating/pressing, which has the large residual imperfection, the stiffness in the axial direction is lower than that in the virgin situation. On the other hand, in the case of the box column after the correction by heating/pressing, whose residual imperfection is small, the stiffness in the axial direction is almost the same as that in the virgin situation. Naturally, the magnitude of the residual imperfection largely affects the stiffness of the steel members under compressive loads.

5.2 *Buckling Mode of Steel Members Corrected by Heating/Pressing*

In the case of the specimens after the correction by heating/pressing, buckling mode under compressive loads is considerably changed comparing with that in the virgin situation. Here, the reason is considered by taking the cruciform column as an example.

The cruciform column after the correction by heating/pressing has large residual imperfection at the centre of the panel. Therefore, the out-of-plane deformation should be largest at that region under compressive loads. It is actually largest not at the center but at the upper side of the panel (refer to Figure 6 (a)). There might be a factor making the panel hard to deform.

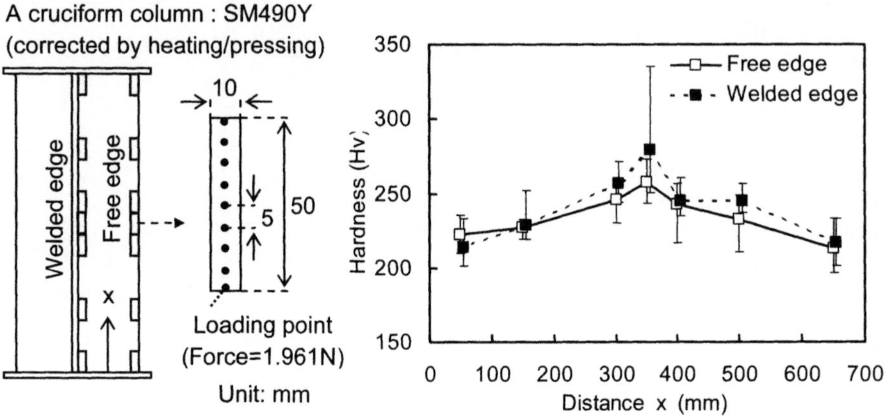

(a) Measuring points of hardness. (b) Distribution of hardness.

Figure 7. Result of Vickers hardness test.

As the factor, increase of yield stress due to work hardening, is noted. That is to say, by buckling in the virgin compressive experiment and its correction by heating/pressing, large plastic deformation is generated at the center of the panel. Therefore, yield stress at that region after the correction by heating/pressing might become higher by work hardening compared with that at another region.

In order to investigate the degree of work hardening, Vickers hardness test is carried out for the panel of the cruciform column, which is buckled by compressive loads and out-of-plane deformation is corrected by heating/pressing. Figure 7 shows the results of Vickers hardness test. It is confirmed that hardness at the center of the panel is larger than that at other region. From this result, it is probable that yield stress becomes higher compared with that at another region as well as hardness, Nakazawa (1987).

In the case of the members corrected by heating/pressing, local buckling under compressive loads does not necessarily appear at the same part as the virgin situation, where the residual imperfection exists, due to increase of yield stress by work hardening.

In any case, increase of yield stress does not decrease ultimate strength of the member. Therefore, it is unrelated with soundness of steel members corrected by heating/pressing.

6 Conclusions

In order to elucidate the effect of correction by heating/pressing on mechanical behaviour of steel members, a series of experiment was carried out.

Obtained main results are as follows:

First of all, compressive experiments for the virgin cruciform and box columns were carried out.

(1) In the case of the cruciform column, local buckling occurred at the center in each projection panel. In the case of the box column, local buckling mode as one cycle of a sine curve occurred. No crack was observed in the welds of the specimens after the experiment.
(2) Local buckling deformation of the cruciform and box columns by compressive loads could be corrected by heating below A_1 transformation temperature and pressing. Although residual imperfection was left according to the degree of buckling deformation, no crack occurred in the welds by the correction.

Compressive experiments for the specimens after the correction were carried out again.

(3) In the case that the residual imperfection was much larger than the initial deflection, the stiffness of the specimen after the correction by heating/pressing was lower than that in the virgin situation. In the case that the residual imperfection was almost the same as the initial deflection, the stiffness of the specimen after the correction by heating/pressing was kept the same as that in the virgin situation.
(4) Ultimate strength of the specimens after the correction by heating/pressing was not lower than that in the virgin situation and no crack occurred during the experiment. This result indicated that mechanical characteristics of the materials have not been influenced by correction with heating below A_1 transformation temperature and pressing.
(5) Buckling mode of the specimens after the correction by heating/pressing was considerably changed compared with that in the virgin situation. The reason might be the increase of yield stress due to plastic deformation by buckling and its correction. It was unrelated with ultimate strength of the specimens after the correction by heating/pressing.

References

Editorial Committee for the Report on the Hanshin-Awaji Earthquake Disaster (1999) *Report on the Hanshin-Awaji Earthquake Disaster, Emergency Repair and Seismic Retrofit*, MARUZEN.

Japan Road Association (2002) *Specifications for Highway Bridge Part II: Steel Bridges*, MARUZEN.

Japan Welding Society (2003) *Welding and Joining Manual*, MARUZEN.

Nakazawa, H. (1987) *Manual for Metal Material Test*, Japanese Standards Association.

8.2 Estimation of Welding Distortions and Straightening Workload Through a Data Mining Analysis*

Nicolas Losseau[1,3,a], Jean David Caprace[1,4,b], Dominique Archambeau [2,c], Amirouche Amrane [1,d], Philippe Rigo [1,4,e]

[1] *University of Liège, Chemin des chevreuils B52, 4000 Liege – Belgium*
[2] *PEPITe ; Rue des Chasseurs Ardennais, 4031 Liège, Belgium*
[3] *Fund for Training in Research in Industry and Agriculture of Belgium (F.R.I.A.)*
[4] *National Fund of Scientific Research of Belgium (F.N.R.S.)*
[a] *n.losseau@ulg.ac.be,* [b] *jd.caprace@ulg.ac.be,* [c] *d.archambeau@pepite.be,*
[d] *a.amrane@ulg.ac.be,*[e] *ph.rigo@ulg.ac.be*

Abstract

This paper will present a way to minimize cost in the shipbuilding industry by using the results of a data mining analysis aiming to improve the cost knowledge of the straightening process. This statistical analysis was based on production data from a shipyard and has the scope to establish an assessment formula of the straightening workload. An intermediate step of the analysis was to estimate the welding distortions appearing in stiffened panels. Those generated formulas are useful to improve the research in the following domains: production simulation, cost assessment of ship hull, structure optimization, design for production, etc.

Keywords: *welding distortions, straightening, shipbuilding, cost assessment, data mining*

1 Introduction

For several years, the big shipyards have need more and more plates of small thickness to build up the stiffened panels in order to decrease the structural weight of ships. The major problem relating to the utilization of thin plates is the appearance of welding distortions that have to be eliminated for esthetical and service reasons. This straightening operation involves non negligible costs and it seems thus important to characterize its economical impact on the global hull fabrication.

The idea to establish the straightening workload formulation was to lead a statistical analysis basing on the production data from a shipyard. This paper gathers the results of two complementary analyses. The first analysis has exploited workload data of 13 passengers' ships and has established a relationship linking the scantling (geometrical characteristics of stiffened panels) to the straightening cost [hour/m²]. The second analysis was led to a measure campaign gathering the welding distortions of one cruise ship; this data mining study exploited those previous data in order to estimate the distortions and to improve the straightening assessment formula.

* A part of this paper results from part of the work performed in sub-project II.1 of InterSHIP, a European R&D project funded under the European Commission's Sixth Framework Programme for Research and Technological Development. (Project n° TIP3-CT-2004-506127)

Figure 1. (1) A distorted panel.
Straightening techniques: (2) by induction, (3) with a 5 nozzles blowtorch.

This paper summarises each stage of the data mining methodology: data feeding, data quality analysis, data exploration, choice of discriminatory attributes and finally generation of the formulae. Some interesting techniques coming from Pepito© software were used in order to perform this analysis like: linear correlations analysis by dendrogram tools, conditioned histograms, conditioned dots clouds, decision tree tools (based on minimisation entropy) and a Neuronal Network Analysis.

Figure 2. (1) panel, (2) section, (3) block, (4) pre-mounting zone and dry dock

2 Fabrication methodology in shipbuilding

The shipbuilding industry utilizes the concept of modular fabrication; the ship is divided in assemblies and sub-assemblies and those elements are fabricated in specialized workshops in order to reach a high productivity.

The basic elements called panels (~15m*15m) are constituted by side plates and longitudinal stiffeners. Then transversal girders and stiffened partitions are assembled on panels to realize sections (~15m*30m*3m); the simple geometric sections are fabricated in production lines while complex ones are built in hall assemblies. The next stage (pre-mounting work) is the realization of blocks (~30m*30m*12m) by assembling several sections. Finally, the blocks are transported in the dry dock and connected together to constitute the ship. The straightening operations are done in the dry dock by experimented workers.

3 Straightening costs assessment: first analysis

The analysis of straightening assessment has followed the Data Mining methodology that is, by definition, the non-trivial process of extracting valid, previously unknown, comprehensible, and useful information from large databases. The successive steps of the analysis were the following ones:

3.1 *Database creation*

A data base was first of all constituted; it gathered, for 13 passenger ships, the characteristics of each section (global geometry, deck thickness, dimensions and inter-distance of stiffeners, section family, deck number, steel grade, section weight, etc.) and the associated straightening workload [hour].

3.2 *Data description stage*

This step consisted in a presentation of the attributes (fields of the data base), with their distribution and other statistical parameters (minimum, maximum, mean and variance).

One of difficulties which arose during the database analysis is that the most structural attributes show a discrete distribution with one or few dominant modes (see Figure 3 (3) and (4)); for instance, the distance between stiffeners has very often the same value. Those attributes are almost "constant" parameters and thus do not constitute a conclusive information source. In order to minimize this effect, we have replaced some attributes. In this scope, we divided the plate weight by the section surface to obtain information similar to the thickness but having the advantage to present a distribution much less discrete.

3.3 *Data quality stage*

This step listed the problematic recordings (strange distribution, missed values, data in conflict with their physical meaning) in order to take care of them in the next stages.

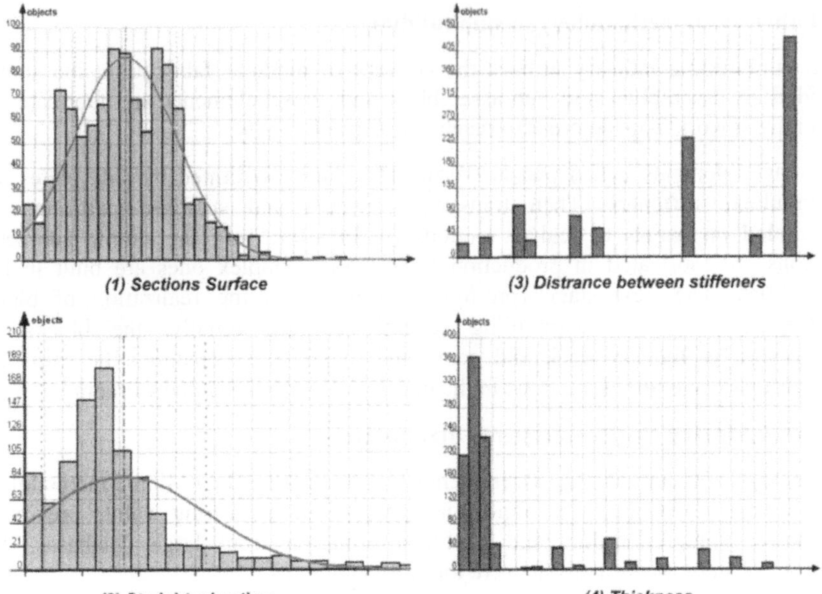

Figure 3. Distribution histograms of attributes

A particular point has been noticed at this stage; the values related to the straightening work realised by sub-contractors are not reliable since this workload corresponds to an estimated time and not a time of strictly effectuated work. Unfortunately, this case concerns more than two third of the records and decreases thus, the quantity of exploitable data.

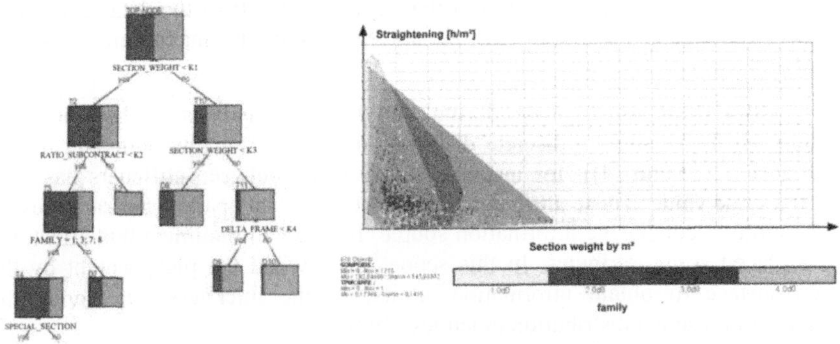

Figure 4. Statistical tools: (1) decision tree, (2) conditioned dots clouds

3.4 *Data exploration stage*

This work stage consisted of using different approaches to visualize the correlations existing between the attributes and the straightening workload in order to finally select the parameters having the most relevant influence on the straightening assessment. In order to fulfil this stage, four different approaches were used: a linear correlation analysis trough dendrograms elaboration, conditioned histograms, conventional dots clouds diagrams and decision trees analyses.

3.5 *Elaboration of the formula*

This stage consisted in building the relation between the straightening cost [hour/m²] and the sections characteristics. The technique selected was the Artificial Neural Networks (ANN) method, which is a powerful technique permitting to elaborate non-linear relations (i.e. hyperbolic tangents) linking several inputs to a unique output.

The input attributes to generate the formula were the following ones: thickness, longitudinal stiffeners spacing, transversal girders spacing, ratio stiffeners spacing/girders spacing, section family, section weight/m², section weight/section length. After having chosen the input parameters, it was necessary to restrict the number of records in order to ignore the sections carrying perturbing information. In this optic, we have ignored the sections where the straightening work was carried out by sub-contractors because time measurements of straightening were less reliable.

The formula was elaborated exploiting 273 records and the correlation between the real value of straightening cost and the value estimated by the formula was 0,838.

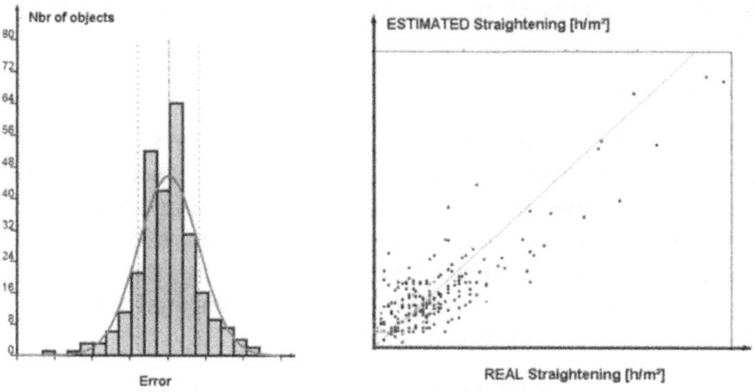

Figure 5. Error diagrams relating to straightening work estimation (from PEPITo ®)

3.6 *Limitation of the formula*

It is necessary to notice that the generated formula has a limit. Firstly, since the recordings were restricted to the works realized by the shipyard workers, the quantity of data exploited was small and thus the robustness of the formula was not excellent. Moreover, when we have constructed the error diagrams, we have voluntarily tested the equation on the same data set than the one used to establish the relation. A consequence of this choice is that the precision given is not representative. Those precisions are optimist compared to the precision obtained if the test set was different than the learning set.

4 Distortions assessment

4.1 *Data description*

Further to a measure campaign gathering the deck distortions of one passenger ship, the data mining analysis relating to welding distortions and straightening has been revived. The deformations were measured after the blocks assembly stage and before the straightening operations. Each recorded value was related to the distortion occurring at the middle of rectangles (about 0,7m*2,5m) delimited by consecutive longitudinal stiffeners and consecutive transversal girders. The campaign covered about 50% of the total deck surface and was associated only to small thicknesses from 5 to 8 mm.

Those distortions correspond thus to the accumulation of welding deformations occurring at each fabrication stage. During the sections fabrication, the welding of plates and longitudinal stiffeners generate firstly a system of bulges that is then modified and sometimes accentuated during the fixing of the transversal girders and bulkheads. The fabrication stage realizing the connection of sections to constitute blocks involves important traction efforts on the plate extremities and modifies again the distortions system. Finally, the resulting decks deformations vary in configuration and in direction; there can be one or several bulges between two consecutive girders and the value of distortions can be positive or negative. Nevertheless, in 70% of the cases, the plates are distorted in the side of the stiffeners and only one bulge is present between successive girders. An example of deformations field is presented in the Figure 6; colours have been applied in order to try to visualise characteristics of the distortions system and eventual correlations between data.

4.2 *Data mining stages*

The database gathers for each rectangle delimited by consecutive stiffeners and girders: plate thickness, stiffeners spacing, girders spacing, steel grade, presence of a plate junction, coordinates of the rectangle inside the ship, deck number, etc. A particular attribute called "additional process" describes the utilisation of a distortion reduction technique (such as application of weld seams onto the plate between stiffeners). Indeed the initial scope of the measure campaign was to evaluate the impact of such distortions reduction techniques.

We added other attributes about the section: section weight by m^2, section workshop (0 for assembly halls; 1 or 2 respectively for the production line of the partially automated workshop). We generate also combined parameters: distortion/thickness in order to have attributes without units.

The successive data mining stages were then applied to: investigate the correlation between attributes, extract meaningful parameters, fix exploitable data sub-sets and finally elaborate a formula trough an ANN analysis.

4.3 *Establishment of the formula*

Numerous attempts have been made to generate an effective estimation formula. It has been revealed that the exploitation of each distortion measure taken as unique record generates too much noise and involves thus a poor quality estimation (correlation around 0,5 for 23 000 elements). By gathering the data by plates (1400 elements) the correlation increased to 0,7 and by gathering the data by sections (108 elements) the correlation reached 0,9. This last solution was thus utilised to estimate the welding distortion in absolute value.

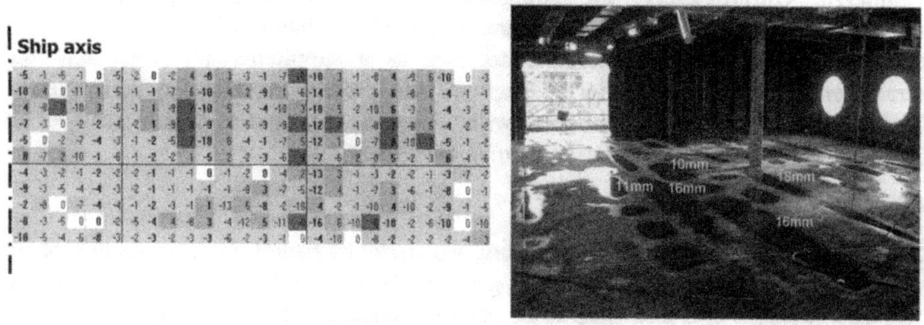

Figure 6. (1) visualisation of a distortions field, (2) photo of a distorted deck

To generate the formula, the 108 sections concerned by the measured campaign were utilised and the following attributes were introduced as inputs: plate thickness, longitudinal stiffeners spacing, transversal girders spacing, ratio stiffeners spacing/girders spacing, section weight by m², workshop of the section, additional process.

The correlation between the real distortions and the values estimated by the formula was 0,947 (see Figure 7). This value is optimist because the data set to test the formula was the same than the set to generate the formula.

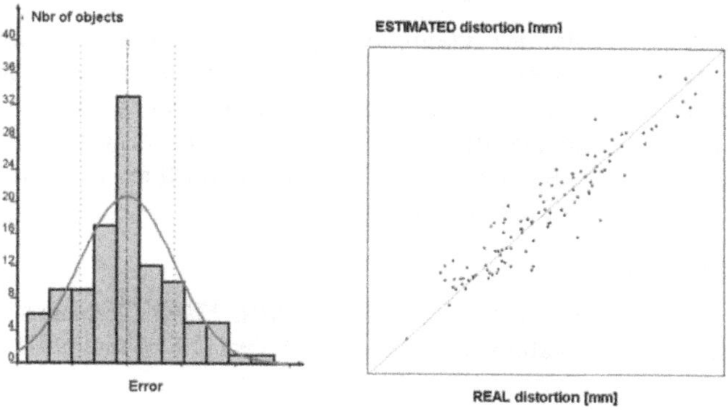

Figure 7. Error diagrams relating to welding distortions estimation (from PEPITo ®)

Here again, the formula has a limitation. The estimation is quite good inside the range values encountered by the attributes in the database but is not excellent outside. For instance, an estimation relating to a thickness lower than 5 mm or higher 8 mm becomes rapidly approximate.

5 Straightening assessment: second analysis

The knowledge of the distortion values seemed interesting to exploit in the straightening assessment. In effect, intuitively, higher are the distortions, longer will be the straightening operations.

The first idea was thus to generate a straightening estimation formula using the real distortions and the real straightening workload of the ship concerned by the measure campaign. Unfortunately, the straightening works of this ship were totally effectuated by sub-contractors, thus, the data was unreliable. The second idea was then to introduce the estimated distortions as a supplementary input into the straightening formula in order to improve the prediction. The Artificial Neural Networks technique was launched with this new parameter; this permitted an increase of the formula correlation from 0,838 to 0,862.

6 Conclusions

The basic structures used in shipbuilding are stiffened panels that are assembled by welded joints. The welding operations generate distortions that accumulate along the fabrication stages and that have to be eliminated for esthetical and service reasons. The common utilization of thin plates increases the straightening operations and it seems thus important to characterize those supplementary works that involve over costs and delivery delays.

This paper summarizes two data mining analyses that exploited production data from the Saint Nazaire shipyard. The first study permitted, trough a statistical analysis of 13 ships data, to establish a relation linking the straightening workload to the sections characteristics. The second one aimed to estimate the welding distortions from the scantling data. A step complementary to the first analysis was then realized; the distortions estimated by the formula were introduced as a supplementary input into the straightening equation.

Those generated formulas are useful to improve the research in the following domains: production simulation, cost assessment of ship hull, structure optimization, design for production, etc. Moreover, the advantages of the distortion formula are the rapidity and the simplicity that are very important points in industry. Indeed, it permits to evaluate rapidly the value of welding distortions without using more complex methods such as Finite Elements.

Acknowledgements

The authors thank University of Liege, AKERYARD FRANCE (Saint-Nazaire) and Pepite (Liège), for the collaboration within sub-project II.1 of InterSHIP.

References

Bruce, G. & Morgan, G. (2006) Artificial Neuronal Networks – Application to freight rates. In: *International Conference COMPIT 2006*, pp. 146-154

Caprace, J.D., Losseau, N. & Archambeau, D. (2007) A Data Mining Analysis Applied to a Straightening Process Database. In: *International Conference COMPIT 2007 pp. 186-197*

Chau, T.T., Jancart, F. & Bechepay, G. (2001) About the welding effects on thin stiffened panel assemblies in shipbuilding. In: *International Conference on Marine Technology Odra 2001*, Szczecin, Poland.

8.3 Using of Local-global Welding Virtual Numerical Simulation as the Technical Support for Industrial Applications

Marek Slováček[1], Jiří Kovarik[2], Josef Tejc[3], Vladimir Diviš[4]
[1]MECAS ESI L.t.d., Technicka 2, 616 69 Brno, Czech Republic, marek.slovacek@mecasesi.cz, [2]MECAS ESI L.t.d., Brojova 2113/16, 326 00 Plzeň, Czech Republic, jiri.kovarik@mecasesi.cz, [3] MECAS ESI L.t.d., Brojova 2113/16, 326 00 Plzeň, Czech Republic, josef.tejc@mecasesi.cz,[4]MECAS ESI L.t.d., Technicka 2, 616 69 Brno, Czech Republic, vladimir.divis@mecasesi.cz

Abstract

Welding as a modern, highly efficient production technology found its position in almost all industries. At the same time the demands on the quality of the welded joints have been constantly growing in all production areas and cause more experimental testing – so called – validation joints. Naturally, these experiments make production more expensive. Numerical simulations supported by experimental measurements can simulate the actual welding process very close to reality. Numerical simulations are very useful tools during production preparation and technology optimization. They are flexible as far as the technology process changes are concerned. Moreover, they reduce the quantity of experimental examinations and consequently make the production less expensive and of a higher quality. The virtual numerical simulations have been done by using SYSWELD CODE, which is being developed by ESI GROUP Company. The SYSWELD CODE work is based on the finite elements method. The new local global solution method is introduced.

Keywords: *Numerical simulation of welding process, SYSWELD, local-global approach, material structure, stress, distortion*

1 Introduction

The welding is the most prevalent metal joining process. The used heat input during welding process creates residual stresses as well as distortion in the steel components. The type and extent of distortion and residual stresses are influenced by a number of different factors as the clamping condition, mechanical and thermal property, type and size of heat source used, welding parameters, preheating temperature, weld joints design, temperature of surroundings etc. Residual tensile stresses have also got a negative influence on the construction lifetime and its brittle fracture resistance. Residual stresses create a balanced system of inner forces, which exists even under no external loading. The numerical simulation of the welding process is one way to determine the level of the residual stresses and distortion.

The calculated residual stresses are used for the prediction of the lifetime including influence of the welding process or finding possibilities of the defect initiation and defect growth under service conditions. The residual stresses are originated mainly in the cases when the steel construction is very stiff. On the other hand, when the construction is free or the movement of some components of the steel construction is allowed, the residual stresses are lower than in fixed cases but the distortions are created. The control of distortion is very important due to manufacturing tolerances for following-up machining and assembly process. It is clear that the compromise between fully fixed and absolutely free construction needs to be found.

The numerical simulation of the welding process can be used during the preparation of welding technology of the new products or during the innovation of the current products and for the assessment of the lifetime of components considering the level of the residual stresses after welding process or after repairs, which have been done with welding technology. Using the welding numerical simulation can decrease number of actual experiments during technology preparation and can also decrease the time needed for the preparation and decrease the cost, as well. The various proposals can be numerically simulated and there, the first idea of the behaviour of the construction during and after welding process arises. Based on the calculated results the most suitable version is chosen. The experiment is done only with the final version of the proposed technology and the calculated and measured results are compared. The optimization of the input welding parameters, welding sequences, clamping condition, support stiffness process is applied to obtain output parameters required.

Based on our experiences the welding numerical simulation can be divided into four following areas:

1. Comparison of several different technologies. Every technological alternative is numerically simulated and the output parameters (quality of the metallurgical structure, residual stresses, distortion etc.) are compared. The aim is to find the best alternative, which corresponds with the customer's requirements.
2. Lifetime assessment. The aim of the analysis is to predict the lifetime of the whole construction or of some components including the consideration of the influence of the welding process. The residual stresses are numerically simulated and considered in the lifetime assessment.
3. Brittle fracture prediction. The calculated level of the residual stresses can be used during the assessment of the brittle construction resistance.
4. Prediction of the distortion during welding assembly process. This type of the prediction is done with the help of the new local global approach.

The verification projects including measurement of the material input data and welding experiments have been prepared and carried out. The numerical simulations of welding experiments have been done and the calculated and measured parameters have been compared. The heat source models for various welding technologies have been found and validated. The main aim of these verification projects is to find the appropriate input parameters and solution methods that the numerical simulations be corresponded as much as possible with the reality. The found parameters and experience have been used in real industrial manufacturing analysis.

The commercial code of SYSWELD has been used. This code is worked-out on the basis of the finite elements method and is developed by ESI Group Company.

2 Numerical simulation of the welding process

The numerical simulations of the welding process are performed with using commercial SYSWELD software (ESI Group Company 2006).

Based on the solution method the numerical simulation of the welding process can be divided into three following groups:

1. Transient welding or step by step method (TW): The heat source model moves according to weld trajectory. The numerical calculation is done at each time increment during welding process. The real welding parameters (current, voltage, velocity and efficiency) are included as parameters of the heat source model. Based on experiences the maximum possible time step is the half length of the molten zone. It comes to that, the transient method is very slow and that it is possible to numerically calculate only small parts of the construction. On the other hand, the results are very complete (temperature fields, hardness, metallurgical structure, residual stresses, distortions). Transient method is suitable for determination of the local effect of the welding joints and optimization of the welding technology (parameters) regarding mainly the metallurgical structure, hardness and residual stresses at weld joint and close surroundings

2. Macro bead deposit method (MBD): The heat source is applied instantaneously in one or several macro areas (elements) in the same time. The real weld trajectory is divided into several macro sections. The energy per unit length of the weld joint is transferred into the structure the same as real welding process. The number of macro sections, time steps are defined based on the welding technology parameters and experience with this method. The MBD methodology is an extent of the TW method. The MBD method decreases calculation time and increases possibility calculation of the big construction with keeping good level of results quality.

3. Local-global method (LG): For very large structure, such as maritime, automotive or heavy industry structures. Standard TW or in some cases MBD methodologies are not feasible because these methods require significant computation time and computer memory size. The idea behind the LG methodology is that welding process is a local modification of stress and strain, the total effect leads to a global state of distortion. The local welding effect is found on the refined local models calculation by using TW or MBD. The stiffness of the local models must be the same or very close to the reality. The results are transferred from local models to the global model and equilibrium linear elastic simulation is performed to ascertain the global distortion. The global model represents often whole structure. The LG methodology enables to simulate a very huge structure with very large number of welded joints. But the results from LG are only distortion, inner force and moments at constraint conditions. The levels of the residual stresses or metallurgical structure are determined by the local modelling, it means by TW or MBD method. The verification of the local-global method is done and published (Slováček et al. 2005).

The TW and MBD method can be also divided into three following stages:

1. During the first stage a complete diagram of anisothermic decomposition (CCT diagram) is entered by special pre-process module. The results of this stage are coefficients describing the kinetic of transformation process depending on cooling rate at individual areas of heat-affected zone. The coefficients depend on the temperature and on the metallurgical phase of particular material and are used as direct input to the second phase.

2. The second stage is a thermo metallurgical solution. This part needs complete thermo physical and thermo metallurgical material properties. A classical equation

of the heat conduction extended with the transformation of latent heat during change of phase and during melting of material is applied. Coupling between phase transformation and heat conductivity is used. The results of the first stage are non-stationary temperature fields, percentage distribution of individual phases, size of primary austenitic grain, hardness.
The temperature analysis is transient calculation in each time interval. The all-welding passes must be simulated.

3. The results of the second stage (mainly non stationary temperature fields) are applied as a loading condition in the third stages, the structural analysis. The complete mechanical properties are needed. The mechanical properties (thermal expansion, yield stress, hardness, Young modulus etc.) depend on the temperature and individual phases. Resulting mechanical properties in welding joint and heat-affected zone are calculated on the basis of individual material structure distribution and their mechanical properties. The results of the second stage are, total deformation (consist of elastic part, thermal part, convectional plastic part, viscoplastic part and transformation plasticity), residual stresses and distortion.

The model of the thermal source is one of the most important input parameters (Slováček et al.2005a). The computation model is loaded only with non-stationary temperature fields. Thermal load represents the thermal energy flow into the material during the welding process. The results (distortion, residual stresses) depend very much on finding the correct model of the thermal source and appropriate temperature distributions. The SYSWELD code has got a special tool „heat source fitting" which enables to find appropriate input computation parameters of heat source in order to simulate reality. The pre-defined double ellipsoidal and conical heat source models can be used by SYSWELD code. The pre-defined translation, rotation and general movement of welding heat source can be used by SYSWELD code as well.

3 Practical use of welding numerical simulation in industry

SYSWELD code has been used for the welding optimisation, for the development of new technological procedures of welding with and without post welding heat treatment, for the simulation of real weld repairs on constructions (Slováček & Junek 1999) for the determination of defect behaviour in the field of residual stresses, prediction of the distortion (Slováček et al 2005b), and comparison between different technologies (Slováček et al 2005b). The possibilities of the SYSWELD code, mainly new solution method based on local-global approach, are shown in following two cases.

3.1 *Distortion prediction of the low pressure turbine part body*

The project has been done for SKODA POWER Company (Tejc & Kovarik 2007). The main aim of the study has been determined the component distortion during and after welding process and possible make a technology modification based on the results of numerical analyses. The very progressive local-global solution method has been used due to requirement for distortion prediction of the huge welded structure. The component is shown in Fig. 1. The welding joints, which have been numerically simulated, are shown in Fig. 2.

The 12 different local models have been prepared and numerically analyzed. Each local model represents different welding technology during the welding process. The each welding step (operation) has been numerically analyzed by using global analyses and the welding distortions have been determined after each welding step. The calculated distortions are shown in Fig. 3, 4 and 5. The reference dimensions and points for measuring distortion are shown in Fig. 6. The Table 1 contains the comparison between calculated and measured distortion.

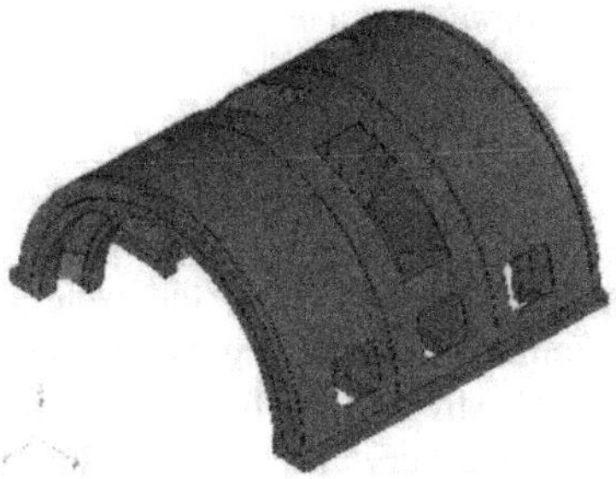

Figure 1. Low pressure part body

Figure 2. Numerically simulated welded joints

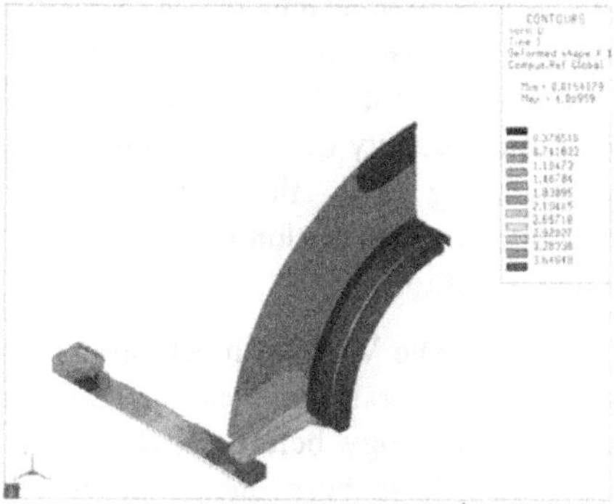

Figure 3. Distortion after 1st step

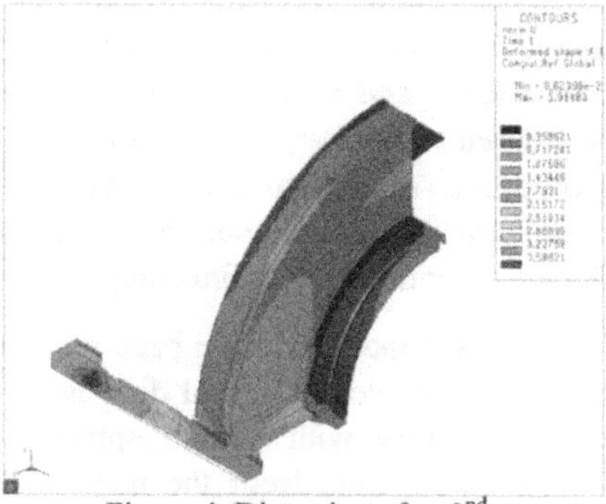

Figure 4. Distortion after 2nd step

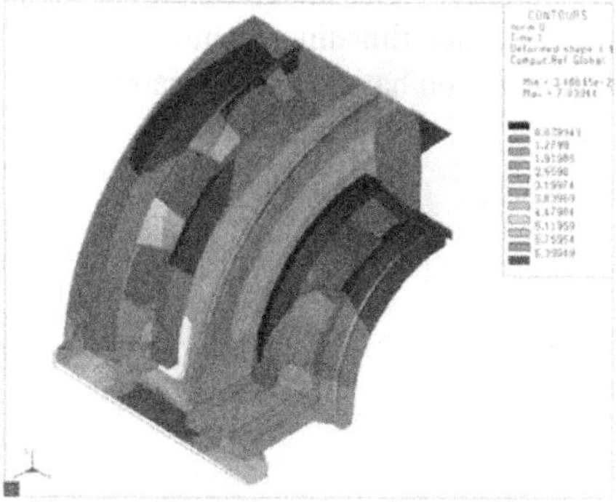

Figure 5. Final distortion after welding

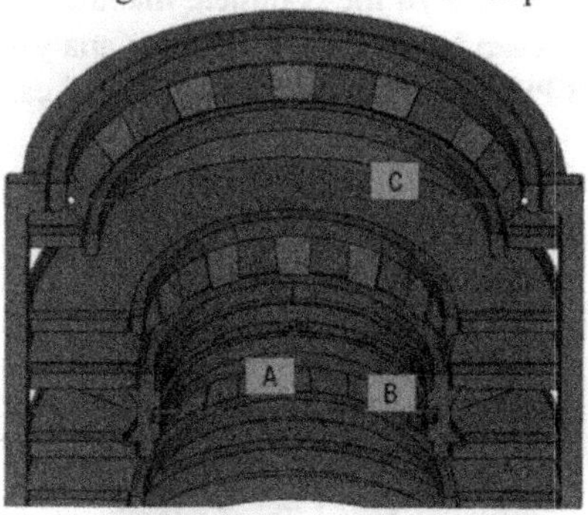

Figure 6. Reference dimensions and points

3.2 *Numerical simulation of the welding of the ITER vacuum reactor vessel*

The new fusion reactor ITER (International Thermo-nuclear Experimental Reactor), Fig. 7, will be built in Cadarache, France. The development of the vacuum vessel reactor welding technology is supported by numerical analyses, which have been done by SYSWELD Code and by using new local-global approach. The projects have been done by Institute of Applied Mechanics Brno and Technical University Ostrava (preparation of small mock-ups testing) (Slováček et al. 2006). The big experimental mock-ups have been prepared and welded by Italy Companies ANSALDO and SIMIC.

Table 1. Comparison between calculated and measured distortion

Reference	*Calculated distortion [mm]*	*Measured distortion [mm]*
A	-0,85	-1
B	-8,8	-7
C	-1,92	-2

The manufacturing of the ITER Vacuum Vessel requires a lot of the welding operations. The weld joints are very long and contain a lot of welding passes. Each welding operation (weld joint) generates residual stresses (deformation) and distortions. The final distortion has to comply with the very strict manufacturing tolerances and requirements for final size and shape of the construction parts. Due to stated facts the several experimental mock-ups have been done and also numerically simulated. The final welding technology will be proposed based on the obtained experimental and calculated results. These mock-ups are used for confirmation of the chosen technology in order to obtain requirement quality of the weld joints and appropriate distortion after welding corresponding with the manufacturing tolerances. The very important aspect for final distortion prediction is removing the clamping condition after finishing of the welding operation.

The VVPSM mock-up have been prepared and done. The VVPSM mock-up is part of the vacuum vessel in real dimension, Fig. 8. The several numerical simulations have been done with aim to optimize welding technology before mock-up real testing. The results from the numerical simulations have been one of bases for finalization of the complete manufacturing technology. Also, the complete VVPSM mock-up has been numerically analyzed as well. After finishing manufacturing of VVPSM mock-up the measured and calculated distortion have been compared.

Figure 7. ITER

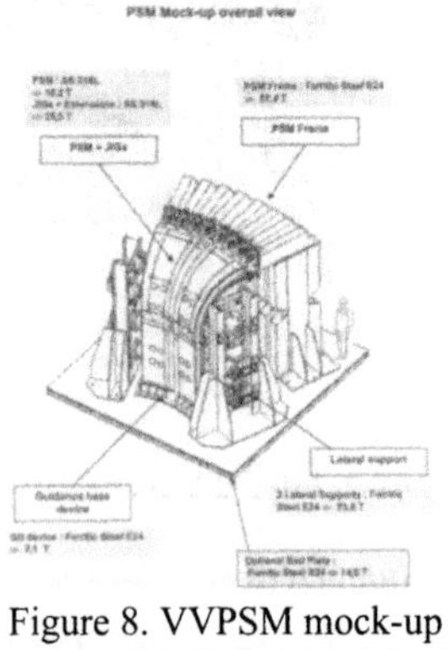

Figure 8. VVPSM mock-up

During the solution of the ITER projects, numerical simulation of the welding of the experimental mock- ups, the 9 global models, 30 local models and approximately 3000 welding beads were numerically simulated. The global and local models of VVPSM numerically simulations are shown in Fig. 9 and 10. The final distortion of some parts of VVPSM mock-ups are shown in Fig. 11 and 12. The Table 2 contains comparison between calculated and measured distortion.

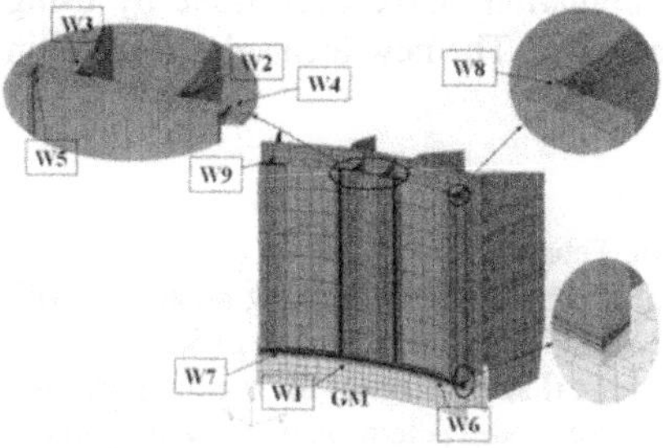

Figure 9. VVPSM global model, PS1 part

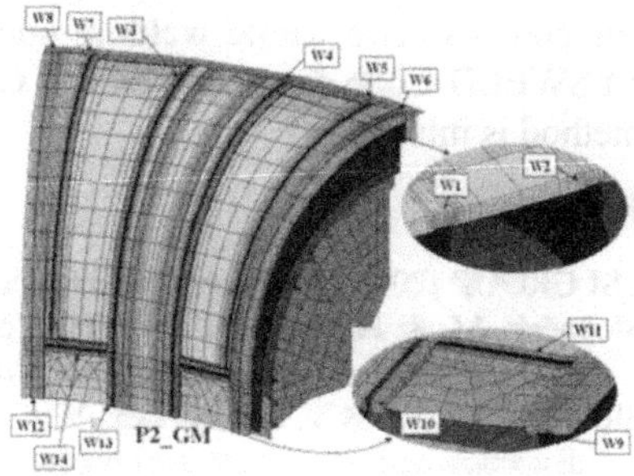

Figure 10. VVPSM global model, PS2 part

Table 2. Comparison between calculated and measured shrinkage

Reference place	*Shrinkage [mm]*		*Reference place*	*Shrinkage [mm]*	
	Measured	Calculated		Measured	Calculated
30	5,4	4,3	35	7,4	4,3
31	4,8	3,6	36	7,6	3,6
32	4,8	3,4	37	6,9	4,4
33	4,8	3,0	38	6,0	4
34	2,0	2,3	39	3,4	3,3
40	3,9	3,9	42	4,0	5,9

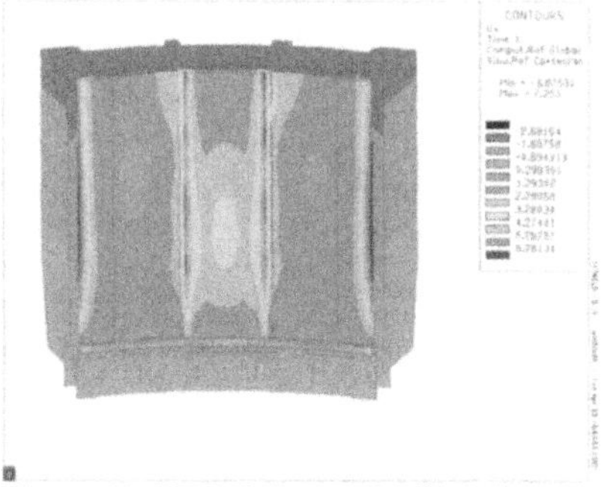

Figure 11. VVPSM,part PS1 final distortion

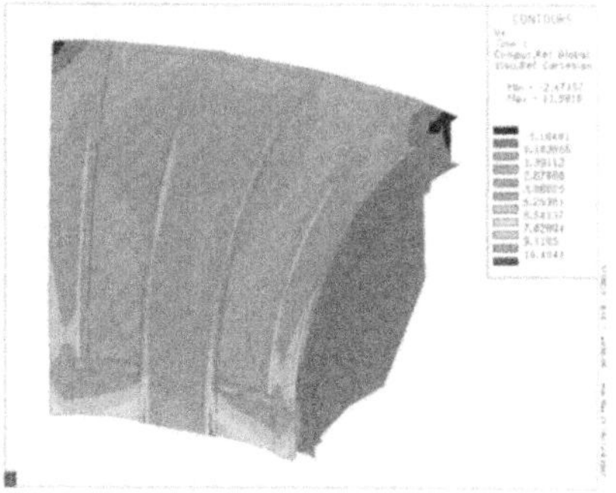

Figure 12. VVPSM,part PS2 final distortion

4 Conclusion

The main aim in the industry is cost reduction with increasing product quality. The very detailed knowledge about the process and product are needed to keep the customer requirements. The reduction of the cost and time for design and technology preparation can be done to decrease the experiments before the finalization of the product and technology. The numerical simulations of the welding process are very modern and productive tool. The recent rapid progress in modelling techniques provides researchers and engineers with more information to achieve a better understanding of the residual stresses and distortion in welded construction. This paper demonstrates how numerical solutions can give some very important and interesting information about welding process and behaviour of the welded construction. The results can be used as one of the basis during the proposal of a new product or new technology and the lifetime prediction of the welded components. The whole welding numerical simulations have been done by using SYSWELD Code developed by ESI Group Company. The new local-global solution method is introduced.

References

ESI GROUP (2006) The welding simulation solution, February.

Slováček, M. & Junek, L. (1999) Effect on residual life time due to welding repairs of NPP components, In: *17th International Conference SMIRT, Prague, Czech Republic*

Slováček, M., Diviš,V. & Ochodek, V. (2005a) Numerical simulation of the welding process, distortion and residua stresses prediction, heat source model determination, *Welding in the World* **49** No. 11/12

Slováček, M., Diviš, V. & Junek, L. (2005b) Support of numerical simulation of the welding process in the production preparation category, In: *IIW conference, Prague*

Slováček, M., Diviš, V., Ochodek, V. & Jones, L. (2006) Welding numerical simulation of real ITER industrial parts, In: *EUROPAM 2006, Toulouse, France*

Tejc, J. & Kovarik, J. (2007) Analýza deformací vrchního dílu NT tělesa, report MECAS ESI, Plzeň (in Czeck)

8.4 Determination of Internal Mechanical Stresses of Large Steel Structures on the Basis of Electromagnetic Reading out of Laser Scribed Marks

János Takács[1], Péter Ozsváth[1], Péter Molnár[2], István Németh[2]
[1]*Budapest University of Technology and Economics Department of Vehicle Manufacturing and Repairing, H-1111, Budapest, Hungary, takacs@kgtt.bme.hu*
[2]*Metalelektro Ltd., Hungary*

Abstract

The loadability of large steel structures (e.g. bridges) is strongly dependent on design and the residual stresses of manufacturing. The developed method is based on distance measurement and it is capable to determine both the residual and operational stresses of large steel structures. Local heat treated marks can be scribed on the surface of steel sheets at a pre determined base distance from each other by using precision laser beam. Electromagnetic detecting of alteration of the base distance of the marks is the fundamental conception of the procedure. The paper presents the development of the process and operation of the system.

Keywords: *Residual Welding Stresses and Distortions,* Structural Safety and Reliability

1 Introduction

The accurately controllable energy input of laser beam makes possible to prepare precisely situated heat treated zones (marks) in the surface of metals. As a result of the laser treatment permanent modification of magnetic properties occurs locally in ferromagnetic materials, which can be used for information storage. The marks can be identified (or read out) by electromagnetic method even if thick paint covers the surface. The variety of special purposes of information storage continuously expands (Ozsváth (2005), Takács (2001)).

If the Hooke-law is valid for the tested material at the current mechanical load, then mechanical stress can be calculated from elongation and Young-module. The deformation of a sheet part of a large steel structure can be determined by the measurement of the distance of two laser scribed marks. The utilization of the method for large steel structures will be presented in Point 2. The mostly used material of large steel frameworks is mild steel which is not sensitive for heat treatment, as a result of this only minor microstructural transformations occur in the tissue of the mark. Electromagnetic detecting of these marks can be difficult especially if thick paint covers the surface. Since the complexity of interaction of laser light and material, laser technological data has to be optimized with scientific approach.

2 The theory of stress determination using laser scribed marks

During manufacturing and operational load of large steel structures like bridges various mechanical stresses arise in the elements. Residual stresses of manufacturing can be significantly high compared to operational loads. Welding is a major source of residual stresses. The residual stresses of manufacturing process can be calculated

if the original stress-free condition of the sheet element is memorized by the distance of two laser marks before it is joined to the other parts of the structure.
Two marks should be scribed into the sheet before welding at a pre-determined base distance (l_0) from each other. When the laser marked element is loaded (residual stress or operational load) average stress level between the marks can be calculated according to the change of l_0 (Δl). For accurate results heat expansion has to be considered in the calculation. What is more, l_0 has to be also compensated with heat expansion during the laser scribing because the base distance is valid for 20°C. Electromagnetic reading out of marks is the main reason for using laser scribing. The distance resolution is hundredth of a millimetre and the marks can be detected also under paint. Furthermore, the laser caused transformation is durable and precisely positioned. The stress level can be monitored from manufacturing process till the end of operation of the structure if the marking is performed before welding of the elements. Theoretically, already operating frameworks can also be marked. In this case only from loading arising deformation and stress can be determined. The principle can be followed in Figure 1.

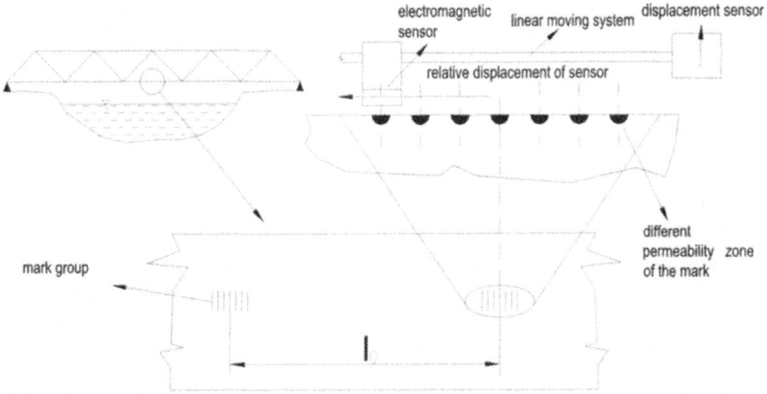

Figure 1. Electromagnetic reading out of laser prepared mark groups for stress determination

For measuring the distance between the two groups marks has to be separated as a function of place. Because of practical reasons a pair of mark groups determine the base distance it means that more identical mark lines form a group. A mark group consists of seven 2 mm wide and 40 mm long heat treated lines. The theoretical centers of the whole groups are used to define the distance as it is displayed in Figure 1. The l_0 base distance is less than 0.5 m.

3 Material properties of the laser scribed marks

Since for bridges used steels contains less than 0.1% carbon, martensite is not forming in the laser affected zone, only the microstructure of grains changes. The paint of bridges is usually thick because of this reliable reading out requires to produce as deep heat treated bands as it is possible without the melting of the surface. Targeted depth of mark zone is 0.5 mm. Melted and rapidly cooled material can be source of micro cracks which results the decrease of fatigue properties. Further key of the reliable detection of a mark is the symmetrical, permanent transformation of material properties (e.g. permeability) of the zone. High mobility potential of laser source is also required for posterior marking of already operating structures.

3.1 *The laser marking process*

The physical properties of a laser treated material is strongly dependent on the applied laser type and technological data. Several experiments were carried out to optimize the laser technology. The microstructural properties and reading out potential will be presented in the following. The dimensions of the transformed zone like depth and width and local modification of magnetic properties are fundamental from point of view of reading out and information durability. Furthermore, the melting of surface during scribing is not favourable because of the increasing risk of micro crack formation. If the material is melted, the surface will be so uneven which can disturb the operation of the reading out sensor. The local thermal expansion during marking of thick steel sheets of large structures cause negligible bending. The scheme of laser marking technology and cross section of resulting heat treated zone can be seen in Figure 2.

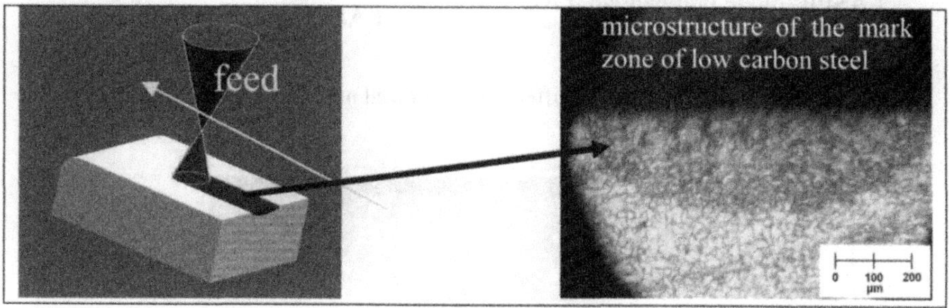

Figure 2. Scheme of the laser marking and structure of marked material (Kalincsák et.al. (2004)

3.2 *Microstructural properties of marking zone*

The laser scribed mark can be systematically characterized in the cross section by metallographic analysis. Due to the chemical interaction with etching reagent is it darker or lighter than the base material. The material of a typical bridge material contains mainly polyhedrical ferrite grains and sparsely situated, more little perlite clusters. The zone differs from the surrounding tissues according to the concrete type of modification. Width and depth can be determined, bubbles, cracks, inhomogenation of the structure can be revealed in the cross section. Important to characterize the boundary of heat treated zone and base structure because relative sudden transition and symmetrical shape is necessary for reliable reading out. In general, if more energy is input, the shape of border line of the treated zone become more even, however it also depends on marked material. The most of the visible microstructural changes are very similar to each other. Figure 3 displays four marks which were optimized with different laser types. "I" means the energy density of the laser beam, laser spot diameter was 2 mm on the surface.

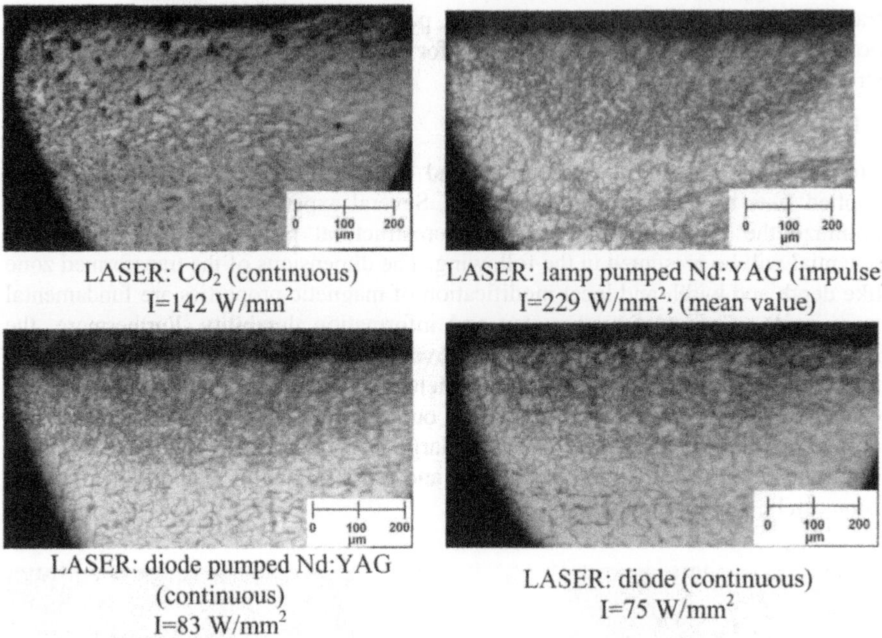

Figure 3. Cross sections of differently optimized marks (polished, nital 5%)

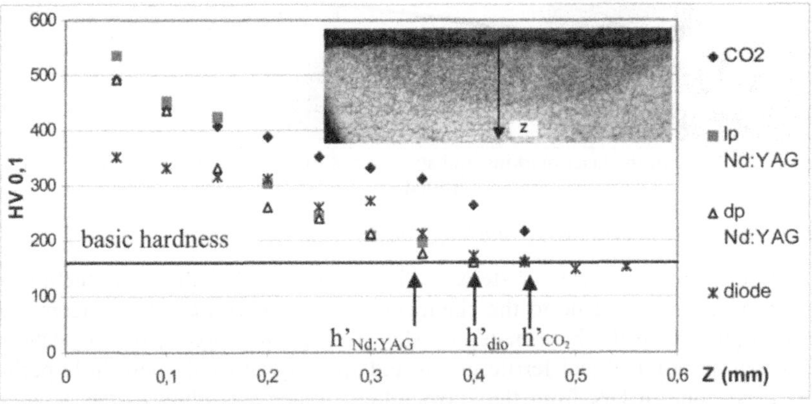

Figure 4. Micro-hardness map of optimal marks of different laser types

Figure 3 represents that quality of Nd:YAG laser marks and diode laser marks are close to each other. All zones are well separated from the base material structure but not as homogeneous, although it seems that micro crack and bubble formation can be avoided. The depth of transformed zones does not seem to be more than 0.2-0.25 mm, which is less than half of the targeted 0.5 mm value. 0.5 mm depth can be achieved but the surface will be too uneven and melted in this case. Optimizing of CO_2 laser technology was less successful than the other laser types. The most promising is the diode pumped Nd:YAG because deep mark can be produced with relative low energy input. Theoretically, with the lamp pumped Nd:YAG laser would be also possible to produce similar marks if impulse frequency was enough high to approximate continuous power output.

The comparison of micro-hardness values of differently prepared marks gives further information about rate of structural transformations and residual stresses in the zone. The following diagram shows that micro-hardness map of the mark. The extension of the transformed zone can be figured out from micro-hardness map (see Figure 4). The displayed hardness values are mean of three mark zones from the group, (which consist of five bands).

4 Reading out potential of laser scribed marks

Figure 6 clearly displays that the detecting potential (signal intensity and character) of visually similar heat treated zones (Figure 5) how much can differ from each other.

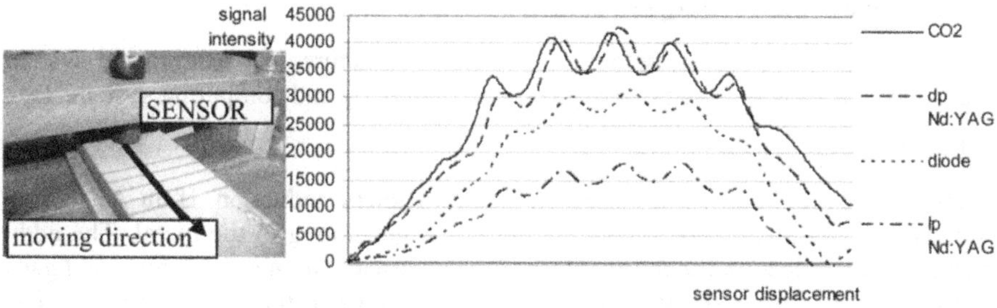

Figure 5. Results of eddy current reading out tests of optimal mark groups

The local maximum places of the diagrams display the marks in the surface of the specimens. The developed software of the reading out system uses the positions of the peaks of the marks. (The differences in the direction of sensor displacement arise from the inaccuracy of positioning of the marked specimen plates in the experimental appliance.) The lift-off of the sensor was 1 mm. These measurements were performed with the experimental set-up of the later prototype of "Stress Reader" device.

5 Summary of laser technology optimization and the marker prototype

According to the several laser experiments more laser type is appropriate for marking of thick mild steel if power output is minimally 200W. At the same time the transformation of treated zone of material is different when different laser types are used. The carried out experiments highlighted that the good reading out has strong connection with the quality of the laser beam. On the basis of different laser test series and analysis of the produced marks the Nd:YAG laser proved to be most suitable for marking low carbon steel. Nd:YAG marks did not have significant disadvantages from the experimented aspects. Practically also Nd:YAG laser is the most appropriate for bridge marking because it can be transmitted in fiber optics which significantly simplifies the scribing of operating bridges on site. A LASAG SLS 200 Nd:YAG laser and ISEL CNC tables were used to build the laser marker prototype. The device is supported with a special software customized for marking.

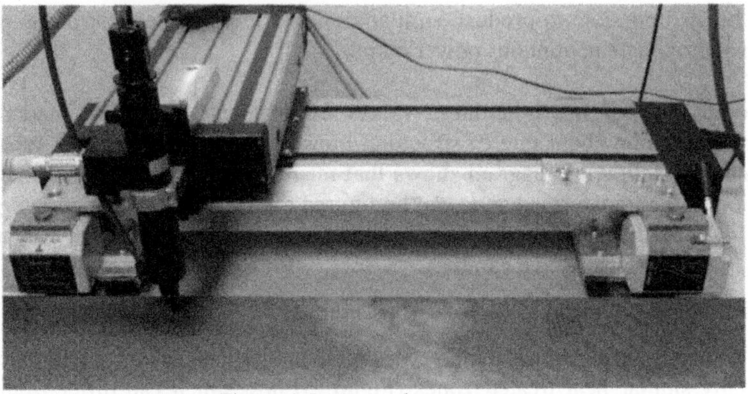

Figure 6. Laser marker prototype

6 Development of reading out system

For determination of small deformation of large structures the reading out system has to operate with sophisticated marker sensor, moving and computing system.

6.1 *Eddy current sensor*

Eddy current measurement is based on electromagnetic induction. If an AC coil is approached to a metal part, then eddy current is generating in the metal. The eddy current is dependent on the tested material, geometrical parameters, defects and gaps in the surface. Eddy current causes impedance change in the generating coil. The reading out method is based on the fact that the eddy current in the laser marked zone will differ from the not marked surfaces. There are several types of eddy current sensors. The main choosing factors are the high sensitivity and narrow view. The distance of mark lines in a group significantly influences the accuracy of the detection. The optimal distance of the lines in a mark group was determined by reading out test series.

6.2 *Reading out system*

Figure 8 shows measured inputs and scheme of the "Stress Reader" device. The reading out system operates with two eddy current sensors. It means that the computing unit has to manage and store more measuring signals and one displacement signal. Temperatures are saved at the beginning and the end of the reading out. The accuracy of stress calculation is ±2,5 MPa which means that the accuracy of the displacement sensor has to be better than ±5μm. The two mark groups are detected simultaneously with two eddy current sensors. Displacement of sensors during the reading out has to be only a few millimetres longer than a mark group. The sensors are moved by servo motors with integrated encoder that supplies displacement data of sensors. The position of a mark group is determined by the average peak position of mark lines. The computer unit is able to ignore damaged marks.

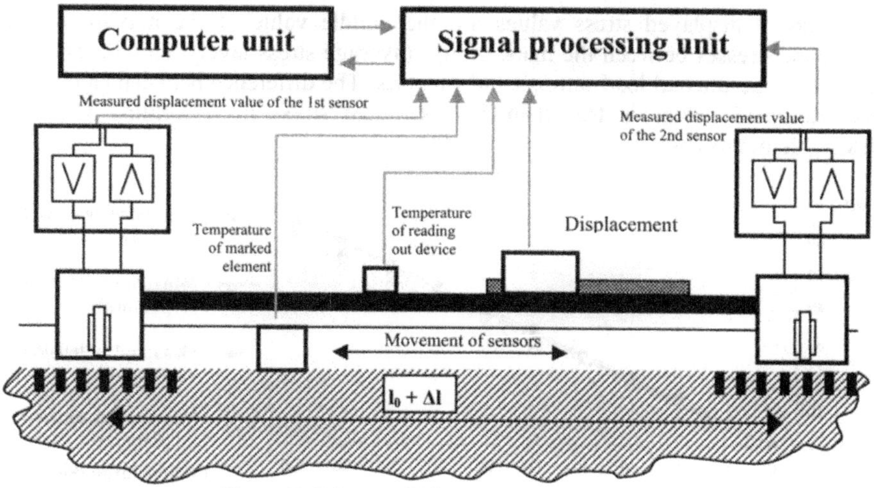

Figure 7. Scheme of "Stress Reader" prototype

6.3 *Verification test of the new method*

The developed stress determination system was tested on an experimental I beam. The inside of plates were marked before welding. The laser marking was made by the lamp pumped Nd: YAG laser marker prototype. The beam was loaded like it is displayed in Figure 8. The Stress Reader was placed in measuring position during the loading and stress calculation was performed between 0 and 100 kN load at every 10 kN steps (Figure 7). The experiment was done on a rail testing instrument.

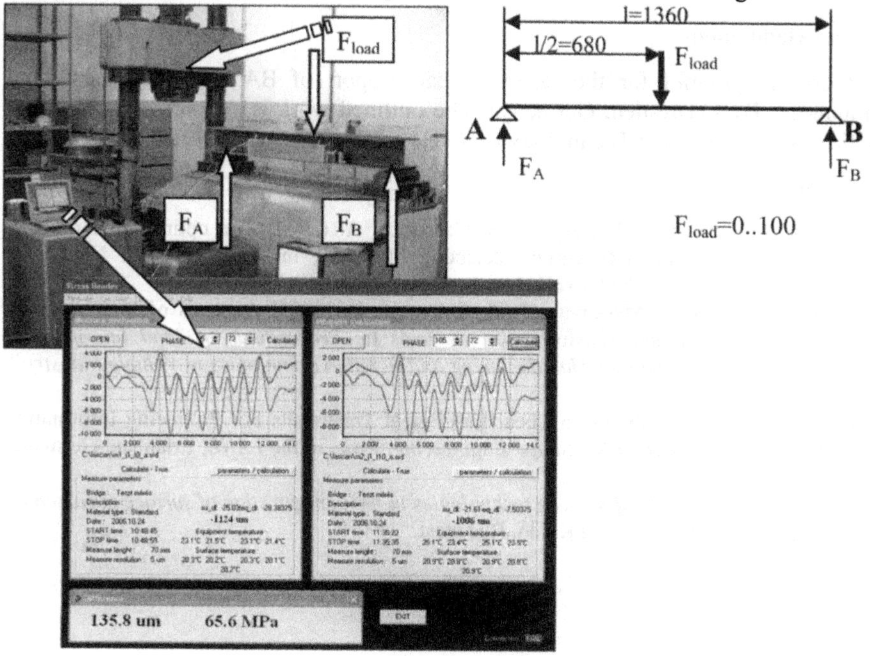

Figure 8: Experimental setup of verification test and theoretical model

In Figure 9 displayed stress values are the middle value of the maximum and minimum stresses between the mark groups (average stress level). This is resulting from the experimental load settings and supports. The difference between calculated and measured values is less than 5%. Residual stress from welding was not detectable in this case.

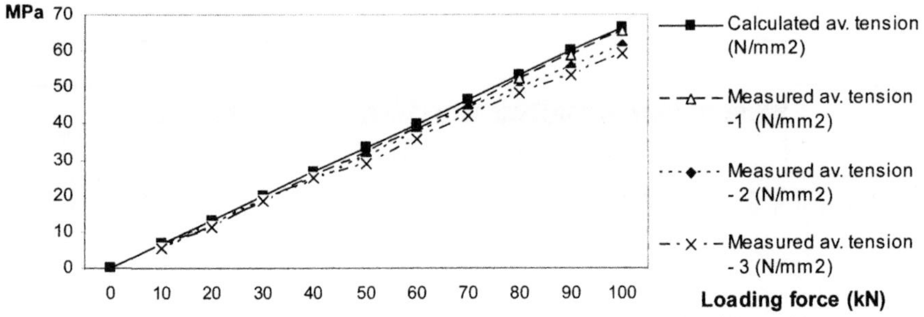

Figure 9. Comparison of theoretical stress calculation and three series of measured values

7 Summary

A complete system for determination of mechanical stresses of large steel structures with electromagnetic detecting of laser scribed marks has been developed. Laser technology and eddy current reading out method of marks was optimized. Laser scribed marks can be read out under paint through several years of operation that facilitates reliable stress monitoring.

Acknowledgement

Authors are grateful for the technological support of BAYATI (Hungary) and Fraunhofer IWS (Dresden, Germany). The outlined work is supported by National Office for Research and Technology (NKTH) project No. AGE-00015/2003.

References

Kalincsák, Z., Takács, J., Vértesy, G. & Gasparics, A. (2004) The optimisation of laser marking signals for eddy current detecting of marks, In: *Laser Assisted Net Shape Engineering, Proc. of the LANE* Erlangen, Germany, **3** 781-788.

Ozsváth, P., Takács, J., Markovits, T. & Kalincsák, Z. (2005) Analysis of laser marking caused microstructural transformation in steels In: *Proceedings of 22nd International Colloquium on „Advanced Manufacturing and Repair Technologies in Vehicle Industry"*, Czestochowa, pp. 73-78.

Takacs, J. et. al. (2001) Precision Local Laser Heat Treatments For Producing Information Input, In: *Laser Assisted Net Shape Engineering, Proc. of the LANE*, Erlangen, Germany, **3** 263-272.

Takács, J. (2004) *The role of modern technologies in the configuration of surface parameters,* (in Hungarian), Műegyetemi Kiadó, Budapest.

8.5 Calculation of Welding Residual Stresses and Distortions under Complex Process Conditions

Markus Urner[1], Tim Welters[1], Klaus Dilger[1]

[1]*Institute of Joining and Welding Technique, Technical University of Braunschweig, Langer Kamp 8, 38106 Braunchweig, Germany, m.urner@tu-bs.de*

Abstract

In a welded construction, residual stresses and distortions develop due to the inhomogeneous heat treatment. Nowadays a decreasing average time to market and an intensified international competition generate a growing demand for reliable and flexible computational methods offering the possibility to predict component properties after welding and integrate a welding simulation into production planning. For instance mounting clamps and other heat sinks have a significant impact on the resulting temperature field in the welded component. Additionally the mechanical deformation of the component during the transient welding process is directly linked to the, often rigid but nevertheless finite, stiffness of the imposed restraints. Several spatiotemporal contiguous welding processes are often as much a challenge as the consideration of a realistic seam formation and phase transformations.

Keywords: *welding, process, residual stresses, residual distortions*

1 Introduction

Most of the components in the machine-building industry are designed as welded constructions. The most appreciable advantages of welded parts are the economical production and the variability in configuration. There are also some disadvantages using welded parts. The inhomogeneous heat treatment of the welding process causes residual stresses and distortions in the component. To keep the requested tolerances in the production it is important to know how the distortion will occur.

The intention of engineering is to avoid distortion already in the development phase. Since a decreasing average time to market the demand for reliable and flexible computational methods to calculate the properties of welded components is growing. The prediction and avoiding of distortions becomes more and more important. One way to compute the residual stresses and distortions in the development phase is, to use a theoretical model for the simulation of the welding process. Today the finite element method is the most chosen way to do these calculations.

The principle procedure of simulating residual stresses and distortions is based on a combination of different physical phenomena. The first step in simulating residual stresses is the calculation of a temperature field. After that, a mechanical analysis is carried out to compute the residual stresses and distortions which occur due to the thermal strains. To describe these connections in the finite element calculation a heat conduction model of welding should be used (Radaj 2002).

2 Finite element calculations

The procedure of finite element simulation is shown in the following example in detail. Intention of the studies was to calculate the welding residual stresses in the used metal sheet. Figure 1 shows a fixture which was used to weld sheets metal out of the steel 1.4301 with a TIG-process without filler metal. First the assembly of the fixture is described. The sheets to be welded lie on a base frame with a copper plate. They are fixed with cover sheets out of the austenitic steel 1.4301 and a downholder made from the steel 1.0570. Figure 2 shows a cross-section schematically.

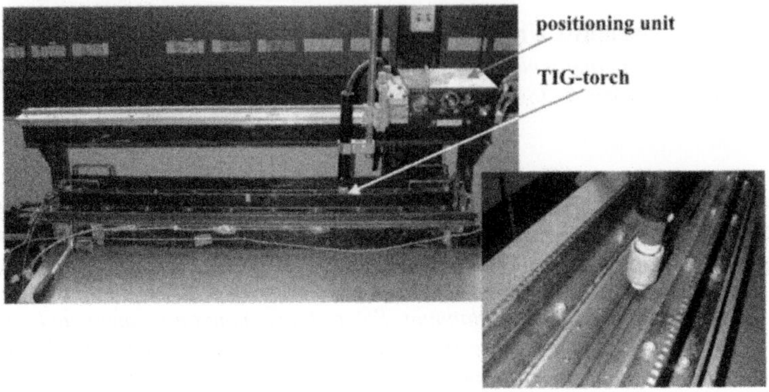

Figure 1. Welding device

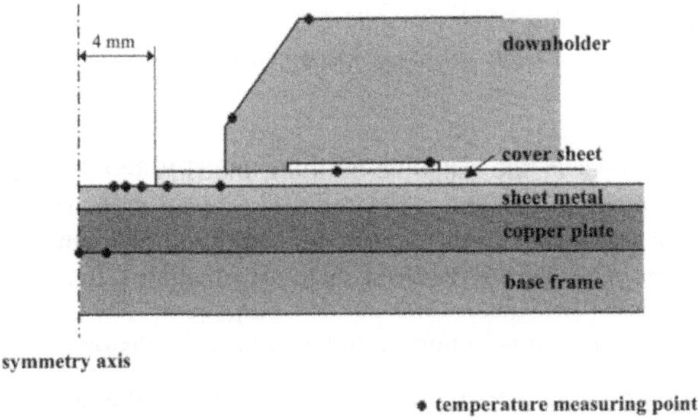

Figure 2. Cross section of the welding device with the position of the thermocouples

The base for the FE-model was a 3D-CAD-model. The complexity of this welding device made it necessary to use a computation-intensive 3D-model to include the heat flow in the fixture. In the shown example first, the dimensions of the clamping fixture were measured and mapped on a CAD-model. Due to symmetry a half-model could be used. The next step was to clean up the geometry and mesh it with 3D-elements. The finite element mesh is shown in Figure 3a. The elements of the welded sheet metal had an edge length of 1 mm in the welding zone. The remaining fixture was meshed with a larger edge length to decrease calculating time.

Measurements with thermo couples showed that the fixture was a strong heat sink which pulls much heat out of the welded sheet metal. Therefore, additional contact elements were added to the model in order to take into account the imperfect heat transfer between the different components of the setup, Figure 3b. The contact elements permitted the definition of heat transfer coefficients (Ansys, Inc. 2005), to manage the temperature flow inside the fixture. To decrease the size of the model a part with a length of only 300 mm was modelled. This length was chosen, because it fulfils both demands, the development of a steady-state temperature field around the moving heat source and a computation time of less than four hours.

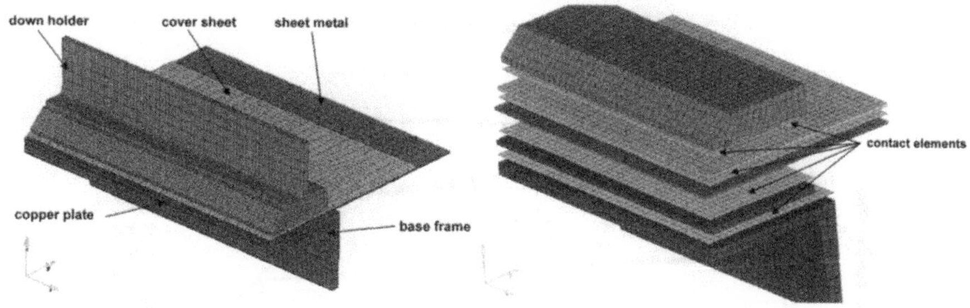

Figure 3. a) FE-model b) contact elements

For the modelling of the TIG-Process, a distributed volume heat source had been chosen. The use of a normally distributed double-ellipsoidal volume heat source after Goldak allowed to create an asymmetrical distribution in movement direction in the case of a travelling welding heat source. Figure 4 shows the characteristic parameters for this source.

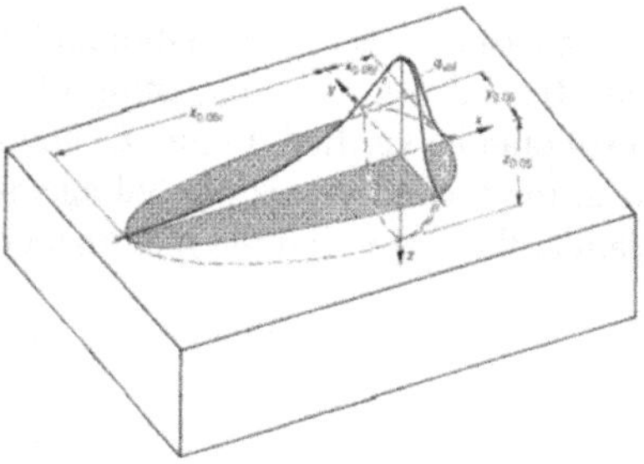

Figure 4. Heat source (Goldak et al.1986)

All calculations were done with the FE-code ANSYS. The calibration of the heat source parameters were performed with temperature measurements at the surface of the sheet metal and the clamping fixture. The temperatures were recorded over the time of the welding process and the cooling face. The used thermocouples of type K were spot welded onto the sheet metal and the welding device. The measuring positions of the thermocouples are shown in Figure 2. By comparing the calculated temperatures with the measured ones the parameters were adjusted. Also the values of the heat transfer coefficients of the boundaries between the different components were adjusted in this way. They were necessary to regulate the heat flow out of the sheet metal in the fixture. The transient calculation of temperature field was done by moving the heat source along the nodes of the groove.

The finite element calculation requires the knowledge of the temperature-dependant mechanical material properties like Young's modulus, yield stress, heat capacity and thermal conductivity. Experimental knowledge about the elastic-plastic behaviour of the used materials is also essential. Figure 5 shows the measured temperature-dependant stress-strain curves of the steel 1.4301. The data points out that the base material is cold hardened by rolling.

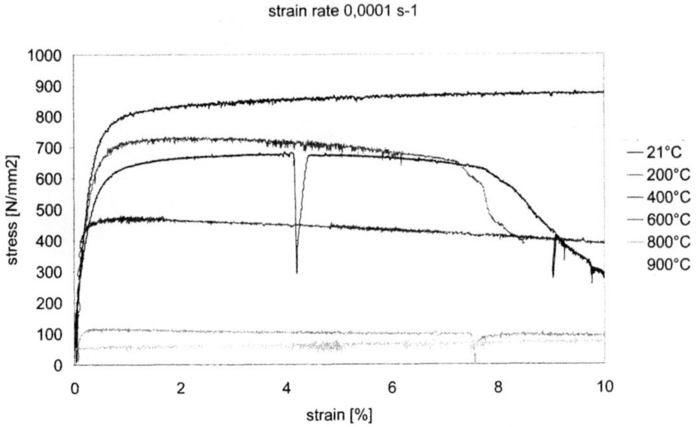

Figure 5. Stress-strain diagram

On the basis of hardness measurements of metallographic cross sections of the melting pool and the heat affected zone a dehardening of the used material by the heat of the welding process could be verified. Figure 6 shows the measured hardness values of four specimens at the second y-axis plotted against the distance from the middle of the welding seam at the x-axis and the maximal occurring temperatures at the first y-axis.

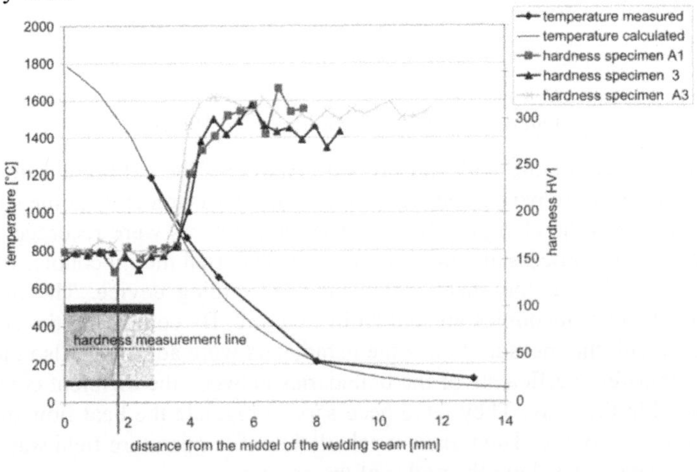

Figure 6. Measurement of hardness

The hardness values drop at a distance from 5 mm to the middle of the weld seam from 300 HV1 in the base material to 150 HV1 in the heat affected zone and the melting pool. This occurs at a temperature of approximately 700 °C. To map these effect to the FE-calculations a second material was defined with the mechanical material properties of the dehardened material. The material properties of the dehardened material were defined by using the EN ISO 18265:2003 on the measured hardness values. The hardness values were correlated with the tensile strength and the yield strength was adjusted.

In each step of the transient calculation the second material was assigned to all elements of the model which had a temperature about 700 °C. Figure 7 shows the elements of the sheet metal influenced by the heat source. The choice of elements with a temperature about 700°C is shown by the first colour in the centre of the heat source. For these elements the material was switched. According to the material switching the calculated residual stresses in this area decrease because the plastic deformation can develop easier than in the base material.

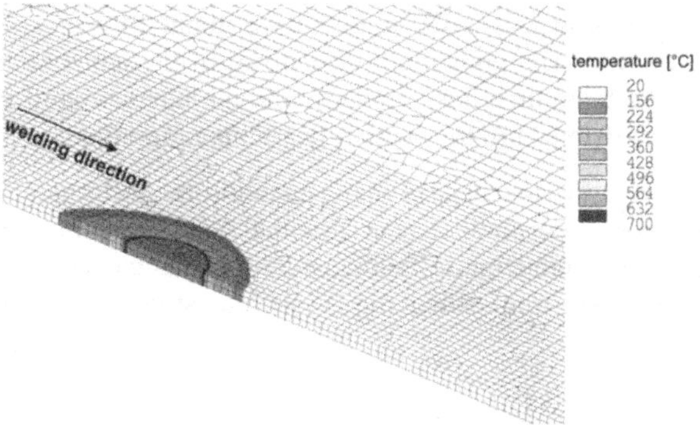

Figure 7. Area of material switching

First, the gap closure was not considered because a model with symmetry constraints is used for the calculations. Therefore, it was necessary to deactivate the elements in front of the heat source by the birth and death capability of the software Ansys: (Ansys, Inc. 2005). These deactivated elements remains in the model but contribute a near-zero stiffness value to the overall matrix. If the heat source moves forward, the elements right in front are activated and integrated into the stiffness matrix. In this way the gap between the welded sheet metal is closed by and by. Figure 8 shows the deactivated elements in a band along the symmetry border in a selected step of calculation

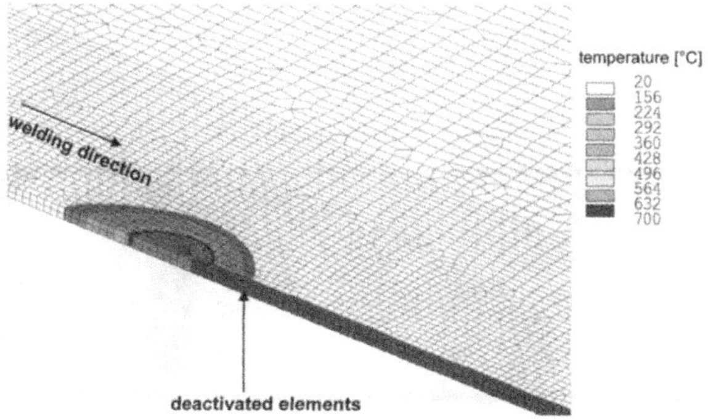

Figure 8. Deactivated elements in front of the heat source

3 Results

Figure 9 shows the measured and calculated temperatures in the sheet metal by comparison. The measured and the calculated curves are displaced in time for better comparability. The curves comply very well. Another control was given by the molten pool. Figure 10 shows the molten pool in a metallographic cross section and the calculated one. Both the dimension and the form match very well.

As a result of the calculation the temperature was now known for every position and time. Figure 11 shows the calculated temperature field in the sheet metal. The displayed colours present a temperature range from 20 to 70°C. All temperatures about 70°C are not displayed. This range was chosen to show the global temperature field. The real welding process with temperatures about 1400°C is localised at the edge of the half model. The typical elliptical form of the temperature field can be recognized in the welding direction. Furthermore, the interaction of the downholder as a big heat sink is visible with the break of the temperature field.

With this information the mechanical analysis could be performed. The calculated element temperatures were now used as thermal loads in a mechanical analysis. Every time step of the temperature field calculation was calculated as load step in the mechanical analysis from which the occurring stresses and distortions were derived.

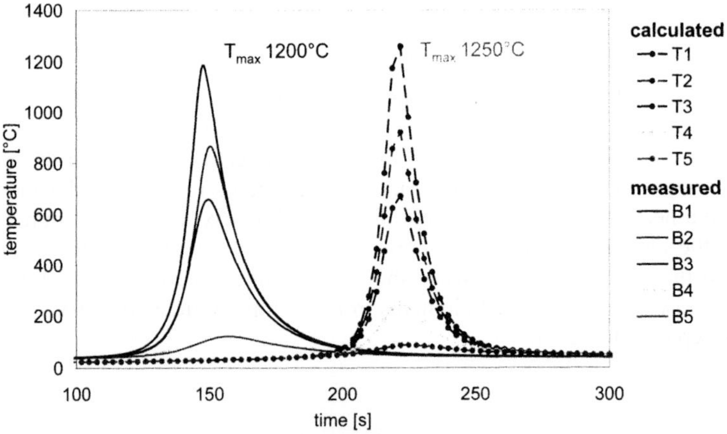

Figure 9. Temperature curves

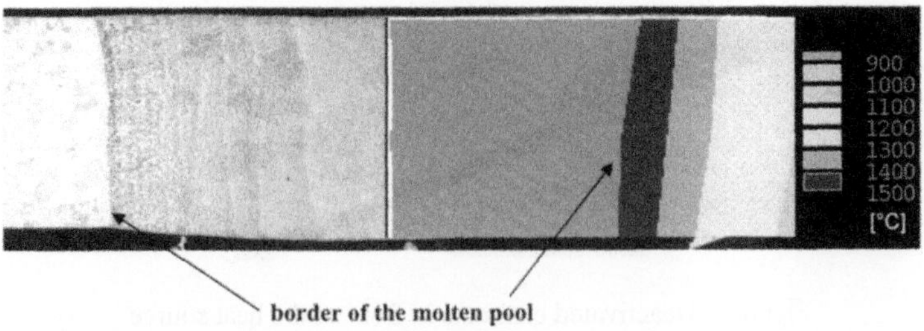

Figure 10. Metallographic cross section

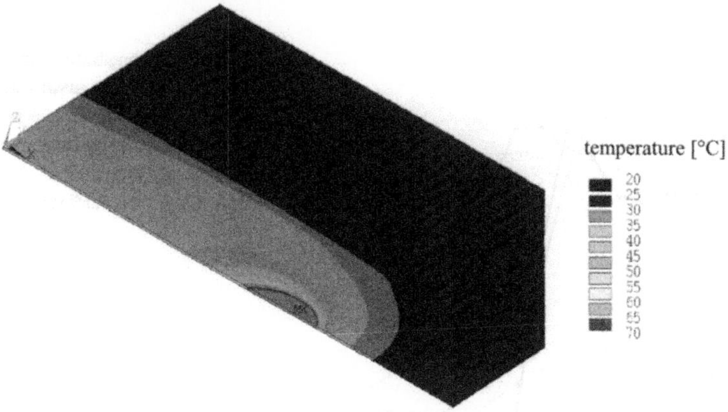

Figure 11. Temperature field

Because of the work hardening caused by the prior cold-rolling process a fundamental change in the temperature-dependant mechanical material properties takes place during the heat treatment. Therefore the material properties of all elements reaching a temperature above 700 °C were changed during the simulation towards a lower yield limit and tensile strength. The last step of the mechanical calculation was to unclamp the sheet metal. Therefore all elements of the downholder are deactivated. The sheet could deform in result of the occurring residual stresses and assume its end shape.

The residual stresses were measured with x-ray diffraction on a line perpendicular to the welding seam. There, a couple of measurements were made. In Figure 12 both measured and simulated values are shown in comparison. The abscissa of the diagram shows the distance to the center of the weld seam in mm, the ordinate specifies the stress in N/mm^2. The measured residual stresses were shifted by 200 N/mm^2 in tension direction. This is a first order approximation for the residual stresses of -200 N/mm^2 which were measured in the base metal prior to welding and which were neglected. The curve progression from the longitudinal as well as from the transverse residual stresses corresponds good to the measured ones. The main peak of the calculated longitudinal residual stresses in 5 mm distance to the centre of the welding seam is the same value as the measured one. The computed transverse residual stresses show a difference of 100 N/mm^2 to the measured ones. This difference can be explained by the preload on the welded sheet metal. These preloads are relieved during the welding process by the influence of the heat. They still have consequences for the welding residual stresses by pushing the high to the measured values. Therefore, particularly the calculated transversal residual stresses appear lower than the measured ones.

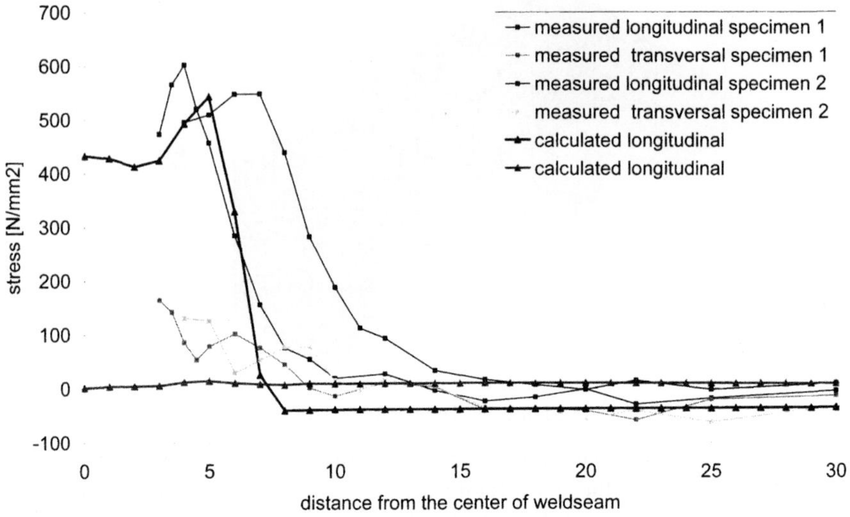

Figure 12. Residual stresses

4 Conclusion

The computation of welding residual stresses in consideration of complex process conditions can be performed. It is possible to model clamping fixture by defining heat flow coefficients and calculate the occurring temperature field. Based on the mechanical analysis the residual stresses and distortions can be calculated. In a next step the welding process can be optimized to decrease the residual stresses in the construction. Through this the preloads on the sheet metal can be reduced and the strength of welded construction can be raised. This leads to the possibility to develop better lightweight constructions.

References

Ansys, Inc. (2005) *Release Dokumentation for Ansys 11.0*

Goldak, J., Chakravarti, A. & Bibby, M. (1986) A new finite element model for welding heat sources. *Metallurgical Trans.* **17B**

Radaj, D. (2002) *Welding residual stresses and distortions*, Düsseldorf: DVS-Verlag.

Wikander, L., Karlsson L., Näsström,M. & Webster, P. (1994) Finite element simulation and measurement of welding residual stresses. *Modelling Simul. Mater. Sci. Eng.* **2** 845-864.

Section 9

Static stresses in welded connections

9.1 Generalised Beam Theory to Analyse Fillet Welded Joints

Zdeněk Kala
Brno University of Technology, Faculty of Civil Engineering, Veveří 95, 602 00, Brno, Czech Republic, kala.z@fce.vutbr.cz

Abstract

The present paper deals with an analysis of the resistance of fillet welded joints used for steel structures. Software products based on the finite element method applied in current design practice are concentrated on stress determination in the basic material of rolled steel beams above all. This stress is often calculated on the FEM beam model. The load–carrying capacity of weld joints in which a complicated triaxial stress is not solved by these programmes at all or is solved only marginally; therefore, the solution of concrete details always depends on the designer's experience. An algorithm will be presented which, in accordance with standard prescriptions, links together the advantages of classical theory with possibilities of contemporary computer technology.

Keywords: *steel, beam, fillet welds, standard, reliability*

1 Introduction

When increasing the level and situation of knowledge in the field of design of steel structures, the step–by–step substitution of formerly used riveting, and/or casting and forging by the technological process of welding place; by this, not only a remarkable decrease in mass but also a saving of manufacturing laboriousness and cost optimisation were obtained Jármai & Farkas (1999). Welded steel structures are commonly used, although welded joints frequently represent the source of crack initiation and of defects which can lead even up to breakdowns with severe consequences. The resistance calculation of a welded joint is very complicated, and its realisation is accompanied by failures caused by human factor. The stress distribution in fillet welded joints is much more complicated than that in butt welds. At present, the triaxial stress of basic material can be solved still by applying the finite elements. However, the calculation model is often very time–demanding, and it always requires a certain reality idealization. The uncertainties are connected, e.g., with the distribution and values of residual stress, with material heterogeneity.

Other uncertainties can be introduced into the calculation by the fact that the welded joint strength depends not only on the type of electrodes used but also on the strength of material welded which, in a welded joint, gets mixed with the electrode metal melted, and further on, that the load–carrying capacity of fillet welded joints with low thickness is higher than that of welds carried out in multiple row spot welding Faltus (1981). Thus, the information mentioned draws attention to the facts that, when specifying the calculation of stress state in welded joints, the calculation should also be in agreement with the conventional theory verified experimentally. When assessing and designing the steel structures both of the Czech nuclear power plant at Temelín, and of the Slovak one at Mochovce, a computation programme was developed enabling practical and accurate assessment of welded joints in generally used rolled profiles, the observance of the articles of the standard EC being emphasized.

2 Design and calculation of welded joints

The calculation of nominal stresses by applying classical mechanics, and the calculation of stresses caused by the macro–geometrical change of structure form are the most practical instruments for a design of welded structures, namely for designers, too. According to the Standards, the load–carrying capacity of a welded joint is evaluated according to Mises stress state, namely for the normal stress $\sigma_\perp$ and shear stresses $\tau_\perp, \tau_\parallel$ in the root of weld, see Figure 1.

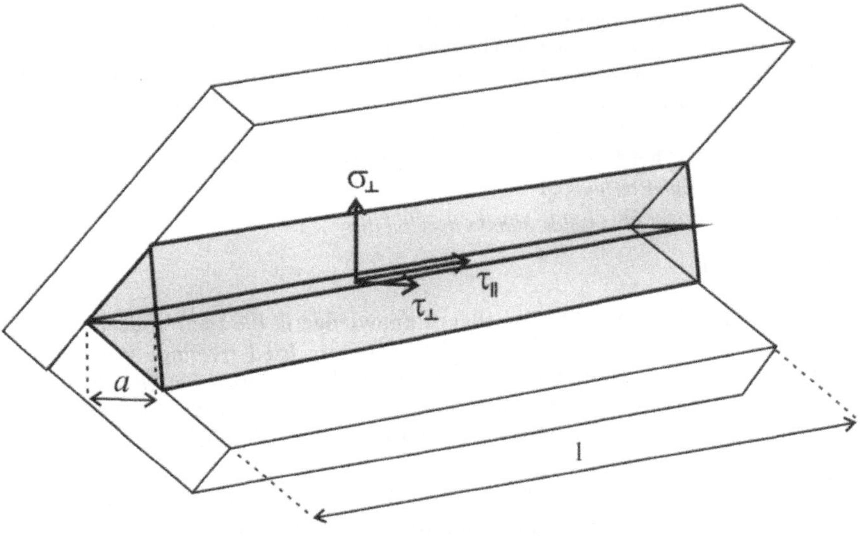

Figure 1. Stress in fillet welded joint

A welded joint must satisfy two conditions at the same time:

$$\sqrt{\sigma_\perp^2 + 3\tau_\perp^2 + 3\tau_\parallel^2} \leq \frac{f_u}{\beta_w \gamma_{Mw}} \tag{1}$$

$$\sigma_\perp \leq \frac{f_u}{\gamma_{Mw}} \tag{2}$$

When calculating the steel structures, mostly components of forces applied to welds in the plane of joined members and perpendicular to them are obtained.

Therefore it is purposeful to convert the formulae into forms enabling the evaluation of Mises stress state based on internal forces of the beam model. Taking into consideration that dimensions of the weld b are, in most cases, substantially smaller than the dimensions of welded members, the stress state in a point assumed can be examined in compliance with the theory of elasticity. Let us consider the loading acting in the root of weld in point M.

In Figure 2, there are marked the forces acting on a very small weld length which will be designated as Δx. Those cross–section areas the external normal line of which has its direction identical with the axes of coordinates are considered to be positive cross–section areas of the section cut through the examined point M. According to this convention, the areas with sides b, Δx of welded diagram in Figure 2 are considered to be negative ones. The normal forces N_y, N_z cause tension and are considered as positive ones. It follows then from the force conditions of equilibrium in the point M that $N_Y = T_{\perp Y}$, $N_Z = T_{\perp Z}$.

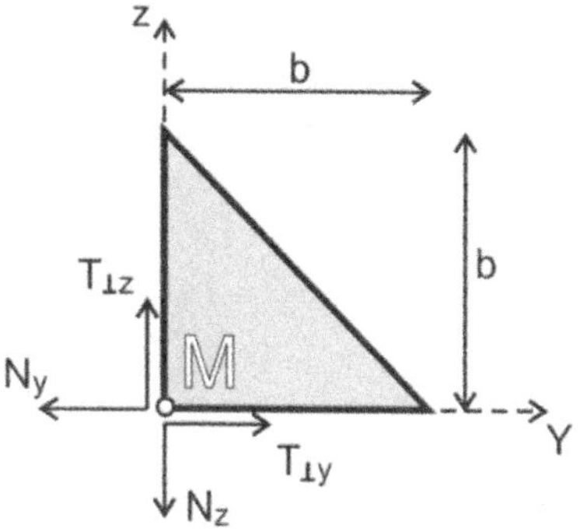

Figure 2. Forces in point M

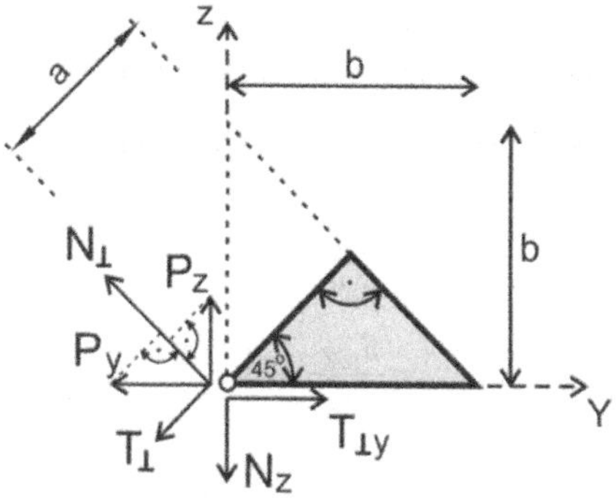

Figure 3. Equilibrium of forces

Let us consider a dangerous welded joint cross–section made at the angle of 45^0, length $a = b/\sqrt{2}$, see Figure 3. By the resolution of forces P_Y, P_Z to the directions parallel with the dangerous welded joint cross–section and perpendicular to it, the normal force $N_\perp$ and shear force $T_\perp$ are obtained. It follows from force conditions of equilibrium:

$$P_Y = T_{\perp Y},$$
$$P_Z = N_Z = T_{\perp Z}. \tag{3}$$

On the chosen cut, the normal and shear resultants will be calculated further on:

$$N_\perp = P_Y \frac{1}{\sqrt{2}} + P_Z \frac{1}{\sqrt{2}} = \left(T_{\perp Y} + T_{\perp Z}\right)\frac{1}{\sqrt{2}}, \tag{4}$$

$$T_\perp = P_Y \frac{1}{\sqrt{2}} - P_Z \frac{1}{\sqrt{2}} = \left(T_{\perp Y} - T_{\perp Z}\right)\frac{1}{\sqrt{2}}. \tag{5}$$

For small values of the cut of cut length a, it can be supposed that the values of bending moments acting on a welded joint will be small, and that they can be neglected when examining the stress state. A similar assumption is applied when calculating the stress of e.g. bolted joints. Under these assumptions, the conventional stress can be determined on the given cut:

$$\sigma_{\perp} = \frac{N_{\perp}}{a\,\Delta x} = \frac{1}{a\sqrt{2}\,\Delta x}\left(T_{\perp Y} + T_{\perp Z}\right), \tag{6}$$

$$\tau_{\perp} = \frac{T_{\perp}}{a\,\Delta x} = \frac{1}{a\sqrt{2}\,\Delta x}\left(T_{\perp Y} - T_{\perp Z}\right), \tag{7}$$

$$\tau_{\parallel} = \frac{T_{\parallel}}{a\,\Delta x}, \tag{8}$$

where $T_{\parallel}$ is the force on the welded joint, Δx is parallel with the welded joint axis. When the designation is introduced

$$\tau_{\perp Y} = \frac{T_{\perp Y}}{a\,\Delta x}, \tag{9}$$

$$\tau_{\perp Z} = \frac{T_{\perp Z}}{a\,\Delta x}, \tag{10}$$

the stress on the given cut can be expressed as:

$$\sigma_{\perp} = \frac{1}{\sqrt{2}}\left(\tau_{\perp Y} + \tau_{\perp Z}\right), \tag{11}$$

$$\tau_{\perp} = \frac{1}{\sqrt{2}}\left(\tau_{\perp Y} - \tau_{\perp Z}\right). \tag{12}$$

With regard to the relation Eq.1, it is evident that it has not any sense to introduce a sign convention for stress $\sigma_{\perp}$ and $\tau_{\perp}$. Taking this fact and the relations Eq.11 and Eq.12 into consideration, only the question is relevant for the problem mentioned whether the quantities $\tau_{\perp y}$ a $\tau_{\perp z}$ have the same signs, or different ones. In this sense, only 4 cases can occur , see Figure 4. For these four cases, the stress actions $\sigma_{\perp}$, $\tau_{\perp}$ are given in Tab. 1:

Table 1. Stress state $\sigma_{\perp}$, $\tau_{\perp}$ according to Figure 4

	A	b	c	d
$\sigma_{\perp} =$	$\frac{\tau_{\perp Y} + \tau_{\perp Z}}{\sqrt{2}}$	$\frac{\tau_{\perp Y} - \tau_{\perp Z}}{\sqrt{2}}$	$\frac{\tau_{\perp Y} - \tau_{\perp Z}}{\sqrt{2}}$	$\frac{\tau_{\perp Y} + \tau_{\perp Z}}{\sqrt{2}}$
$\tau_{\perp} =$	$\frac{\tau_{\perp Y} - \tau_{\perp Z}}{\sqrt{2}}$	$\frac{\tau_{\perp Y} + \tau_{\perp Z}}{\sqrt{2}}$	$\frac{\tau_{\perp Y} + \tau_{\perp Z}}{\sqrt{2}}$	$\frac{\tau_{\perp Y} - \tau_{\perp Z}}{\sqrt{2}}$

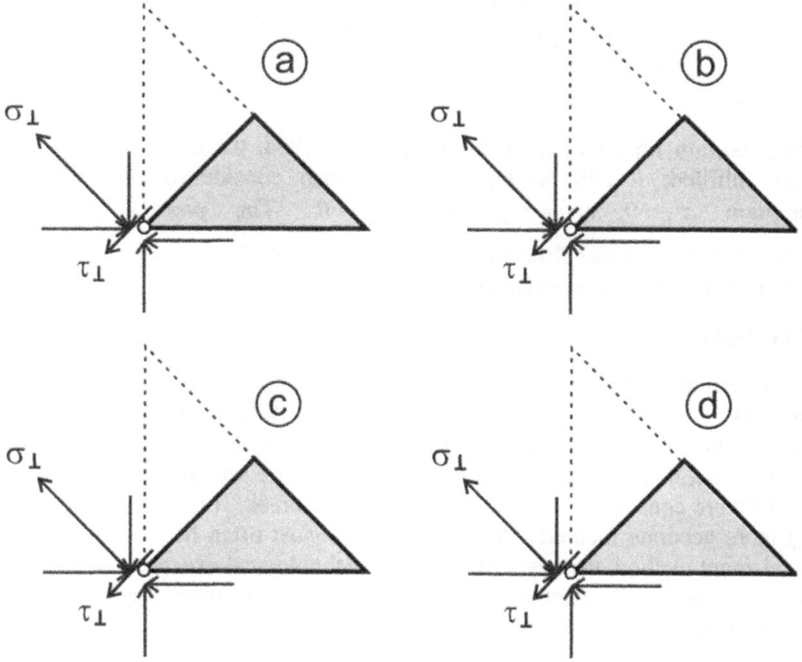

Figure 4. Welded joint stress variants

Essentially, only two cases can occur:

a) $$\sigma_{\perp} = \frac{1}{\sqrt{2}}\left(\tau_{\perp Y} + \tau_{\perp Z}\right), \qquad \tau_{\perp} = \frac{1}{\sqrt{2}}\left(\tau_{\perp Y} - \tau_{\perp Z}\right) \tag{13}$$

b) $$\sigma_{\perp} = \frac{1}{\sqrt{2}}\left(\tau_{\perp Y} - \tau_{\perp Z}\right), \qquad \tau_{\perp} = \frac{1}{\sqrt{2}}\left(\tau_{\perp Y} + \tau_{\perp Z}\right) \tag{14}$$

By inserting Eq.13 and Eq.14 into the relation Eq.1, the following expression is obtained:

$$\sqrt{\frac{\left(\tau_{\perp Y} \pm \tau_{\perp Z}\right)^2}{2} + 3\left[\frac{\left(\tau_{\perp Y} \mp \tau_{\perp Z}\right)^2}{2} + \tau_{\parallel}^2\right]} \leq \frac{f_u}{\beta_w \gamma_{Mw}}. \tag{15}$$

The relations Eq.1 and Eq.15 are practically based on the Mises plasticity condition. The relation Eq.15 enables to evaluate a welded joint based on the intenal forces from a beam model analogously as it is usual for the beam cross–section based on components of forces acting on welded joints in the plane of joined members or perpendicularly to them.

Stresses in absolute values are introduced into the Eq.15, for binomials in brackets, it is possible to consider, in a conservative manner, the signs: –, +:

$$\sqrt{\frac{(\tau_{\perp Y}-\tau_{\perp Z})^2}{2}+3\left[\frac{(\tau_{\perp Y}+\tau_{\perp Z})^2}{2}+\tau_{\parallel}^2\right]}\le\frac{f_u}{\beta_w\gamma_{Mw}} \quad (16)$$

If the condition Eq.1 written in the Eq.16 is fulfilled, the condition Eq.2 will be always fulfilled; it can be proved, e.g., when considering the conservative assumption $\tau_{\parallel}=0,\ \tau_{\perp Y}=\tau_{\perp Z}\Rightarrow\tau_{\perp Y}-\tau_{\perp Z}=0$. The process proposed is conservative; in case that fillet welded joins do not satisfy the condition Eq.16, it is necessary to carry out a more accurate calculation.

3 Conclusion

The calculation method presented allows a quick and – with the use of computer technology – also relatively simple assessment of fillet welds. When the load–carrying capacity of a weld joint is higher than the load–carrying capacity of the basic material, this assessment can be considered to be satisfactory. Conversely, in case of severe consequences in form of a possible break, it is necessary to proceed using more accurate methods. A concrete detail must often be solved, e.g., by the finite element method, namely with regard to technological execution of the weld, to stresses caused by local structural discontinuity, to residual stresses caused by irregular cooling down, etc.

Acknowledgement

The article was elaborated within the framework of project GAČR 103/07/1067, KJB201720602 AVČR and research MSM0021630519 and GAČR 103/08/0275.

References

Jármai, K. & Farkas, J. (1999) Cost calculation and optimisation of welded steel structures, *Journal of Constructional Steel Research*, **50** No. 2, 115–135, ISSN 0143–974X.

Junek, L. & Vejvoda, S. (2000) Navrhování a posuzování svařovaných strojních součástí, konstrukcí a tlakových nádob z hlediska únavového poškozování, sborník z II. konference Velkostroje a těžební technika, Teplice, (Design and Assessment of Welded Machine Components, Structures, and Pressure Vessels From the Point of View of Fatigue Damage, Proceedings of the Conference on Giant Machines and Extraction Technique), pp. 61–71, ISSN 1212–4613.

Kala, Z. (2003) Verification of the partial reliability factors on a case of a frame with respecting random imperfections, Int. Colloquium, In: Proc. of Int. Conference on Metal Structures, Miskolc, Edited by K. Jarmai & J. Farkas, Proceedings, pp.19–22, 3–.5. April 2003, Millpress Science Publishers, Rotterdam, ISBN 90 77017 75 5.

Faltus, F. (1981) Spoje s koutovými svary (Joints With Fillet Welds), Academia, nakladatelství ČSAV (ČSAV Publishing House).

Svoboda, J. & Kala. Z. (2000) Metodika zpřesnění konvenčního výpočtu svarových spojů v aplikaci i na dopravní stavby, Spolehlivost a diagnostika v dopravní technice 2000 (Methodology of Precision of the Conventional Solution of Weld Joints in Application Also to Transport Structures), Reliability and Diagnostics in Transport Technology 2000, Pardubice, ISBN 80–7194–303–7.

Strauss, A., Kala, Z., Bergmeister, K., Hoffmann, S. & Novák, D. (2006) Technologische Eigenschaften von Stählen im europäischen Vergleich, Stahlbau, 75, Januar 2006, Heft 1, ISSN 0038–9145.

9.2 Investigations of Strength and Ductility of Welded High Strength Steel (HSS) Connections

Ulrike Kuhlmann, Hans-Peter Günther, Christina Rasche
University of Stuttgart, D-70569 Stuttgart, Germany,
e-mail: sekretariat@ke.uni-stuttgart.de

Abstract

The developments of steel structures aim at light and slender constructions. Therefore, high strength steels (HSS) with good welding characteristics and a high ductility in addition to higher strength have been developed by the steel industry. With increasing strength of the steel also the loads which have to be transferred in the welded connections, grow. In the building construction industry fillet and partial penetration connections are commonly used. When using HSS it is very important to ensure the strength as well as sufficient ductility of these welded connections. This paper presents results of a research project analysing the strength and ductility of welded high strength steel connections. Especially for S460 existing design rules show a conservative approach which now maybe revised based on experimental as well as numerical tests.

Keywords: *High strength steel, welded connections, fillet welds, filler metal, mismatch.*

1 Introduction

High strength steels (HSS) are a continuously winning market share in the construction industry as they can bring significant savings due to the reduced material consumption. Advantages also result from reduced weld sizes, smaller weight and faster fabrication. With increasing strength of the steel also the loads which have to be transferred in the welded connections, grow. For the building construction industry fillet and partial penetration connections are commonly used. When using HSS it is very important to ensure the strength as well as sufficient ductility of these welded connections.

The calculation method in Eurocode 3 Part 1-8 (2005) requires that the nominal stress $\sigma_{w,Ed}$ within the area of the weld throat section has to be smaller than the weld resistance denoted $f_{vw,d}$:

$$\sigma_{w,Ed} \leq f_{vw,d} \tag{1}$$

It is known (Treiberg T. 1991) that in fact the ultimate strength of the welded connection depends on the weld metal strength. However, within Eurocode 3 the resistance is simply expressed by a function of the tensile strength f_u of the base metal divided by the factor β_w that correlates the weld strength to the base metal strength:

$$f_{vw,d} = \frac{f_u}{\beta_w \cdot \gamma_{Mw}} \quad \text{with} \quad \gamma_{Mw} = 1.25 \tag{2}$$

The correlation factor β_w is given in EN 1993-1-8 (2005) rising from 0.8 for mild steel to 1.0 for high strength steel. Table 1 summarises the design weld resistance $f_{vw,d}$ for fillet welds depending on the structural steel grades according to

EN 10025 (2005). From this table it appears that currently the design resistance of the higher strength steel grade S460 is somewhat lower than of the lower strength steel grade S355. The main reason for the low design resistance of steel grade S460 in comparison to S355 is the limited number of available test results and the large scatter in those (Background Documentation D.03 1990), (Gresnigt 2002).

Table 1. Values for β_w and $f_{vw,d}$ according to EN 1993-1-8 for steel grades according to EN 10025 in case of fillet welds, t ≤ 40mm

Steel grade		S235	S355	S460N	S690
f_y	[N/mm^2]	235	355	460	690
f_u	[N/mm^2]	360	510	540	770
β_w	[-]	0.8	0.9	1.0	1.0
$f_{vw,d}$	[N/mm^2]	360	453	432	616

The rules in EN 1993-1-8 (2005) furthermore require that for all steel grades the strength and ductility of the filler metal (electrodes) might to be at least matching the strength and toughness of the base metal. This requirement is especially for high strength steels unfortunate as undermatching electrodes lead, besides advantages regarding weldability and quality, to a better ductility and redistribution of stresses and thereby leading almost to the same strength as for matching filler metal. Allowing undermatching filler metal for fillet welds would thus enable the designer to choose an optimal combination of base metal, filler metal and welding parameters to comply with the requirements for ductility and strength (Gresnigt 2002), (Collin & Johansson 2005). In order to abolish this unfavourable requirements particular for high strength steels, first efforts have started by (Collin & Johansson 2005) on steel grade S690 using matching and undermatching filler metal. Based on these investigations a first approach concerning the strength of undermatching fillet welds have been included within EN 1993-1-12 (2007) that gives additional rules for extending the scope of Eurocode 3 up to steel grades S700. According to EN 1993-1-12 (2007) it is allowed to use undermatching filler metal for steel grades higher than S460, when calculating the weld resistance according to Eq. 2 by replacing the tensile strength of the base metal f_u by the tensile strength of the filler metal f_{eu}.

2 Experimental Test Programme

2.1 *Scope and overview*

The absence of funded strength functions for steel grade S460 and S690 according to EN 10025 (2005) as well as the restrictions of EN 1993-1-8 (2005) concerning the allowance of using undermatching filler metals was the starting point of a large German national research project realised by four partners (AiF-Vorhaben-Nr. 14195 BG 2007). Aim of this project was the investigation of strength and ductility of welded high strength steel connections by means of experimental and numerical investigations.

A total of 328 weld tests were conducted to investigate the effect of the following parameters: steel grade of base metal S355, S460 and S690, matching level between base metal and filler metal, welding procedure and welding parameters, low temperature, weld geometry concerning thickness and length, joint type, electrode manufacturer and steel fabricator.

2.2 *Test Specimens and Fabrication Procedure*

Table 2 gives an overview on the three general joint types of the test specimens used. Most experimental tests were performed at room temperature on fillet welded lap joints where the welds are orientated parallel to the load direction, as it is well known from other research work (Background Documantation D.03 1990), (Sedlacek & Stangenberg 2002), that this case is the most severe configuration and thus governs the strength function for fillet welds. Additional test series comprised 107 tests on cruciform joints with welds transverse to the loading direction and 108 tests on butt joints with partial penetration welds (Double-Vs) from both sides. A limited number of 34 specimens were tested at low temperature (-40°C). The experimental design included duplicated tests in all cases.

Table 2. General test specimens determining weld resistance

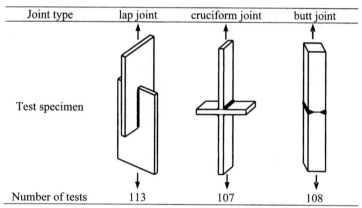

Joint type	lap joint	cruciform joint	butt joint
Test specimen	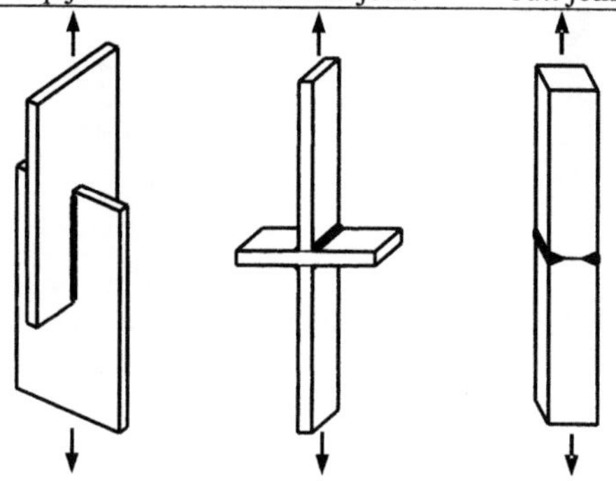		
Number of tests	113	107	108

Most test specimens were fabricated using MAG welding. Welding parameters, cooling rate etc. were chosen following the recommendations for welding of metallic materials according to EN 1011-2 (2001). All welding procedure parameters e.g. wire speed, ampere, voltage, time etc. were well documented in a large data base. This paper will concentrate on the investigations on lap joints. More information on the research work can be found in (AiF-Vorhaben-Nr. 14195 BG 2007), (Hildebrand et al. 2007).

2.3 *Test Procedure for Lap Joints*

Prior to testing detailed measurements of the weld profile, thickness and length as well as non destructive tests such as magnetic particle tests were conducted.

The test set-up used for testing of the lap joints is shown in Figure 1a. The tests were performed in standard tension testing machines with capacity of 1.0 and 5.0 MN. In order to get precise information about the deformation behaviour along the weld an optical displacement measurement system was used in addition to standard strain gauges. For the optical displacement measurement system best results have been achieved by preparing the surface with a kind of a cross pattern.

During the tests the fracture face mostly in the weld itself (see Figure 1b), only in a few cases a failure occurred partially in the weld and partially in the heat affected zone (HAZ) of the base metal. No pure failure of the base metal was observed. After the fracture of the welds, the specimens were carefully removed from the testing machine to avoid any damage to the fracture surface. Details of the fracture surface, including the weld root penetration and the angle of fracture were measured at various locations along the weld length. The fracture surfaces of some of the weld specimens were examined under a scanning microscope in order to identify the nature of the fracture process (ductile or brittle).

a)

b)
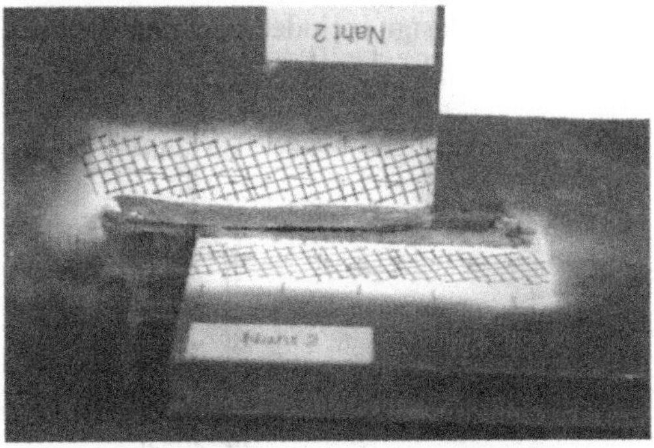

Figure 1. Test set-up and marking of test specimens for optical measurements

2.4 *Test Results of Lap Joints*

For lap joints where the welds are orientated parallel to the loading direction Figure 2 shows a comparison of the test results for steel grades S355J2, S460M and S690Q in terms of the ultimate shear stress τ in the throat section of the weld determined by testing. For all steel grades a matching condition was chosen, meaning the nominal strength of the filler metal is the same as for the base metal.

By comparing the ultimate loads it seems, that for the same nominal geometrical dimensions the connection made of steel grade S460M has a higher strength than S355J2. As shown in Table 1, the design resistance for S355 and S460 is nearly the same. The test results show the conservative approach of the design rules. By comparing the ultimate loads it seems, that the connection made of steel grade S690Q has only a slightly higher strength than S460M. However, this slight increase in strength is related to a loss of ductility as it can be seen in Figure 3. Herein, the load-displacement behaviour is shown, where Δu indicates the relative displacement between the weld ends u_1 and u_2 oat the base metal.

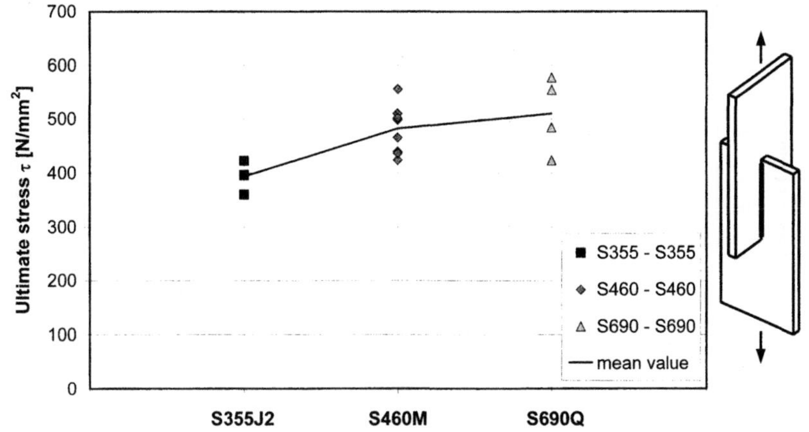

Figure 2. Comparison of ultimate stresses S355J2, S460M and S690Q, matching condition

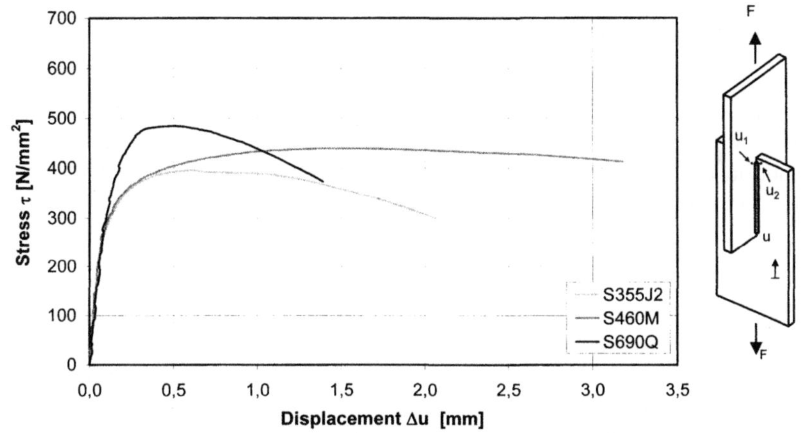

Figure 3. Load–displacement behaviour of steel grade S355J2, S460M und S690Q, matching condition

Figure 4 shows an example where for a connection of steel grade S690 the influence of the filler metal strength on the ductility and strength of the connections has been investigated. Two different strength classes of filler metal G46 with 460 MPa yield strength and G69 with 690 MPa yield strength were chosen. In relation to the S690 base metal, the filler metal G69 represents the matching and G46 the undermatching condition. The load-displacement behaviour in Figure 4 clearly indicates that in case of using undermatching filler metal the connection behaves much more ductile while having almost the same ultimate load resistance.

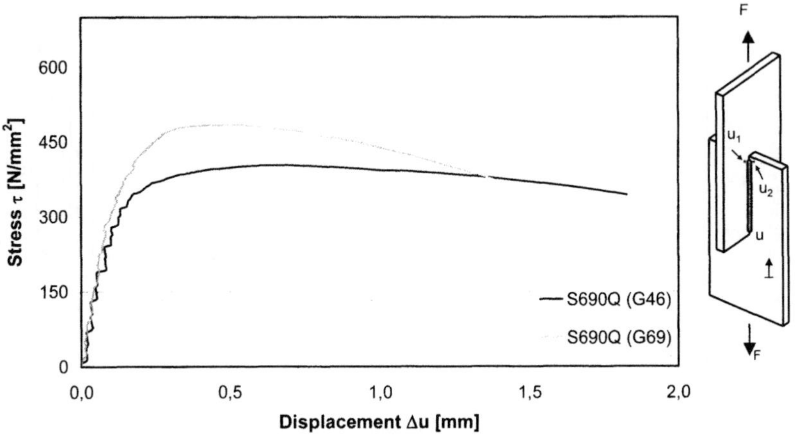

Figure 4. Influence of filler metal strength on load–displacement behaviour of steel grade S690)

3 Numerical simulations

Aside of the experimental tests, numerical investigations have been performed. The numerical simulations were subdivided among the different project partners and include welding simulations providing information upon e.g. cooling rates, distribution of microstructure and hardness and residual stresses after the welding process, strength simulations to determine the ultimate load and deformation behaviour and fracture mechanic calculations in order to specify the safety against brittle fracture at low temperature.

First results of ultimate load simulations performed by the authors are presented in Figure 5. The Finite Element (FE) simulations were performed using the commercial software package ANSYS® 11.0 and were verified against the load-displacement and load-strain measurements received by the experimental tests. The FE simulations use a geometrical non-linear level with a rate independent multi-linear *von Mises* elastic-plastic isotropic hardening model. Two different material properties (weld metal and base metal) are covered where the single properties are derived by standard tension test coupons. Figure 5 shows the load-displacement diagram obtained by tests and FE simulation of a standard test specimen of steel grade S460M with matching filler metal. Based on the multi-linear material properties the ultimate load is predicted with an accuracy of about 7%. The yielding and stress redistribution were analysed and show a good agreement between simulation and test.

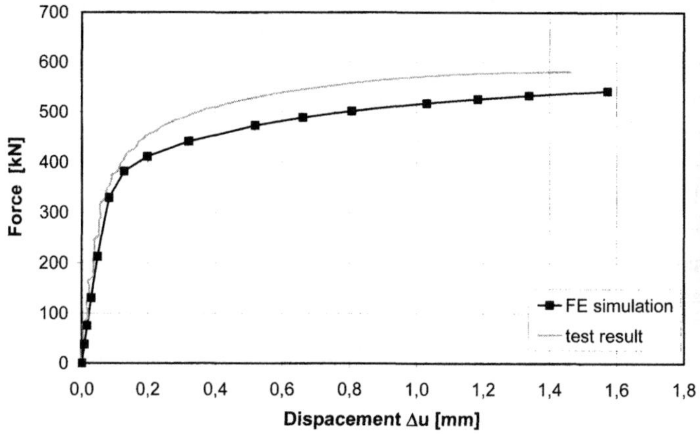

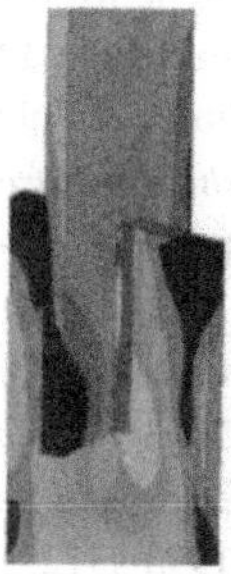

Figure 5. Load–displacement diagram, FE simulation vs. test result

4 Conclusions

The paper presents results of a research project dealing with the strength and ductility of welded high strength steel connections. It is shown, that the present design rules in Eurocode 3 (steel structures) present a rather conservative approach especially for S460M and are inadequate to cover e.g. mismatch effects. The described experimental and numerical investigations indicate the need to update the existing design rules in terms of improved strength functions taking directly into account the strength of the filler metal. For fillet welded connections of high strength steels it has been shown that undermatching filler metals are especially favourable in order to gain more ductility through a better redistribution of stresses. Thus, the results may lead to an improvement of the design and safety of welded high strength steel connections in the construction industry.

5 Acknowledgements

This project is realised by four German research partners: Universität Stuttgart, Bauhaus-Universität Weimar, Technische Universität Darmstadt and the Günter-Köhler-Institute, IFW Jena. Detailed information about the research project may be found under: *http://www.uni-stuttgart.de/ke/AiF_14195.html.* Special thanks are given to AiF and FOSTA for the financial support and to all the industry partners for the supply of material and the welding work done preparing the test specimens.

6 References

AiF-Vorhaben-Nr. 14195 BG (2007) *Wirtschaftliche Schweißverbindungen höherfester Baustähle,* Arbeitsgemeinschaft industrieller Forschungsvereinigungen e.V. (AiF), FOSTA P652, Runtime 2005-01-01 until 2007-01-31 (in German).

Background Documantation D.03 (1990), *Evaluation of Tests Results on Welded Connections made from Fe E 460 in order to obtain Strength Functions and suitable Model Factors*, Eurocode 3 Editorial Group.

Collin P. & Johansson, B. (2005) Design of welds in high strength steel. In: *Proceedings of the 4th European Conference on Steel and Composite Structures,* Maastricht, Volume C, pp. 4.10-89 – 4.10-98.

EN 1011-2 (2001) *Welding – Recommendations for welding of metallic materials, Part 2: Arc welding of ferritic steels*, European Standard, CEN, Brussels.

EN 1993-1-8: Eurocode 3 (2005) *Design of steel structures – Part 1-8: Design of joints*, European Standard, CEN, Brussels.

EN 1993-1-12: Eurocode 3 (2007) *Design of steel structures – Part 1-12: Additional rules for extension of EN 1993 up to steel grades S700*, European Standard, CEN, Brussels.

EN 10025 (2005) *Hot rolled products of structural steels,* European Standard, CEN, Brussels.

Gresnigt A.M. (2002) Update on Design Rules for Fillet Welds. In *Proceedings of the 3rd European Conference on Steel Structures, Coimbra –Portugal*, pp. 919-927.

Hildebrand J., Wudtke I., Werner F., Rasche C. Günther H.-P., Kuhlmann U., Versch C., Vormwald M. (2007) Wirtschaftliche Schweißverbindungen aus höherfesten Stählen im Bauwesen. In *2. Konferenz – Gestaltung und Konstruktion, SLV Halle, Halle –Germany*, pp. 50-55 (in German).

Sedlacek, G. & Stangenberg, H. (2002) Design Philosophy of Eurocodes – Background Information. *Journal of Constructional Steel Research* **54** 173-190.

Treiberg T. (1991) "Influence of Base and Weld Metal Strength on the Strength of Welds" Connections in Steel Structures II. Behaviour, Strength and Design. In: *Proceedings of the Second International Workshop held in Pittsburgh*, pp. 46-53, AISC

9.3 Design of Welded Joints - Determining Stress State

László Molnár[1]; Károly Váradi, [1]; Balázs Czél[1]; László Oroszváry[2] and Csaba Molnár[2]

[1]Budapest University of Technology and Economics, H-1111 Budapest, Hungary, e-mail: mol@eik.bme.hu, [2]Knorr Bremse Hungária Kft, H-1202 Budapest, Hungary

Abstract

The VDI Richtlinien titled "Rechnerischer Festigkeitsnachweis für Maschinenbauteile" was published by Forschungskuratorium Maschinenbau (FKM) in 1998; it provides designers with a design method of a standardized approach on a new basis, even for the stress analysis of welded structures. One of the essential criteria for designing welded seams for static stress and fatigue is knowledge of the stress state in the seam and its surroundings. Erwin Haibach proposes to determine the so-called structural stress characterizing stress state of welded seams along the Haibach line. In our presentation, a brief introduction to the designing of welded structures according to FKM will be followed by a presentation of results of FE analyses to demonstrate the use of stresses interpreted along the Haibach line as structural stresses. Our research and development activities were performed on assignment by and in cooperation with Knorr Bremse Hungária Kft.

Keywords: *welded joints, structural stress, Haibach line*

1 Introduction

Basically, **FKM** design guidelines are calculation algorithms built up in a standardized manner for various cases of application, which consist of rules, calculation procedures, correlations, and factors used for calculations. The calculation steps of static stress analysis and fatigue analysis are substantially concurrent with each other. The suitability of the component / welded joint examined is demonstrated by the **degree of utilization**. The degree of utilization is the ratio of the stress (σ) and the limit stress (R) modified by a safety factor:

$$a = \frac{\sigma}{R / j_{erf}} \leq 1 \, , \tag{1}$$

where j_{erf} is a safety factor depending on the structure and the mode of stress. The highest value of the utilization factor can be 1. If the utilization factor is higher than 1 (or 100 %), then the component or welded joint is not suitable in terms of static or fatigue strength.

The following is a discussion of determining the stress state of welded joints.

2 Stress state of welded joints

Welded components should be considered as 2D, surface-like components as the highest stresses are generated on the surface. For surface-shaped structural components, stress analysis / design for fatigue must be performed both for normal stress in directions x and y as well as for shearing stress in xy direction. Therefore, characteristic stress components are:

- in case of *nominal stress*: S_x, S_y, T_{xy} ;

– in case of *structural stress*: $\sigma_{x,max}$, $\sigma_{y,max}$, $\tau_{xy,max}$;
– in case of *effective notch stress (total stress)*: $\sigma_{Kx,max}$, $\sigma_{Ky,max}$, $\tau_{Kxy,max}$.

Figure 1 shows the interpretation of *nominal*, *structural* and *effective notch stresses.*

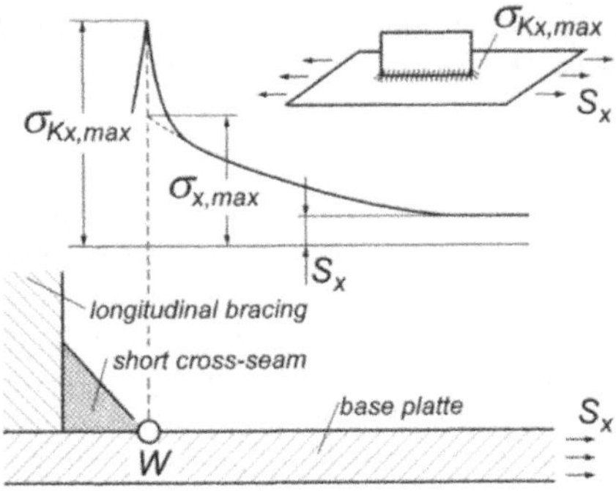

Figure 1. 2D welded component. Longitudinal belt sheet with bracing (FKM 1998).

According to the marks in the figure:

S_x — *Nominal stress.* Average stress, calculated in analytical way, using the dimensions of the structure;

$\sigma_{x,max}$ — *Structural stress.* Maximum stress generated directly before the welded seam. Extrapolated from the stresses generated before the welded seam.

$\sigma_{Kx,max}$ — *Effective notch stress.* A stress taking the impact of stress concentration into consideration: maximum stress directly in the welded seam, calculated with a fictitious rounding radius of $r = 1$ mm at weld toe or weld root (It can only be used in design for fatigue.)

An example is provided for the interpretation of *nominal*, *structural* and *effective notch stresses* by the welded sheet with a hole shown in *Figure 2*.

The example shown in *Figure 2* basically corresponds to a uniaxial stress state, where *nominal stress* (S_x) can be calculated from the actual cross-section:

$$S_x = \frac{F}{(b-d)s}, \qquad (2)$$

Structural stress ($\sigma_{x,max}$) can be determined by a numerical method, e.g. by a 2D FE method, taking into consideration the impact of the stress concentration at the hole, but specifying stress distribution on a non-welded component.

Effective notch stress ($\sigma_{Kx,max}$) can also be determined by an FE method, but with a 3D model which models the welded seam as well, with the generally accepted fictitious rounding of $r = 1$ mm. Let us note again that this stress can only be used for design for fatigue strength.

Based on a literature review, it can be established that the majority of researchers agree that the *effective notch stress state* of a welded seam can be determined by a so-called notched model. In this procedure, the geometry of the component is taken into consideration together with the seam in the course of the calculation, modelling the impact of seam transition by a notch radius resembling the actual welded joint.

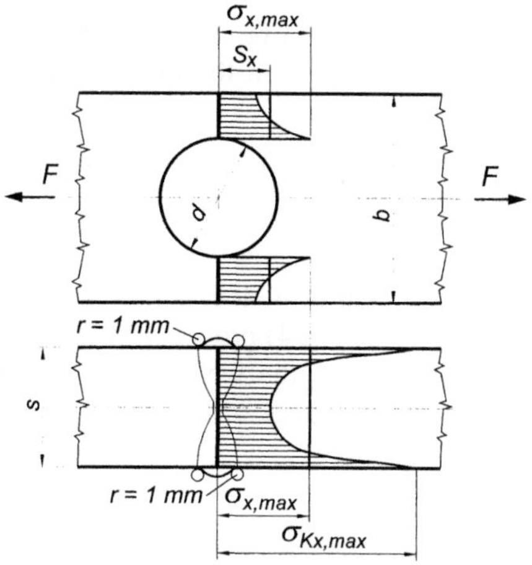

Figure 2. Welded sheet with a hole to interpret *nominal*, *structural* and *effective notch stresses* (FKM 1998).

Practical experiences show that real conditions can be properly approximated by modelling the transition with a radius of r = 1mm (*Figure 3*). Modelling the impact of the stress concentration of the weld toe by means of this radius, the peak stresses arising can be calculated by a numerical method.

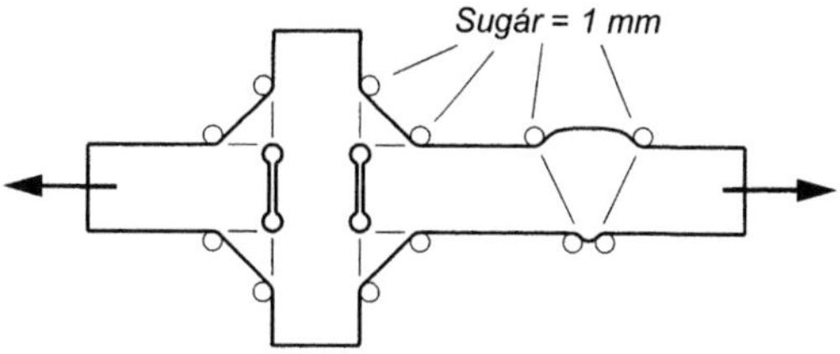

Figure 3. Modelling of weld toes and roots with radius r = 1 mm (IIW 2003).

Benefits of the procedure include high calculation accuracy and versatility of use. This procedure can be used for determining the *effective notch stress peak* (total stress) at the weld toe.

A disadvantage of this method is that an extremely fine FE mesh is required for calculating the actual stress distribution, which implies a considerable demand of calculation time and disk space, and this may cause serious difficulties when modelling larger-size structures.

Difficulties arising from the application of the *effective notch stress* concept led to the development of the *structural stress concept.* Structural stress is determined by taking into consideration the geometry of the real component, by neglecting the impact of the welded seam – more specifically, by taking it into consideration in a highly simplified manner. The *structural stress state* is determined by an FE model, using a relatively coarse mesh, and stress peaks at critical locations are approximated by the extrapolation of stress distribution.

Structural stress (or hot spot stress) can be determined by the extrapolation of the stresses calculated at given distances from the point of hot spot – at the so-called reference points (*Figure 4*).

Figure 5 shows the interpretation of structural stress in case of a butt seam. The hot spots at the top and the bottom side, respectively, do not fall into a single vertical line.

Literature suggestions differ as regards the location of reference points. International Institute of Welding (IIW 2003 and 2004) proposes the distances of 0.4*t* and 1.0*t* for the distance of reference points from the weld toe in the case of finer FE mesh; however for larger elements the distances of 0.5*t* and 1.5*t* (t is the thickness of the sheet) is recommended.

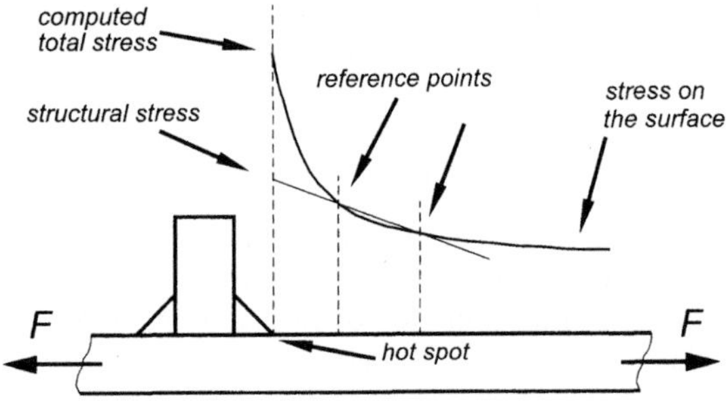

Figure 4. Determining the structural stress in case of an fillet seam (IIW 2003).

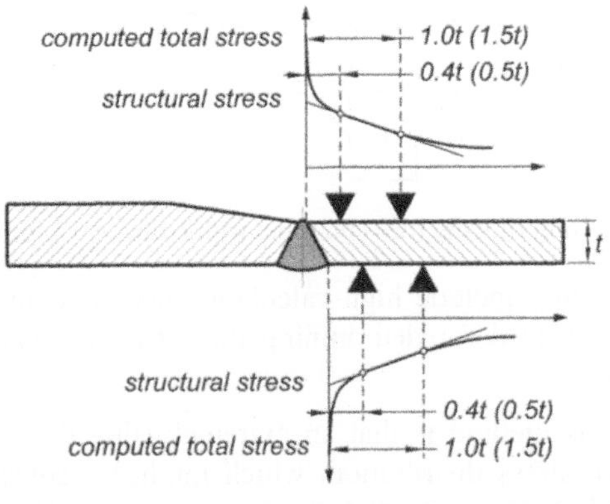

Figure 5. Interpretation of structural stress for a butt seam (Ibso 2006)

As regards designing for structural stress, the shaping of the seam, its technology – meaning, in effect, the relationship between effective notch stress and structural stress – are taken into consideration with an FAT (fatigue class) qualification defined by a number of fatigue tests.

In order to determine structural stress by extrapolation, stresses must be selected at two or three reference points (or along such lines parallel with the seam). *Haibach* proposes a point (or line) from the weld toe where the calculated stress yields directly the structural stress. *Figure 6* shows the location of the Haibach line compared to the welded seam (Haibach 1968).

The distance D_H of structural stress location from the toe of the seam – the so-called *Haibach* distance – is

$$D_H = D_H^* \approx 2{,}5 \text{ mm}. \tag{3}$$

In the case of shell models, there is a discussion between the researchers, if the *Haibach* distance is measured from the weld toe (D_H), or from the so-called singularity point (D_H^*). The former is lower by 5-10 %, the latter is higher by 5-10 % than stress calculated by solid model.

A benefit of introducing the Haibach line is that instead of stress selection at two reference points (along two lines) and stress extrapolation, it is only necessary to perform a stress selection at one point (along one line).

An important task of our investigation is to demonstrate the applicability of the Haibach approach.

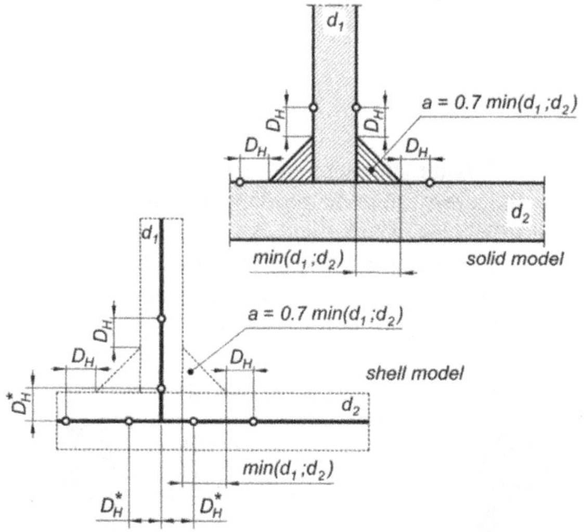

Figure 6. Position of the Haibach line compared to the weld toe in case of a solid and a shell model

3 Structural stress along the Haibach line

The investigation was performed by calculating the stress state of two sheets perpendicular to each other in the vicinity of the welded seam on the basis of various FE models, and by comparing the stresses calculated along the Haibach line with each other, on one hand, and with traditional methods for determining stress, on the other hand. It was also examined how the element type, the element size, and the boundary conditions affect the stress values calculated.

Figure 7 shows the geometry, load and boundary conditions of the model examined.

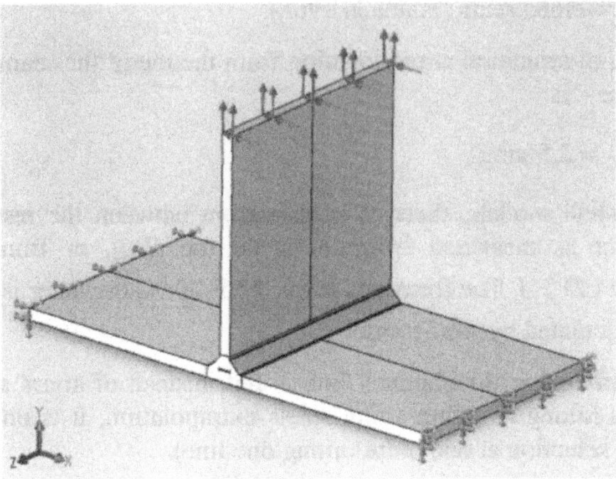

Figure 7. Geometry, loading and boundary conditions to examine an fillet seam.

Table 1 summarizes the models examined. The factor S_D is used for comparing the models (FKM 1998):

$$S_D = \frac{FAT}{225} \cdot \frac{92}{\sigma_{max}} , \tag{4}$$

where fatigue class *FAT* is the characteristic fatigue strength of the detail at $2x10^6$ cycles; *FAT=225* is the stress range or double-amplitude of the fatigue strength at the reference number of cycles of $2x10^6$ and determined experimentally according to IIW; *92* is the specific fatigue limit value of welds in steel for completely reversed normal stress at a number of cycles $5x10^6$ ant it corresponds to *FAT=225* ; and σ_{max} is the highest stress of the line examined (weld toe or Haibach line).

Table 1. Comparison of FEM models based on Haibach stress

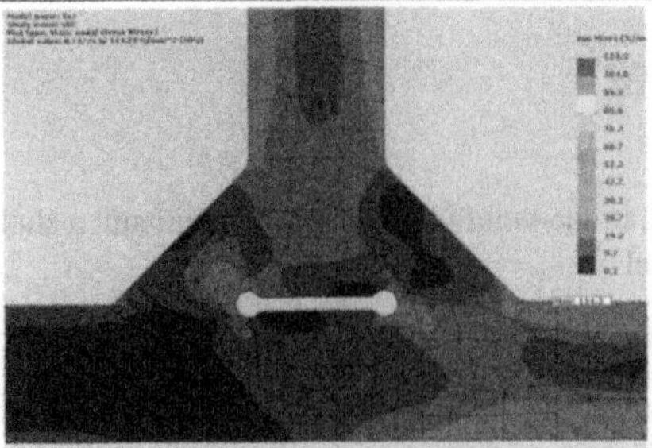	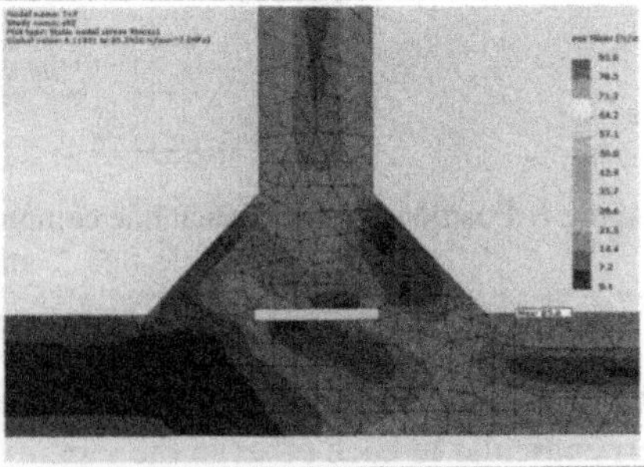
Tv1 – S_D = 225/225*92/140 = 0.65	Tv2 – S_D = 100/225*92/65 = 0.63

Table 1 continues

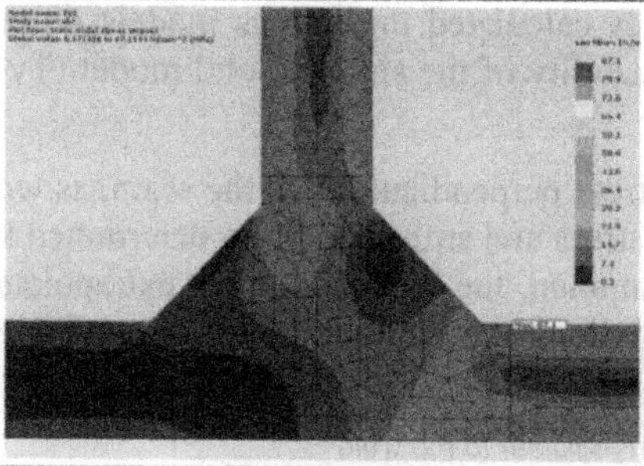

Tv3 – S_D = 100/225*92/65 = 0.63

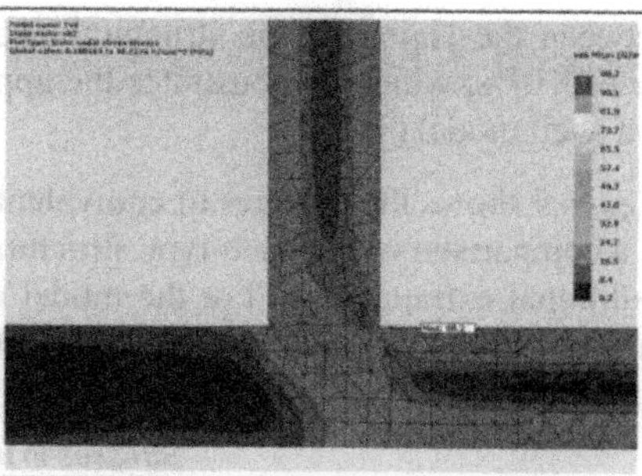

Tv4 – S_D = 100/225*92/60 = 0.68

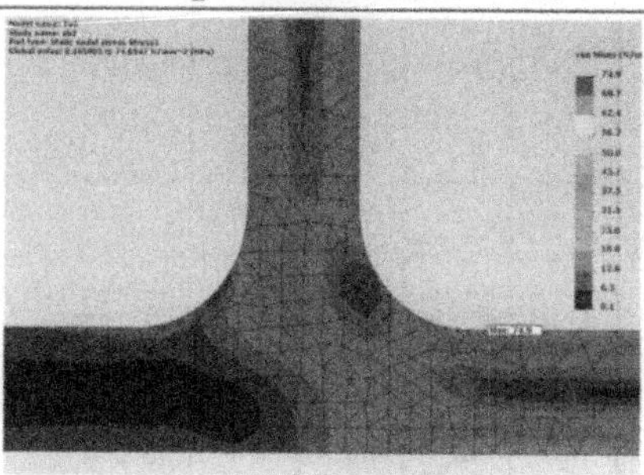

Tv5 – S_D = 100/225*92/62 = 0.66

TvS – S_D = 100/225*92/59 = 0.69

The first model in the table serves for determining the effective notch stress, while, for the sake of comparison, stresses along the Haibach line were also determined in each case. The last shell model is intended to demonstrate the applicability of shell elements for structural stress calculations.

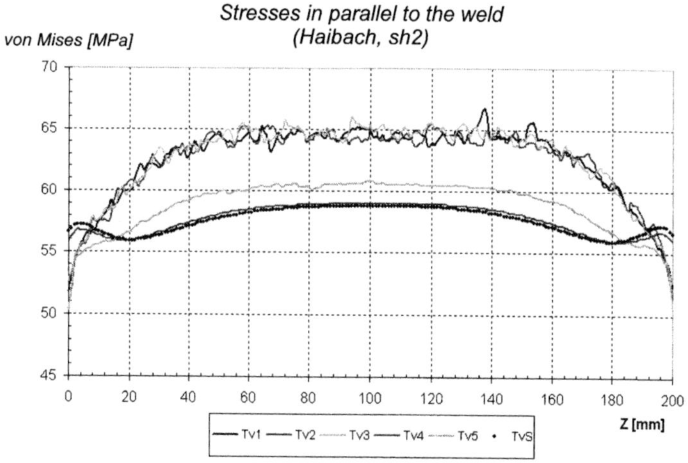

Figure 8. Stress of the six kinds of models along the Haibach line.

Figure 8 shows the equivalent stress distribution of the six kinds of FE models along the Haibach line parallel with the seam. The figure clearly shows that discrepancies between the Haibach-type structural stresses calculated on various models do not exceed 10%, which demonstrates the applicability of the simpler solid model (Tv4) and shell model (TvS).

Figure 9 shows the changes of equivalent stress perpendicularly to the seam, as well as a comparison of Haibach-type structural stress and structural stress determined by traditional extrapolation. For the model examined, the stress value for extrapolated case was 64 MPa (blue and red lines), and along the Haibach line it was 65 MPa (green line).

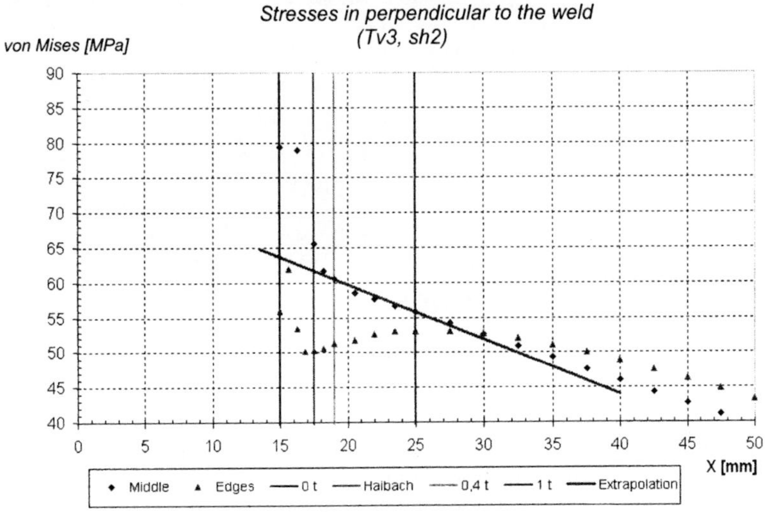

Figure 9. Determining structural stress.

Our analysis performed demonstrated that stress calculated along the Haibach line can be considered as structural stress and that the shell model is also suitable for determining structural stress.

References

Common Structural Rules for Bulk Carriers. (2006) Chapter 7. Direct Strength Analysis.

FKM (1998) Rechnerischer Festigkeitsnachweis für Maschinenbauteile. *Forschungskuratorium Maschinenbau Richtlinie.*

Haibach E. (1968) Die Schwingfestigkeit von Schweissverbindungen aus der Sicht einer örtlichen Beanspruchungsmessung. *Laboratorium für Betriebsfestigkeit. Darmstadt.* Bericht Nr. FB-77

IIW (2003) Recommendations for fatigue design of welded joints and components. *International Institute of Welding.* IIW document XIII-1965-03 /XV-1127-03

IIW (2004) Recommendations on fatigue of welded components. *International Institute of Welding.* IIW document XIII-1539-94 / XV-845-94

Ibso J. B. (2002) Fatigue design of offshore wind turbines and support structures. DNV.

Morgenstern C. (2006) Kerbgrundkonzepte für die schwingfeste Auslegung von Aluminiumschwessverbindungen. Darmstadt

NORSOK Standard N-004 (1998) „Design of Steel Structures" 1.Dec. 1998.

9.4 Finite Element Modelling of Fillet Weld Connections

Lucjan Ślęczka, Aleksander Kozłowski
Rzeszów University of Technology, ul. Poznańska 2, 35-082 Rzeszów, Poland,
sleczka@prz.edu.pl

Abstract

The paper describes the finite element modelling method employed to analyse welded connections. The strength phenomena of fillet welds is based on ductile crack initiation and propagation, so finite element model should include proper modelling of stress-strain relationship, including hardening and softening effects, incorporation of geometrical non-linearity and different material models for base and weld metals and also for heat affected zone. The occurrence of large plastic strains requires the use of the true stress-strain relationship. Non-linear stress-strain curve gives possibility to model local yielding and redistribution of stresses. "Element dead" option available in the FE programs traces initiation and propagation of ductile crack in these places, where ductility of welds is exhausted.

Keywords: *Steel structures, connections, fillet welds, FE-modelling.*

1 Introduction

Fillet welds are one of the most common components of steel connections. In structural application, they are usually designed with some level of overstrength, due to their small ductility, compared with other joint parts. The strength criteria for statically loaded connections used in various codes were developed in last century, based on results of extensive tests, (Design rules 1976). Such design rules in many cases are likely conservative because of the large scatter of the test results and limited number of tests. Now, finite element simulations give possibility to model accurate behaviour of fillet weld connection up to failure. In many cases such failure analysis estimating the real strength is necessary. The emphasis in such analysis is usually placed on the assessment of overall behaviour, Jamshidi & Birkemoe (2005), Rodriguez et al. (2004), Mellor et al. (1999). Only several researches have focused on ductile failure of welded connection. Such studies, e.g. Wang et al. (2006), have been performed mainly for aluminium joints, where fracture occurs indeed in heat affected zone (HAZ). Failure analyses of welded connections are usually performed with special user defined constitutive material model, which makes them difficult to perform for non-highly advanced users. "Element dead" option in the commercially available FEM programs gives possibility to model initiation and propagation of ductile crack of welds in easy way, provided that some basic phenomena are included in numerical model.

2 Basic phenomena in FE modelling of welded joints

Finite element modelling of weld connections has to take into account some basic phenomena to achieve accurate results.

2.1 *Material heterogeneity*

The welded joint consists of three different zones: weld metal, base metal and heat affected zone. The weld and base metal are usually chemically and structurally compatible, but in typical steel building structures the nominal material properties of weld metal are usually higher then those of base metal. The weld metal strength level and their ductility is also influenced by welding conditions. The heat affected zone (HAZ) is area adjacent to the weld, where properties of base metal are changed by heat input. For medium strength steel grade the yield strength and ultimate strength of HAZ are usually increased. Moreover, the specific regions exist in HAZ zone with evidently different microstructure and mechanical properties. The range of HAZ depends strongly on welding thermal cycle. Normally, the width of HAZ zone is narrow, usually has less than 2÷3 mm, after application of modern welding technologies. Figure 1 shows zones of weld metal, base metal and HAZ in fillet weld connection. Such simplified division can be used in FEM modelling.

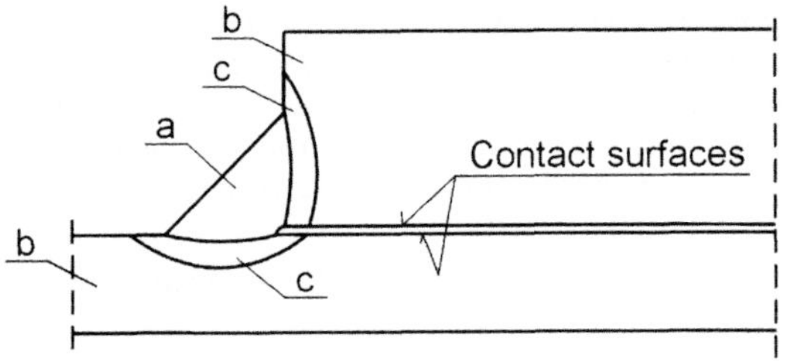

Figure 1. Zones and contact surfaces in fillet weld connection: a) weld, b) base metal, c) heat affected zone (HAZ).

2.2 *Stress – strain relationship*

The behaviour of fillet welds is affected by yield strength f_y, but in particular the strength is based on ultimate strength f_u and ultimate strain ε_u, so stress-strain relationship should be modelled in true (not engineering) values. Figure 2 shows simplified multilinear true stress – true strain σ-ε curve, which can be implemented in finite element models, where large plastic deformation occurs. Last branch of this line in σ-ε relationship, where values of stress exceed f_u, simulates the behaviour of material after the necking. According to Faella et al. (2000) Table 1 gives suggested values of σ-ε curve parameters for constructional steels.

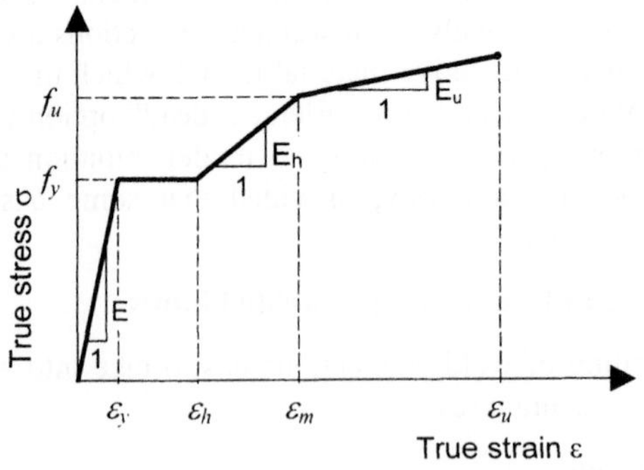

Figure 2. σ-ε multilinear constitutive law.

Table 1. σ-ε curve parameters for constructional steels (Faella et al. 2003)

Steel grade	f_y MPa	f_u MPa	$\varepsilon_h/\varepsilon_y$ [-]	E/E_h [-]	E/E_u [-]	$\varepsilon_u/\varepsilon_y$ [-]	ε_u [%]
S235	235	360	12.3	37.5	523.2	$\geq$ 357	$\geq$ 40
S275	275	430	11.0	42.8	447.6	$\geq$ 425	$\geq$ 56
S355	355	510	9.8	48.2	381.7	$\geq$ 479	$\geq$ 81

2.3 *Geometrical nonlinearities and process of fracture*

FE analysis of ductile fracture requires consideration of the material properties at the necking phenomenon, which can be reached by stress – strain curve presented in Figure 2, and also by application of the true ultimate strain ε_u. The ultimate strain ε_u corresponds to the last point input for the strain-stress curve. When such strain is reached in given element integration point, the corresponding finite element is removed from FE model. This "element dead" option based on ultimate strain, gives possibility to model initiation and propagation of ductile crack in easy way, in the commercially available FEM programs. The kinematic description in such nonlinear analysis should take into account large displacement/large strain formulations.

2.4 *Contact surfaces*

The friction forces between plates in lap, welded connections are small enough to neglect them. But modelling of frictionless contact conditions between plates (Figure 1) can be beneficial. It will prevent the adjoining plates from penetration, when their deformations occur, during load increase.

2.5 *Residual stresses*

It is well known that after welding residual stresses occur in the connection. They usually produce tension in the weld metal and compression in the surrounding base metal, in case of one-pass welding. Because the resultant of residual tensile stresses is compensated by resultant of residual compressive stresses, overall capacity of welded joint is the same as for joint without residual stresses, so there is tendency to omitting them in numerical simulations. But in case of large three dimensional connections, residual stresses cause biaxial or triaxial stress state, and ductility of the joint can be decreased.

3 Validation of finite element modelling of ductile fracture

All numerical calculations were performed using the FE package ADINA 8.3, (ADINA R&D, Inc. (2005)). Figure 3 shows comparison of FE modelling results of rounded specimen with 4 mm diameter with test results. The main aim of this analysis was comparing numerical results of ductile fracture with data from tests. The specimen was cut out from weld metal, done using E42 4 B42 electrodes. Its mechanical characteristic were as follow: yield strength f_y=425 MPa, ultimate strength f_u=550 MPa and ultimate strain ε_u=ln(A_0/A_f)=1,38, where A_0, A_f are initial section area of the specimen and final section area (after the fracture) respectively. Shape of stress-strain curve was modelled according to Table 1. Material model is based on the Huber-von Mises yield condition, an associated flow rule using the Huber-von Mises yield function and on an isotropic hardening rule. One half of the specimen was modelled, using the 2D, 9 nodes axisymmetric elements. The size of mesh was as 0,2x0,2 mm.

In spite of simplified σ-ε curve modelling as in Figure 1, quite good accuracy is achieved in post critical behaviour (Figure 3 and 4). The difference in elongation measured on 60 mm base is 18% between test and FEM results, and the difference between ultimate force (when total fracture occurs) is only 0,1%.

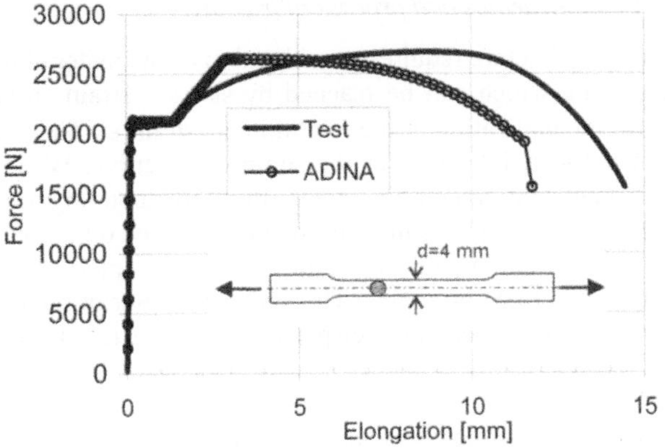

Figure 3. Modelling of tension test of rounded specimen.

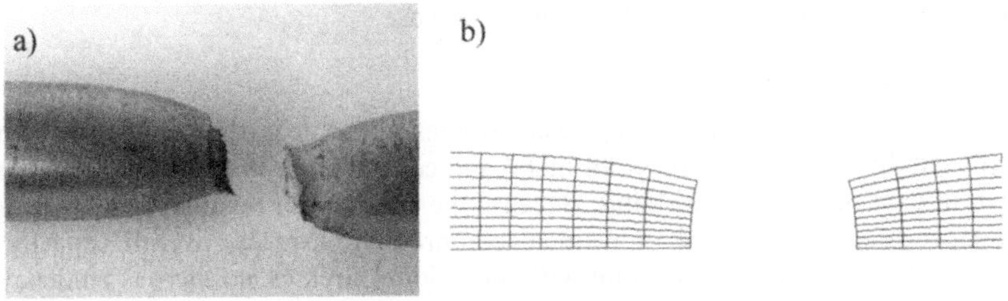

Figure 4. Necking of the specimen a), and necking of FE model b) after the fracture.

4 Parametric study of lap connections with transverse fillet welds

4.1 *Effect of weld thickness*

A view of investigated lap connections with fillet welds is shown in Figure 5. The joint consist of three plates, which are connected using two fillet welds. There were analysed connections with varying weld thickness: a=0,2t, a=0,3t, a=0,4t, a=0,5t and a=0,6t, where thickness of the plate t is equal to 12 mm. Due to symmetry only half of the joint was modelled, Figure 6. One multilinear σ-ε material model was used to simulate weld, base and HAZ metal, with parameters described in previous chapter (in joint with weld thickness a=0,6t central plate was modelled using elastic material model to ensure weld failure). Average size of FE elements in weld zone was equal to 0,25x0,25 mm. 2D, plane strain, 9-nodes elements were used with thickness equal to 1,0 mm. The frictionless contact was assumed between upper and the central plate. No residual stresses have been modelled.

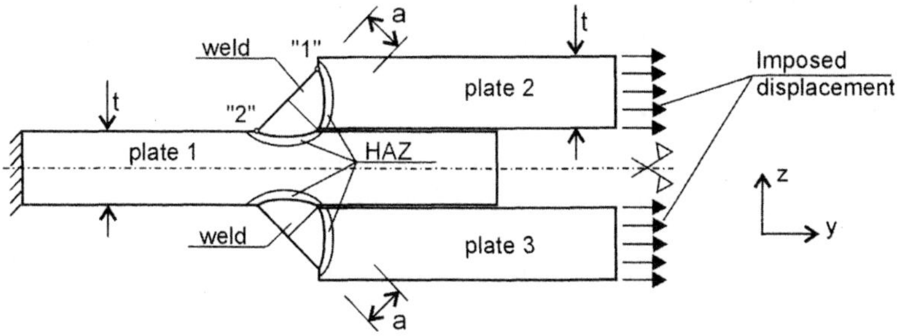

Figure 5. General view of analysed fillet weld connection.

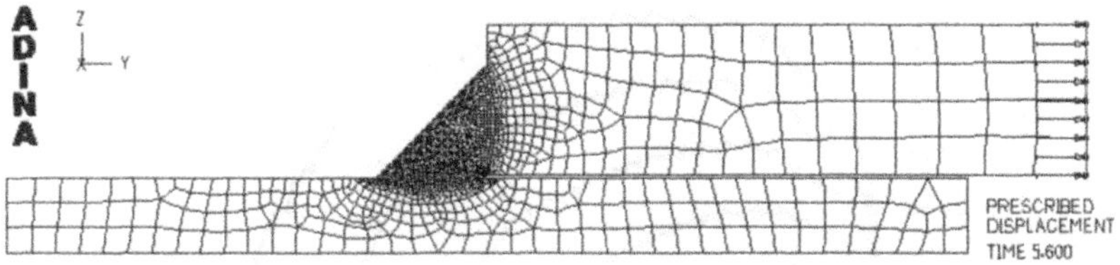

Figure 6. FE model of the joint.

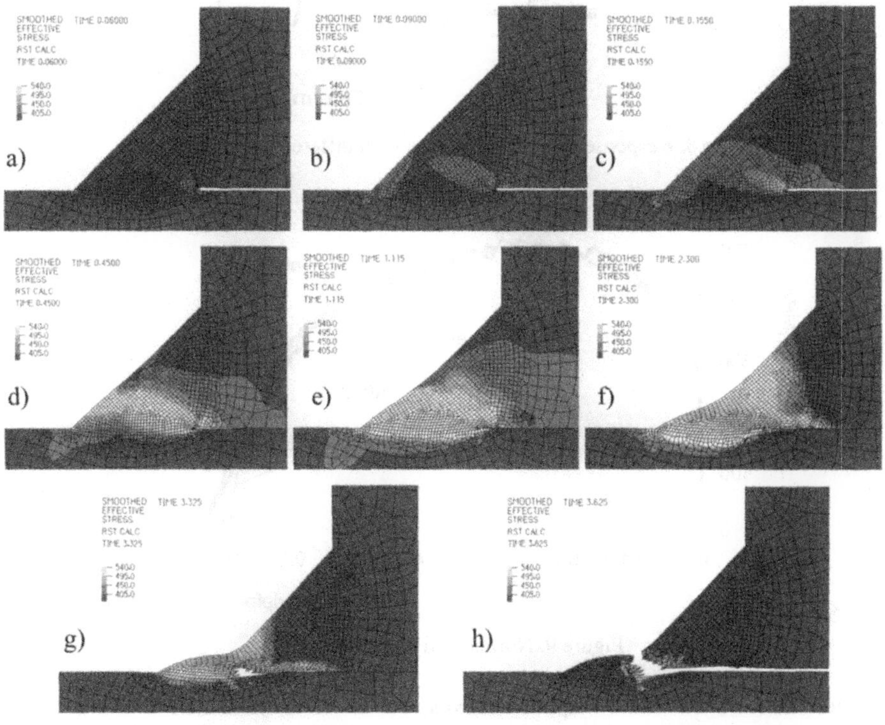

Figure 7. Evolution of yielding and fracture of weld a=0,4t with applied load: a) elastic stage F=1004 N, b) elastic-plastic stage F=1445 N, c) F=1673 N, d) F=2170 N, e) $F=F_{max}=2639$ N, f) propagation of fracture F=2292 N, g) F=1268 N, h) total fracture F=0 N.

The weld failure mode obtained from FE simulation is shown in Figure 7. It is based on tearing the parent metal along the weld contour, parallel to direction of loading. This form of fracture is very similar to that observed experimentally, Grondin at al (2002), Callele at al. (2005).

Figure 8 shows joint response curves with different weld thickness, where weld deformation Δ is assumed as measured difference between displacement of point "1" and "2" in direction of loading (Figure 5). It can be noticed that analysis show the same initial stiffness of the welds, approximately equal 50000 N/mm^2 in elastic stage.

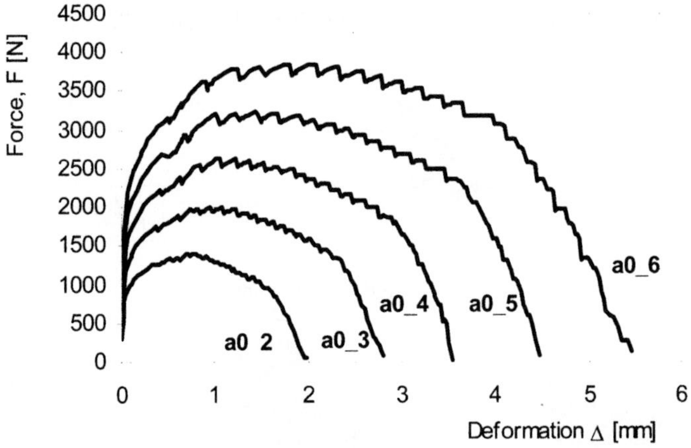

Figure 8. Response of the connections with different weld thickness.

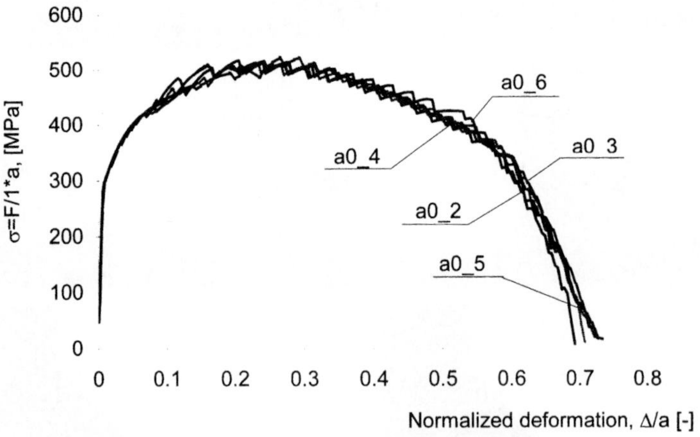

Figure 9. Normalized response curves.

Figure 9 shows the same response curves, but in normalized values. Normalized deformation is obtained by dividing measured deformation of weld Δ by their thickness *a*. Values at ordinate axis show stress at throat of the weld, $\sigma = F/1 \cdot a$. As can be seen, average normalized deformation Δ/a at ultimate load is equal approximately to 0,2, which is value quite close to this obtained by Grondin et al. (2002).

4.2 *Effect of HAZ modelling*

The joint with weld thickness $a=0,4t$ was selected for further comparison. Figure 10 shows the joint response curves with, and without HAZ modelling. Model "a0_4" has only one material constitutive model for base, weld and HAZ metal. Model "a0_4_HAZ_1" has one material model for weld and base metal, with mechanical properties as described in chapter 3, and the second one with $f_{y,HAZ}$=625 MPa, $f_{y,HAZ}$=750 MPa and $\varepsilon_{u,HAZ}$=0,9 for HAZ metal. Also models with decreasing value of $\varepsilon_{u,HAZ}$, equal to $\varepsilon_{u,HAZ}$=0,7 and 0,5 respectively were analysed. But in these cases, there was no difference in total joint response compared to "a0_4_HAZ_1" model. As can be seen, increase of yield strength of HAZ decreases a little deformation of weld. The weld failure pattern, is independently of HAZ modelling.

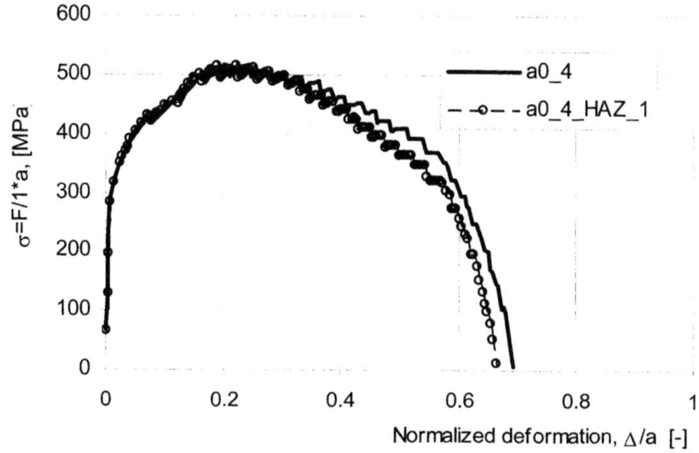

Figure 10. Effect of HAZ modelling.

4.3 *Effect of weld metal ductility*

The next analysed parameter was influence of weld metal ductility on behaviour of connection. Also the joint with weld thickness $a=0,4t$ was selected for comparison, marked as "a0_4" (Figure 11). The next models, "a0_4_weld_1" ÷ "a0_4_weld_3", have the same geometry, and also one material model for weld, base and HAZ metal properties. The shape of σ-ε material constitutive law was constant, but values of ultimate strain was decreasing from ε_u=1,0 for "a0_4_weld_1", ε_u=0,8 for "a0_4_weld_2" and ε_u=0,7 for "a0_4_weld_3". It can be seen, that along with decreasing weld metal ductility, also ultimate capacity (maximum stress) is decreasing.

5 Conclusions

The ultimate capacity of fillet weld connections is governed by the strength of weld metal f_y, f_u and its ultimate strain ε_u. Described above methods of FE modelling give possibility to analyse fillet weld connections in fast and easy way. Average total calculation time of described above FE models was equal approximately to 400s on the desktop computer.

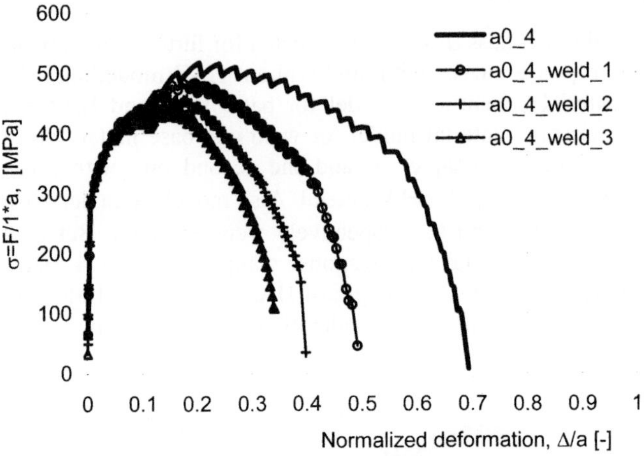

Figure 11. Effect of weld metal ductility.

References

ADINA R&D, Inc. (2005) *ADINA Theory and Modeling Guide*. Report ARD 05-7. Watertown, MA 02472 USA

Callele, L. J., Grondin, G. Y. & Driver, R.G. (2005) *Strength and behaviour of multi-orientation fillet weld connections*. Structural Engineering Report No. 255, University of Alberta

Design rules for arc-welded connections in steel submitted to static loads. (1976) *Welding in the World*. **14** No. 5/6

Faella, C., Piluso, V. & Rizzano. (2000) *Structural steel semirigid connections. Theory, Design and Software*. CRC Press

Grondin, G.Y., Driver R.,G. & Kennedy D. J. L. (2002) *Strength of transverse fillet welds made with filler metals without specified toughness*. Research report, University of Alberta

Jamshidi, A.K. & Birkemoe, P.C. (2005) Strength analysis of fillet welded steel connections. In: *4th European Conference on Steel and Composite Structures, Maastricht, June 8-10*, Volume B, paper 1.12.

Mellor, B.G., Rainey, R.C.T. & Kirk, N.E. (1999) The static strength of end and T fillet weld connections. *Materials and Design* **20** No.4, 193-205.

Rodriguez, J. L., Alvarez, E. R. & Moreno, F. Q. (2004) Study of the distribution of tensions in lap joints welded with lateral beads, employing three dimensional finite elements. *Computers and Structures* **82** No.15-16, 1259-1266.

Wang, T., Hopperstad, O. S., Larsen, P. K & Lademo, O.G. (2006) Evaluation of a finite element modelling approach for welded aluminium structures. *Computers and Structures* **84** No.29-30, 2016-2032.

Section 10
Application

10.1 Welding Works on Floodplain-bridge of Danube Bridge at Dunaújváros

László Érsek, Vasile Hodrea, László Léber
GANZ Manufacturing Company for Bridges, Cranes nad Steel Structures, Budapest, Hungary

Abstract

Manufacturing of the floodplain bridge of the new Danube bridge in Hungary is described. Welding technology, preassembly, site erection and welding of trapezoidal stiffeners is detailed. Dimensions optimal for manufacturing and erection are selected. Hungarian standards and prescriptions for highway bridges are used.

Keywords: *Steel construction, bridges, welding technologies, manufacturing*

1 Introduction

The Danube bridge at Dunaújváros is the largest highway bridge of our country. Total length of the bridge is 1680 m, length of river bed arch bridge is 307,9 m. Mass of steel structures is nearly 25.000 t. Furthermore it is the first application in Hungary of the thermo mechanically rolled high yield strength structural steels (S460M/ML) in field of bridge construction (Domanovszky 2005).

Welding technological problems in connection with the building of the right-side North floodplain bridge are discussed.

2 Applied welding technologies

2.1 *Parent metals*

Materials applied for the floodplain bridges are standard steels of bridge construction, according to standard MSZ EN 10025: for heavy duty applications S355J2G3, or S355K2G3 and even S235 for the secondary structures.

2.2 *Welding consumables*

Applied welding consumables in accordance with the welding processes are: welding wires for gas metal arc welding (GMAW), welding wires for submerged arc welding (SAW), flux powders and electrodes for shielded metal arc welding (SMAW).

2.3 *Welding processes*

During building of bridge structures following processes were used:
- submerged arc welding (SAW) with solid electrode,
- active-gas metal arc welding (GMAW),
- shielded metal arc welding (SMAW) et manual metal arc welding

2.4 *Welding equipment, welding fixture accessories*

Regarding to high volume of welding works and the tight deadlines we purchased the most update and best accepted high-tech welding equipment of the industrial practice for each area of the work. For the manufacturing we used water cooled GMAW and SAW welding machines, but at the erection we used relatively low weight inverter type machines for SMAW, lower capacity and lower weight machines for GMAW and automatic SAW machines. For positioning and welding of the main structures we constructed adequate accessories even before starting of manufacturing.

2.5 *Welding procedure tests*

Before the manufacturing/erection we executed all welding procedure tests according to relevant prescriptions and standards which can prove capability of the applied welding technologies.

2.6 *Welders certificates*

During the execution works we employed welders having valid certifications according to standard. Even operators of SAW machines allowed working with valid standard certificate. Welders (operators) were allowed to make welding seams which were in accordance with their certificate. Even tack welds made by certified welders were used.

2.7 *Welding supervision during manufacturing/erection*

Welding technology was treated with high priority in both manufacturing and site erection works during the whole execution work. Therefore, each contractor company employed responsible welding engineers for manufacturing-, preassembly- and site erection works. Activities of these responsible welding engineers were supported by welder foremen and gang bosses. Working of the above mentioned persons were coordinated by chief welding engineer of the consortium.

3 Main periods in execution works of right bank floodplain bridge

Right bank floodplain-bridge consists of 65 totally preassembled bridge sectors having different lengths and masses:

- supporting spreader sector	h= 15 m	Mass: 91,2 t
- 5 spreader sectors	h= 16,9 m	Mass: 83,9 t
- 4 spreader sectors	h= 16,85 m	Mass: 77,9 t

Welding and one-by-one assembling of 65 bridge sectors were executed in Csepel and erection site of company GANZ Company, then after descaling and painting they were shipped to the site.

4 Manufacturing

The main steps of workshop manufacturing are as follows.

4.1 *Preassembly of track/bottom/ridge panels*

The first step is welding of longitudinal stems (trapezoidal stems), independently from the panels. Welding of longitudinal welding seams of trapezoidal stems were executed by special two head automatic welding machines. The occasional elongation welding seams were made by GMAW.

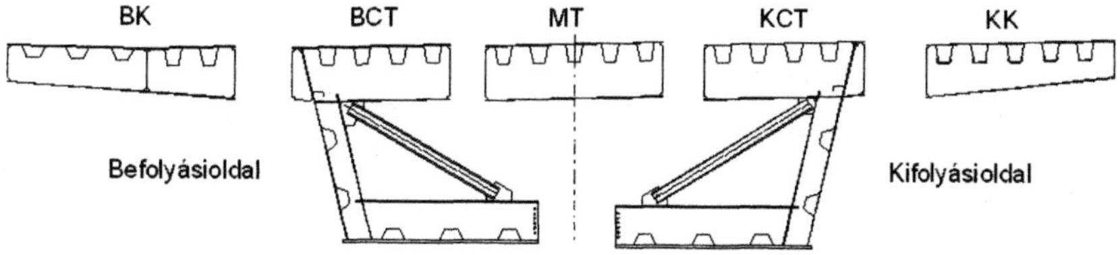

Figure 1. Normal truss holder divided into transportation units

Explanations: BK – inlet pavement bracket; BCT – inlet C-module; MT – intermediate module; KCT – outlet C-module; KK – outlet pavement bracket
Befolyási oldal: Flow- inside; Kifolyási oldal: Flow- outside

4.2 *Manufacturing of cross trusses*

The cross trusses were made in two types:
- normal - and
- reinforced design.

The total cross section was dismantled into 5 units - considering the transportability by truck (dimensions, masses):
- 2 pieces so called C- unit (it consists of one side ridge, the lower half girdle and the track sector above the ridge),
- intermediate track (which will be inserted between the C–units),
- track bracket and
- pavement bracket (which are built outside of the body segment).

Figure 1 shows the division of a normal cross truss. According to constructional division there were averagely 16,5 m long and 10-30 t weight units. Figure 2 shows typical steps of the workshop manufacturing.

4.3 *Preassembly*

There were a lot of welding seams (some of them are very important from structural point of view) must have been made. These welding seams are as follows (Below list is a welding sequence meantime.).

1. Longitudinal seam of bottom plate.
2. Seam of cross truss girdle plate on bottom plate.
3. Seam of cross truss ridge plate on bottom plate.
4. Seam of cross truss girdle plate under the track plates.
5. Longitudinal seam of the track plate (body segment).
6. Seams of ridge plates of cross truss under the track plate.
7. Seams on girdle plates of pavement bracket.
8. Longitudinal seams of the track plate (bracket).
9. Seams of ridge plates on pavement bracket.

Figure 2. A typical step of the manufacturing

The most applied welding processes was usually SMAW; for longitudinal seams of track plates was a combination of processes GMAW and SAW. The whole section assembly from 5 units (mentioned at manufacturing) arrived by truck was made up to 100 t mass according to lifting capacity of the cranes. Preassembly and welding (of longitudinal and transversal seams) were performed in special purposed welding accessories (carriages). Figure 3 shows the preassembly workstation. Transportation of larger units to the erection site was performed via shipping.

4.4 *Site erection*

The similar structured Danube bridge at Szekszárd still has cross sectional settings with high strength bolts on border of the shipped units. The bottom girdles of each joints are welded, even in case of the last two units made by "free erection". Execution of cross sectional fitting forms the essential part of the work.

4.5 *Application of ceramic backings*

The workshop patches (for plates with different wall thickness) were very rare, but at preassembly and site erection they needed very frequently. Rule for roads ÚT 2-3.413 also prescribes execution and inspection of patches at the track plates. In the tender according to the special technical specification, every patches should be done with ceramic backings except the trapezoid ribs elongations.

4.6 *Welding processes increasing the productivity*

Among our applied methods increasing the productivity there has to be underlined the plates, sections (trapezoidal ribs) were cut to size (in longitudinal and cross direction) according to the drawing. It was advantageous for decreasing of cutting-bevelling- and welding works, but some times caused problem in selection of the adequate dimension plate due to the pure storing area. Application of two torches SAW automatic welding machines because of assurance of even quality were necessary to be mentioned.

Figure 3. Preassembly of manufactured modules in Csepel

4.7 *Application of welding fixture accessories*

Manufacturing of welded structures with correct dimension tolerances can be realised by application of fixtures accessories only. For manufacturing of the main structural units we designed and fabricated capable accessories. We made manufacturing benches with the necessary dimensions for example for making the panels made of trapezoidal ribs, compact ribs and base plates (track, bottom and ridge), which ensured the shape tolerances and made possible the required preliminary straining.

We made large endorsers (approximate diameter is 8 m) for welding the so-called „C"-elements (It can be seen in background of Figure 2).

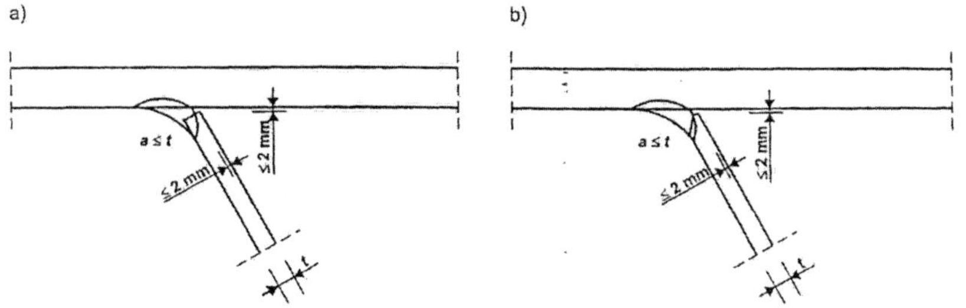

Figure 4. Edge preparation of trapezoidal rib for welding a) automatic welding b) manual welding

4.8 *Welding of trapezoidal ribs*

Welding joints connecting longitudinal trapezoidal ribs on to ortotrop track plates have majority among welding joints of vehicular bridges because of the type of loading and the difficult execution (one side approaching) and testing. This problem is regulated in departmental standard for vehicular bridges (ÚT 2-3. 413). Accordingly, in one side neck seams of trapezoidal ribs on track plates (usually 8 mm thick) the maximum lack of penetration is 2 mm. In case of SMAW the edge of ribs should be bevelled. In case of automated process, it is not required. (Figure 4a and 4b).

GANZ Company applied two of the most updated torches SAW machines from the beginning. During the preliminary welding experiment, we made a lot of welding specimens with - and without bevelling. For introduction of this, we attached some photos (Figure 5a and 5b). One can see that the requirement can be satisfied even without bevelling.

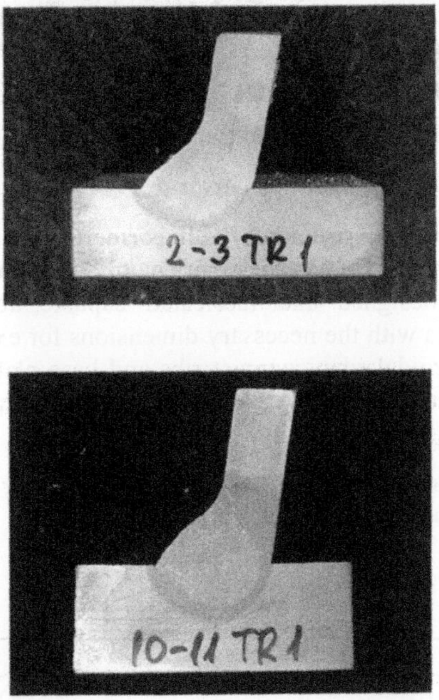

Figure 5. Macro photo of neck joint on trapezoid rib a) submerged arc welding (SAW) without bevelling; b) submerged arc welding with bevelling

The most frequent type are the longitudinal seams of the trapezoidal ribs. We had to weld nearly 110.000 m length from this type of joint at the floodplain-bridges.

The practice abroad is also similar, for example at the Greabaelt- bridge (Rouvillain 1995). The position of welding joint in the structure is more important than automation of the welding. Consequently, at track plates the trapezoidal ribs should be bevelled according to the standard, meanwhile, at other panels bevelling is not necessary.

5 Tests of welding seams, tests during manufacturing and erection as well as control tests

5.1 *Destructive tests*

Welding plans used to be worked-out by the potential contractors and to be worked out during the manufacturing/erection including control tests of manufacturing and erection.

5.2 *Non-destructive tests*

Method and extension of non-destructive tests was determined according to relevant standards (for example road standard ÚT 2-3-413) and the participant contractors and QC supervisors agreement.

6 Selection of dimensions optimal for manufacturing/erection

From economy reason every contractor would like to realize as large structure as possible. From this viewpoint the primary limit is the vehicular transportability and secondary one is the lifting capacity of the cranes. The primary limit is the width dimension for both vehicular and railway transport. Height of the structures usually does not exceed the overhead bridges height. In our case the manufacturing was arranged in 16,5 long units. In other countries for example at Greatbaelt-(Storebaelt) bridge between Denmark and Sweden applied 20 m long units were applied (Rouvillain 1995). We have to remark that this trend has reached our country: at HÉV (suburban railway) bridges manufactured recently there were even 22,5 m long units.

Acknowledgement

We would like say special thanks to dr. Sándor Domanovszky, winning of the Széchenyi prize, chief engineer of consortium DunaÚjhíd for helping in issue of present paper and for his photos (Figure 2 and Figure 3).

References

Domanovszky, S. (2005) Thermomechanically rolled S460M/ML steels in steel construction (in Hungarian). *Hegesztéstechnika*, No. 4. 13 – 21.

Rouvillain, F. (1995) High strength steel in a major construction over Storebaelt (the Great Belt) *Svetsaren* No. 1. 15 – 20.

Standards and prescriptions (in Hungarian)

ÚT 2-3.404.2002: Building of highway bridges II. Production and erection of steel bridges.

ÚT 2-3.413:2005. Design and complementary prescriptions of highway bridges.

Special technical prescriptions: Nemzeti Autópálya Rt – Budapest.

10.2 Lessons from Failures and Design Errors of Thin-walled Steel Structures

József Farkas

University of Miskolc, H-3515 Miskolc, Hungary, altfar@uni-miskolc.hu

Abstract

The failure possibilities of thin-walled steel structures are summarized. Examples of buckling, torsion, fatigue, freezing and corrosion are described in details. These analyses show for designers, manufacturers and engineers in maintenance how to prevent these failures or wrong designs.

Keywords: *thin-walled structures, buckling, fatigue, torsion, corrosion*

1 Introduction

Modern steel structures consist of welded plated and tubular components. Since the behaviour of rolled section rods are similar to plated components, it can be stated that all the steel structures used nowadays are thin-walled ones. The most effective way to decrease the structural mass is to decrease the plate thicknesses. Thinner structures have not only smaller mass, but also lower cost, since the welding cost is proportional to second power of thickness.

Besides of these advantages thin-walled structures have some problems, which can sometimes cause catastrophic failures. In the present study these problems are illustrated by some failure cases. Lessons from these failures show, that the design system of structural optimization should be applied, which includes all the important engineering aspects.

2 Problems of welded thin-walled structures

- Residual welding stresses and distortions arise. Compression stresses decrease the buckling strength, tension stresses accelerate the fatigue crack propagation, distortions can exceed the allowable fabrication tolerances.

- Compressed plates and shells can buckle.

- Torsion of thin-walled open section rods can cause additional normal stresses and large deformations.

- High stress concentrations in the connections of thin elements can lead to fatigue failures.

- Welded beam-to-columns connections in frames can be damaged by earthquake.

- Thin-walled components are very sensitive to fire.

- Closed section rods can be destroyed by freezing of water accumulated in them.

- Corrosion can damage thin-walled sections in dangerous measure.

In order to strengthen plates and shells against buckling and vibration stiffening should be used, but the stiffening needs additional material and welding cost. A question arises whether a thicker unstiffened or a thinner stiffened component is more economic. To answer this question a systematic study is necessary, since the economy depends on more parameters. Our answer can be found in a series of studies published recently.

3 Collapse of a belt-conveyor bridge due to shell buckling

The bridge has been designed with a skew part as shown in Figure 1. Coal-powder settlings inside and outside of the circular cylindrical thin-walled shell – mainly in the vicinity of the connection of the horizontal and the skew bridge part – caused overloading and the bridge collapsed due to local buckling of 4 mm thin shell.

The failure could have been prevented by a better design taking into account the possibility of overload by coal-powder and by using a better bridge shape.

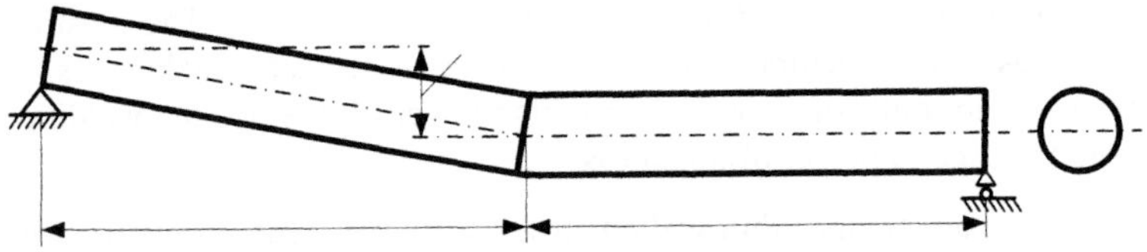

Figure 1. Main dimensions and cross-section of a belt-conveyor bridge of circular shell structure

4 Collapse of a transit silo hopper due to inadequate torsional stiffness of the transition ring-beam

The transition ring-beam is an important structural component of transit silos, and is loaded by compression, bending, shear and torsion. In a transit silo the designer has used only a horizontal plate for this ring-beam (Fig.2) (Farkas 1986). The variable

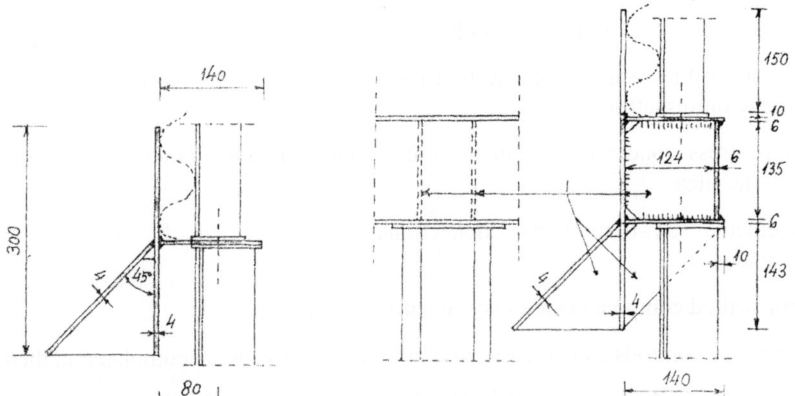

Figure 2. The original and the stiffened transition ring beam structure

load caused a low cycle fatigue in the fillet welds connecting this plate to the silo and the hopper collapsed. To ensure a suitable torsional stiffness a welded box section is used instead of the simple plate (Fig.2).

5 Fatigue fracture of the Kielland offshore platform

The semi-submersible offshore platform capsized suddenly on March 1980, 123 persons died while 89 persons were rescued. (Almar-Naess et al. 1982). The detailed fracture analysis has shown that the catastrophe was caused by fatigue failure of one of horizontal tubular bracings in which a hydrophone has been mounted by cutting a hole and use double fillet welds (Fig.3).

Welding caused initial cracks and lamellar tearing because of the poor through thickness properties of the material in the hydrophone tube. The fatigue crack propagation was caused by high wave loads. The other bracings fail due to overloading.

The catastrophe could have been prevented by using suitable steel for the hydrophone tube and better fillet weld quality. This collapse showed that the fatigue failure is very dangerous for welded structures, since the local initial cracks can propagate suddenly.

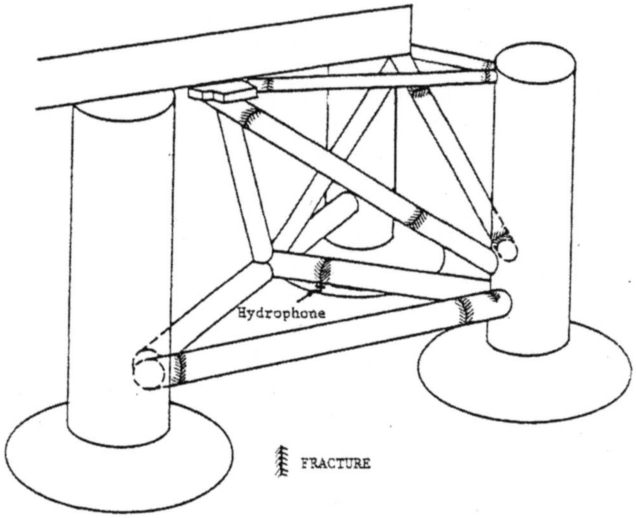

Figure 3. Tubular structure of the Kielland offshore platform

6 Freezing failure of a truss chord of welded box section

A sport stadium roof is supported by four steel trusses of 42 m span length (Farkas 1979). The lower chord has a box section profile. For stiffening of the shape of the box section inner diaphragms are used welded to the section in three sides by fillet welds. Since it is impossible to weld the fourth side from inside, a backing plate is used connecting to the upper flange by small diameter plug welds (Fig.4). The holes for these welds have not been welded in some places. This fabrication error caused the accumulation of rain water inside of the box.

Since the construction has been performed in winter, accumulated water froze and caused very dangerous cracks and large local deformations. The damage was so dangerous that the whole lower chord should have been changed.

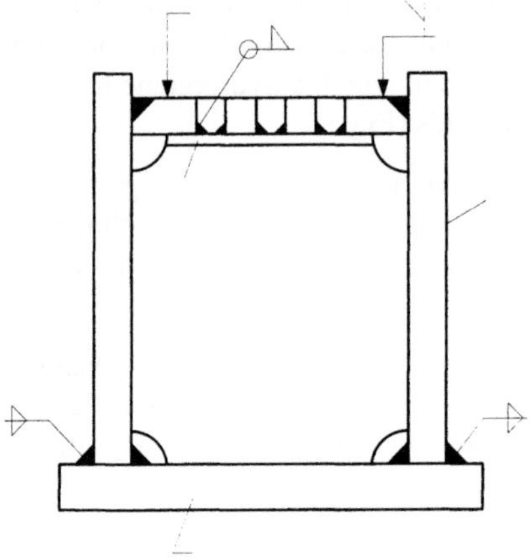

Figure 4. Welded box cross-section of the central part of the lower flange

The lesson from this case is that the accumulation of water inside of box sections should be avoided by using some small holes, which do not cause any weakening of the cross section. It should be mentioned that the welding of the diaphragms to the upper flange was superfluous.

7 Local buckling of stiffening columns of a silo caused by corrosion

Thin-walled profiles can loss their load-carrying capacity by corrosion, since their thickness can be decreased significantly. In a silo with a bin of corrugated circular cylindrical shell stiffening vertical columns of flanged channel section are used (Fig.5). The bad maintenance in which these columns have not been repainted during 15 years, they corroded in so great measure that they have locally buckled in their bottom part. Since the columns have been fixed in concrete bases, instead of their change new columns should have been constructed to support the bin.

It can be concluded that it is important to protect thin-walled steel components continuously against corrosion.

8 Unsafe design of supporting frames for pressure vessels

In a chemical plant U-profile side beams have been designed for support of pressure vessels (Figure 5). (Farkas 1989). The connections of these beams to the main I-girders have been realized by simple bolting of the webs of the connecting beamwebs (Figure 6.). The flanges of the U-profile side beams have not been connected to the I-girders.

In such connections the reaction forces act in the vertical plane of the U-profile web (Figure 6). Since the shear center of U-profile is outside of this plane, additional torsional moment arises and causes warping normal stresses and torsional deformations. These additional effects can cause vibrations during the operation, since the pressure vessels are loaded dynamically. This fact has been observed during the dynamic water proof of pressure vessels. Therefore, the designer decided to strengthen these side U-beams.

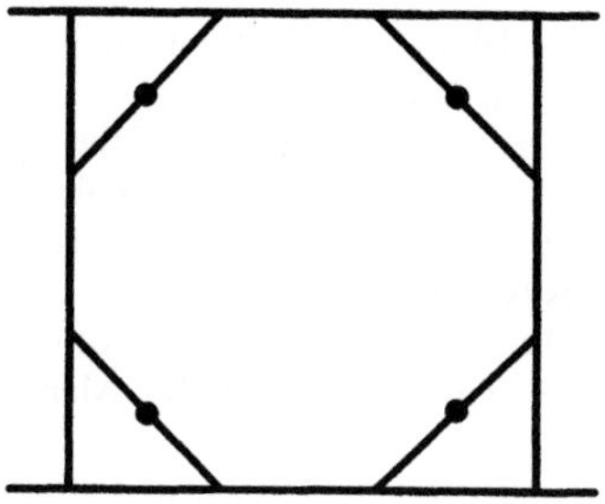

Figure 5. Plan view of the supporting structure with main girders and side beams, S – support points of a pressure vessel

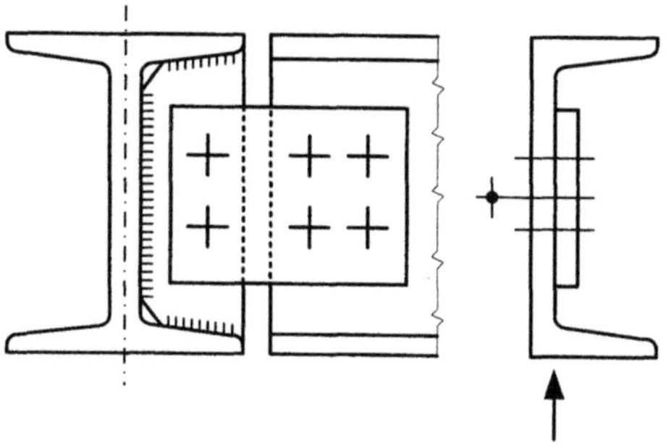

Figure 6. Bolted connection of a side channel-beam to a main I-girder. T - the torsion (or shear) center. The reaction force is marked by a vector sign

The correct design of these supporting frames should use rectangular hollow section or welded box beams to avoid large torsional deformations, since the open-section beams are very sensitive to torsion.

It should be mentioned that Augustyn & Śledziewski (1976) as well as Scheer (2001) have collected many failure cases of engineering structures in their books.

9 Conclusions

Thin-walled structures are sensitive to many modes of failure such as buckling, torsion, fatigue, freezing of box beams, corrosion. Examples of buckling of circular shell structure for a belt-conveyor bridge, torsion of a silo transition ring beam, warping torsion of supporting beams for pressure vessels, fatigue of a tubular offshore platform, corrosion of supporting columns of a silo are described in details. Lessons learnt from these examples can help engineers to avoid these failure possibilities.

References

Almar-Naess,A., Haagensen,P.J., Lian,B. & Simonsen,T. (1982) Metallurgical and fracture analyses of the Alexander L. Kielland platform. International Institute of Welding, IIW-Doc. XIII-1066-82.

Augustyn,J. & Śledziewski,E. (1976) Awarie konstrucji stalowych (Failures of steel structures, in Polish). Warszawa, Arkady.

Farkas,J. (1979) Opinion from the main truss girder of the Sport Stadium in Zalaegerszeg. Manuscript in Hungarian, Miskolc, University of Miskolc.

Farkas,J. (1986). Opinion from the strengthening of a TS10 transit silo. Manuscript in Hungarian, Miskolc, University of Miskolc.

Farkas,J. (1989). Opinion from the load-carrying capacity of the supporting steel beams of the MDI investment of Borsod Chemical Plant (BVK). Manuscript in Hungarian, Miskolc, University of Miskolc.

Scheer,J. (2001) Versagen von Bauwerken. Ursachen, Lehren. Band 1. Brücken. Band 2. Hochbauten und Sonderbauwerke. Ernst & Sohn, Weinheim.

10.3 Cost Optimal Design of Composite Floor Trusses

Uroš Klanšek, Stojan Kravanja
University of Maribor, Faculty of Civil Engineering, Smetanova 17, SI-2000 Maribor, Slovenia, e-mail: uros.klansek@uni-mb.si, stojan.kravanja@uni-mb.si

Abstract

This paper presents the cost optimization of composite floor trusses. The composite structure was constructed of a reinforced concrete slab and welded steel Pratt trusses consisting of hot rolled channel sections. The structural optimization was carried out by the non-linear programming approach, NLP. An accurate objective function of the manufacturing material, power and labour costs was defined for the optimization. The objective function was subjected to structural analysis constraints. The design constraints were defined according to Eurocode 4 for the conditions of both the ultimate and the serviceability limit states. A numerical example of the optimization of the composite floor trusses presented at the end of the paper shows the advantages of the proposed approach.

Keywords: *Manufacturing costs, Structural optimization, Non-linear programming, Composite structures, Welded structures*

1 Introduction

This paper presents the cost optimization of composite floor trusses. The composite structure was constructed of a reinforced concrete slab and welded steel Pratt trusses consisting of hot rolled channel sections. The structural optimization was carried out by the non-linear programming approach, NLP. An accurate objective function of the manufacturing material, power and labour costs was defined for the optimization, Klanšek & Kravanja (2006). The objective function was subjected to structural analysis constraints. The design constraints were defined according to Eurocode 4 (1992) for the conditions of both the ultimate and the serviceability limit states. A numerical example of the optimization of the composite floor trusses presented at the end of the paper shows the advantages of the proposed approach.

2 Composite floor trusses

The composite structure was constructed of a reinforced concrete slab and welded steel Pratt trusses consisting of hot rolled channel sections, see Figure 1.

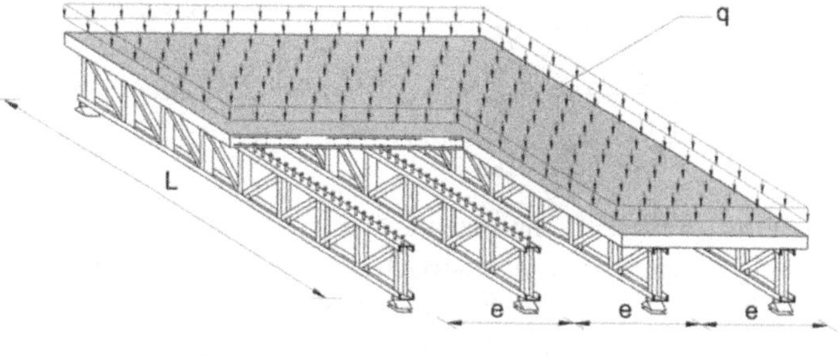

Figure 1. Composite floor truss system.

The full composite action between the concrete and the steel truss is achieved by the cylindrical shear studs, welded to the top chord of truss and embedded in concrete. The composite trusses were designed according to Eurocode 4 (1992) for the conditions of both the ultimate and the serviceability limit states. The design loads were calculated with regard to Eurocode 1 (1995). The concrete slab was designed according to Eurocode 2 (1992) as a one way spanning continuous slab. The optimization of steel elements was performed with respect to Eurocode 3 (1995).

3 NLP optimization

3.1 *NLP problem formulation*

The structural optimization was performed by the non-linear programming approach, NLP. The general NLP optimization problem is formulated as:

$$\begin{aligned} &\text{Min } z = f(\boldsymbol{x}) \\ &\text{subjected to:} \\ &\boldsymbol{h}(\boldsymbol{x}) = 0 \\ &\boldsymbol{g}(\boldsymbol{x}) \leq 0 \\ &\boldsymbol{x} \in \boldsymbol{X} = \{\boldsymbol{x} \mid \boldsymbol{x} \in \boldsymbol{R}^n, \boldsymbol{x}^{LO} \leq \boldsymbol{x} \leq \boldsymbol{x}^{UP}\} \end{aligned} \qquad \text{(NLP)}$$

where $\boldsymbol{x}$ is the vector of the continuous variables, defined within the compact set $\boldsymbol{X}$. Functions $f(\boldsymbol{x})$, $\boldsymbol{h}(\boldsymbol{x})$ and $\boldsymbol{g}(\boldsymbol{x})$ are the (non)linear functions involved in the objective function z, the equality and inequality constraints, respectively. All the functions $f(\boldsymbol{x})$, $\boldsymbol{h}(\boldsymbol{x})$ and $\boldsymbol{g}(\boldsymbol{x})$ must be continuous and differentiable.

In view of the optimization of composite floor trusses, the continuous variables define dimensions, forces, stresses, strains, cost parameters, etc. The equality and inequality constraints and the bounds of the variables represent a rigorous system of design, load, resistance and deflection functions known from the structural analysis. The objective function was proposed to minimize the structure's manufacturing costs.

3.2 *Cost objective function*

The optimal design of composite floor trusses was determined by the minimum of the manufacturing costs. In this way, the cost objective function was defined as:

$$\begin{aligned} \text{min: } Cost = \{ & C_{M,s,c,r} + C_{M,sc} + \sum_{i,j} C_{M,e_{i,j}} + \sum_{i,j} C_{M,ac,fp,tc_{i,j}} + C_{M,f} \\ & + \sum_{i,j} C_{P,c,hs_{i,j}} + \sum_{i,j} C_{P,c,gm_{i,j}} + \sum_{i,j} C_{P,w_{i,j}} + C_{P,sw} + C_{P,v} \\ & + \sum_{i,j} C_{L,c,hs_{i,j}} + \sum_{i,j} C_{L,g_{i,j}} + C_{L,p,a,t} + \sum_{i,j} C_{L,SMAW_{i,j}} \\ & + C_{L,sw} + \sum_{i,j} C_{L,spp_{i,j}} + C_{L,f} + C_{L,r} + C_{L,c} + C_{L,v} + C_{L,cc} \} / (e \cdot L) \end{aligned} \qquad (1)$$

where *Cost* [€/m^2] represents the manufacturing costs per m^2 of the usable floor surface; $C_{M,...}$, $C_{P,...}$ and $C_{L,...}$ are the material, power and labour cost items calculated in €; $\sum_{i,j}$ denotes the sum of all the individual truss element cost contributions; subscripts i, j are the end joints of a single truss member; e [m] is the intermediate distance between the trusses and L [m] is the span of the composite truss. The material, power and labour cost items are defined in the following equations.

Material costs

Steel, concrete and reinforcement:

$$C_{M,s,c,r} = c_{M,s} \cdot \rho_s \cdot \sum_{i,j} A_{i,j} \cdot l_{i,j} + c_{M,c} \cdot d \cdot e \cdot L + c_{M,r} \cdot \rho_s \cdot A_s \cdot l_s \cdot L \qquad (2)$$

where $c_{M,s}$ [€/kg], $c_{M,c}$ [€/m^3] and $c_{M,r}$ [€/kg] are the prices of the structural steel, the concrete and the reinforcement; ρ_s is the steel density 7850 kg/m^3; $A_{i,j}$ [m^2] and $l_{i,j}$ [m] are the cross-section area and the length of a single truss member; d [m] is the depth of the concrete slab; A_s [m^2/m^1] is the cross-section area of steel reinforcement per m^1 and l_s [m] is the length of the reinforcing steel.

Cylindrical shear studs:

$$C_{M,sc} = c_{M,sc} \cdot n_{sc} \qquad (3)$$

where $c_{M,sc}$ [€/stud] and n_{sc} are the price and the number of cylindrical shear studs.

Electrode consumption, see Creese et al. (1992):

$$C_{M,e_{i,j}} = c_{M,e} \cdot \rho_s \cdot A_{w_{i,j}} \cdot l_{w_{i,j}} / EMY \qquad (4)$$

where $c_{M,e}$ [€/kg] is the price of the electrodes; $Aw_{i,j}$ [m^2] is the cross-section area of the weld; EMY is the electrode metal yield and $lw_{i,j}$ [m] is the length of the weld.

Anti-corrosion, fire protection and top coat paint:

$$C_{M,ac,fp,tc_{i,j}} = \left(c_{M,ac} + c_{M,fp} + c_{M,tc}\right) \cdot \left(1 + k_p \cdot k_{sur} \cdot k_{wc}\right) \cdot A_{ss_{i,j}} \qquad (5)$$

where $c_{M,ac}$ [€/m^2], $c_{M,fp}$ [€/m^2] and $c_{M,tc}$ [€/m^2] are the prices of the anti-corrosion, the fire protection and the top coat paints per m^2 of painted surface; k_p, k_{sur} and k_{wc} are the paint loss factors which take into account the painting technique, the complexity of the structure's surface and the weather conditions in which the structure is painted; $Ass_{i,j}$ [m^2] is the steel surface area of the truss member.

Formwork floor-slab panels:

$$C_{M,f} = c_{M,f} \cdot e \cdot L / n_{uc} \qquad (6)$$

where $c_{M,f}$ [€/m^2] is the price of the formwork floor-slab panels per m^2 of the concrete slab panelling surface area and n_{uc} is the number, how many times the formwork panels may be used before they have to be replaced with the new ones.

Power costs

Sawing the steel section:

$$C_{P,c,hs_{i,j}} = c_P \cdot \left(P_{hs} / \eta_{hs}\right) \cdot k_{am} \cdot T_{c,hs} \cdot b_{i,j} \qquad (7)$$

where c_P [€/kWh] is the electric power price; P_{hs} [kW] and η_{hs} are the machine power and the machine power efficiency of the hacksaw; k_{am} is the factor which considers the allowances to machining time; $T_{c,hs}$ [h/m] is the time for steel cutting and $b_{i,j}$ [m] is the overall web width of the truss member.

Edge grinding the steel section:

$$C_{P,c,gm_{i,j}} = c_P \cdot \left(P_{gm}/\eta_{gm}\right) \cdot k_{am} \cdot T_g \cdot l_{g\,i,j} \tag{8}$$

where P_{gm} [kW] and η_{gm} are the machine power and the machine power efficiency of the grinding machine; Tg [h/m] is the time of edge grinding and $lg_{i,j}$ [m] is the grinding length of the individual truss member.

Shielded metal arc welding, see Creese et al. (1992):

$$C_{P,w_{i,j}} = c_P \cdot \rho_s \cdot \left(I \cdot U/\eta_w\right) \cdot A_{w_{i,j}} \cdot l_{w_{i,j}} / DR \tag{9}$$

where I [kA] and U [V] are the welding current and the voltage; η_w is the machine power efficiency of the arc welding machine and DR [kg/h] is the deposition rate.

Stud arc welding:

$$C_{P,sw} = c_P \cdot \left(I_{sw} \cdot U_{sw}/\eta_w\right) \cdot n_{sc} \cdot T_{sw} \tag{10}$$

where I_{sw} [kA], U_{sw} [V] and T_{sw} [h/stud] are the current, the voltage and the time required for stud welding.

Vibrating the concrete:

$$C_{P,v} = c_P \cdot \left(P_v/\eta_v\right) \cdot T_v \cdot e \cdot L \tag{11}$$

where P_v [kW] and η_v are the power and the machine power efficiency of the concrete vibrator; T_v [h/m^2] is the time required for consolidation of the concrete.

Labour costs

Sawing the steel section:

$$C_{L,c,hs_{i,j}} = c_L \cdot k_{am} \cdot T_{c,hs} \cdot b_{i,j} \tag{12}$$

where c_L [€/h] denotes the labour cost per working hour.

Edge grinding of the steel section:

$$C_{L,g_{i,j}} = c_L \cdot k_{am} \cdot T_g \cdot l_{g\,i,j} \tag{13}$$

Preparation, assembly and tacking of the steel structure to be welded:

$$C_{L,p,a,t} = c_L \cdot T_{p,a,t} \tag{14}$$

where $T_{p,a,t}$ [h] is the time for the preparation, assembling and tacking.

Manual shielded metal arc welding:

$$C_{L,SMAW_{i,j}} = c_L \cdot k_d \cdot k_{wp} \cdot k_{wd} \cdot k_{wl} \cdot k_r \cdot T_{SMAW} \cdot l_{w_{i,j}} \tag{15}$$

where k_d is the difficulty factor which reflects the local working conditions; k_{wp} is the factor which considers the welding position; k_{wd} considers the welding direction; k_{wl} considers the shape and the length of the weld; k_r considers the chamfering of the root of the weld; T_{SMAW} [h/m] is the time for manual shielded metal arc welding.

Semi-automatic stud arc welding:

$$C_{L,sw} = c_L \cdot T_{swp} \cdot n_{sc} \tag{16}$$

where T_{swp} [h/stud] denotes the time needed for stud welding, placing/removal of a ceramic ferrule and cleaning the connection.

Steel surface preparation and protection:

$$C_{L,sppi,j} = c_L \cdot k_{dp} \cdot \left(T_{ss} + n_{ac} \cdot T_{ac} + n_{fp} \cdot T_{fp} + n_{tc} \cdot T_{tc}\right) \cdot A_{ssi,j} \tag{17}$$

where k_{dp} is the difficulty factor related to the painting position; T_{ss} [h/m^2], T_{ac} [h/m^2], T_{fp} [h/m^2] and T_{tc} [h/m^2] are the times required for the sand-spraying, the anti-corrosion resistant painting, the fire protection painting and the top coat painting of the steel surface, respectively; n_{ac}, n_{fp} and n_{tc} are the numbers of layers of the anti-corrosion resistant paint, the fire protection paint and the top coat paint.

Placing the formwork (panelling, levelling, disassembly and cleaning):

$$C_{L,f} = c_L \cdot T_f \cdot e \cdot L \tag{18}$$

where T_f [h/m^2] represents the time necessary for panelling, levelling, disassembly and cleaning a formwork.

Cutting, placing and connecting the reinforcement:

$$C_{L,r} = c_L \cdot \rho_s \cdot k_{rh} \cdot k_{ri} \cdot T_r \cdot A_s \cdot l_s \cdot L \tag{19}$$

where k_{rh} and k_{ri} are the difficulty factors related to the structural height and inclination of the concrete slab; T_r [h/kg] is the time required for the cutting, placing and connecting of the reinforcement.

Concreting the slab:

$$C_{L,c} = c_L \cdot T_c \cdot d \cdot e \cdot L \tag{20}$$

where T_c [h/m^3] represents the time required for placement of the pumped concrete.

Concrete consolidation:

$$C_{L,v} = c_L \cdot T_v \cdot e \cdot L \tag{21}$$

Curing the concrete:

$$C_{L,cc} = c_L \cdot T_{cc} \cdot d \cdot e \cdot L \tag{22}$$

where T_{cc} [h/m^3] is the time required for the curing of the concrete.

A detailed interpretation and the ranges of values for the parameters included in the cost objective function may be found in reference Klanšek & Kravanja (2006).

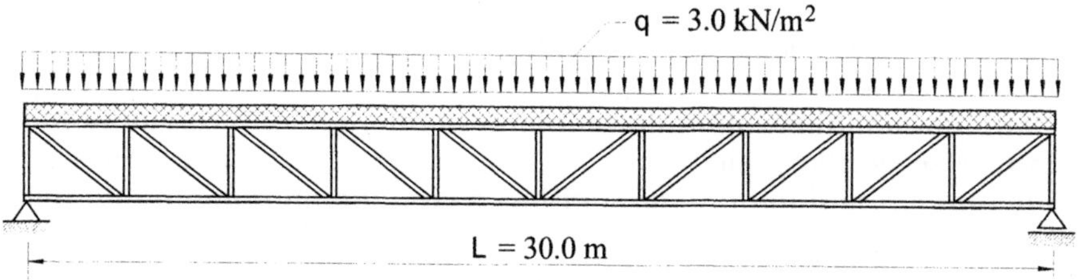

Figure 2. Composite floor trusses.

3.3 *Structural analysis constraints*

The objective function was subjected to structural analysis constraints defined according to Eurocode 4 for both the ultimate and the serviceability limit states. The constraints for the presented structure may be found in paper Klanšek et al. (2006).

4 Numerical example

The paper presents the cost optimization of a 30 m long simply supported composite truss, subjected to self-weight and the variable load of 3.0 kN/m^2, see Figure 2.
The material, power and labour cost parameters used in the optimization are shown in Table 1. The fabrication times and the approximation functions for the fabrication times are shown in Table 2 and Table 3. All other input data are listed in Table 4.

Table 1. Material, power and labour costs parameters

$c_{M,s}$	Price of the structural steel S 235 – S 355:	1.00 – 1.07	€/kg
$c_{M,c}$	Price of the concrete C 25/30 – C 50/60:	85.00 – 120.00	€/m^3
$c_{M,r}$	Price of the reinforcing steel S 400:	0.70	€/kg
$c_{M,sc}$	Price of the cylindrical shear studs:	0.50	€/piece
$c_{M,e}$	Price of the electrodes:	1.70	€/kg
$c_{M,ac}$	Price of the anti-corrosion paint:	0.85	€/m^2
$c_{M,fp}$	Price of the fire protection paint (F 30):	9.00	€/m^2
$c_{M,tc}$	Price of top coat paint:	0.65	€/m^2
$c_{M,f}$	Price of the prefabricated floor-slab panels:	30.00	€/m^2
c_P	Electric power price:	0.10	€/kWh
c_L	Labour costs:	20.00	€/h

Table 2. Fabrication times

$T_{c,hs}$	Time for sawing the steel sections: 1.337 h/m
T_g	Time for edge grinding of the steel sections: 33.333×10^{-3} h/m
T_{sw}	Time for stud welding: 2.333×10^{-4} h/stud
T_v	Time for consolidation of the concrete: 0.200 h/m^2
T_{swp}	Time for welding, placing/removal of a ferrule and cleaning: 55.555×10^{-4} h/stud
T_{ss}	Time for sand-spraying: 0.050 h/m^2
T_{ac}	Time for anti-corrosion resistant painting: 0.050 h/m^2
T_{fp}	Time for fire protection painting: 0.050 h/m^2
T_{tc}	Time for top coat painting: 0.050 h/m^2
T_f	Time for panelling, levelling, disassembly and cleaning the formwork: 0.300 h/m^2
T_r	Time for cutting, placing and connecting the reinforcement: 0.024 h/kg
T_{cc}	Time for curing the concrete: 0.200 h/m^3

The optimization was performed in two successive steps. The first step included the NLP optimization, where the continuous variables were calculated inside their upper and lower bounds. At this stage, the structure was fully exploited considering either ultimate or serviceability limit state conditions. In the second step, the calculation was repeated/checked for the fixed variables rounded up, from continuous values obtained in the first stage, to their nearest upper standard values. CONOPT (Generalized reduced-gradient method) was used for the optimization, Drud (1994). The obtained optimal structural design is presented in Figures 3 and 4.

Table 3. Approximation functions for fabrication times

$T_{p,a,t}$*	Time for preparation, assembling and tacking: $T_{p,a,t} = C_1 \cdot \Theta_d \cdot (\kappa \cdot \rho_s \cdot V_s)^{0.5}/60$ [h]; $C_1 = 1.0$ min/kg$^{0.5}$; $\Theta_d = 3.00$; $\kappa = 23$ elements; $\rho_s = 7850$ kg/m^3 and V_s [m^3].
T_{SMAW}	Time for manual shielded metal arc welding: Fillet welds: $T_{SMAW} = a_2 \cdot a_w^2 + a_1 \cdot a_w + a_0$ [h/m]; $a_2 = 1.2653\times10^{-2}$; $a_1 = 1.3773\times10^{-3}$; $a_0 = 1.6111\times10^{-2}$ and a_w [mm]. ½ 60° V welds: $T_{SMAW} = b_6 \cdot a_w^6 + b_5 \cdot a_w^5 + b_4 \cdot a_w^4 + b_3 \cdot a_w^3 + b_2 \cdot a_w^2 + b_1 \cdot a_w + b_0$ [h/m]; $b_6 = -1.7138\times10^{-8}$; $b_5 = 1.7372\times10^{-6}$; $b_4 = -0.5576\times10^{-4}$; $b_3 = 4.1851\times10^{-4}$; $b_2 = 1.0805\times10^{-2}$; $b_1 = -0.7401\times10^{-1}$; $b_0 = 2.8286\times10^{-1}$ and a_w [mm].
T_c	Time for placement of pumped concrete: $T_c = c_2 \cdot d^2 + c_1 \cdot d + c_0$ [h/m^3]; $c_2 = 2.4000\times10^{-3}$; $c_1 = -5.4000\times10^{-2}$; $c_0 = 9.9500\times10^{-1}$ and d [cm].

* Fabrication time proposed by Jármai & Farkas (1999).

Table 4. Input data

ρ_s	Steel density: 7850 kg/m^3
ρ_c	Concrete density: 2500 kg/m^3
EMY	Electrode metal yield: 0.60
k_p	Paint loss factor – painting technique: 0.05 for brush painting
k_{sur}	Paint loss factor – complexity of the structure: 1.00 for large surfaces
k_{wc}	Paint loss factor – weather conditions: 1.00 for brush painting
n_{uc}	Number, how many times the formwork floor-slab panels may be used: 30
k_{am}	Factor – allowances to machining time: 1.09 for the machining process
P_{hs}	Power of the hacksaw: 2.20 kW
η_{hs}	Machine power efficiency: 0.85 for the hacksaw
P_{gm}	Power of the grinding machine: 1.10 kW
η_{gm}	Machine power efficiency: 0.85 for the grinding machine
I	Welding current: 230 A
U	Welding voltage: 25 V
η_w	Machine power efficiency: 0.90 for the arc welding machine
DR	Deposition rate: 3.7 kg/h
P_v	Power of the internal vibrator ø 48 mm: 3.10 kW
η_v	Machine power efficiency: 0.85 for the internal concrete vibrator
k_d	Difficulty factor – working conditions: 1.00 normal conditions
k_{wp}	Difficulty factor – welding position: 1.00 flat position 1.10 vertical, overhead position
k_{wd}	Difficulty factor – welding direction: 1.00 for flat position and vertical welds
k_{wl}	Difficulty factor – welding length: 1.00 for long welds
k_r	Difficulty factor – root of the weld: 1.00 for welds without treatment of root
k_{dp}	Difficulty factor – painting position: 1.00 for horizontal painting
k_{rh}	Difficulty factor – structural height: 1.00 for structural height less than 6 m
k_{ri}	Difficulty factor – inclination of the concrete slab: 1.00 for horizontal slab

The optimal result of 7529.53 € per single composite truss (or 88.69 €/m^2 of useable floor surface) was obtained in the second NLP stage. The optimal steel sections are listed as follows: top chord (UPE 270) bottom chord (UPE 240); diagonals D_1 (UPE 160), D_2 (UPE 140), D_3 (UPE 120), D_4 (UPE 100), D_5 (UPE 100); verticals V_1 (UPE 160), V_2 (UPE 160), V_3 (UPE 140), V_4 (UPE 120), V_5 (UPE 100), V_6 (UPE 100).

5 Conclusions

This paper presents the cost optimization of the composite floor trusses composed from a reinforced concrete slab and from steel trusses made from hot rolled channel

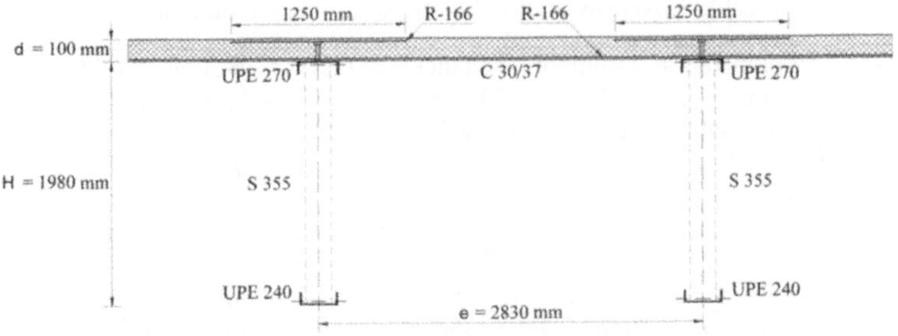

Figure 3. Optimal design and materials of composite floor trusses.

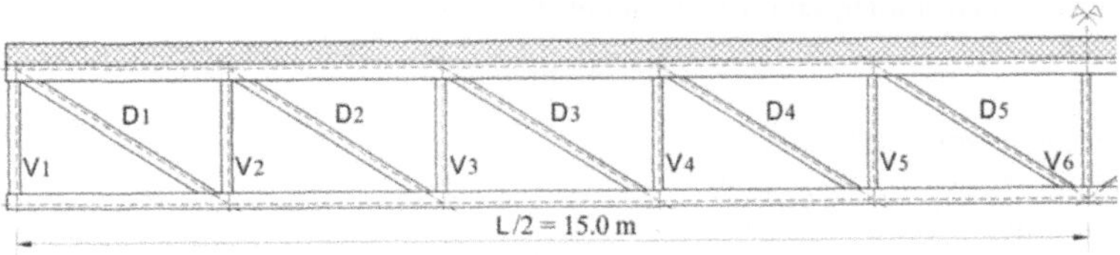

Figure 4. Arrangement of steel truss members.

sections. The optimization was performed by the non-linear programming approach, NLP. The objective function of the manufacturing material, power and labour costs was defined for the optimization. The objective function was subjected to a rigorous system of equality and inequality constraints, defined according to Eurocode 4 to satisfy both the ultimate and the serviceability limit states. Since the cost function is detailed and formulated in an open manner, it can be easily adopted and used for any specific data in different economical and technological conditions. The numerical example presented at the end of the paper shows the advantages of the proposed approach.

References

Creese, R.C., Adithan, M. & Pabla, B.S. (1992) *Estimating and costing for the metal manufacturing industries*, New York: Marcel Dekker.

Drud, A.S. (1994) CONOPT – A Large-Scale GRG Code. *ORSA Journal on Computing* **6** No.2, 207–216

Eurocode 1 (1995) *Basis of design and actions on structures*, Brussels: European Committee for Standardization.

Eurocode 2 (1992) *Design of concrete structures*, Brussels: European Committee for Standardization.

Eurocode 3 (1995) *Design of steel structures*, Brussels: European Committee for Standardization.

Eurocode 4 (1992) *Design of composite structures*, Brussels: European Committee for Standardization.

Jármai, K. & Farkas J. (1999) Cost calculation and optimization of welded steel structures. *Journal of Constructional Steel Research* **50** No.2, 115–135

Klanšek, U. & Kravanja, S. (2006) Cost estimation, optimization and competitiveness of different composite floor systems. Part 1, Self-manufacturing cost estimation of composite and steel structures. *Journal of Constructional Steel Research* **62** No.5, 434–448

Klanšek, U., Šilih, S. & Kravanja, S. (2006) Cost optimization of composite floor trusses. *Steel & Composite Structures* **6** No.5, 435–457

Initial Shape of Cable-suspended Roofs

Kazimierz Myslecki and Piotr L. Sawinski
Wroclaw University of Technology, Poland
e-mail: kazimierz.myslecki@pwr.wroc.pl, piotr@sawinski.org

Abstract

A method is presented for obtaining initial shape of cable-suspended roofs. The background to the method is Hellinger-Reissner fuctional, which is adapted to the problem of net structures. It is expressed in matrix form what makes it suitable to use while writting software. As examples three net-structres are solved with different boundary conditions (hyperbolic paraboloid rigid edge with and without columns inside, flat edge with columns inside and different anchorage of columns).
Keywords: *cable-suspended roofs, Hellinger-Reissner functional, net structures*

1 Introduction

1.1 *Background*

Nowadays architects more often design valiant shape of roofs, giving to the structure own aesthetical thought and showing a diversity of engineering solutions. Designing coverage of public utility buildings, which are often a star-turn of many cities architects likely strive for cable-suspended roofs.

The first such structures were built in tsarist Russia (Nizhny Novgorod) at the end of the XIXth century and the pioneer was Vladimir Shukhov. Of course from the beginning there has been a necessity of cope with computing them and many problems connected with assurance of being tensile at all time and in every part. It is worth to mention the name of Frei Otto, the German architect, who developed methods of computing cable structures and design a lot of such constructions, i.e. Olympiastadion in Munich (Figure 1). Today the name of Santiago Calatrava, who design i.e. The Milwaukee Art Museum (Figure 2) and Entrance to the Olimpic Stadium in Athens (Figure 3) can not be omitted. Nevertheless, in many cities a lot of less famous architects and civil engineers design, compute and construct cable-suspended roofs.

Due to the popularity of cable-suspended roofs engineers have to gain a simple tool which allows them to calculate the structures. According to this there is created a fucnctional, which joins all the necessary conditions and mechanics laws in one formula. The solution is obtained under the iterations algorithm in order to large displacements and so geometrically non-linear character of work of the structures.

1.2 *Why to design cable-suspended roofs?*

There are some aesthetical, engineering and economic reasons according to which cable-suspended roofs are constructed. Undoubtly this structires have

Figure 1. Olympiastadion in Munich

Figure 2. The Milwaukee Art Museum

Figure 3. Entrance to Olimpic Stadium in Athens

interesting, eye-focusing shapes. Smooth curvature with willowy piles on which the tent is spread catches the eyes of visitors.

Despite of visual pleasure, there are also more rational arguments:

- possibility of full exploatation of durability of material. It occurs because of the fact that cable-suspended roofs are only tensile structure and so there is no buckling,
- lightness of structure,
- large spans,
- no necessity of using scaffoldings nor boardings.

However we have to face and solve some problems:

- cables should not be compressed,
- the net has no shape until forces or loads are applied (Figure 4),
- the vibrations occurs eagerlier at flat parts of net, so we have to provide great forces there or try to design nets with constant curvature (i.e. hyperbolic paraboloid)

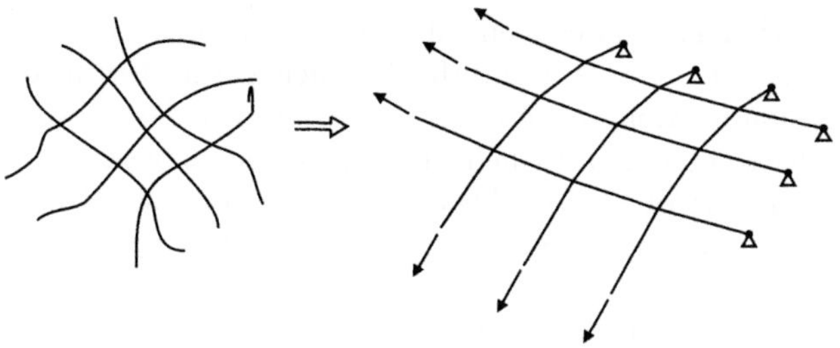

Figure 4. A net without and with applied forces

Moreover it is worth to remain some other facts:

- nets take no shear (have no shear rigidity). Under applied load the roof is forced to change its shape in such manner that in new form it will be able to carry the load without shear,
- large displacements occur,
- great elongation of cable due to applied forces e.g. for 35m long cable with diameter ϕ = 20mm under the force of 333kN (80% of destroing force) the elongation is 0.23m,
- small displacements of supports cause large increase of sag i.e. for 35m long cable with beginning sag of 1m due to 5cm displacement of support the increase of sag is 0.33m.

1.3 *What does the initial shape mean?*

Initial shape of a cable-suspended roof is a shape, which roof obtains under applied prestressing edge forces. There is no other load acting on structure. This shape depends on the shape of edge ring, existence of piles, applied forces and the way of application (to all cables or not).

2 Modyfied Hellinger-Reissner functional as a method of solution

2.1 *The Hellinger-Reissner functional*

Equation 1 shows well known Hellinger-Reissner functional:

$$\begin{aligned}\Pi(\sigma^{ij}, u_k) &= \int_V \left[-\frac{1}{2} D_{ijkl}\sigma^{ij}\sigma^{kl} + \frac{1}{2}(u_{i,j} + u_{j,i})\sigma^{ij} - f^i u_i \right] dV + \\ &\quad - \int_{S_\sigma} u_i \overline{p}^i \, dS_\sigma - \int_{S_u} (u_i - \overline{u}_i)\sigma^{ij} n_j \, dS_u \qquad (1)\end{aligned}$$

where:
σ^{ij} – stress tensor, u_i – displacement tensor, D_{ijkl} – compliance,
f – volume loads, V – volume, p – edge loads,
$\overline{u}$ – known displacements, n – normal to the surface,
S_σ, S_u – surface with known stress and displacements respectively

The above functional is very convenient to adapt for solving cable net structures due to its arguments. Stress σ and displacements u in described case are the vectors of tension in cables $\mathbf{S}$ and displacements of nodes $\mathbf{u}$. We postulate all boundary nodes to be fully attached (no displacements possible) and so we can omit the proper part of Eq. 1. Also carring out analysis of initial shape we assume there is no loads at nodes and then the part $(f^i u_i)$ is also neglected.

2.2 *Modified Hellinger-Reissner functional*

After the modifications described in subsection 2.1, adding new boundary conditions with Lagrangian multipliers and expressing new formula in matrix notation as in Myslecki et al. (1993) we obtain a functional suitable for net structure analysis:

$$\Pi_R(\mathbf{S},\mathbf{u},\lambda) = -\frac{1}{2}\mathbf{S}^T\mathbf{D}\mathbf{S} + [\mathbf{L}(\mathbf{u}) - \mathbf{L_0}]^T\mathbf{S} + \lambda^T(\overline{\mathbf{S}} - \overline{\mathbf{S}}_0) \tag{2}$$

where:
$\mathbf{D}$ – diagonal matrix of compliance,
$\mathbf{L}(\mathbf{u}), \mathbf{L_0}$ – actual and initial cable length respectively,
$\overline{\mathbf{S}}$ – forces in edge cables,
$\overline{\mathbf{S}}_\mathbf{0}$ – fixed tension in some of the edge cables,
λ – Lagrangian multipliers (physically interpreted as edge cables stretches equivalent to initial constrains)

The stationary conditions of Eq. 2 are:

$$\frac{\partial\Pi}{\partial\mathbf{S}} = 0 \qquad -\mathbf{D}\mathbf{S} + [\mathbf{L}(\mathbf{u}) - \mathbf{L_0}] + \Lambda\lambda = 0 \tag{3a}$$

$$\frac{\partial\Pi}{\partial\mathbf{u}} = 0 \qquad \mathbf{G}^T(\mathbf{u})\mathbf{S} = 0 \tag{3b}$$

$$\frac{\partial\Pi}{\partial\lambda} = 0 \qquad \overline{\mathbf{S}} - \overline{\mathbf{S}}_0 = 0 \tag{3c}$$

where:
Λ — one-zero matrix which chooses suitable $\overline{\mathbf{S}}_0$ among $\overline{\mathbf{S}}$,
$\mathbf{G}^T(\mathbf{u})$ — matrix consists of direction cosines of cable axis, which are joined in nodes.
In general form we can write Eqs. 3a–3c as follows:

$$\mathbf{f}(\mathbf{x}) = 0 \tag{4}$$

where $\mathbf{x}$ — vector of all state parameters $\{\mathbf{S},\mathbf{u},\lambda\}$ and Eq. 4 is a non-linear vector equation due to displacements $\mathbf{u}$. Using Newton-Raphson method for this equation we get an iteration algorithm:

$$\mathbf{x}^{i+1} = \mathbf{x}^i + \Delta\mathbf{x}^i \tag{5}$$

where increment $\Delta\mathbf{x}^i$ can be calculated by solving the equation below:

$$\mathbf{A}^i\Delta\mathbf{x}^i = \mathbf{f}(\mathbf{x}^i) \tag{6}$$

with matrix $\mathbf{A}^i = \left.\frac{\partial\mathbf{f}(x)}{\partial\mathbf{x}}\right|_i$ and i – step of iteration.

3 Examples

3.1 *Assumptions*

In the examples shown below it is assumed that:

- any cable has the same longitudinal stiffness $EA = 100\text{MN}$ (like for a cable with Youngs modulus $E = 205\text{GPa}$ and diameter $\phi \cong 2.5\text{cm}$),
- there is a constant edge load along side,
- edge ring and columns inside are rigid.

A net structure is considered with general parameters shown at Figure 5. The dash-lined arrows show possibility of applying edge loads from 4 sides.

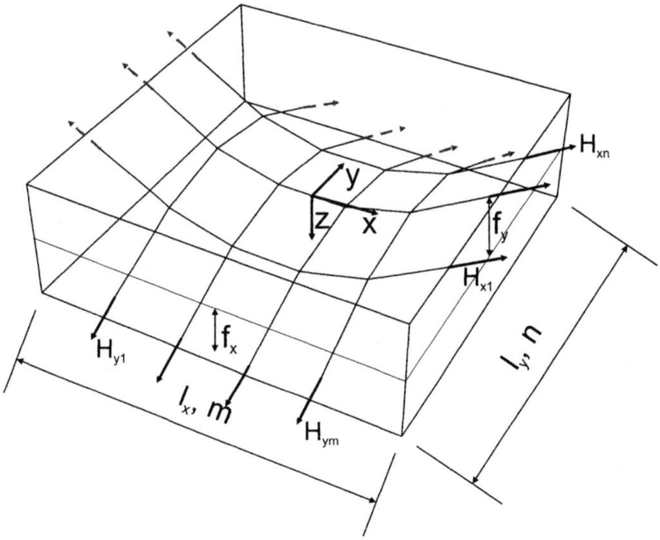

Figure 5. Parameters decribing considered net

The iterations in solving algorithm start from position occupied by nodes according to coordinates calculated as for hyperbolic paraboloid with respect to formula 7:

$$z(x, y) = \frac{f_x}{l_x^2}x^2 + \frac{f_y}{l_y^2}y^2 \tag{7}$$

3.2 *Example 1*

The 5×5 cables net is analized. The edge is a 10m×10m square in plane with $f_x = 2\text{m}$ and $f_y = -2\text{m}$. Prestressing forces are applied along four edges: in case ❶ $H_x = 50\text{kN}$ and $H_y = 5\text{kN}$ and in case ❷ $H_x = 5\text{kN}$ and $H_y = 50\text{kN}$. The solutions show how we can influence on the initial shape of net structure changing edge loads. Obtained shapes are shown at Figures 6 and 7. One

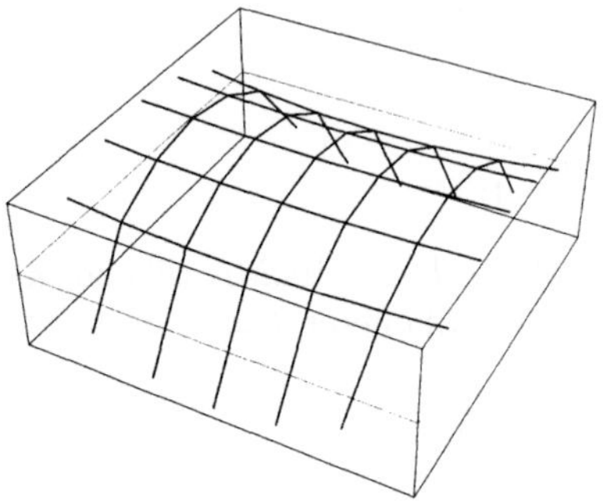

Figure 6. Case ❶

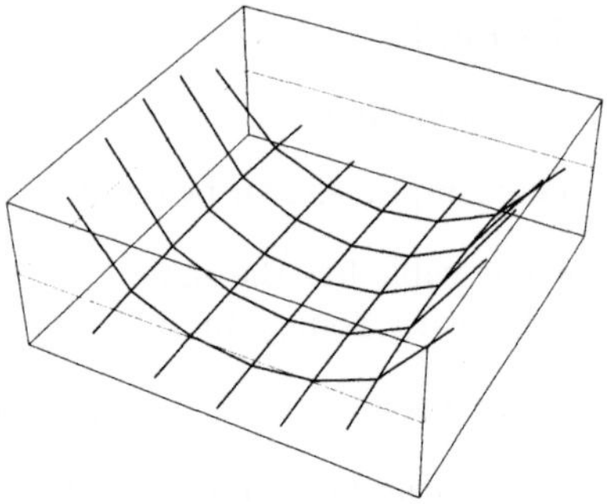

Figure 7. Case ❷

can see from those figures that manipulating the value of prestressing force influence the initial shape of cable net structure in significant way and due to this also on the stiffness and the possibility of applying loads to nodes (Hajduk & Osiecki (1970)).

3.3 *Example 2*

In this example we calculate a net square in plane but spread on flat edge. Additionally four columns are added inside. The length of edge side is 10m and the height of columns is 2.5m. There are 5×5 cables. The prestressing forces are applied only to two perpendicular edges and have the value $H_x = H_y = 3\text{kN}$. There are two different cases: in the ❸ the columns are fully attached to the ground and in the ❹ the rotations are unblocked in anchorage.

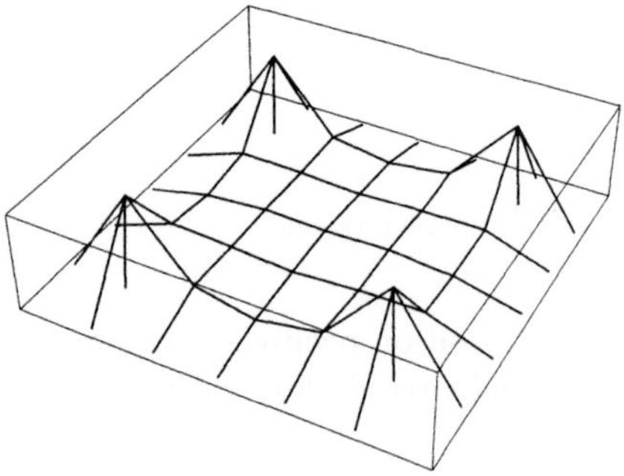

Figure 8. Case ❸

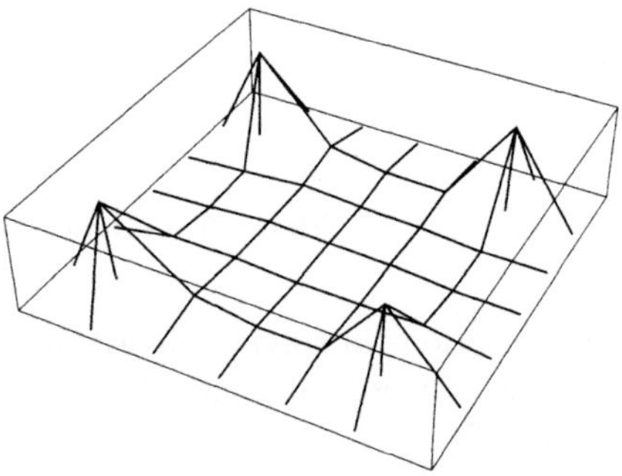

Figure 9. Case ❹

Starting iteration with this same conditions and doing this same number of steps we can notice that the overall height of nodes in case ❹ is lower than in case ❸. It is the obvious result of inclining of the columns due to horizontal displacements of nodes under applied edge forces. This shows that choosing the proper anchorage of columns is one of the problem during designing such structures.

3.4 *Example 3*

In this example we point only to show variety of form which can be used by designer. The rectangular in plane net is spread on hyperbolic paraboloid (vertically) edge (as in subsecton 3.2) with additional ten columns. The columns have different height and anchorage. Due to that ita mountain-shaped net shown at figure 10 is constructed .

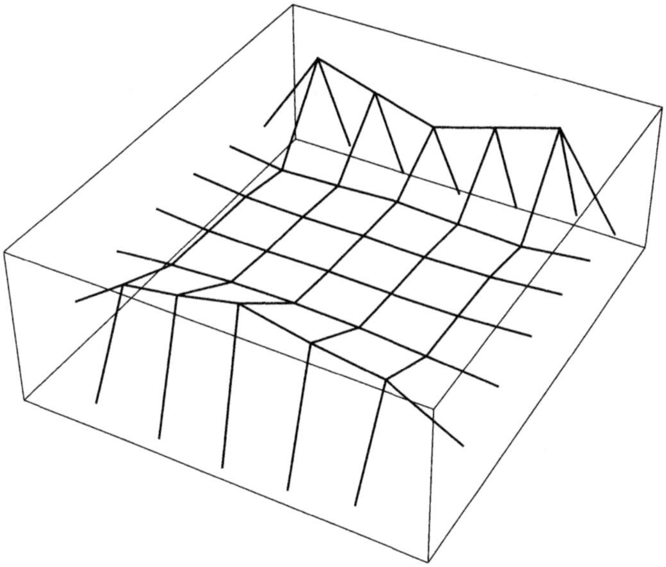

Figure 10. A mountain-shaped net

3.5 *Conclusions*

Presented method of solutions of cable-suspended roofs on the field of calculating initial (erection) shape has an easy adaption to use in software. Due to that it can be strongly adviced for engineers or designers of such structures. Shown examples prove a huge variety of possible shapes. The other publications such as Myslecki (1993), Sawinski (2006) and (2007) reveal wide possibility of using the modified Hellinger-Reissner functional in complex calculating cable-suspended roofs with respect to its optimization. The futher researches are going to extract in evident form part of functional responsible of bending and compressing the edge ring and the part connected with fulfillment of cells between cables.

References

Frackiewicz, H. (1970) *Mechanics of net structures* (in Polish), Instytut Podstawowych Problemow Techniki Polskiej Akademii Nauk, Warszawa, Panstwowe Wydawnictwo Naukowe

Hajduk, J. & Osiecki, J. (1970) *Tensile constructions - theory and calculation*

methods (in Polish), Warszawa, Wydawnictwo Naukowo-Techniczne
Myslecki, K., Kasprzak, T. & Wasniewski, G. (1993) Erection shape analysis of prestressed nets with elastic filling of cells and elastic joints. In: *Gesellschaft fur Angewandte Mathematik und Mechanik. Wissenschaftliche Jahrestagung'93.* Dresden, Germany
Sawinski, P. (2006) *Static analysis of cable-suspended roofs* (in Polish), Masters thesis, Wroclaw
Sawinski, P. (2007) Numerical sensitivity analysis in the field of statics of cable-suspended roofs. In: 6th *International Conference of PhD Students.* Miskolc, Hungary
Szabo, J. & Kollar, L. (1984) *Structural design of cable-suspended roofs*, Budapest, Akademiai Kiado

Section 11

Welding technology I

11.1 Study of the Weld Zone of Friction Welded C45/1.3344PM Dissimilar Steel Joints

Erdinc Kaluc[1], Emel Taban[2]
[1] *Welding Research Center of Kocaeli University, 41200, Kocaeli, Turkey*
e-mail:ekaluc@kou.edu.tr, erdinc.kaluc@yahoo.com
[2] *Kocaeli University, 41200, Kocaeli, Turkey,*

Abstract

In this study, the dissimilar friction welding of a medium C45 (AISI 1040) carbon steel and powder metallurgically produced tool steel 1.3344PM (AISI M3:2) (Vanadis 23) was investigated. Round bars of 11,3 mm diameter in hot-rolled and normalized condition were used. The welding experiments were carried out by an industrial friction welding machine with different friction welding parameters determined from pre-experimental studies and literature. Post weld heat treatment (PWHT) at 650 °C for 4 hours was also applied. Variations of microhardness were measured along the rod axis with 50g test load in all joints and the welding zones were examined with light optical microscope (LOM) and scanning electron microscope (SEM) including both as-welded and PWHTed samples.

Keywords: *Dissimilar welding, friction welding, 1.3344PM, Vanadis 23, C45, AISI 1040.*

1 Introduction

Joining of dissimilar materials is required for several situations arised in industrial applications. The materials are employed in the same structure for effective and economical utilisation of the special properties of each one. In order to take full advantage of the properties of different metals, it is essential to produce joints with high integrity. Welding of dissimilar metals is generally more challenging than that of similar metals due to the difference of the chemical, mechanical and microstructural properties of the parent metals to be joined (Rao et al. (2006).
Joints of dissimilar metal combinations are employed in different applications requiring a combination of properties to save costly and scarce materials. Solid state welding processes are widely employed for joining dissimilar metals because in such situations, fusion welding is usually not feasible owing to the formation of brittle and low melting intermetallics because of the metallurgical incompatibility (Meshram et al. (2007).

As a mass production process, friction welding finds widespread industrial use particularly for high quality joining of dissimilar materials with little or no post weld machining since it has significant economical and technical advantages such as low energy consumption for joining, higher welding exactness than arc welding, efficiency and easy control of welding parameters. Also, the submelting temperatures and short weld times allow joining many combinations of metals (Rao et al (2006), Jayabharath et al.(2007), Ozdemir et al. (2007), Sathiya et al. (2007), Lee et al (2004), D'Alvise et al (2002).

There have been numerous studies related to wrought dissimilar welding, metal matrix composite welding and bulk metallic glasses welding by means of friction welding process (Meshram et al. (2007), Uwaba et al. (2007), Ozdemir et al. (2007),

Hascalik & Orhan. (2007), Arivazhagan et al. (2006), Fu & Du (2006), Hascalik et al. (2006), Kimura et al. (2005), Sahin (2005), Ozdemir (2005), Lee et al. (2004), D'Alvise (2002), Yilmaz et al. (2003), Yilmaz et al. (1996).

Innovation in materials and manufacturing processes make PM components increasingly competitive in replacing wrought or cast materials for various practises. However for some applications, PM parts may have to be joined to one another or with other wrought materials as integrated components to achieve the required task (Jayabharath et al.(2007). 1.3344PM steel, commercially known as Vanadis 23 is a powder metallurgical (PM) high alloyed steel containing (wt. %) 1,28 C, 4,20 Cr, 6,40 W and 3,1 V. For some applications, tool industry might encounter welding of the steels produced by powder metallurgy and carbon steels. There have been few articles found that are originally related to the friction welding of PM steels with other metals such as carbon steel and copper (Rao et al (2006), Jayabharath et al.(2007), Atsushi et al. (1985), Atsushi et al. (1986), Nakahara et al. (1990).

Taking this into account, a study about the microstructural investigation of the weld zones of friction stir welded 1.3344PM and C45 (AISI 1040) steels were realised. For friction welding, friction times of 3 s, 4 s, 5 s, 6 s, 7 s and 8 s were used. Post weld heat treatment (PWHT) at 650°C for 4 hours was applied to the joints. Microstructural properties of the weld zones were investigated including heat treated and as well as non heat treated samples.

2 Experimental procedure

In this study, a powder metallurgically produced tool steel commercially known as Vanadis 23 (1.3344PM) and C45 (AISI 1040) medium-carbon steel were used as base metals.
Chemical composition of the related materials is shown in Table 1.

Table 1. Chemical composition of the base metals (wt %)

	C	Mn	Si	P	S	Mo
C45 (1040)	0,40	0,76	0,25	0,03	0,03	0,003
	C		W	Cr		V
(1.3344PM)	1,28		6,40	4,20		3,1

For friction welding, round steel bars in hot-rolled and normalized condition of 55mm and 70mm lengths respectively for PM and C45 steels with 11,3mm diameter were used.
The friction welding was carried out by an industrial friction welding machine with a hydraulic pressure capacity of 150MPa. The welding has been performed by using optimum parameters, with friction times of 3s, 4s, 5s, 6s, 7s and 8s, determined from pre-experimental investigations and literature, see Table 2.
For each friction time, 7 welds, in total 42 welds, have been produced. PWHT was applied at 650°C for 4 hours to every 4 sample from each group of friction time. The remainder 18 welds were left as welded.
Cross sections were removed from the joints prepared, polished, etched and examined by means of LOM and SEM. Macro- and micrographs of the as welded and PWHTed samples were taken and microhardness measurements with 50g test load were done.

Table 2. Friction welding parameters for 1.3344 PM and C45 steels.

Joint code	Friction time (s)	Rotational speed (rev min^{-1})	Forging time (s)	Friction pressure (MPa)	Forging pressure (MPa)
F3	3	1400	1	60	110
F4	4	1400	1	60	110
F5	5	1400	1	60	110
F6	6	1400	1	60	110
F7	7	1400	1	60	110
F8	8	1400	1	60	110

3 Results and discussion

After welding, quality control with bend testing was applied in industrial conditions and it was observed that the welds with more than 3s friction time exhibit enough strength.

In the visual inspection of all welds, higher flash was observed with increasing friction time. At long friction times, especially at 8s, the flash was observed to form more from C45 side while PM steel did not participate much in the flash formation while for shorter times, such as 4s, flash formed was much less than the one with 8s. Taking into account the microstructural characteristics, the welds with 4s and 8s were chosen as representative of short and long friction times.

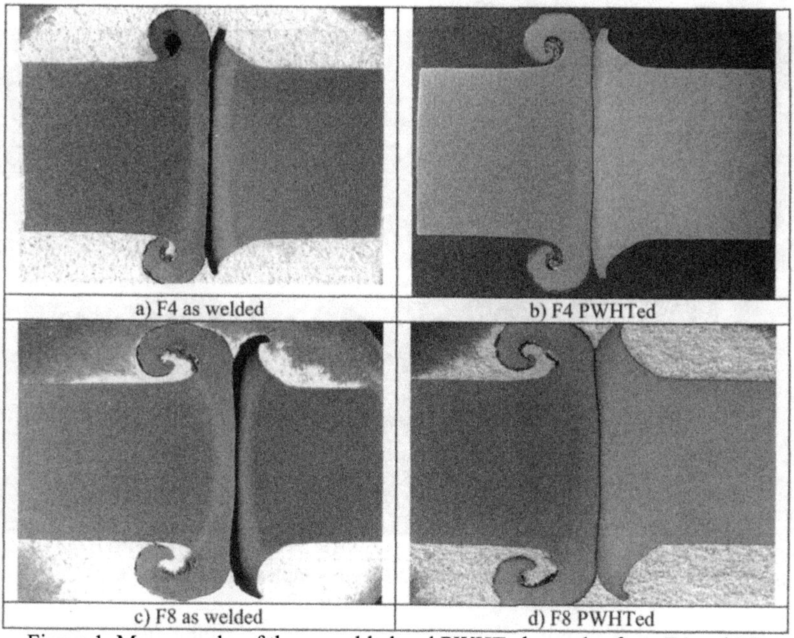

Figure 1. Macrographs of the as welded and PWHTed samples from F4 and F8.

Figure 1, shows the macrographs of the specimens prepared from the joints F4 and F8 which were welded with friction times of 4s and 8s respectively. Since, PWHT was also applied, Figure 1 includes both as welded and PWHTed samples.

The micrographs of the related samples can be seen in Figure 2 for the joint welded with 4s and in Figure 3 for the one with 8s.

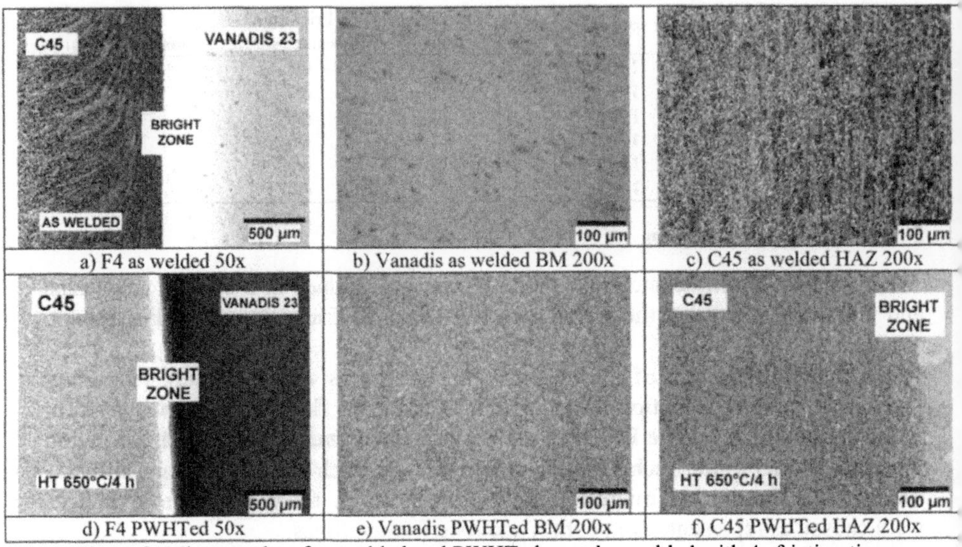

Figure 2. Micrographs of as welded and PWHTed samples welded with 4s friction time.

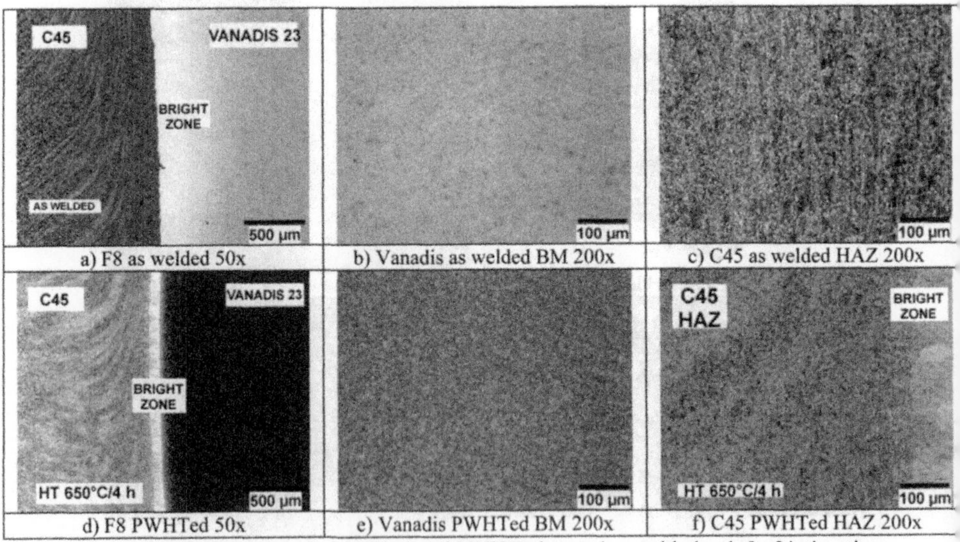

Figure 3. Micrographs of as welded and PWHTed samples welded wth 8s friction time.

A bright zone which is not affected by etching hence giving bright image when examined by microscopy, has been observed for the samples, see Figure 2a, 2d, 3a & 3d. For the joint welded with 4s (F4) friction time has shown a wider bright zone than that of welded with 8s in as welded condition. Bright zone has significantly been narrower after PWHT compared to as welded ones, see Figure 2d and 3d. However at the bright zone, some grain coarsening has been observed after PWHT.

For both joints- F4 and F8, Vanadis base metal showed finer and equiaxed microstructure after PWHT while some improvement was seen at the HAZ of C45 side.

SEM photographs of the weld zones taken from the joint F4, in as welded and PWHTed condition are presented in Figure 4.

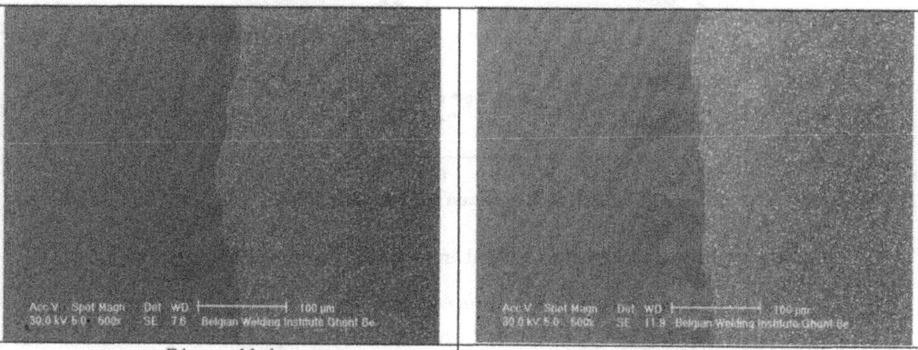

F4 as welded	F4 PWHTed

Figure 4. SEM photographs of the weld zones of F4 as welded and PWHTed.

Hardness plots of the cross sections are presented in Figure 5 and Figure 6 respectively for F4 and F8 both for as welded and PWHT.

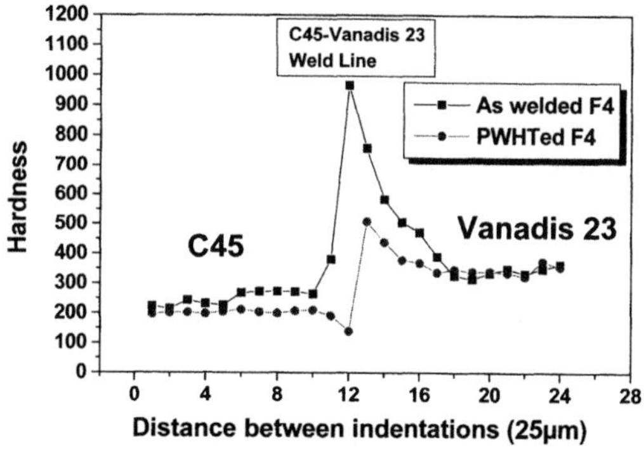

Figure 5. Hardness graphs of as welded and PWHTed samples from the joint F4.

It has been observed that all the weldments have the same characteristic hardness variation curves. Figure 5 shows the hardness profile on the as welded and PWHTed specimens which has been welded with friction time of 4s while the hardness plot of those welded with a friction time of 8s is shown in Figure 6. The distance between indentations is 25μm. In general, the hardness variation at the side of C45 is approximately constant. Minimum hardness value for both cases is thought to depend on the carburization which occurred from C45 side to Vanadis side similar to the research by Yilmaz et al. (1996). It can clearly be seen that hardness generally decreases due to the PWHT at 650°C for 4hours for both types of joints.

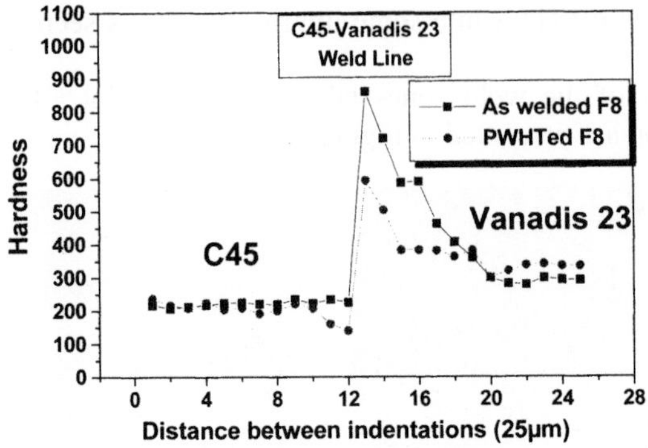

Figure 6. Hardness graphs of as welded and PWHTed samples from the joint F8.

The microstructural studies earlier (Yilmaz et al (1996), Yilmaz (1993) have shown that in the friction welded C45/S6-5-2 dissimilar steel joints, C profile is very similar to the hardness profile showing that the hardness is mainly determined by carbon due to its solid solution on precipitation hardening by carbides. A possible explanation of the bright zone would be in the temperature increase caused by the joining process. The matrix is soluated with carbon and carbide forming elements due the carbide dissolution. And during cooling very fine carbides precipitate which still remain during PWHT. This precipitation causes the hardness to increase at the Vanadis 23 side of the joint.

Acknowledgements

The authors would like to acknowledge Kocaeli University Research Fund in scope of research project (Project No: 2004/51) for financial support. Thanks to Korkmaz Celik for providing the materials and to Supsan Co. for the friction welding. Acknowledgements to Prof. Dr. M. Yilmaz from Kocaeli University and E. Babiloglu for their contribution. Colleagues at the Belgian Welding Institute are acknowledged for the additional microstructural examinations and SEM photos.

References

Arivazhagan, N. Singh, S. Prakash, S. & Reddy, GM. (2006) High temperature corrosion studies on friction welded dissimilar metals. *Materials Science and Engineering B* **132** 222-227.

Atsushi, H. Der-Ming, L. Yoshio, N. & Toru, K. (1985) Friction welding of sintered steel and carbon steel. *Quarterly Journal of the Japan Welding Society* **3** No. 4 696-702.

Atsushi, H. Der-Ming, L. Yoshio, N. & Toru, K. (1986) Friction welding of sintered steel and carbon steel. *Transactions of the Japan Welding Society* **17** 65-70.

D'Alvise, LD. Massoni, E. Walloe, SJ. (2002) Finite element modelling of the inertia friction welding process between dissimilar materials. *Journal of Materials Processing Technology* **125-126,** 387-391.

Fu, F. & Du, SG. (2006) Effects of external electric field on microstructure and property of friction welded joint between copper and stainless steel. *J of Materials Science* **41** 4137-4142.

Hascalik, A. Orhan, N. (2007) Effect of particle size on the friction welding of Al2O3 reinforced 6160 Al alloy composite and SAE 1020 steel. *Materials and Design* **28,** 313-317.

Hascalik, A. Unal, E. & Ozdemir, N. (2006) Fatigue behaviour of AISI 304 steel to AISI 4340 steel welded by friction welding. *Journal of Materials Science* **41,** 3233-3239.

Jayabharath, K. Ashfaq, M. Venugopal, P. & Achar, DRG. (2007) Investigations on the continuous drive friction welding of sintered powder metallurgical (P/M) steel and wrought copper parts. *Materials Science and Engineering A* **454-455,** 114-123.

Kimura, M. Nakamura, S. Kusaka, M. Seo, K. & Fuji, A. (2005) Mechanical properties of friction welded joint between Ti6Al4V alloy and Al-Mg alloy (AA5052). *Science and Tech. of Welding and Joining* **10,** No.6, 666-672.

Lee, DG. Jang, KC. Kuk, JM. Kim, IS. (2004) Fatigue properties of inertia dissimilar friction welded stainless steel. *Journal of Materials Processing Technology* **155-156** 1402-1407.

Meshram, SD. Mohandas, T. & Reddy, GM. (2007) Friction welding of dissimilar pure metals. *Journal of Materials Processing Technology* **184,** 330-337.

Rao, SS. Jayabharath, K. Ashfaq, M. & Krishna, BV. (2006) Preliminary studies on friction welding of sintered P/M steel preforms to wrought mild steel. *Science and Tech. of Welding and Joining* **11,** No.2, 183-190.

Ozdemir, N. (2005) Investigation of the mechanical properties of friction welded joints between AISI 304L and AISI 4340 steel as a function rotational speed. *Materials Letters* **59,** 2504-2509.

Ozdemir, N. Sarsilmaz, F. & Hascalik, A. (2007) Effect of rotational speed on the interface properties of friction-welded AISI 304L to 4340 steel. *Materials and Design* **28,** 301-307.

Sahin, M. (2005) Joining with friction welding of high speed steel and medium carbon steel. *Journal of Materials Processing Technology* **168,** 202-210.

Sathiya, P. Aravindan, A. & Haq, AN. (2007) Effect of friction welding parameters on mechanical and metallurgical properties of ferritic stainless steels. *Int J Adv Manuf Technol* **31,** 1076-1082.

Seiji, N. Tomio, S. & Hideo, K. (1990) Friction welding of sintered steels to carbon steel. In the case of bar type of sintered compact steel. *Journal of Mechanical Engineering Laboratory* **44,** No.1, 1-7.

Uwaba, T. Ukai, S. Nakai, T. Fujiwara, M. (2007) Properties of friction welds between 9Cr-ODS martensitic and ferritic-martensitic steels. *Journal of Nuclear Materials* **367-370,** 1213-1217.

Yilmaz, M. (1993) Investigation of the weld zone of the dissimilar friction welded tool steels. PhD thesis, Yildiz Technical University.

Yilmaz, M. Kaluc, E. Tulbentci, K. & Karagoz, S. (1996) Investigation into the weld zone of friction welded C45/HS6-5-2 dissimilar steel joints. *J of Materials Science Letters* **15,** 360-362.

Yilmaz, M. Col, M. & Acet, M. (2003) Interface properties of aluminium/steel friction welded components. *Materials Characterization* **49,** 421-429.

11.2 Solid Wire vs. Flux Cored Wire – Comparing Investigations for GMA-laser-hybrid Welding

Stephan Lorenz[1], Thomas Kannengiesser[1], Gerhard Posch[2]
[1]*Federal Institute for Materials Research and Testing, 12205 Berlin, Germany*
e-mail: stephan.lorenz@bam.de, [2]*Böhler Welding Austria GmbH, 8605 Kapfenberg, Austria*

Abstract

Application of up to 20kW high-power fibre lasers enables joint welding of high-alloyed steels with wall thicknesses < 10 mm at high welding speed. Notably GMA-laser-hybrid welding opens up new applications in industry, since, for example, it allows gap bridging and increases the overall process option availability for crack-resistant welds. Comparative laser welding experiments using a 20kW fibre laser as well as GMA-laser-hybrid welding experiments with conventional solid wires and new types of tubular cored wire were performed. Particularly due to the high achievable deposition rate in connection with the appearing deposition characteristics, tubular cored wires may offer a new alternative for laser-hybrid welding. Various tubular cored electrodes, as for example with rutile and metal powder filling, have been investigated. Each wire type exhibits special process-specific properties influencing the welding process stability as well as the metallurgy. This study demonstrates the potential of advanced filler materials in the field of laser-GMA-hybrid welding.

Keywords: tubular cored electrode, solid wire electrode, GMA-laser-hybrid welding

1 Introduction

The "arc augmented laser welding" exists as a basic procedure since the end of the 1970ies and was developed by Steen (1980) for the first time. At the beginning of the 1990's when laser beam sources of adequate beam power and quality were available for advanced industrial applications this technology became more important for welding plates above 5 mm wall thickness, Matsuda et al. (1988). Referring to this a specific kind of procedure is represented by the Laser-GMA-hybrid welding. It requires high deposition rates by the arc process and accordingly it features good gap bridgeability as well as deep penetration because a focussed laser beam and an arc of a constant melted wire operate in the same area. Nowadays solid wires are primarily used for Laser-GMA-hybrid welding processes Wieschemann (2001), Thomy et al. (2003), Coutouly et al. (2006), Liu et al. (2006).

Mild and high strength (StE 690V) steel hybrid welding welded with a matching filler material and a basic cored wire accompanied to a CO_2-Laser beam source, were investigated by Neuenhahn (1999). The CO_2-Laser-GMA-hybrid welding gave a good toughness of weld metal. The higher toughness can be reached if the Laser-GMA-hybrid welding is used instead of the laser beam-cold wire process.

Cored wires own special properties like efficient and economic modification of the filling for an adaptation to new welding tasks and procedures. Therefore, it is possible to develop new alloying concepts in small batch sizes for a short time Perteneder et al. (2001). Regarding a special shape of cored wire and an additional alloying metal powder inside of the core wire, this produces a specific arc characteristic. This causes high current density in the outer thin walled tube and generates a broader arc especially by using a rutile cored wire. As shown in Figure 1 the weld seam geometry produced by solid wire welding presents a lower penetration than welded with a metal cored wire, Widgery (1994).

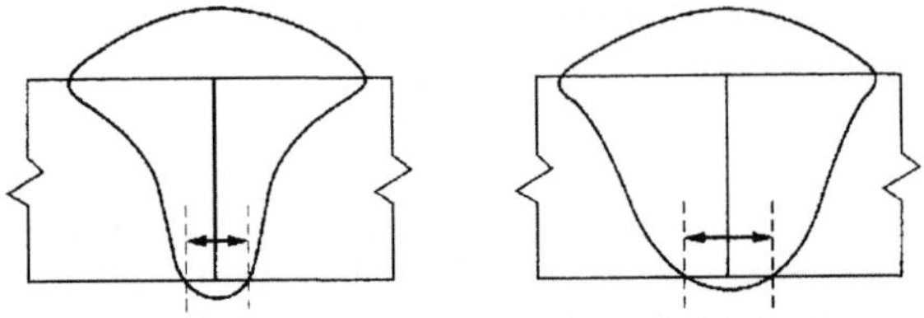

Figure 1: penetration section of solid wire (left hand side) and tubular cored wire weld (right hand side), Widgery (1994)

The mentioned properties of cored wires provide a scientific approach for research. The phenomena of welding different wire types are discussed in the following.

2 Test equipment and materials

The division "Safety of Joined Components" features modern welding equipment like a solid state laser. The nominal output power P_{nom}, beam parameter product *BPP* and output fiber core diameter d_{fc} are listed in
Table 1. The welding speed was varied in a broad range due to the high laser beam power. The wires were welded with a 500 ampere GMA welding power source. For each type of wire a specific available welding program was chosen to start the experiments with optimal conditions.

Table 1: specifications of the solid state laser beam source

P_{nom} in kW	*BPP* in $mm \cdot mrad$	d_{fc} in μm
20	11.5	200

Base material selected to use in this investigation is an austenitic stainless steel X2 Cr Ni 19 11 (AISI 304L). An extraction of the chemical composition analysed by spark source spectroscopy is shown in
Table 2.

Table 2: Chemical composition of the base material, wt-%

Fe	*Cr*	*Ni*	*Mn*	*Cu*	*Si*	*Mo*	*Co*
69.50	18.13	9.92	1.09	0.15	0.49	0.26	0.20

2.1 *Parameter setting*

In order to support an economic demand on a high power welding procedure a maximum wire feed rate of 15 m/min was used. Another important point in order to

perform a comparison basis in welding four different filler materials are used. Solely in the case of the rutile cored wire with slowly solidifying slag T 19 9 L R M (C) 3 (FCW (R)) the deposition rate was varied because of investigations concerning the arc behaviour at lower wire feeder velocities. As shown in
Table 3 maximum arc power P_A is different in each wire type. The reason is the fact that arc power is a unique characteristic of welding properties. The other parameters, i.e. laser beam power P_L, welding speed v_s, focal position Δz and distance a_t between laser beam axis and the end of the wire were adapted to a specific value during welding. In order to better understand short terms represented in
Table 3, the following terms are used. The metal cored wire T 19 9 L M M 1, the rutile cored wire with rapidly solidifying slag T 19 9 L P M 1 and the solid wire G 19 9 L Si are marked by MCW, FCW (P) and SW, respectively. The diameter of all wire is 1.2 mm.

Table 3: Parameter setting range

	P_L *in kW*	P_A *in kW*	Δz *in mm*	a_t *in mm*
FCW (P)	12	8.6 - 10	0	4
FCW (R)	5 - 12	5.8 - 10.1	-8, -3, 0	2.5, 5
MCW	10 - 12	10 - 11.3	-3, 0	4
SW	8 - 12	10.3 - 11.8	0	4

The shielding gases were selected by supplier recommendation. Welding with rutile cored wires was applied with 18 % CO_2 in argon (M21). The metal cored wire as well as the solid wire were welded with 2 % CO_2 in argon (M11). An additional trailing nozzle with argon sprinkling was employed to minimise the oxidation on the typically elongated weld pool, DIN EN 439 (1994).

3 Results

3.1 *Arc position*

Based on the test results all wire types exhibit a uniform tendency. The Laser-GMA-hybrid process in leading arc position can be applied in broad welding speed range without any instability. A stable arc process with less spatter generation notably can be observed. The arc has a detrimental effect on the weld pool in leading arc position that is characterised by an increased spatter generation.

For clean welding processes using rutile cored wires it is recommended to adjust a slightly leading arc position with an inclination angle of 10°. Otherwise slag ingredients would be rinsed in front of the weld pool and finally included under the liquid weld metal. But due to constructive restrictions a minimum inclination angle of 21° can be adjusted. The consequence is a larger weld seam elevation and pronounced notch penetration for welding with metal cored wire.

In trailing-arc position peculiar increased pore shaped slag inclusions are visible in the cross sections of rutile cored wire welded specimens. It is significant, as illustrated in

Figure 2 a), that pores are mostly developed in the laser induced area of the weld metal. Furthermore a higher resolution might uncover the spherical forming of the slag in the pores surrounded by a thin area of gas (black circle), as shown in

Figure 2b). In some cases the pore is not completely filled up with slag due to a rinse out effect during metallographic preparation. A root shape penetration of the dendritic solidified slag into the weld metal did not occurred in any of the micrographs. Typical examination accompanied to radiographic tests were used to reveal microstructures of weld in particular for a pore size. But, critical pore size was not found.

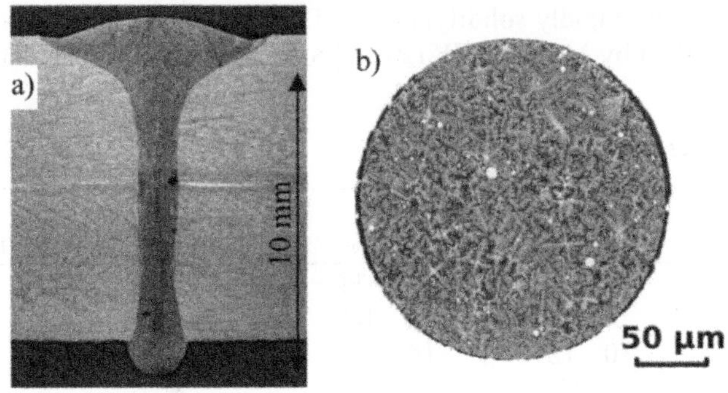

Figure 2: a) a cross section of a Laser-GMA-hybrid weld produced with rutile cored wire (etchant according to Lichtenegger Bloech I), b) a cross section of a slag-containing pore (picture generated by backscattered electron)

3.2 *Wire characteristics*

There are distinct differences regarding the slag development between both used rutile cored wire types. Compared to the rutile cored wire with slowly solidifying slag welding with the wire with rapidly solidifying slag leads to easier slag separability, as illustrated in

Figure 3 a) and b). It must be deducted that the chemical composition of the applied rutile slag systems has influenced a generation of the slag during welding.

For the rutile cored wire with slowly solidifying slag, it is found that partially adherent slag still remains on the weld metal surface, which is not a proper characteristic of inclusion removal. As presented in

Figure 3 b) slag remainders, so called slag islands, appeared at the weld metal centreline of the top surface along the weld seam. During the formation of these adherences the influence of the focussed laser beam is essential because of its high heat input affecting the centre of the process zone. Oladipupo (1987) points out that due to the formation of chromium spinels (1600 °C) as well as calcium titanate (1900 °C) badly detachable slag remainders are formed. In this case the adverse influence of aluminium on slag detachability, as Killing (1996) refers, is negligible. In the base material, a minimum content of 0,007 wt-% by spark source spectroscopy was measured.

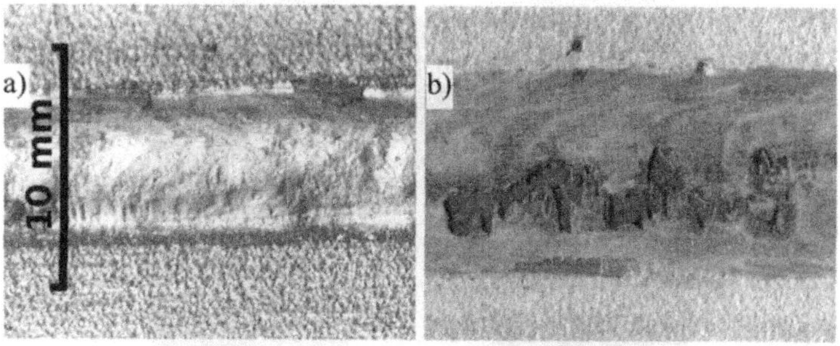

Figure 3: a) hybrid weld seam top side surface using rutile cored wire with rapidly solidifying slag directly after welding in brushed condition and b) using rutile cored wire with slowly solidifying slag, pickled for one hour after brushing

More importantly, the effect of chromium on slag detachability was investigated by an electron probe micro analysis (EPMA) as illustrated in Figure 4 a) and b). The investigated location was depicted from a longitudinal section of a hybrid weld seam. In the weld metal and the slag layer chromium was found. This implied that cohesion of this element can be found even if the spacing between weld metal and slag is narrow as shown Figure 4 a).

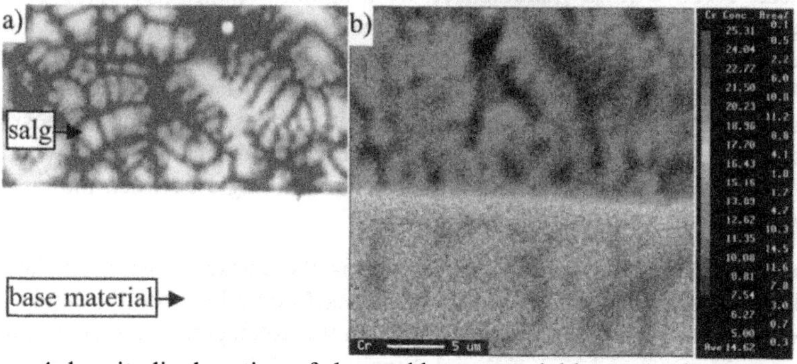

Figure 4: longitudinal section of slag and base material layer a) a backscattered electron (BSE) photograph and b) chromium mapping of the same surface by EPMA

As another important point, some drops of a slag-metal mixture transferred from the torch to the weld pool can be easily captured by an advantage of a high speed camera, as shown in Figure 5.

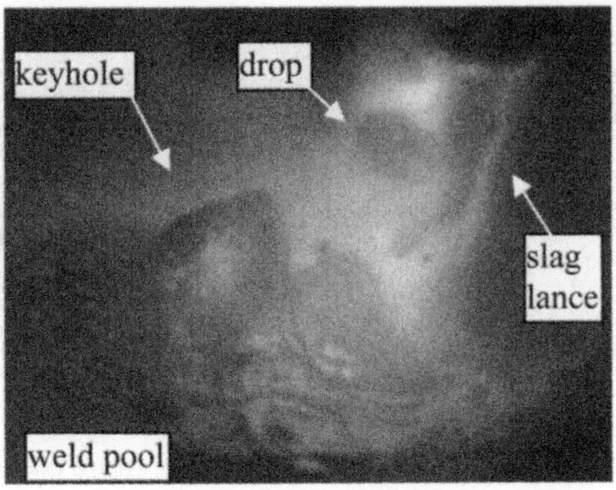

Figure 5: Hybrid-process area with the end of a rutile cored wire and a lance shaped slag

The solid wire was welded with pulsed arc in order to avoid degradation of alloying element from burning process. Moreover, a controllable drop changeover is developed to get a stable arc process and it only marginally affects weld pool. In leading arc position a narrow weld seam and less of spatter can be produced with this wire type. Otherwise in trailing arc position no stable parameter setting range could be observed.

The metal cored wire is characterised by a smooth spatter-free spray arc for all parameter settings due to the specific arc behaviour. For the most part the current is conducted in the thin shell and the current density is higher compared to a solid wire. Additionally, to melt the powdered filling a lower energy is used. This means that a metal cored wire seems to be high deposition rates. In leading arc position the weld seam profile shows a deeper penetration compared to the solid wire. Slight notch penetration can be detected, if the inclination angle between the laser beam axis and wire axis is up to 20°.

Another reason for the undercut appearance is the surface tension on the molten pool. Feng et al. (1998) observed that the contact angle between the liquid surface and the fusion line has a significantly effect on the notch generation at the transition between weld metal and base material. Finally, notch penetration becomes lower if the weld pool volume is increased and the molten width is decreased. There are still some geometrical factors which can be changed to minimize the undercut occurrence (e.g. torch configuration).

For the trailing arc position the results show that the rutile cored wires are applicable in a narrow parameter setting range because of their smooth arc behaviour. In case of the metal cored wire a stable parameter setting range above 3m/min welding speed is predicted.

4 Conclusion and outlook

In this study four different wire types were investigated for the Laser-GMA-hybrid welding process. Distinct differences between trailing and leading arc position could be regarded. Only the rutile cored wires can be applied in the trailing arc process above 2 m/min welding speed. For the metal cored wire a stable process is expected above 3 m/min welding speed. The solid wire can only be welded in leading arc position.

Concerning the two different rutile cored wire types in both arc positions pores including slag were generated. Almost slag inclusions appeared at the middle height and the bottom part of the weld seam. The slag is mainly formed like a sphere. The detected pore size does not exceed critical values which could be measured by radiographic tests.

Furthermore, badly detachable slag remainders were generated on the weld seam surface by using the rutile cored wire with slowly solidifying slag. With an electron probe micro analysis could be proven that this phenomenon might be indicated by the formation of chromium spinels when the slag is exposed to high temperatures affected by laser beam energy. The rutile cored wire welding with rapidly solidifying slag results in easy detachable slag directly after welding due to a different composition of the filling.

For the further studies the formation of undercuts during hybrid welding with metal cored wire will be investigated. This wire type features high deposition rate ability above 15 m/min. There is a scientific demand to weld such materials up to the upper limitation of welding machine.

The improved slag removability of the rutile cored wire with rapidly solidifying makes this type of filler metal reliable for the positional Laser-GMA-hybrid welding.

References

Coutouly, J. F., Deprez, P., Demonchaux, J. & Koruk, A. I. (2006) The Optimisation of Laser Welding and MIG/MAG - Laser Hybrid Welding of Thick Steel Sheets. *Lasers in Engineering*, **16** 399 – 411.

DIN EN 439 (1994) Welding consumables - Shielding gases for arc welding and cutting. German version

Feng, L., Chen, S., Li, L. & Yin, S. (1998) Static equilibrium model for the bead formation in high speed gas metal arc welding. *China Welding*, **7** No. 1, May, 22 – 27.

Killing, R. (1996) Angewandte Schweißmetallurgie - Anleitung für die Praxis. *DVS-Verlag GmbH*, ISBN 3-87155-130-9

Liu, Z., Kutsuna, M. & Sun, L. (2006) CO_2 Laser-MAG Hybrid Welding of 590 MPa High Strength Steel. *Quarterly Journal of Japan Welding Society*, **24** February 1 – 9.

Matsuda, J., Utsumi, A., Katsumura, M., Hamasaki, M., Nagata, S. (1988) TIG or MIG arc augmented laser welding of thick mild steel plate. *Joining & Materials*, **1**, July 31 – 34.

Neuenhahn, J. C. (1999) Hybridschweißen als Kopplung von CO_2-Hochleistungs-lasern mit Lichtbogenschweißverfahren. *Fraunhofer –Institut für Lasertechnik/ Institut für Schweißtechnische Fertigungsverfahren, Aachen,* Dissertation, © Shaker Verlag, 1999.

Oladipupo, A. O. (1987) Slag detachability from submerged arc welds. *Welding Research Council - Progress Reports*, **42** 1/2, 23 – 24

Perteneder, E., Dörfler, R., Posch, G., Tösch, J. & Ziegerhofer, J. (2001) MAG-Schweißen unlegierter und hochlegierter Stähle mit Fülldrahtelektroden - dargestellt aus der Sicht eines Schweißzustzherstellers. *Schweiß- und Prüftechnik* 09, pp. 130 – 135.

Steen, W. M. (1980) Arc augmented laser processing of materials. *Journal of Applied Physics,* **51** No. 11, 5636 – 5641.

Thomy, C., Schilf, M., Seefeld, T., Vollersten, F., Sepold, G. & Hoffmann, R. (2003) CO_2-Laser-MSG-Hybridschweißen in der Rohrfertigung, *DVS Berichte 225*, pp. 167 – 173.

Widgery, D. (1994) Tubular Wire Welding. *Woodhead Publishing Limited,* ISBN 1 85573 088 X

Wieschemann, A. (2001) Entwicklung des Hybrid- und Hydraschweißverfahrens am Beispiel des Schiffbaus, *Aachener Berichte Fügetechnik*, Band 2001, 1, ISBN 3 8265 8852 5

11.4 Precision Joining of Al-Al Micron Foils by Improved Ultrasonic Friction Welding

XueChao LU, Rong LI, XinChun LAI

China Academy of Engineering Physics, Mianyang, P.O.Box:919, 621900, Sichuan, P.R.China.e-mail:thirty2006@yahoo.com

Abstract

Precision joining of 20μm~25μm thick Al-Al foils with no introduction of different to parents materials was realized by improved Ultrasonic Friction Welding(UFW) method, since the conventional UFW method puts more focus on large scale of bulk metal materials. The monomer and post-joining Al foils were performed to the analysis or comparison analysis including surface roughness, thickness homogeneity, interface and bonding strength by using White Light Interferometer (WLI),Optical Microscope (OM) and Scanning Electron Microscope(SEM) etc.. As experimental results indicate, high quality Al-Al foils components and one kind of precision device applied in ICF micro-targets were well prepared by our improved UFW.

Key words: *Ultrasonic Friction Welding ; Precision joining ; Surface roughness ; Thickness homogeneity*

1 Introduction

Precision joining of the same or dissimilar micron foils with fine surface roughness parallel interface in nanometer size levels homogeneous thickness, and no introduction of different parents materials are essential needs in many aspects like Inertial Confinement Fusion (ICF) micro-targets, ULSI, MEMS etc., since the conventional methods like glued joint, diffusion bonding and solder brazing can not meet the stringent requirements well. The friction welding, as one kind of high-quality、high-efficiency and low- cost solid state joining spread out in the area of aircrafts、automobiles、and cutting tools of interest (Fuji et al. 1995, 1996, Yibas et al. 1995) is expected to realize the precision joining of metal foils. However, it is almost large scale of bulk metal materials reported as the friction welding objectives instead of micron foils like Al Copper (Sahin et al. 1998, 1996, Okuyucu 2007). In this study, emphasis is put on the about 20μm and 25μm thick Al foils as objectives and the Ultrasonic Friction Welding (UFW) method with improvement is employed to realize the precision joining of Al foils, in which the post-joining quality of components or devices is characterized by White Light Interferometer (WLI), Optical Microscope (OM) and Scanning Electron Microscope(SEM).

2 Experimental

2.1 *Al foils and the UFW*

Rolled and annealed pure Al foils were commercially provided with the nominal 20μm and25μm in thickness respectively. The ~2mm×~2mm area size of monomer Al as the sidestep foils and another group of a little greater of Al as the substrate foils are together subjected to the UFW process, subsequently in half-lapping components. The fundamental principle of UFW is that output ultrasonic through the energy-exchange components brings out high-frequency vibration of the up force head, which acts on the materials on anvil bed perpendicularly and quickly, thus the joining completed owing to the friction heat occurrence in interface. Before the UFW application, the Nicle-35 UFW equipment was developed to reduce the load force from $50N/mm^2$ to5 N/mm^2 due to the high ductility of Al and realize the continuous driving force. Also, re-machining was executed for the force head and anvil bed so that high surface accuracy was expected. In this experiment, the UFW frequency is about 3500HZ and the welding and delay time are both optimized in the ranges of 0.02s to 0.06s.

2.2 *Measurement and analysis*

The surface roughness and thickness of both of the Al foils and the post-joining Al-Al components are measured by using the contact less type Taylor CCI6000 White Light Interferometer with 100μm of scope in thickness. High accuracy of 0.01nm resolution on Z-axis and 0.003nm repeatability of the R_q (Root Mean Square) value can meet the precision characterization. The monomer Al foils about ~2mm×~2mm area size are necessary, previously measured on standard gauges, then as sidestep foils, subjected to the UFW with another group of greater area size of Al substrates foils. The surface roughness R_q of gauge is about 20nm. The point is that measurement of the extraction lines on monomer Al foils are localized in order to ensure the datum comparison in corresponding line-line relations before and after the UFW, in which the measurement localization is completed through the pre-made micro-mark on the monomer Al foils. In this study, three measurement lines with 100μm space distance between two neighbouring lines are extracted as the characterization of surface roughness and thickness of Al foils.
The cross section, interface and configuration micrographs of the post-joining Al-Al components were observed by SEM(KYKY-1010B) and OM(OLYMPUS) respectively.

3 Results and discussion

3.1 *Thickness homogeneity*

Two groups of 25μm /25μm and 20μm /25μm of Al-Al components were prepared respectively. For instance the 25μm/25μm components are introduced as follows. Figure 1 shows 3D diagram of the monomer Al foil and thickness measurement result and Figure 2 displays the results of Al-Al components. Both of the curves represent the measurement of 1st line according to Table 1. As can be seen from the measurement results, the post-joining Al sidestep foil has no obvious deformation and is of homogeneous thickness in the Max./Mea. of 25.35μm/24.88μm by comparison to the previous monomer of 24.55μm/24.49μm. The thickness datum is negligibly changed. In addition, although thickness of the substrate Al foil is not initially measured, after UFW its thickness in the Max./Mea. of 24.15μm /24.10μm closer to the monomer one still can be seen in Figure 2, which in turn indicates the capacity of UFW for the precision joining of micron metal foils. In fact, all of the 25μm thick Al foils were obtained from same one sheet in area size of 50mm×50mm. The measurement results of extracted three lines for Al foil before and after UFW are listed in Table 1. The thickness change of Max./Mea. values of the rest two lines after joining are both slightly greater than that before and also can

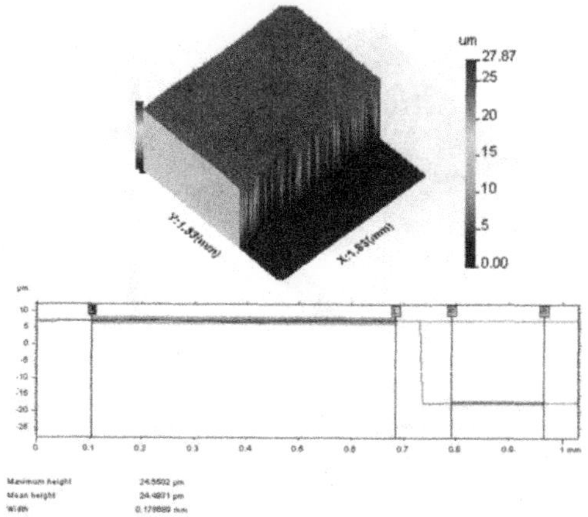

Figure 1. WLI measurement results of the Al monomer foil

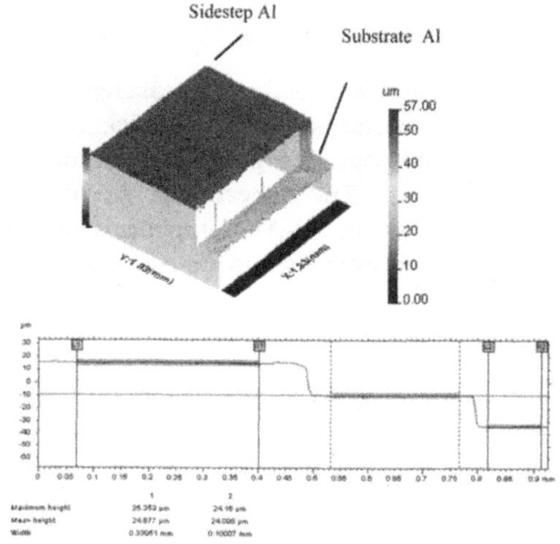

Figure 2. WLI measurement results for 25μm/25μm Al-Al components after UFW

Table 1. The thickness results of extracted three lines on 25μm Al foil (μm)

Lines extraction	as monomer foil before UFW		as sidestep foil after UFW	
	Max.	Mea.	Max.	Mea.
1st line	24.55	24.49	25.35	24.88
2nd line	24.89	24.21	25.41	24.50
3rd line	24.12	24.11	25.13	24.38

be neglected. The main reason is that the measurement datum plane is different, in which the gauge as datum plane is for monomer Al foil and the substrate Al as that is for the sidestep Al foil after UFW(seen in Figure 2.).Clearly, the surface roughness of the standard gauges is much better than that of the rolled Al foils.

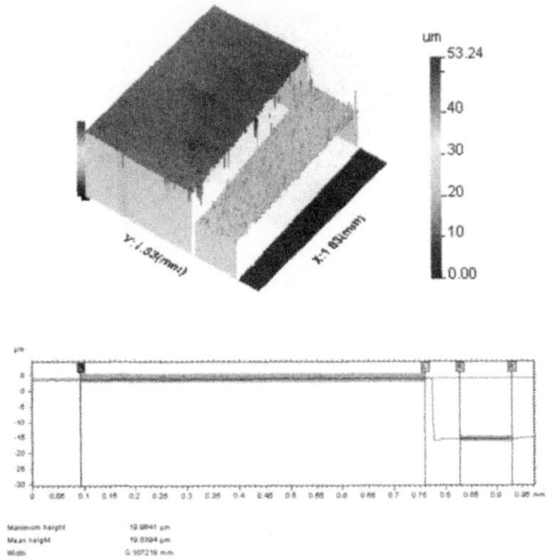

Figure 3. WLI measurement results for 20μm/25μm Al -Al component after UFW

Figure 3 displays the 3D diagram and thickness measurement of 20μm/25μm Al-Al component. Slight deformation and homogeneous thickness are still present much like the joining of 25μm/25μm ones. In Figure 3 the thickness of the sidestep Al is in the Max./Mea. of 19.96μm /19.64 respectively. According to the results, the highly homogeneous thickness in Al-Al components without introduction of different to parent materials in interface can be achieved by our improved UFW method.

3.2 *Surface roughness*

For all monomer and sidestep Al foils, the conditions and parameters of WLI measurement are the same. In this study of surface roughness, the cutting off wavelengths for extraction lines are defined as 80μm. Three extraction lines for surface roughness analysis are actually as the same as the corresponding ones for thickness homogeneity analysis mentioned previously, but each R_a , R_q, R_p and R_v values for surface roughness is averaged before and after UFW. As can be seen in Table 2, the arithmetical average R_a, R_q and the lowest wavelength valley R_v are

Table 2. The surface roughness analysis of Al foils (μm)

	20μm/25μm component				25μm / 25μm component			
	R_a	R_q	R_p	R_v	R_a	R_q	R_p	R_v
As monomer before UFW	0.045	0.071	0.25	0.12	0.028	0.051	0.10	0.13
As sidestep after UFW	0.055	0.082	0.20	0.18	0.037	0.062	0.044	0.31

all increased after UFW, but the highest peak R_p are both decreased more or less. Owing to the friction between force head and Al surface, the ductile Al is easily subjected to damage on surface. On the other hand, the R_p value is decreased due to the impact of force head acting on the Al surface. Generally, all of the influences are not obvious.

3.3 *Micrographs of components*

Figure 4. (a) and (b) show the cross-sectional micrographs of 25μm/25μm and 20μm/25μm Al-Al components respectively. Obviously both of Al-Al components are dense and parallel in interface (joints), thus it is possible to make the homogeneous thickness after UFW. Also the surface morphologies of the joining faces of Al foils were observed as can be seen in Figure 5. (a). By comparison to the non-joining Al surface displayed in Figure 5.(b) the surface of the former is more smoothly and have some dispersive defects, whereas the latter presents the original texture of rolled Al foils. According to the UFW principle the defects can be considered the highly plastic deformation and atoms diffusion points in the course of welding. Almost of these defects are hundreds nanometers in depth, so the original thickness and homogeneity of the monomer foils can be kept too much extent after UFW, which is actually required in many precision-joining aspects of interest.

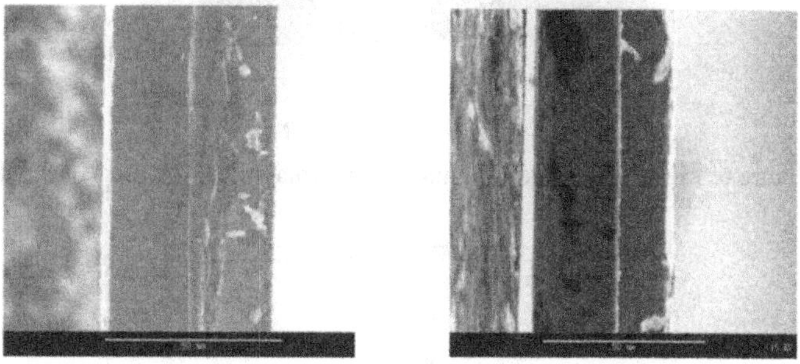

(a) (b)

Figure 4. SEM cross-sectional micrographs of Al-Al components

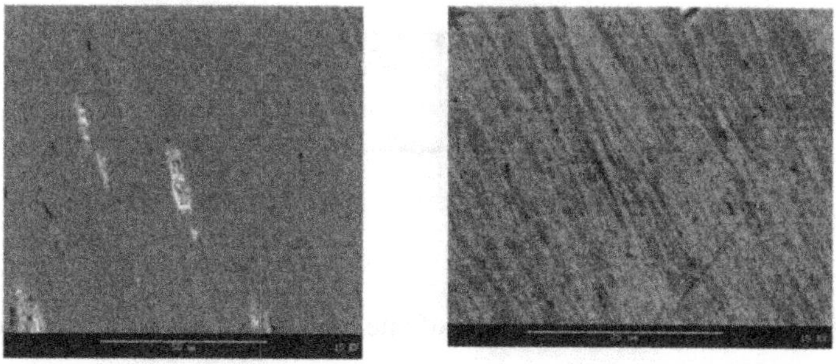

Figure 5. SEM morphologies of joining and non- joining faces of Al foils

Due to the very small components, it is hard to observe the bonding strength of Al-Al components. But because of the local atoms diffusion in interface (Meshram 2007) the ratio of bonding strength to mass can be accepted in many micro-assembly scopes.

3.4 *One kind of precision device*

By using the improved UFW, one kind of precision device applied in ICF micro-targets were well prepared in one-step process. Figure 6 displays the device was assembled with bi-sidestep Al foils and one substrate Al foils. One sidestep Al is 20μm thick and another is 25μm thick, between which the rectangle groove is about 117μm in width. The homogeneous thickness、low deformation and fine surface roughness etc. are also required in the device. Figure 7 shows the homogeneous thickness of two sidesteps of one random extraction line on the device.

Figure 6. Mesh figure and OM micrograph (magnification 80×) of the device

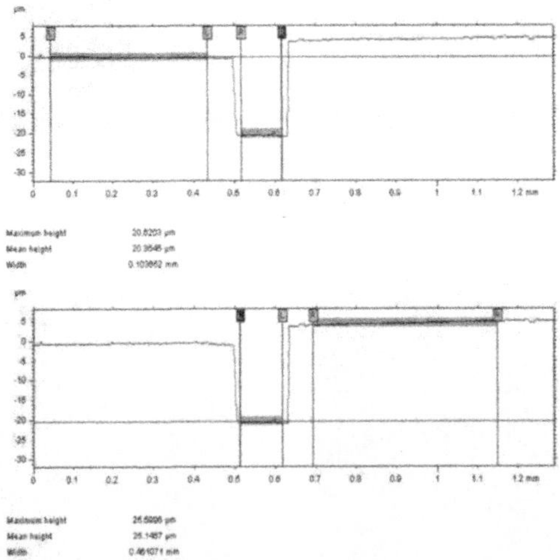

Figure 7. WLI measurement results of one random extraction line on the device

3 Conclusions

Precision joining of Al-Al micron foils have been realized by our improved UFW, although the conventional UFW is usually employed to the joining of bulk metal materials. In the experiments the monomer micron Al foils and the same as sidestep ones in Al-Al components are negligibly changed in thickness, in which the slight deformation and homogeneous thickness are also obtained from the datum analysis and indication of the parallel interface. One kind of well-prepared device applied in ICF micro-targets indicates that the UFW with special improvement is capable of precision joining for Al foils.

References

Fuji, A., Ameyama,K. & Futamata, M.(1995) Improving tensile strength and bend ductility of titanium/AISI304L stainless steel friction welds. *Materials Science and Technology* **8** 219-235

Fuji, A., Ameyama. K. & North, T.H.(1996) Improved mechanical properties in dissimilar Ti-AISI 304L joints. *Journal of Materials Science* **31** 829-827.

Meshram, S., Mohandas, D. & Madhusudhan, T.G. (2007) Friction welding of dissimilar pure metals. *Journal of Materials Processing Technology* **18**

Okuyucu, H., Kurt.E. & Arcaklioglu, A. (2007) Artificial neural network application to the friction stir welding ofAl plates. *Materials & .Design* **28** 78-84.

Sahin, A.Z. ,Yilbas, B.S. & Garni, A.Z.(1996) Friction welding of Al-Al, Al-steel, and steel-steel samples. *Journal of Materials Engineering* **5** 89-99.

Sahin, A.Z., Yilbas, B.S.& Ahmed, M.(1998) Analysis of the friction welding processin relation to the welding of copper and steel bars. *Journal of Materials Processing Technology* **82** 127-136.

Yibas, B.S., Sahin, A.Z. & Coban. A. (1995) Investigation into properties of friction welding of Al bars. *Journal of Materials Processing Technology* **54** 76-81.

11.4 Experiments and Simulations of Welded Joints with Post Weld Treatment

Peter Schaumann, Christian Keindorf
Leibniz University Hannover, Appelstr. 9A, 30167 Hannover, Germany,
schaumann@stahl.uni-hannover.de, keindorf@stahl.uni-hannover.de

Abstract

This paper provides detailed information about experimental investigations of an arc welding process carried out for seam butt welded joints followed by a process of needle peening. The objective is to analyze the difference in fatigue strength between as welded and treated conditions. The results are compared to numerical simulations including contact-algorithm for post weld treatment. Furthermore the fatigue life is calculated numerically with notch strain approach to consider the different residual stress states for both conditions. As result the increase of fatigue resistance due to needle peening known from experimental investigations could now be comprehended also by numerical simulation.

Keywords: experiment, *simulation, needle peening, post weld treatment, notch strain concept*

1 Introduction

For welded joints subjected to dynamic loads fatigue resistance of the base material and the weld is very important to design steel constructions in fatigue limit state. In most cases residual stresses due to the welding process reduce the fatigue resistance. However, different methods of post weld treatment can be chosen to enhance the fatigue strength of welded structures. One of these methods is the Ultrasonic-Impact-Treatment (UIT), which introduces beneficial compressive stresses at the weld toe zones and reduces stress concentrations by improving the weld toe. (Statnikov et al., 1977). The equipment consists of a handheld tool and an electronic control box (Figure 1). The tool is easy to handle during application. It operates at the head movement with a mechanic frequency of 200 Hz overlain by an ultrasonic frequency of 27000 Hz. The method involves post-weld deformation by impacts from single or multiple indenting needles excited at ultrasonic frequency, generating mechanic impulses on the work surface (Statnikov et al., 1997).

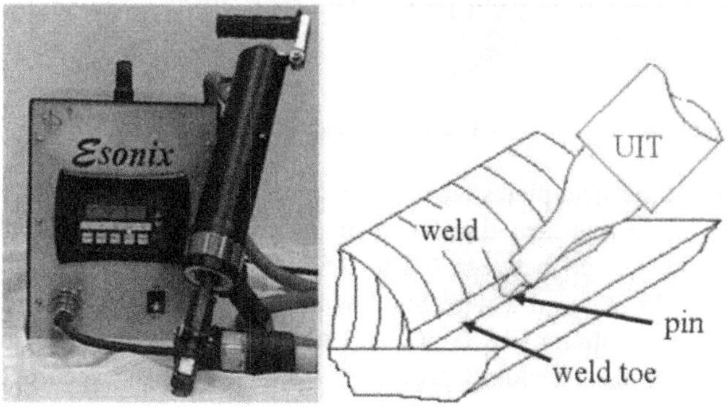

Figure 1. Ultrasonic Impact Treatment (UIT) [Applied Ultrasonics Europe].

The objective of the treatment is to introduce beneficial compressive residual stresses at the weld toe zones and to reduce stress concentrations by improving the weld toe profile. Furthermore, the area being treated is highly plastically deformed which has the effect of work hardening. Figure 1 (right) shows a weld seam which is treated by UIT with one pin along the weld toe line.
For detailed numerical analyses of UIT-effects the process of welding was coupled with the post weld treatment process for two steel plates with a single butt weld. Additionally, experimental investigations of welding were carried out for steel plates to verify the numerical model concerning thermodynamics and structure mechanics. Finally, the crack initiation life was calculated based on the notch strain approach. Residual stresses and the different elastic stress concentration factors were taken into account for both conditions (as welded and treated). With this demonstrated process coupling critical zones of welds can be identified in steel constructions and can be optimized in fatigue limit state especially for productions in series.

2 Experimental investigations

A test setup shown in Figure 2 (left) was chosen to realize an arc welding process. The wire feeder and also the welding head were handled mechanically. The welding head was fixed at a track beam, which was displacement controlled to ensure a constant welding velocity. Furthermore, the head could be positioned exactly with the track beam in x, y and z direction. The coordinate system is shown in Figure 2 (right) with the point of origin at one edge of the plates. Only the boundaries' conditions at x = +/-150 mm were fixed.

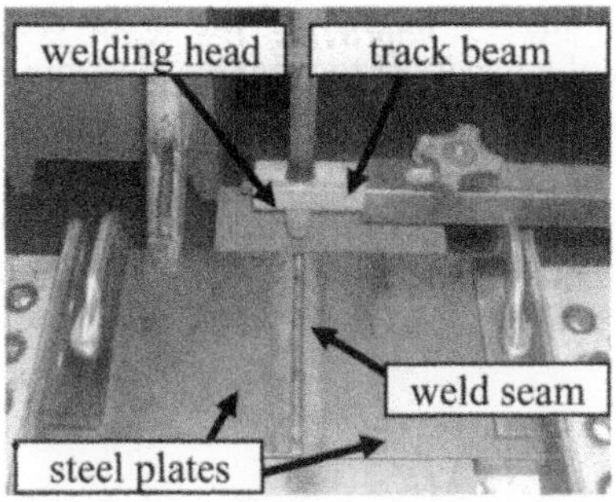

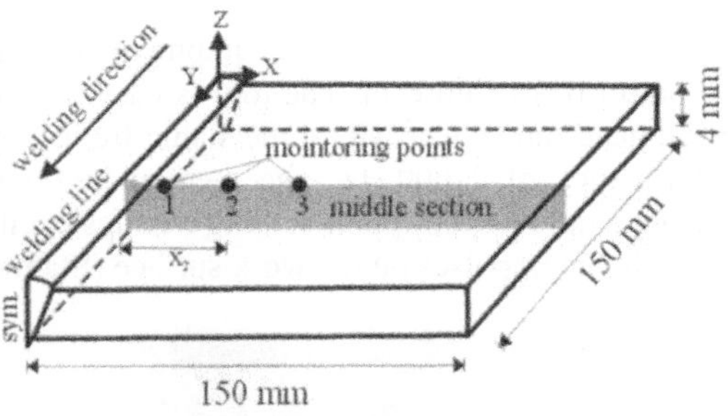

Figure 2. Test setup (left) and dimensions of test specimen (right).

The dimensions of the steel plates fabricated from S355 J2G3 are 150 x 150 x 4 mm. The seam preparation was carried out with an angle of 22.5° at every flank. The gap at the welding line varied between 0.2 to 0.5 mm for the bottom surfaces of the plates. For the weld material a wire SG 2 with a diameter of 0.8 mm and for the shielding gas a composition of 18 % CO_2 and 82 % argon was used. The butt weld is generated as single pass welding process. Together with the recorded time the welding velocity was estimated to v = 255 mm/min. During the welding process the electric input was measured with U = 18.3 V and I = 125 A. Furthermore, the temperature was measured with three thermo elements from type Newport Omega

TC-GG-KI-30-40 with a range of -90°C to 1370°C. The thermo elements were positioned at the right plate with distances $x_1 = 5$ mm, $x_2 = 30$ mm and $x_3 = 50$ mm transverse to the welding line (s. Figure 2 (right)). Additionally, three strain gauges from type HBM 3/120LY11 were oriented at the left plate exactly with the same distances as the thermo elements at the right plate. The strain gauges were necessary to measure the residual strains during and after the welding process. All measured signals were recorded online with the data logger HBM Spider 8 and analyzed with the software Diadem. The whole testing time included the welding time of 35 s and after then the cooling time, which was terminated at nearly t = 2000 s. The results of experimental investigations will be compared in the following chapters with solutions of numerical simulations.

3 Numerical investigations

The symmetric half model, implemented in ANSYS®, consists of three zones: the weld, the heat affected zone and the base material of the steel plate (Figure 3). Material properties are implemented depending on temperature (Wichers, 2006). The notch at the weld toe has a variable radius to consider the actual weld geometry for as-welded and treated conditions. The smallest dimension of an element at the notch has been chosen to 0.2 mm in direction of thickness. The model is driven by displacement boundary conditions like in the test setup. For the simulation of the post weld treatment the head of a 3 mm diameter pin of UIT is modelled including contact elements between the plate and the pin (Figure 3 right).

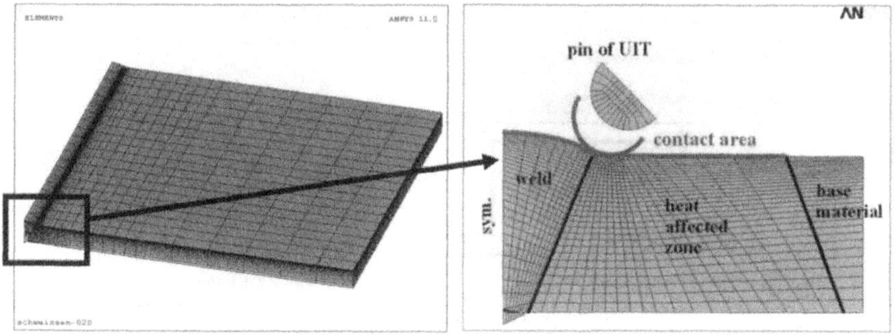

Figure 3. Numerical model of the steel plate with a butt weld.

The simulation of welding consists of thermal and structural analyses. The coupling for the multiphysic problem is realized with a multifield solver (MFS) available in ANSYS®. It is an automated tool for solving sequentially coupled field problems and is built on the premise that each physic is created as a field with an independent solid model and mesh. Surfaces or volumes can be identified for coupled load transfer. The solver automatically transfers coupled loads across dissimilar meshes. The field 1 is defined for transient thermal analysis and modelled with 3D-thermal elements Solid70. These elements have eight nodes with a single degree of freedom, temperature, at each node. The field 2 modelled with Solid45 elements, is for structural analysis to estimate the residual strains and stresses. As load for the structural model, the nodal temperatures are transferred von field 1 to field 2. For thermal symmetry the heat flux passing across the surface of symmetry shown in Figure 2 is assumed to be zero, and, for structural symmetry, the translation in the x-direction of the same surface is also zero. It is important to note that the automatically estimated time increment in the analysis drops to a very small values due to the vast difference in the temperatures. This effect can be dramatic for small element sizes of the weld pool at the fusion surface. But the meshing in the weld pool must be fine enough to account for the high temperature gradient calculation. The transient temperature profile during the welding process is analyzed using Goldak's formulation of the double-ellipsoidal heat source (Goldak, 1984). At the beginning of the simulation all elements for the weld are deactivated with a birth & death function in ANSYS®. These elements are activated step by step when the head source is moved along the welding line. Parametric meshing is used in order to easily track the results along a certain predefined path in any direction. Due to the multifield solver strategy the kinetics of phase transformation could be also implemented in the numerical model. For every time step the actual content of each phase like austenite or ferrite is estimated. After the process of welding and cooling the process for post weld treatment follows and finally the macro for the calculation of fatigue life is used. The whole path of process coupling is shown in Figure 4.

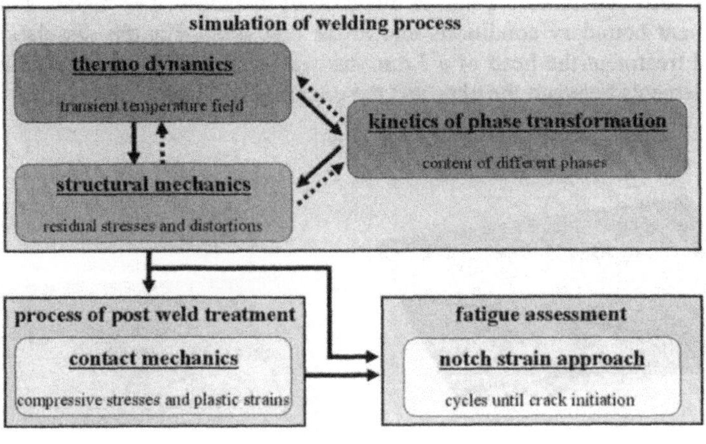

Figure 4. Multiphysic process coupling.

4 Thermodynamics

The numerical simulation of the transient temperature field was verified with two different measuring techniques. On the one hand with thermo elements (1D) and on the other hand with thermograph cameras (2D). The Figure 5 (a) shows a plot for the field of temperature at the top surface at t = 42 s. The welding started from left side and ended at t = 35 s at the right side. Since the welding head disturbed the view of thermograph camera, a plot during the cooling time without the welding head was chosen. It can be seen that the temperatures over 80°C have a smaller width for the left edge as for the right one. This comparison between numerical simulation and experiment shows a good agreement for the qualitative distribution of the temprature field. The zones with the same temperature levels have obviously the same range.

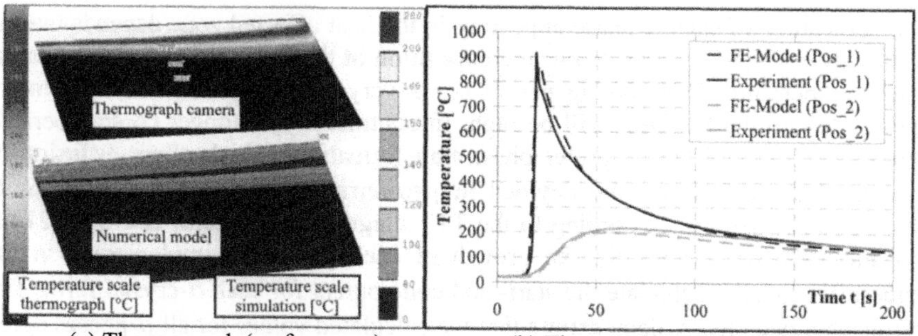

(a) Thermograph (surface area) (b) Thermo elements (two points)
Figure 5. Comparison of temperature between experiment and simulation.

Additionally, temperature curves over time have been measured at two different monitoring points. The thermo elements were positioned at $x_1 = 5$ mm, $x_2 = 30$ mm on the top surface in the mid section (s. Pos_1 and Pos_2 in Figure 2). The comparison with the numerical results in Figure 5 (b) shows a very good agreement. The maximum temperature for monitoring point 1 is 900°C at nearly t = 18 s, which is the time, when the head source passes the mid section. Of course, the maximum temperature of point 2 with 200°C is due to the longer distance to the welding line lower as for point 1. With these comparisons it can be concluded that the numerical model is verified concerning the transient temperature field.

5 Kinetics of phase transformations

The multifield solution for the welding process includes the effects of solid state phase transformation. The calculations are based on a TTT-diagram for the used base material S355 according to [S-ZTU from Seyfarth et al., 1992]. Therefore, two different types of crystal forms are considered for phase transformations. On the one hand the γ-phase, also called austenite, exist as the only one for temperatures over A_{c1} = 718°C in the case of the used S355. On the other hand different α-phases (martensite, ferrite, perlite and bainite) exist which start to grow during the cooling time when the notch temperatures decrease under A_{c1}.

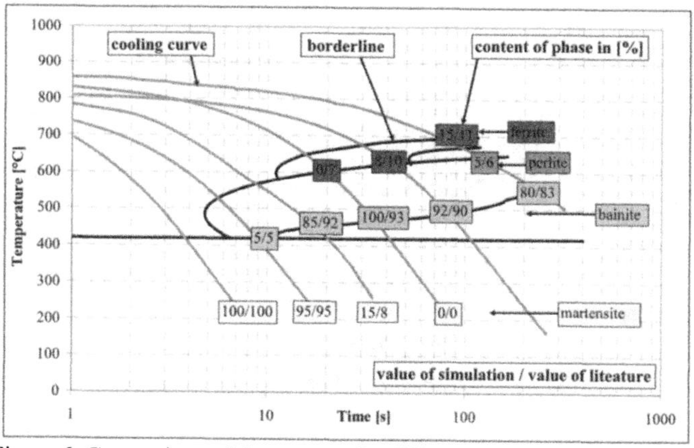

Figure 6. Comparison of the content for different phase transformations.

The percentage of each α-phase appearing in the heat affected zone depends mainly on the cooling velocity, the chemical composition of the base material and the actual notch temperature. The cooling period $t_{8/5}$ is very short, therefore the content of martensite and the hardness will be high. In contrast to this longer cooling periods between 800°C and 500°C prefer phase transformations which allow diffusion of carbon. These effects can be devided in γ-to-ferrite, γ-to-perlite and γ-to-bainite phase transformations. According to the TTT-diagram in (Seyfarth et al., 1992) the 1% and 99% borderlines for the growth of each phase are implemented in the numerical model to calculate the start- and end-content for each α-crystal form. The results obtained clearly demonstrate that the calculated values for all α-phases agree very well with the values in literature (s. Figure 6). The quality of these values depends mainly on the polynominal function which should approximate the borderlines 1 % and 99 % of each phase as good as possible (Börnsen, 1989). For example if an element shows a cooling behaviour like the slowest cooling curve in Figure 6 the α-phase ferrite will grow at first with a content of 15 %. The value of literature (Seyfarth et al., 1992) in this case is 11 %. After then the perlite phase starts to grow and ends with a content of 5 % (6 % in literature). Finally the bainite phase develops up to 80 % (83 % in literature). The content of martensite is here 0 % because the cooling time $t_{8/5}$ is to slow for this phase transformation.

6 Structure mechanics

The structural analysis was carried out simultaneously to the thermal analysis and phase transformations. Therefore, the nodal temperatures over time were the load for the structural elements. The simulation finished at t = 2000 s to estimate the final residual stresses and strains after cooling. For example, the residual strain curves over time and longitudinal to the welding line are shown in Figure 7 for two tests (experiments 26 and 28). The results belong to the monitoring point 2 which was located at x_2 = 30 mm transverse to weld (s. Figure 2 right). Furthermore, the curve for this point (notch) calculated with the FE-model is plotted as dotted line.

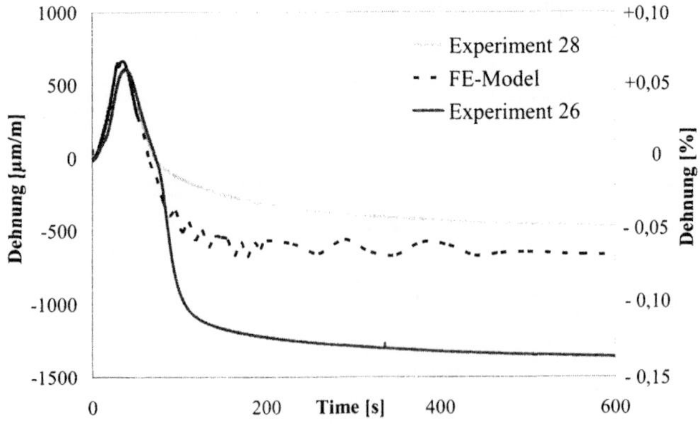

Figure 7. Comparison of phase transformation.

At first all curves show strains in tension because of the heating during the welding process the material expands. After welding, the strains change to compressive values when the plates cool down. The comparison of the curves shows that the numerical result is located in the spectrum of experimental results which have a spreaded strain range like in Figure 7.

7 Contact mechanics

The process of post weld treatment by UIT was implemented using a contact algorithm with element types Conta173 and Targe170. For the simulation only the top of the UIT-pin shown in Figure 3 was modelled. The pin was positioned in the mid section from y = 40 – 60 mm with an initial gap to the base plate. After restarting the transient structural analysis the pin moved in an angle of 60° to the weld toe and deformed this surface. The residual stress state transverse to welding line (x-direction) for as welded condition is shown in Figure 8 (left) and the changed stress distribution after post weld treatment in Figure 8 (right).

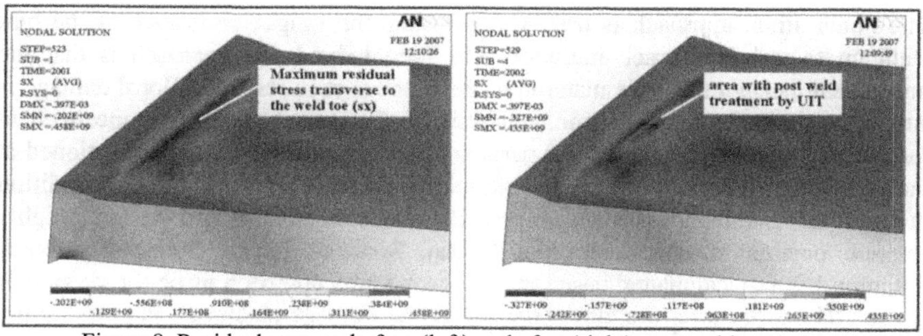

Figure 8. Residual stresses before (left) and after (right) post weld treatment.

The transverse stress in the treated area changed from tensile residual stresses (σ_{R,T,as_welded} = +458 N/mm²) to compressive residual stresses near the yield stress ($\sigma_{R,T,uit}$ = -327 N/mm²). Additionally, the weld geometry was rounded by the pin and the toe notch radius increased.

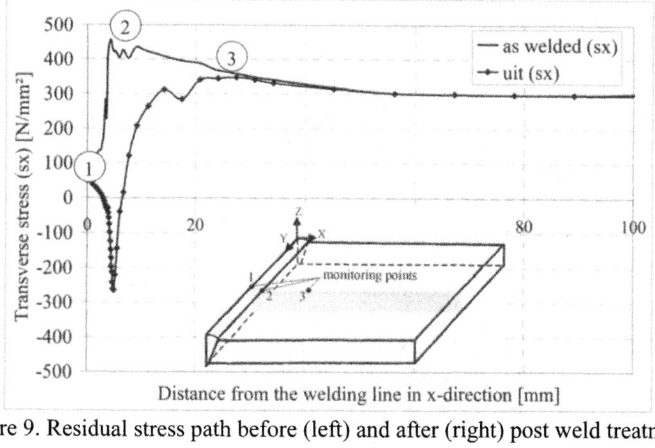

Figure 9. Residual stress path before (left) and after (right) post weld treatment.

Furthermore, the residual stress distribution transverse to the weld toe (x-direction) was compared in Figure 9 for a path in mid section. The stress path for the as-welded condition rises up from the welding line (monitoring point 1) to the maximum value at the toe (monitoring point 2) and decreased slowly to a stress level of 300 N/mm² in base material (monitoring point 3). In contrast to the as-welded condition the stress distribution after post weld treatment had a negative peak located at the weld toe. The compressive stress reached a minimum value of -280 N/mm². In the near of the peak the residual stress were also reduced. But with more distance to the welding line the curve reached the stress level like in the as-welded condition.
Finally it can be noticed that the post weld treatment by UIT has a great influence on the local stress state at the weld toe which is often the location for crack initiation. The numerical results are suitable to assess the fatigue strength under consideration of the modified residual stress state at the weld toe.

8 Fatigue assessment

The notch strain approach is used for assessing the fatigue resistance of the butt weld up to technical crack initiation. The idea behind this approach is that the mechanical behaviour of the material at the notch root in respect of local damage is similar to the behaviour of a miniaturized, axially loaded, unnotched specimen in respect of global damage. The comparison specimen is assumed to be positioned at the notch root. It should have the same microstructure, the same surface condition inclusive of residual stresses and, if possible, the same volume and the same highly stressed material at the notch root (Radaj, Sonsino, Fricke, 2006). The crack initiation life was calculated based on the notch strain approach using the version of Seeger. The S-N curve was determined with the damage parameter P_{SWT} assuming the material properties of the base material. Elastic-plastic behaviour, residual stresses and the different elastic stress concentration factors for both conditions were taken into account. The predicted S-N curves for crack initiation of butt weld are presented in Figure 10.

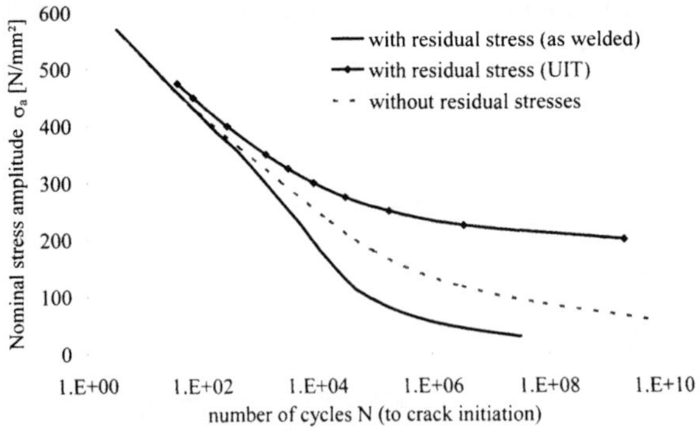

Figure 10. Comparison of phase transformation.

In Figure 10 the numbers of cycles N is plotted against the nominal stress amplitude σ_a for alternating loading (R = -1). The curve for as-welded condition has the lowest fatigue strength because of the tensile residual stresses. In contrast to as welded condition the butt weld with the treated toe has higher fatigue strength. The post weld treatment introducing compressive stresses in the notch root is highly effective in this case. The analysis indicates that the influence of residual stresses at the weld toe is significant for the crack initiation life. The results for both conditions are compared with the S-N curve without considering residual stresses which S-N curve is obviously between these both cases. Finally, the results are shown graphically as plots using the „fatigue tool" which is implemented in ANSYS®.

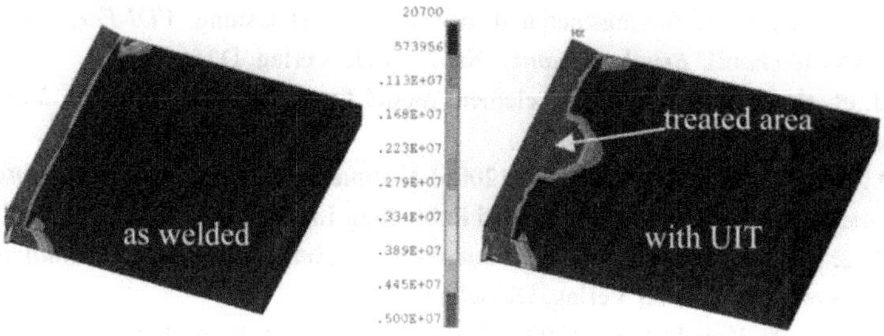

Figure 10. Cycles until crack initiation for as welded (left) and treated condition (right).

For the application of this user specified fatigue tool, it has to be mentioned critically that the iterations for the elastic-plastic stress-strain behaviour at notch and finally the one for getting life cycle is very time-consuming in the calculation. To save computing time the accuracy for the iteration of stress-strain path at notch is set to $\Delta\sigma = 0.1$ N/mm² and for the iteration of life cycle to $\Delta N = 500$. This means in every step of iteration the number of cycle is raised by this delta value. The amplitude is set to $\sigma_a = 0.5 \cdot f_{y,k} = 180.0$ N/mm² with a stress range of $R = -1$ and $\sigma_m = 0.0$ N/mm². Furthermore, a fine mesh especially at crack critical zones is necessary. However, these plots can be used for qualitative illustrations and detections of areas with number of life cycles which are minimal and therefore critical zones for crack initiation.

9 Conclusions

The fatigue strength of welded joints represents the weak point of every very dynamically loaded steel construction. Induced residual stresses or changes in the crystalline structure due to heating are only two of many notch effects. In the underlying research project an arc welding process was analysed in experiments and simulations. The comparisons concerning thermal and structural analysis show a good agreement. Due to the use of multifield solution technique in ANSYS the kinetics of solid phase transformations could be considered. Additionally a following process of post weld treatment by Ultrasonic Impact Treatment (UIT) was investigated. The objective of the numerical investigations was analysing the influence of residual stresses on the fatigue life for welded joints with and without post weld treatment. Finally, the fatigue resistance until crack initiation was assessed for both conditions based on the notch strain approach. As result, after UIT-treatment of the weld toe in the midsection, the predicted location of crack initiation moved from the weld toe towards the heat affected zone. The increase of fatigue resistance due to UIT post weld treatment known from experimental investigations could now be comprehended also by numerical simulation.

References

Bäumel, A. & Seeger, T. (1990) *Materials Data for Cyclic Loading*, Suppl. 1, Amsterdam, Elsevier Science.

Börnsen, M. (1989) Zum Einfluss von Gefügeumwandlungen auf Spannungen und Formänderungen bei thermischer und mechanischer Belastung, *VDI-Fortschrittberichte, Reihe 18 (Mechanik/ Bruchmechanik)*, Nr. 72, VDI-Verlag, Düsseldorf.

Goldak, J. et. al. (1984) A new finite element model for welding heat sources, *Metallurgical Transactions B*, 15B, pp. 299-305.

Radaj, D.; Sonsino, C.M. & Fricke, W. (2006) *Fatigue assessment of welded joints by local approaches*, Second edition, Woodhead Publishing Limited and CRC Press LLC.

Radaj, D. & Sonsino, C.M. (2000) "*Ermüdungsfestigkeit von Schweißverbindungen nach lokalen Konzepten*", DVS Verlag, Düsseldorf.

Schaumann, P. & Keindorf, C. (2007) Enhancing fatigue strength by ultrasonic impact treatment for welded joints of offshore structures, *Third International Conference on Steel and Composite Structures* (ICSCS), pp. 921-926, Manchester, UK.

Schaumann, P.; Keindorf, C. & Kirsch, T. (2007) Multiphysikalische Prozesskopplung am Beispiel einer stumpf geschweißten und nachbehandelten Stahlplatte, *SYSWELD Forum 2007*, Weimar.

Seyfarth, Meyer, Scharf (1992) Großer Atlas Schweiß-ZTU-Schaubilder, *Fachbuchreihe Schweißtechnik*, Band 110, DVS Verlag, Düsseldorf.

Statnikov, E.S. et al. (1977) Ultrasonic impact tool for strengthening welds and reducing residual stresses, *New Physical Methods of Intensification of Technological Processes.*

Wichers, M. (2006) Schweißen unter einachsiger, zyklischer Beanspruchung - experimentelle und numerische Untersuchungen, *Dissertation*, Institute for steel construction, TU Braunschweig.

Section 12
Welding technology II

12.1 Application of Laser Peening on Welded Rib-plate of Mild Steel

Yoshihiro Sakino[1], Yuji Sano[2] and You-Chul Kim[1]
[1]*Joining and Welding Research Institute, Osaka University*
11-1, Mihogaoka, Ibaraki, Osaka, Japan, 567-0047 sakino@jwri.osaka-u.ac.jp
[2] *Power and Industrial Systems Research and Development Center, Toshiba Corporation*
8, Shinsugita-cho, Isogo-ku, Yokohama, Japan, 235-8523

Abstract

Laser peening is an innovative surface enhancement technology to introduce a compressive residual stress on metallic materials. Experimental results showed that laser peening was effective to prevent SCC and enhance fatigue strength. However, effects of laser peening to steel for structure and its welded zone are not clarified. In this paper, residual stress of welded toes in rib-plates and fatigue life with laser peening was investigated by comparing to that without laser peening. X-ray diffraction was used to measure the residual stress. The main results are summarized as follows. 1) Laser peening can change the tensile residual stress to large compressive residual stress in welded toes. The nearer from the welded toe, the larger the effect by laser peening was. 2) Fatigue crack was not initiated from peened parts, even though stress amplitude of the peened part is twice stress amplitude of crack-initiated parts. 3) Fatigue life of the rib-plates with laser peening was prolonged at least three times or more compared to that without laser peening.

Keywords: *Laser peening, Fatigue life, Residual stress, Fillet weld, Steel for structures*

1 Introduction

Laser peening is an innovative surface enhancement technology to introduce a compressive residual stress on metallic materials. (Y. SANO (2002)) Fundamental The process of laser peening is summarized as follows. When an intense laser pulse is focused on the material, the surface absorbs the laser energy and a submicron layer of the surface evaporates instantaneously. Water film confines the evaporating material and the vapor is immediately ionized to form plasma by inverse bremsstrahlung. The plasma absorbs subsequent laser energy and generates a heat-sustained shock wave, which impinges on the material with an intensity of several gigapascals, far exceeding the yield strength of most metals. The shock wave loses

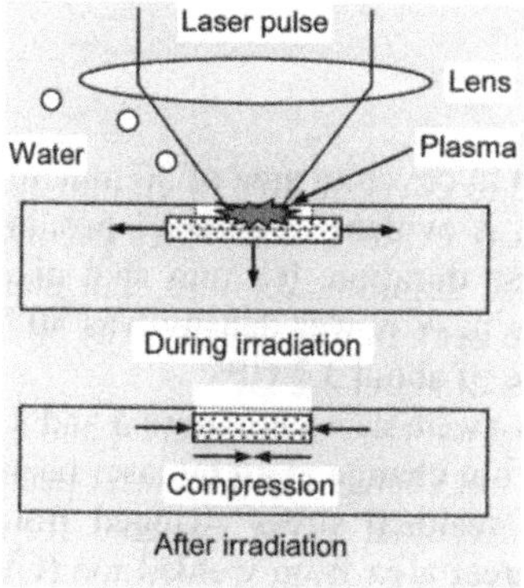

Figure 1. Basic process of laser peening

energy as it propagates to create a permanent strain. After the shock wave propagation, the surface is elastically constrained to form a compressive residual stress on the surface. X-ray diffraction study showed that the compressive residual stress nearly equal to the yield strength was imparted on the surface of the material. Laser peening was effective to prevent stress corrosion cracking (SCC). (Y. SANO (2005))

Laser shock peening can change tensile residual stress to compressive residual stress. So it seems that laser shock peening is a very effective to enhance the fatigue strength, because tensile residual stress is one of the most important factor of fatigue strength.

In this study, the residual stress and the fatigue life of welded rib-plates with laser shock peening was investigated by comparing the rib-plates with and without laser shock peening. Mild steel (SM490A) was used for the plates and fillet weld was used to join the ribs to the plates.

2 Residual stress

As shown in Figure 2, fillet-welded rib-plates of SM490 mild steel were prepared by welding a 9 mm thick rib to a 12 mm thick plate with a length of 180mm and a width of 50 mm. Carbon-dioxide (CO_2) gas shield welding was used with a JIS Z 3312 YGW11 filler wire.

Figure 3 shows measuring points of the residual stress in the fillet-welded rib-plates. X-ray diffraction was used to measure the residual stress.

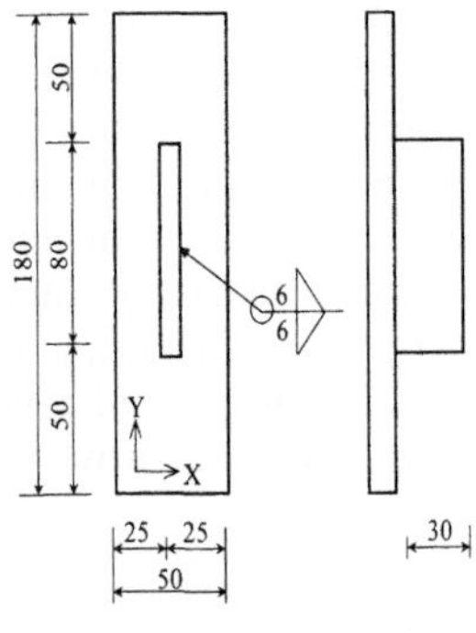

Figure 2. Specimen for

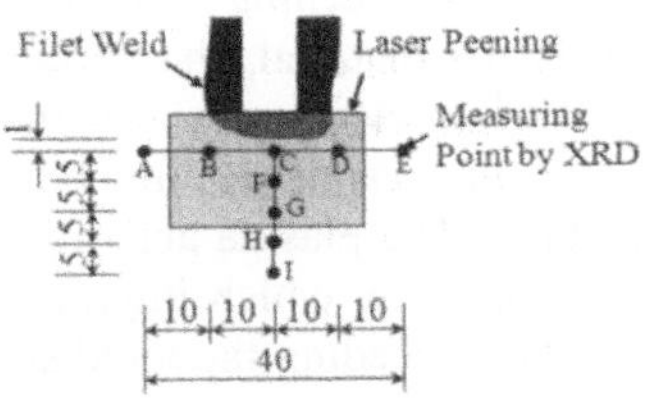

Figure 3. Measuring points of residual stress by XRD

Laser peening was performed to cover an area of 20 mm 30 mm around the weld toe where stress concentration was evident. The laser peening condition was 200 mJ laser pulse energy, 8 ns pulse duration, 0.8 mm spot diameter and 36 pulse/mm^2 irradiation pulse density. The peak power density was 50 TW/m^2, which generated plasma with the peak pressure of about 3.2 GPa.

Residual stress distributions (σ_y) were shown in Figure 4 and 5. In unpeened areas (A, E, H, I), the Residual stress did not change at all by laser peening. But in laser peened areas (B, C, D, F, G), the residual stress changed from tensile stress to large compressive stress. The nearest area from welded toe (C) changed larger than the other laser peened area (B, D, F, G). And the compressive residual stress of area C

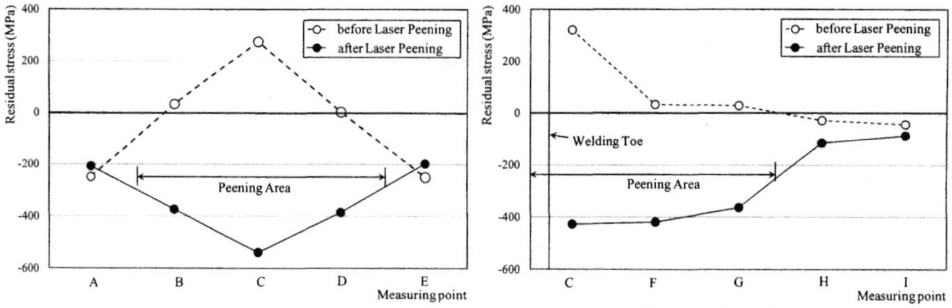

(a) X-direction (b) Y-direction

Figure 4. Residual stress distribution of upper side of lib (σ_y)

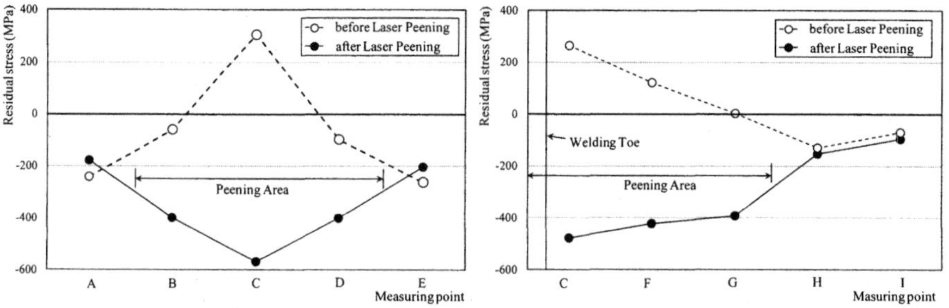

(a) X-direction (b) Y-direction

Figure 5. Residual stress distribution of lower side of lib (σ_y)

Figure 6. Crack initiation points

was nearly equal to or more than the yield strength. So it can be said that the nearer from welded toe, the larger the effect to the residual stress by laser peening was. (Y. SAKINO et al. (2007))

3 Fatigue life

Fillet-welded rib-plates of SM490A mild steel were prepared by welding a 9 mm thick rib to a 12 mm thick plate with a length of 450 mm and a width of 98 mm. Carbon-dioxide (CO_2) gas shield welding was also used. Laser peening was performed to cover an area of 20 mm 30 mm around the weld toe where stress concentration was evident. The laser peening condition was the same in Chapter 2. High-cycle fatigue properties were studied through a series of fatigue tests under

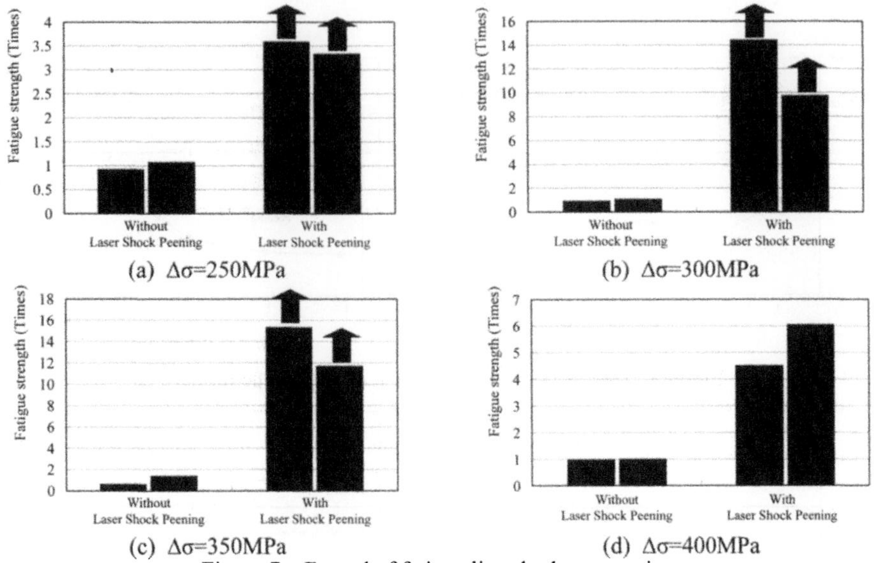

Figure 7. Extend of fatigue lives by laser peening

Figure 7 illustrated the fatigue test results. The vertical axes are normalized so that the average fatigue lives of two unpeened samples would be a unity for each stress amplitude. The fatigue strength or lives could not be deduced properly for specimens with laser peening, since some specimens still survived after available fatigue cycles and cracks broke out from unexpected area other than the toe for some other specimens. However, the overall results showed that laser peening drastically prolonged the fatigue lives of the fillet-welded rib-plate specimens. (Y. SAKINO et al. (2006))

4 Conclusions

(1) Laser peening can change the tensile residual stress to large compressive residual stress in welded toes. The nearer from welded toe, the larger the effect by laser peening was.
(2) Fatigue crack was not initiated from peened parts, even though stress amplitude of the peened part is twice stress amplitude of crack-initiated parts.
(3) Fatigue life of the rib-plates with laser peening was extended at least three times or more compared to that without laser peening.

Acknowledgments

This study was financially supported in part by the Grant-in-Aid for Encouragement Research (A) from the Ministry of Education, Culture, Sports, Science and Technology, and by the research grant from the Japan Iron and Steel Federation.

References

SAKINO Yoshihiro, SANO Yuji & KIM You-Chul (2007) Effect of laser peening on residual stress of steels and fillet welded zone, *Journal of Constructional Steel,* No.15, 419-424.

SAKINO Yoshihiro, SANO Yuji & KIM You-Chul (2006) Improving fatigue strength with laser peening, *Proceedings of National Symposium on Welding Mechanics and Design 2006*, No.2, 605-608.

SANO Yuji (2002) Residual stress improvement on metal surface by underwater irradiation of high-intensity laser, *Journal of Japan Laser Processing Society,* No.9, 163-170.

SANO Yuji (2005) Residual stress improvement by laser peening without coating and its applications, *Proceedings of the 65th Laser Materials Processing Conference*, 111-116.

12.2 A New Potential for Welded Structures in Modified X2CrNi12: Characterization of Dissimilar Arc Weld with S355 Steel

Emel Taban[1], Eddy Deleu[2], Alfred Dhooge[2,3], Erdinc Kaluc[4]
Kocaeli University, 41200, Kocaeli, Turkey, e-mail:emelt@kou.edu.tr
Research Center of the Belgian Welding Institute, B-9000, Gent, Belgium,
University of Ghent, B-9000, Gent, Belgium,
Welding Research Center of Kocaeli University, 41200, Kocaeli, Turkey

Abstract

In this study, modified X2CrNi12 ferritic stainless steel conforming in composition to grades UNS S41003 (ASTM A240) and Wr. Nr. 1.4003 (EN 10088-2 and EN 10028-7) has been welded to non-alloy S355 steel by means of flux cored arc welding (FCAW) process. A dissimilar butt welded joint was subjected to a series of tests in order to investigate all aspects of the weld properties. Mechanical testing was carried out by means of Charpy impact, tensile and bend tests. Microstructural examinations including hardness surveys, ferrite content measurements and grain size analyses of various weld regions were realised. Salt spray and blister tests were also applied as exposure testing. Rather promising results have been obtained yielding a correlation between microstructure and impact toughness.

Keywords: *ferritic stainless steel, 12%Cr, S355, welding, microstructure, toughness, resistance again atmospheric exposure.*

1 Introduction

12%Cr ferritic stainless steels have widely been used as low cost utility stainless steels and were developed to fill the gap between stainless and the rust-prone carbon steels providing an alternative which displays both the advantages of stainless steels and engineering properties of carbon steels. The former generation of these steels is known as 3Cr12 stainless steel with 0,03%C conforming to UNS S41003 of ASTM A240 (Greef & du Toit (2006), du Toit et al. (2006), Kotecki (2005), Meyer & du Toit (2001), Marshall & Farrar (2000), Gooch & Ginn (1988).

3Cr12 has good corrosion resistance in mild environments, however, its weldability is limited. To improve the weldability, EN 1.4003 steel can be modified from conventional 12% Cr stainless steel by decreasing the C-content below 0,03%, which is regarded as the limit for low-carbon steels. Advanced steel making technology now enables fabricating the modified X2CrNi12 ferritic stainless steel complying with EN 10088-2 and EN 10028-7 with C-levels below 0,015% and with reduced amount of impurities improving weldability and mechanical properties consequently. Initial applications of these steels were materials handling equipment in corrosive/abrasive environments while they are now commonly used in the coal and gold mining, for sugar processing equipments, road and rail transport, power generation, for petrochemical, metallurgical and paper industries, but also in structural applications (Taban (2007), Greef & du Toit (2006), du Toit et al (2006), Kotecki (2005), Meyer & du Toit (2001), Marini & Knight (1995), Bennett (1991), Thomas & Hoffmann (1982).

Compared to carbon steels regarding long term maintenance costs, modified X2CrNi12 requires less coating renewals offering a substantial economic and environmental advantage. For other applications, compared to higher alloyed stainless steels, the use of modified 12Cr could be more economical (Taban (2007), Taban et al. (2007), Deleu et al. (2006), Taban et al. (2006), Dhooge & Deleu (2005), Meyer & du Toit (2001), Marini&Knight (1995).

Due to the fact that there have not been many studies on the weldability and the properties of this modified steel and considering the tendency to expand the structural applications, dissimilar welding was taken into account. In this paper, modified X2CrNi12 stainless steel and S355 steel plates were welded by means of the FCAW process. Mechanical, toughness, microstructural and atmospheric exposure properties of the joint was evaluated and reported.

2 Experimental procedure

Chemical composition and transverse tensile properties of the 12 mm thick base metals are given in Table 1.

Table 1. Properties of modified 12Cr stainless steel and S355 plates (data from steel producer)

Chemical composition of modified 12Cr and S355 respectively (wt %)							
C	Si	Mn	P	S	Cr	Ni	N (ppm)
0,012	0,26	0,95	0,035	0,0010	12,45	0,51	0,008
0,122	0,33	1,47	0,009	0,0007	0,11	0,09	0,005
Mechanical properties of modified 12Cr and S355 respectively							
R_e (MPa)			R_m (MPa)			Strain at fracture (%)	
362			501			31	
379			504			37	

Although matching consumables are commercially available for welding the 12Cr stainless steel, they are not recommended in applications where any form of non-static loading such as impact or fatigue is anticipated. Reported weldability studies have shown that austenitic, especially 309 type of consumables are recommended to produce tough welds required for structural purposes (du Toit et al. (2006), Marshall & Farrar (2000), Gooch & Ginn (1988), Tullmin & Robinson (1988), Pagani & Robinson (1998), van Lelyveld, & van Bennekom (1995), Bennett (1991). In accordance with literature, AISI 309 wires were used for this study.
Plates of dissimilar metals provided with a V-shaped weld preparation with an angle of 50° were FCAW-welded in four passes. An E309LT-1 wire with a diameter of 1,2 mm and EN 439-M21 oxidising gas were used. The heat input varied between 1,13kJ/mm and 1,78kJ/mm.

After welding, chemical analysis samples entirely located at the weld metal were prepared. The measurements were done by GDOES. Nitrogen was determined by melt extraction. Also full thickness tensile specimens transverse to the weld and cylindrical test samples completely positioned at the weld metal in longitudinal direction were machined and tested as well as transverse face and root bend test specimens. Bending was executed till 180° unless cracking was observed before. A nominal specimen width of 30 mm and a mandrel diameter of 55 mm were used. A series of notch impact test samples were extracted from both face and root sides and prepared with notches positioned at the weld metal (WM), fusion line (FL) and at the HAZ 2 mm from fusion line (FL+2). Charpy impact toughness test was done at the temperatures of -20°C, 0°C and 20°C. Cross sections were removed from the joint and etched in order to make macro- and microphotographs. According to EN 1043-1 standard, HV5 traverses were made at sub-surface from the face and the root sides of the welds. Ferrite content of the WM was determined by means of Ferritscope. Depending on the toughness test results, ASTM grain size numbers were measured on the existing macro sections to investigate for a possible correlation between toughness and microstructure. Salt spray and blister exposure tests were executed to assess the resistance against atmospheric attack. The salt spray test was done according to ASTM B117 on uncoated and coated samples positioned with S355 steel side downwards in a 5%NaCl aqueous solution with a fog volume of 24 ml to 28 ml per 24 hours, a pH of 6,5 to 7,2 and at a temperature of 35°C for 350 and 1000 hours of exposure. Coating consisted of a two layer protection system which is used in practice and artificially scratched. This allowed to estimate the resistance of the welds when the coating is accidentally damaged prior to or during operation. Blister test was executed on coated samples prepared similar to those for the previous test. Samples were exposed to real atmospheric conditions for 3120 hours at the centre of Gent Belgium.

3 Results and discussion

The chemical composition of the weld deposit is summarised in Table 2. The weld contained more Si, V and N than the base metals due to the filler metal.

Table 2. Chemical composition of the weld deposit of dissimilar joint (wt%).

C	Si	Mn	P	S	Cr	Cu	Ni	Mo	Ti	V	Al	Nb	N
0,03	0,57	1,36	0,02	0,004	22,8	0,07	11,0	0,15	0,03	0,09	0,03	0,002	0,02

Respective transverse tensile strength values of 485MPa and 502MPa have demonstrated the actual matching strength of the weld versus the base metals. Fracture in both cases occurred at the structural steel although the actual strength of both steels was very similar. All-weld metal tensile properties determined on the cylindrical test samples varied between 531MPa and 569MPa for R_m and 342MPa and 360MPa for R_p. Moreover, none of the bend samples failed during testing while the welds were found to be sound. Notch impact test results, expressed in J are illustrated in Figure 1. Interpreting the data based on 27J as required mean toughness level irrespective of test temperature, weld failed at the 12Cr-side at FL and FL+2.

Furthermore, the 12 mm thick dissimilar FCAW-weld showed a normal weld profile with some angular distortion, see Figure 2. Micrographs reveal some grain coarsening and martensite islands at the high temperature heat affected zone (HTHAZ) of the 12Cr-side. ASTM grain size numbers between 2 and 3 were measured at the HTHAZs of the corresponding samples.

Hardness measured over the entire weld cross sections is illustrated in Figure 3. Values for locations 0,7 mm respectively above (HAZu) and below (HAZd) the line of indentations for the left (12Cr-side) and right HAZ (S355-side) are shown with open symbols.

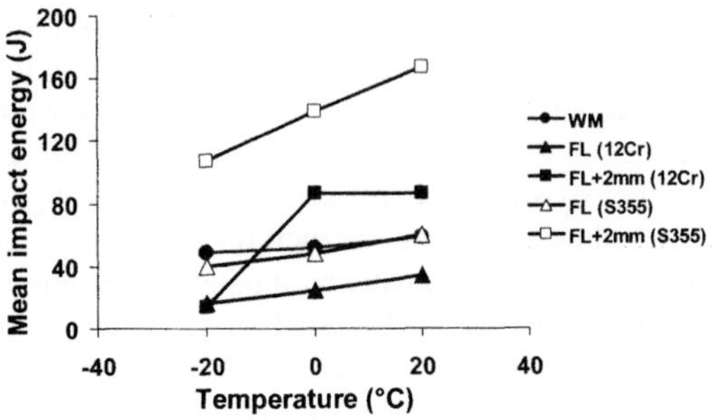

Figure 1. Notch impact toughness of the dissimilar weld.

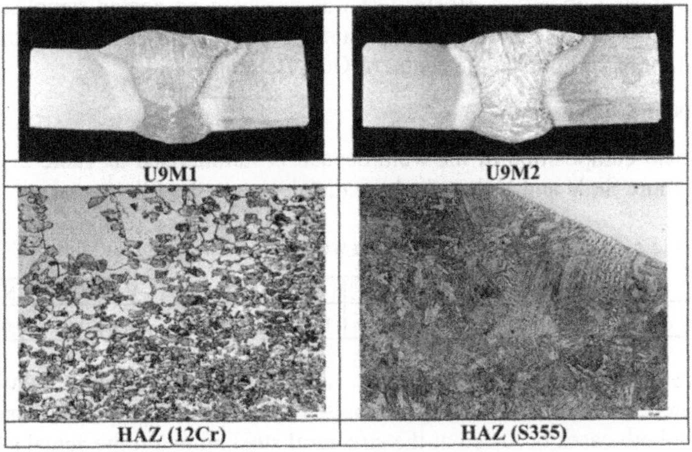

Figure 2. Macro- and microphotographs of the dissimilar welded joint.

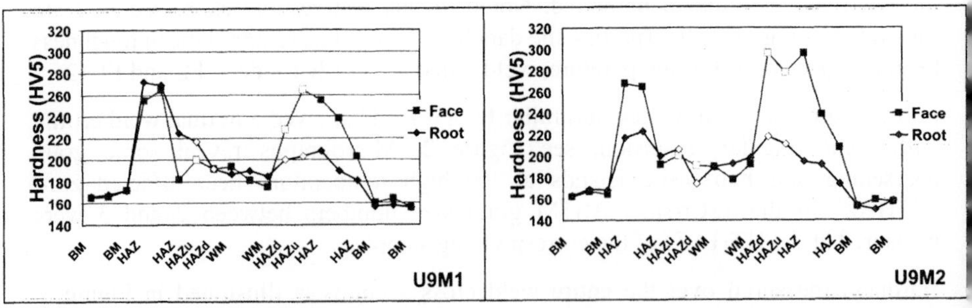

Figure 3. HV5 hardness across macro sections of the dissimilar weld.

Weld metal hardness data varied between 178HV5 and 198HV5. Maximum HAZ hardness instead was about 272HV5 from 12Cr-side. Maximum hardness at the S355 steel in the investigated dissimilar weld was about 296HV5 and was measured at the HAZ of the face side. Weld metal ferrite content of the respective samples varied between 11% and 21%.

Low carbon 12%Cr steels with ferritic-martensitic structure have the tendency to transform to ferrite in the HTHAZ of fusion welds resulting in grain coarsening and toughness reduction (du Toit et al. (2006), Greef & du Toit (2006).

A correlation between impact toughness and grain size of the welds was examined with ASTM grain size number measurements at the HAZs. Coarse grained microstructures which are identified by small ASTM grain size numbers coincided with low impact test results. This situation is also in accordance with the papers by Meyer & du Toit (2001) and Gooch & Ginn (1988) as DBTT of 12%Cr steel increases with ferrite grain size. Thus, efforts to limit grain size help to enhance toughness properties. To restrict grain growth, heat input should be kept as low as possible (Meyer & du Toit (2001), du Toit et al. (2006), Greef & du Toit (2006).
Photographs of uncoated and coated samples are shown in Figure 4 after an exposure time of 350 and 1000 hours for salt spray and of 3120 hours for blister tests.

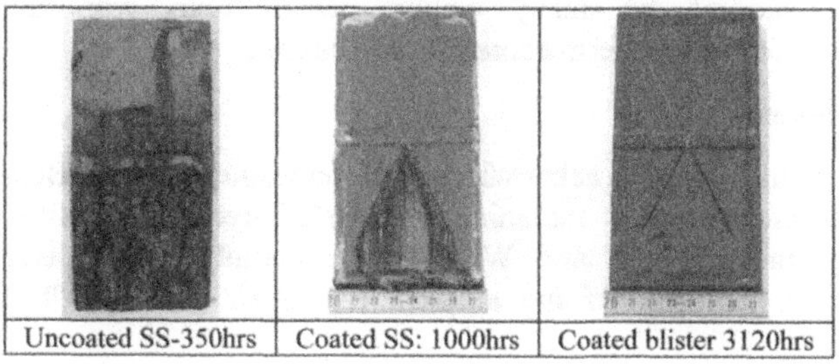

Figure 4. Photographs of uncoated and coated samples of salt spray (SS) and blister tests.

350 hours of exposure of the uncoated salt spray test sample showed some deterioration at the 12Cr-side (top) and heavily corroded S355-side (bottom). Long-term (1000 hours) salt spray corrosion behaviour of the coated sample heavily scratched across the weld revealed some corrosion products just at the scratch on the 12Cr-side while more damage was observed on the S355-part. Obviously the coating provided a good protection for the weld, as in general only, the scratched region had deteriorated. For the coated blister test sample, some small spots at the scratch were noticed, so the weld can be considered as having a good resistance against atmospheric attack over a period of 3120 hours even when severely scratched across the entire welded joint.

4 Conclusions

The following conclusions can be drawn concerning the weld properties of a dissimilar FCAW joint made between modified X2CrNi12 stainless and S355 steels. Manufacturing 12Cr ferritic stainless steel conforming to EN10088:X2CrNi12 but with reduced amounts of carbon and impurities is possible with reasonable production costs so that it possesses tensile and toughness properties complying with non alloy structural steel EN10025:S355J2. In general, sound dissimilar welds can be made by FCAW by means of highly alloyed AISI 309 type of consumables. This means that joining of this stainless steel can be accomplished by welding under economical conditions producing weldments with attractive properties. The major withdrawal of 12Cr is the tendency to grain coarsening at the HAZ unless the heat input during welding is properly controlled. Grain coarsening does not affect tensile nor bend properties, however the HAZ toughness for sub-zero temperatures may be disappointing. Due to possible grain growth, care should be taken and the heat input should be restricted. In general Charpy impact toughness of the joint with notch positions at FL and FL+2 of the 12Cr-side were found to be less than those for S355-side. This can be related to the heat input sensitivity of the modified 12Cr steel. The weld metal in the present welds was overmatched in tensile strength while the bend tests revealed no defect.

Resistance against atmospheric attack in coated condition of modified X2CrNi12 stainless steel welds is promising even when evaluated under severe circumstances, i.e. artificially damaged. Under pure atmospheric conditions, the weld demonstrated the possibility to prevent undue development of corrosion products, once initiated. The innovative aspect is that, in case welding can be done properly, this stainless steel can be applied for many welded constructions working under mild environments yielding long term economic advantages.

Acknowledgements

The authors would like to acknowledge all colleagues at the Belgian Welding Institute. IWT and members including Industeel, Buyck, University of Gent, Air Liquide, Lincoln Smitweld and WTCM is gratefully acknowledged for their contribution and support. Of the authors, E. Taban thanks TUBITAK for the scholarship.

References

Bennett P. (1991) The weldability of 12% chromium ferritic corrosion resisting steels. *Mater Australia* June, 15-17.

Deleu, E., Dhooge, A., van Haver, W. & Taban, E. (2006) Stainless Steel Type X2CrNi12 for Structural Applications- Weldability, Welding Technology and Properties of Welded Joints. Final Report, BWI Document No. EDM06062 (to be issued).

Dhooge, A. & Deleu, E. (2005) Ferritic stainless steel X2CrNi12 with improved weldability for structural applications. *In: Proc. Stainless Steel World 2005 Conference&Expo, Maastricht, Netherlands*, pp. 160.

du Toit, M., van Rooyen, GT. & Smith, D. An overview of the HAZ sensitization and SCC behaviour of 12 % chromium type 1.4003 ferritic stainless steel. IIW Doc IX-2213-06, IIW Doc IX-H-640-06.

Gooch, TG. & Ginn, BJ. (1988) HAZ toughness of MMA welded 12%Cr martensitic-ferritic steels. TWI Welding Institute Members Report 373/1988 3.

Greef, ML. & du Toit, M. (2006) Looking at the sensitization of 11–12% chromium EN 1.4003 stainless steels during welding, *Welding Journal* **85** No. 11, 243s- 251s.

Kotecki, DJ. (2005) Stainless Q&A, *Welding Journal* **84** 14-16.

Marini, A. & Knight, D.S. (1995) The use of 3Cr12 for corrosion- abrasion applications in the mining industry. Corrosion and Coating SA March 4-12.

Marshall, AW. & Farrar, JCM. (2000) IIW Doc: IX-1975-00, IXH-494-2000.

Meyer, AM. & du Toit, M. (2001) Interstitial difussion of carbon and nitrogen into HAZs of 11-12% chromium steels. *Welding Journal* **80** 275s- 280s.

Pagani, S.M. & Robinson, F.P.A. (1998) Microstructure and mechanical and electrochemical properties of martensitic weld deposits developed for welding of a 12%Cr duplex stainless steel. *Materials Science and Technology* **4** 554-559.

Taban, E. (2007) Weldability and Properties of Modified 12 Cr Ferritic Stainless Steel for Structural Applications, PhD thesis, University of Kocaeli, 515 pages.

Taban, E. Deleu, E., Dhooge, A. & Kaluc, E. (2006) Mechanical and microstructural properties of welded X2CrNi12 ferritic stainless steel. *In: Proc. DVS GST 2006, Schweissen und Schneiden,* Germany, pp. 74- 79.

Taban, E. Deleu, E., Dhooge, A. & Kaluc, E. (2007) Gas metal arc welding of modified X2CrNi12 ferritic stainless steel. *Kovove Materialy-Metallic Materials* **45** No. 2, 67- 73.

Thomas, CR & Hoffmann J.P. (1982) Metallurgy of a 12 % chromium steel, *In: Proc. Speciality steels and hard materials conference, Pretoria*, pp. 299-306.

Tullmin, M.A.A. & Robinson, F.P.A. (1988) Effect of molibdenum additions on the corrosion resistance of a 12% chromium steel. *Corrosion-Nace*, **44** 9.

van Lelyveld, C. & van Bennekom, A. (1995) Autogenously welded 3Cr12 tubing for use in the sugar industry. *Stainless Steel* September/October, 16 - 18.

12.3 Friction Stir Welding of T-joints Fabricated with Three Parts

S. M. O. Tavares, P. M. G. P. Moreira, P. M. S. T. de Castro
Faculty of Engineering, University of Porto, Porto, Portugal
sergio.tavares@fe.up.pt

Abstract

Compared with conventional welding processes, friction stir welding (FSW) requires different structural designs to produce T-joints. The friction stir welded T-joint is a type of welding connection with important or even critical applications. Among those being considered by industry or already in industrial practice, the reinforcement of skins using stiffeners for aircraft fuselages, or the body of railroad passenger cars can be mentioned as examples. New geometries to produce this type of joint are presented. For one of these solutions a metallurgical and mechanical characterization was performed using stress/strain and bending tests comparing the behaviour of base material and FSW butt joints. Numerical simulations of the bending tests of T-joints using the non-linear finite element code ABAQUS were performed and their results compared successfully with experimental data.

Keywords: *friction stir welding, T-joints, metallographic defect analysis, mechanical characterization, numerical simulation.*

1 Introduction

Friction stir welding (FSW) is a solid-state welding process developed and patented by W. Thomas in the The Welding Institute - TWI (Thomas et al., 1992). This process has a high industrial interest with several advantages compared with traditional welding processes. This process is suitable to weld different types of materials including welds with dissimilar materials.

The friction stir weld is promoted by the interaction of a non consumable rotating tool, composed by one shoulder and one pin, and the parts to be joined.

Aluminium alloys are attractive materials in mechanical design because of their low density, high strength, good corrosion resistance and good weldability. Some of aluminium alloys can be welded by fusion processes with good efficiency; however the high strength alloys and most of the 2xxx alloys are not recommended for fusion welding because of crack sensitivity during welding. FSW is capable of producing better welds than fusion processes, and in addition it is capable to weld the alloys that cannot be welded by fusion processes.

The T-joint is an important geometry because it can increase significantly the inertia and strength of thin skins or plates without significant weight increases. This shape can be obtained using welding processes where the web (the vertical element) is joined to the flange or to the plate (horizontal element). This geometry is commonly used in the reinforcement of aircraft fuselages, in ships, trains and many other applications where the weight of the structure is especially important. This type of geometry can be obtained using extrusion processes, commonly applicable to small components. When large reinforced panels are needed the extrusion is not a solution, and there is a need for riveting or welding processes. The FSW is a promising process to do this task, already used in several industrial companies.

Just a few studies concerning friction stir welded T-joints can be found in the literature. Erbsloh et al. (2003) studied T-joints in aluminium alloy 6013-T4 using 4 mm thick plates and plunging the tool through the skin and the stiffener, in order to study the influence of backing plates. Fratini et al. (2006) presented a comparison of the performance of T-joints obtained with metal inert gas welding, extrusion and FSW. Another study concerning A36 mild steel T-joints produced using FSW is presented by Steel et al. (2005); these authors discuss attempts to avoid defects using chamfers in the backing plates.

In the present paper different solutions to produce T-joints are presented. The geometries presented allow variations in geometry and choice of materials in order to optimize the mechanical properties.

The purpose of the present paper is to present some experimental results of friction stir welded T-joints of aluminium alloy 6082-T6. These T-joints were manufactured using three components and a different geometrical arrangement. The T-joint specimens were produced from three flat plates, all of them 3 mm thick, using a rig that was designed and manufactured for this purpose. Different welding parameters were used to select be best tool and parameters, in order to obtain good mechanical properties without root defects or cracks in the T-joint corners. Firstly, manual bending tests from small parts of the welds were performed, and the specimens presenting good behaviour were then subjected to metallographic analyses. Tensile tests of specimens cut from the welded structures were carried out and the results were compared with FSW butt joints and base material data. Also, experimental bending tests were carried out measuring the load versus displacement and these results were compared with base material behaviour. Numerical simulations of the bending tests of T-joints using the non-linear finite element code ABAQUS were performed and their results compared with experimental data, in order to improve the understanding of the behaviour of this type of joint.

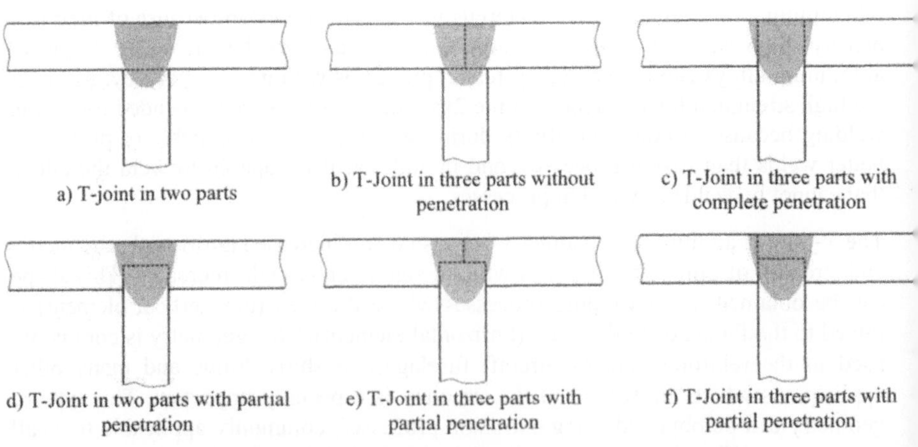

a) T-joint in two parts

b) T-Joint in three parts without penetration

c) T-Joint in three parts with complete penetration

d) T-Joint in two parts with partial penetration

e) T-Joint in three parts with partial penetration

f) T-Joint in three parts with partial penetration

Figure 1. Different alternatives to produce T-joints using FSW.

2 Geometry and welding process

Different solutions of geometrical arrangement of the parts composing the T-joint can be used, aiming at achieving sound welds and at the same time preventing defects in the T-corners. Examples of these geometries are presented in Figure 1, where c) to f) are the object of a pending patent, Tavares et al. (2007).

The solution with three parts where the web is positioned in the middle of both parts that compose the flange, Figure 1.c) is studied in the present paper. This geometry simplifies the process and facilitates the production of T-joints reducing the initial preparation of the components and producing welds with few defects. The more common solution is presented in Figure 1.a), though cracks are found in the T-corners caused by the lack of mixture in the bottom of the plates. A solution to solve this problem is to use some chamfer or fillet in the backing plates in order to improve the mixture of the materials.

The cross section of the geometry used for this study is presented in Figure 2. All components are in aluminium alloy 8082-T6.

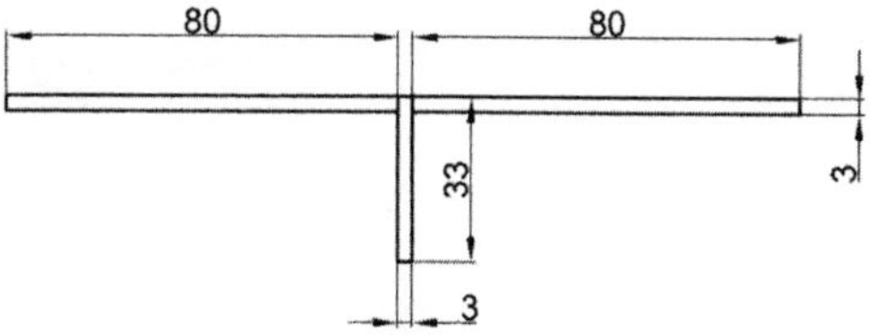

Figure 2. Cross section of geometry adopted for the tests.

Several tests were carried out using different parameters. From these tests 6 specimens were chosen to perform metallurgical and mechanical characterizations. The welding parameters and pin geometries for these specimens are presented in Table 1. In the first tests it was observed that if the pin ends with a plane surface, this creates a transition plane along the welding line, with low mechanical properties. Hence in the next tests a pin with conical geometry was adopted, but still the mechanical properties did not improve with this solution.

Table 1. Welding paramenters and pin geometries of each test.

	Test 1	Test 2	Test 3	Test 4	Test 5	Test 6
Angular speed [rpm]	1500	1500	1500	1500	1500	1500
Linear speed [mm/min]	360	216	216	290	216	216
Tilt angle [°]	2.5	2.5	2.5	0	0	0
Fillet radius	< 1 mm	< 1 mm	< 1 mm	1 mm	1 mm	1 mm
Tool	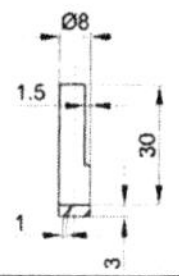		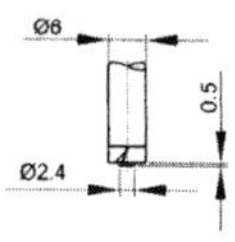		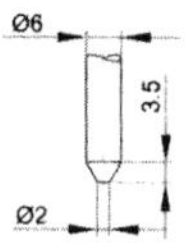	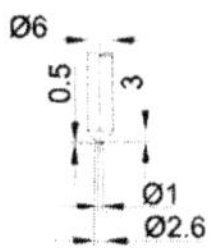

3 Metallurgical characterization

A metallurgical characterization was performed to study the microstructural changes due to the FSW process. The joints were cross-sectioned perpendicularly to the welding direction and etched with HF reagent and Picral reagent, ASM Handbook (1985). In addition, these analyses help to identify the root defects, the presence of cracks in the T-corners, porosities and lack of mixture. Figure 3 shows a macrostructure of each test and Figure 4 shows the microstructures of some areas of the weld joint in order to visualize the type of defects that can be found using FSW.

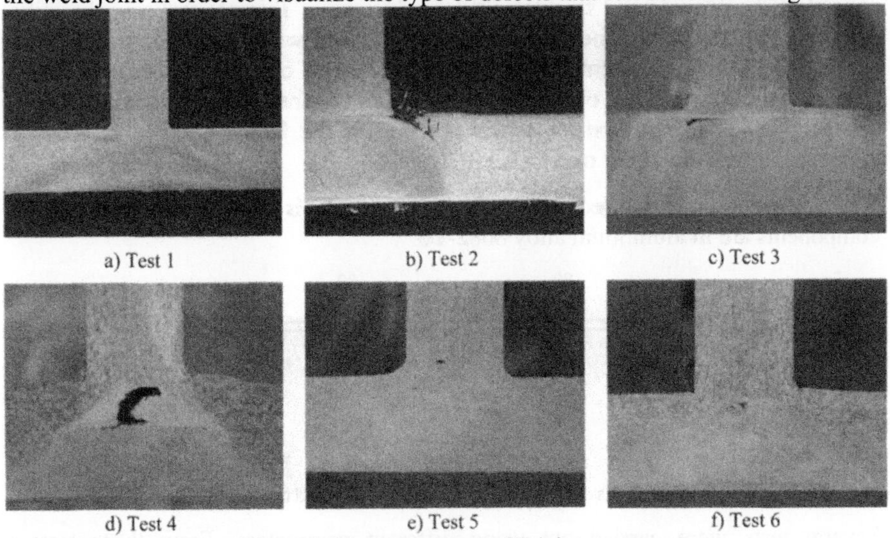

Figure 3. Macrostructures of T-joints tests.

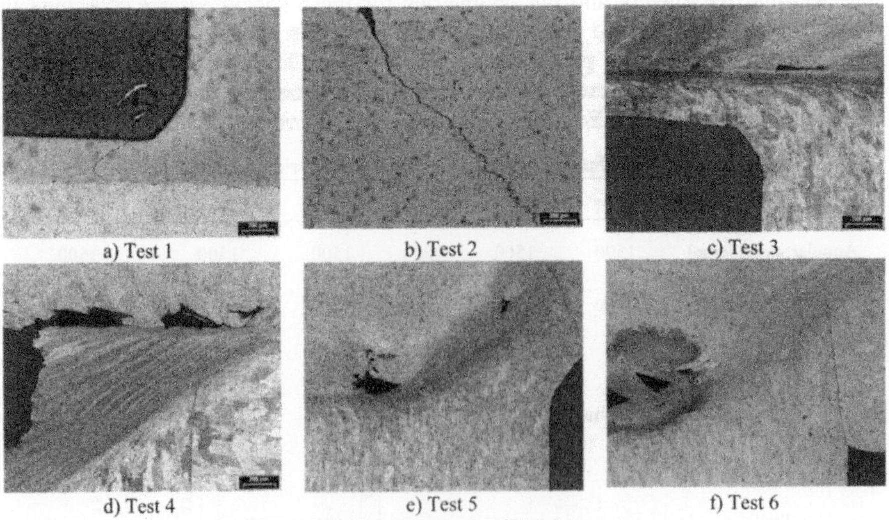

Figure 4. Microstructures of T-joints tests.

Several defects can be easily detected in the macrostructures. The first and fifth tests present fewer defects, however some porosity was found. Analysing the microstructures, Figure 4, small cracks in the limit of the pin region are visible. In the second test the pin had excessive penetration and in some locations it scrabbled in the backing plates. The cross sections examined were taken from damage zones. A possible cause of the bubble of material in the middle of the weld (eg. Figure 3 b) may be the reduction of linear speed which promoted an imperfect mixture.

The third test shows the effect of cutting the pin tip creating a flat surface, and the poor results obtained may also be attributed to the low linear speed used (approximately 210 mm/min). Finally, in the fourth and sixth tests the lack of penetration is visible; this creates small cracks in the T-corners deteriorating the mechanical properties. It was concluded that the complete material mixture is obtained if the pin is distanced about 0.1 mm from the back plates.

4 Mechanical characterization

In order to characterize mechanically these joints two types of mechanical tests were performed: tensile and bending tests.

4.1 *Tensile tests*

Tensile specimens were made from the welds, crosswise to the weld line. For each weldment the average of three stress strain tests was used in order to determine the main mechanical properties (Yield Tensile Strength - s_{YS}, Ultimate Tensile Strength – s_U and elongation at fracture– ε). The tensile stress strain curves were obtained using a servo hydraulic MTS testing machine using a 25 mm gage length and a load cell of 250 kN.

One curve of each weldment is plotted in Figure 5 and the averages of mechanical characteristics are presented in Table 2.

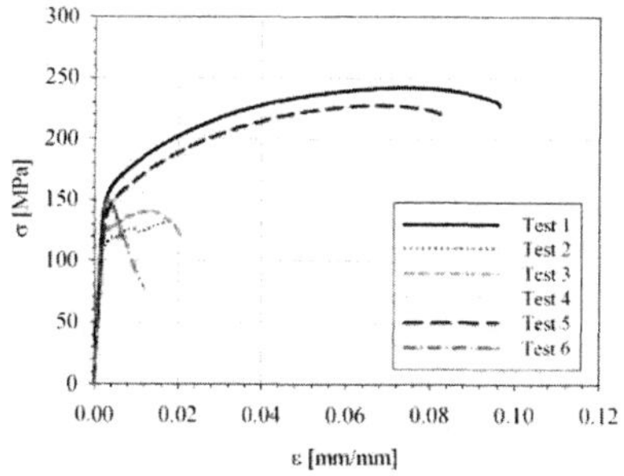

Figure 5. Stress strain curves for each wedment.

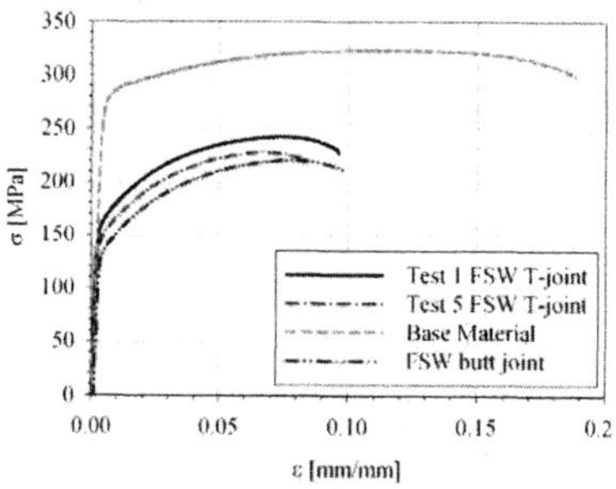

Figure 6. Stress strain curves comparison with base material and FSW butt joints.

From the stress strain curves obtained, only the first and fifth tests display complete integrity of the specimen since the rupture occurred in the limit of the thermo mechanical affected zone. In the other tests the rupture occurred in the proximity of the T-corners caused by some defects in this zone.

The stress strain curves of specimens 1 and 5 were compared with the base material and with a butt joint of the same material (Al6082-T6). This comparison is presented in Figure 6.

Table 2. Mechanical properties obtained from stress strain curves.

	Test 1	Test 2	Test 3	Test 4	Test 5	Test 6
σ_{YS} [MPa]	161.09	114.20	127.17	134.47	147.74	140.19
σ_U [MPa]	241.81	128.74	138.84	155.70	225.99	154.34
ε	8.61%	4.49%	5.56%	5.56%	9.56%	5.36%

4.2 *Bending tests*

The mechanical integrity of the weld zone in weldments was carried out by bending tests. With these tests it is possible to detect defects and their direction in the T-corners. These tests were performed in the MTS machine mentioned before, using displacement control with a velocity of 1mm/min, Figure 7. Typically three specimens of each weldment were tested. These specimens were 150 mm width and 20 mm thick. For each one the displacement vs. load force in the punch was recorded. One curve representative of each weldment is presented in Figure 8.

Figure 7. Bending test setup.

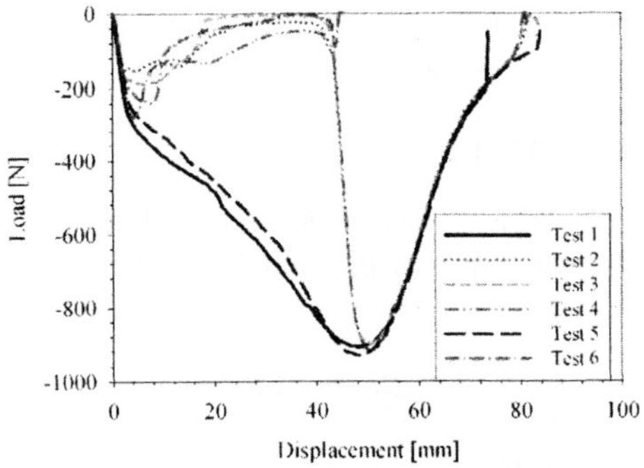

Figure 8. Bending test results – load vs. displacement

The first and fifth tests present the greater bending test energy, confirming the results obtained in the stress strain tests and from the metallographic analyses. The others tests presented unsatisfactory behaviour with partial rupture in the beginning of the test; the same maximum load was attained at 50 mm displacement but this is a result of type of test which is unrelated to the integrity of the joint.

4.3 *Numerical analyses*

Numerical analyses of successful bending tests were performed using a finite element model in order to identify potential differences between the theoretical and experimental results. As an approximation, the stress strain curve of base material was used in order to include the plasticity effect in the model.

The finite element solver ABAQUS was used to model the load *vs.* displacement curves using material and geometrical non-linearities. The rollers of the test device were defined by analytical rigid surfaces and the frictionless contact between these surfaces and the specimen was modelled. In this complex model large displacement theory, contact theory and multiple stress-plastic strain data were used.

In Figure 9 presents the mesh with the rigid analytical surfaces and the deformed shape obtained after the test. In Figure 11 a comparison between finite element results, data of weldments 1 and 5, base material, and FSW butt joint is presented.

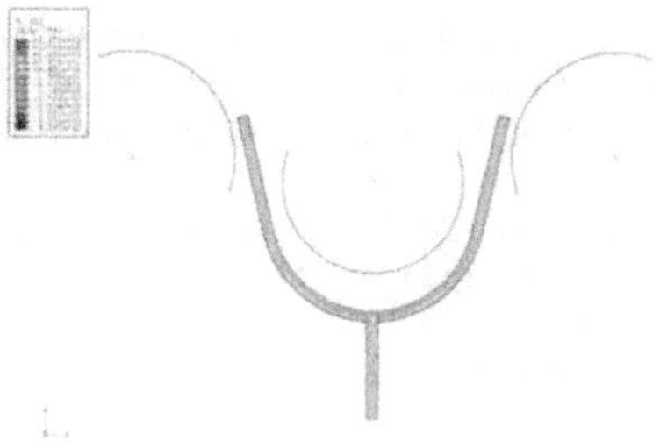

Figure 9. Bending test – deformed model.

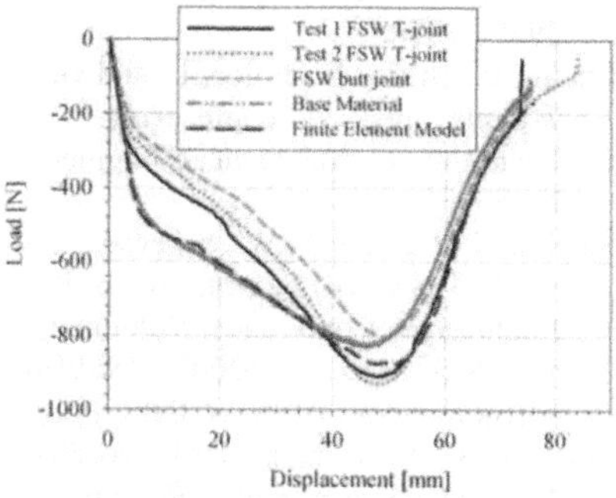

Figure 10. Bending test – results comparison.

A good agreement between base material and the finite element model was found. The FSW T-joints have less energy resistance that the base material (up to 40 mm); this fact can be caused by the softening or/and the loss of the T6 heat treatment condition in the welding zone. In the final part of the bending test a good agreement among all numerical models and experimental results is found.

5 Conclusions

New T-joints designs manufactured with FSW, with great potential to improve the mechanical characteristics of these joints were discussed. Friction stir welded T-joints were manufactured using three components and the vertical member tangential to the top surface of the horizontal members. Some specimens with and without defects were cut from the T-joints and tested in order to understand the better solutions for the pin geometry and process parameters. Metallurgical and mechanical characterization was performed. Several defects were detected in the T-corners, eg. cracks and excess of pin penetration. Some loss in yield and ultimate tensile strength was found when comparing base material with the weldments. Predictably the bending tests of weldments without defects presented the higher energy values. A finite element model using geometric and material non-linearities was performed in order to model the bending test and to understand the effect of the welding line in the mechanical behaviour of the specimens.

Acknowledgments

The work was partially supported by FCT through contracts SFRH/BD/19281/2004 and SFRH/BD/29004/2006. Experimental work of J. R. Almeida, R. Silva, M. Figueiredo and F. Oliveira is acknowledged.

References

ASM Handbook (1985) – *Vol. 9 Metallographic and Microstructures*, ASM International, Materials Park, OH.

Erbsloh, K. C., Donne, D. and Lohwasser, D. (2003) *Friction stir welding of T-joints*, Materials Science Forum -Trans Tech Publications Inc.; Thermec'2003, Pts 1-5, 426-4. 2965 - 2970.

Fratini, L., Buffa, G., Filice, L. and Gagliardi, F. (2006) *Friction stir welding of AA6082-T6 T-joints: process engineering and performance measurement,* Proceedings of the Institution of Mechanical Engineers Part B-Journal of Engineering Manufacture, **220** No.5. 669-676.

Steel, R. J., Nelson, T. W., Sorensen, C. D. and Packer, S. M. (2005) *Friction stir welding of steel T-joint configurations* In Proceedings of the International Offshore and Polar Engineering Conference, Volume 2005, Seoul, Korea.

Tavares, S. M. O., et al (2007), *"Soldadura por fricção linear - Soldaduras em T"*, pending patent number 103867 S, INPI, Portugal.

Thomas, W.M., Nicholas, E.D., Needham, J.C., Murch, M.G., Templesmith, P. & Dawes, C. J. (1992) *Improvements Relating to Friction Welding*, WO/1993/010935, International Patent Number PCT/GB92/02203.

12.4 Economic Benefit of Upgrading the In-service Inspection System of Pressurized Components at Paks Nuclear Power Plant

Péter Trampus[1], Sándor Rátkai[2] and Dénes Szabó[2]
[1]*Trampus Consulting & Engineering, Bicske, Hungary, trampusp@trampus.axelero.net*
[2]*Paks Nuclear Power Plant, Paks, Hungary, ratkai@npp.hu, szabod@npp.hu*

Abstract

Nuclear power plant pressurized components (primarily their welded joints) are periodically inspected by non-destructive examination (NDE) methods during the plant service period. As a substantial part of the Paks Nuclear Power Plant ongoing life extension project, the in-service inspection (ISI) program is being upgraded to meet the worldwide accepted ASME Code requirements. This will allow to reduce the length of planned outage periods which contributes to a more cost-efficient operation and maintenance. Implementation of the qualification of NDE systems may provide a quantifiable benefit due to the proper planning and preparation of NDE, a lower probability of repetition of examinations and a higher probability of flaw detection. Introduction of risk-informed ISI focuses on inspection of high risk components and, thus, may decrease inspection efforts.

Keywords: *ISI, outage time, inspection qualification, NDE, risk-informed ISI.*

1 Introduction

Nuclear power plants (NPPs) are hazardous technological systems due to the existence of ionizing radiation as the inherent feature of the technology. The pressurized components and piping that transport the generated heat from the reactor core via the cooling medium, are welded structures. Their structural integrity must be maintained and justified throughout the whole service life of the plant. Therefore, the pressurized components and piping are periodically inspected by non-destructive examination (NDE) methods to provide data on possible flaws to their structural integrity assessment. This activity is called in-service inspection (ISI).

Paks, Hungary's sole NPP consisting of four VVER-440 model 213 units, was commissioned in the mid-eighties with a design life of 30 years (VVERs are Russian designed pressurized water reactors, 440 refers to the nominal electric capacity in MW, model 213 is the second generation of VVER-440s). The owner is now preparing the operational life extension up to 50 years. Preparing the comprehensive technical justification for the long term operation of the units has provided a unique opportunity for the ISI program upgrading. The current ISI program in Hungary is based on the former Soviet normative technical documents. Although the program went through in a quite permanent development, it still shows fundamental deficiencies which can not be changed easily.

The objective of the ISI program upgrading is basically twofold. On the one hand, it should significantly contribute to safety enhancement as a consequence of using a more effective ISI program, see Engl & Trampus (2002). On the other hand, it is expected to provide substantial economic benefit. This paper intends to present the outline of the ISI program upgrading project with a special focus on its economic aspects.

2 Main features of the ISI program upgrading

Figure 1 shows the scheme of the ISI program upgrading project. It can be seen that basically two sets of documents were created: (1) background studies, and (2) documents of the new, i.e. upgraded ISI program. The background studies cover the comparative assessment of the existing Hungarian ISI program and those according to ASME because the requirements of Section XI of the ASME Boiler and Pressure Vessel Code (BPVC) have been decided to adapt. The ASME BPVC is a comprehensive, worldwide accepted and proven system for the NPP pressurized components. The feasibility study deals with the introduction of a risk-informed ISI (RI-ISI); the other study focuses on the further development of inspection qualification. The new ISI program documents include the modified ISI plans, the new acceptance standards, a set of qualification target flaw configurations, and the NDE procedures applied. In the following, the three major areas of the ISI upgrading are presented. After a brief introduction of each the economic aspects will be discussed.

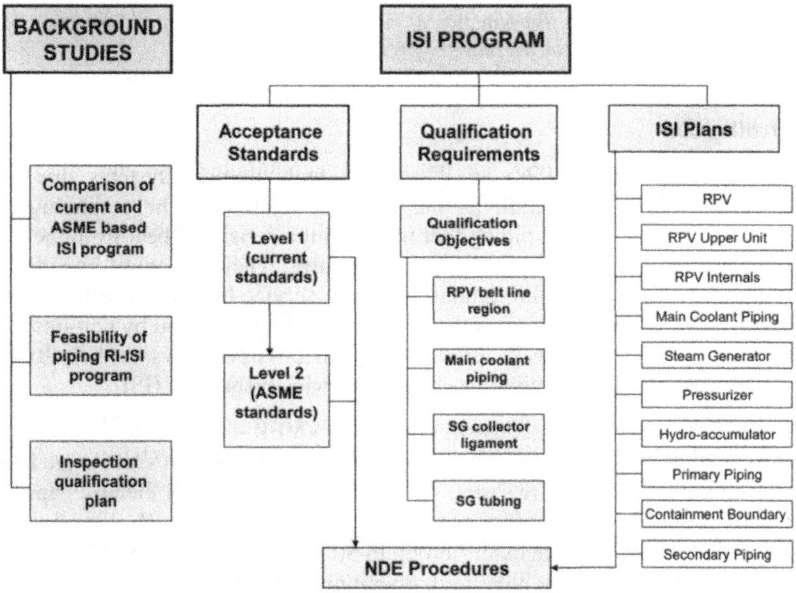

Figure 1. ISI program upgrading project

2.1 *Extension of ISI interval*

The comprehensive adaptation of ASME BPVC at Paks NPP could allow to increasing the ISI interval of the Safety Class 1 components. For these components the Russian based ISI interval was set to four years, while the ASME Section XI applies a ten-year inspection interval. Paks NPP considered justifying a two times four years period. It is less rigorous than the ASME one and to which the transition from the current system may easily be done. Since a prerequisite for application of

Section XI is the meeting of Section III requirements, a design review of selected components is being done to justify compliance with the Section III requirements. The design review will confirm the applicability of the eight year ISI interval.

Parallel with the adaptation of ASME BPVC Section XI requirements, the tasks were supplemented with the incorporation of the recently issued new requirement and guidelines for the ageing management of Safety Class 1 components. Ageing management is one of the key elements of long term operation. This work results in an approximately thirty percents larger volume of the necessary inspections compared to the previous practice. Pragmatically, we can conclude that the new ISI system will require more effort.

Meanwhile, this increased volume of the inspections will be distributed on a longer period (two times of the original ones), which finally results in an approximately 65% volume of the inspection in any eight years time frame. This 35 % saving could have a significant role in the economic operation of Paks NPP. At least two aspects of this feature can be mentioned:

(1) The direct costs of the operation could be reduced, because of the laggard inspection. This cost type is proportional to the effective volume of the conducted inspections. Therefore, in the case of any reduction in this activity results in saving money directly.
(2) If the volume of the inspection is less than the usually applied one, the plant outage period could be reduced, which results in more effective operational days a year. Taking this aspect into account we can assume that 5-10 days reserve can draw up in every year at all nuclear units. The economic advantage of this reserve can be estimated using the following assumptions:
    Number of units = 4;
    Average remaining operational lifetime (assuming the success of the operational life extension) = 25 years;
    Average electric capacity of each unit = 500 MW;
    Revenue after 1 kWh = USD cent 0.05;
    Average operational revenues per unit per year: 500 x 1000 x 24 x 0.05 = USD 0.6 M

Taking this data we can estimate the total extra incoming, which is 4 x 0.6 x 25 x (5-10) = USD 300 – 600 M. This is a very huge amount and does not require more explanation.

2.2 *Inspection qualification*

Results of large scale international projects have demonstrated that Ultrasonic Testing (UT) during ISI of NPP components could be quite effective. However, use of standardized procedures has been proved to be inadequate for some flaw types and geometries, see PISC (1992). This recognition has led both to Performance Demonstration Initiative (PDI) in the USA, and to the inspection validation concept introduced first in the UK. From the latter one, the European Network for Inspection and Qualification (ENIQ) as a European wide inspection qualification framework came into existence. The objective of both the PDI and ENIQ is to establish confidence that the NDE procedure, equipment and personnel are capable of meeting the inspection requirements in real circumstances. Details on the European methodology are in ENIQ (2007), and a VVER specific guidance in IAEA (1998).

The two major elements of inspection qualification are the technical justification and practical assessment. Technical justification is a collection of all the information which provides evidence about the reliability of an NDE technique as applied to a specific component. It could include physical reasoning, feedback from field experience, relevant round robin trials, mathematical models, etc. Practical assessment may involve test pieces replicating the component under test in size and geometry. The defective condition may also be accurately replicated.

The first steps in the field of inspection qualification were made in Hungary in the late 1990s. Paks NPP together with other VVER operating countries participated in both the International Atomic Energy Agency (IAEA) and European Commission (EC), supported sponsored projects to establish inspection qualification infrastructure in these countries. In 2000, the Hungarian Nuclear Regulatory Authority commenced the inspection qualification activities at Paks NPP. Due to the lack of experience the plant proposed to perform first a pilot qualification. During the pilot qualification exercise Hungary developed its national qualification strategy, involved leading NDE experts in the qualification process, which established an independent inspection qualification body based on experience gained, determined the future qualification goals and agenda. In the period of 2003 to 2007 a series of qualifications have been performed, see in Somogyi et al. (2006).

The main benefit of the inspection qualification is that it improves the nuclear safety by giving confidence that the qualified inspection system is capable of meeting requirements. Nevertheless the qualification can also bring economic benefits to offset costs. One the one hand, the cost of qualifications cannot be ignored sometimes even it can be on the large side. The cost factors typically include the charges of fracture mechanics calculations of the target flaw size, the operation of the independent qualification body, the manufacturing of qualification test pieces and also the costs of inspection organization related to qualification (optimizing the inspection procedure, performing laboratory trials, etc). Especially the manufacturing of qualification test pieces can be very expensive, since the test piece used for qualification must be representative for the component to be inspected in terms of geometry, material properties, welding and fabrication procedure used during manufacturing etc. Moreover, the defects implanted into the qualification test piece usually desired to be realistic.

The expenses of qualification can be saved by different ways, for instance by using qualification levels, since the rigour of qualification represented by the qualification levels is directly proportional to the costs. For some inspections a lower qualification level can be sufficient taking into consideration the qualification requirements. In this case the application of a simpler test piece can be accepted that can be significantly cheaper than a 1:1 scale mock up required when using higher qualification levels.

Sometimes it is possible to save costs by renting qualification test piece from other plant operator, which is much less expensive than manufacturing. Today a qualification test piece database is available created jointly by ENIQ and IAEA, in

2003, to help nuclear utilities to save cost by sharing information and finding existing and applicable test piece for qualification goals.

Some of the economical benefits of inspection qualification are not quantifiable or at least difficult to measure. The qualification allows the adaptation of new technology by showing that it meets the required performance criteria. Similarly, it can allow the use of simpler inspections than those prescribed by the codes, again if the performance criteria are met. Thus, qualification can save costs by justifying the use of simpler and hence less time-consuming and expensive inspections.

If the detection capability and sizing accuracy is proven by the qualification, the inspection intervals can be determined better to ensure that no defects will grow to serious levels before the next inspection. Intervals determined in such an objective way are frequently longer than those set on the basis of pessimistic assumptions of inspection capability, leading to less inspection without any drop in confidence in structural integrity of the component.

Another benefit of using qualified NDE system is that the decision on defect repair can be taken at an earlier stage. If the detection capability and the sizing accuracy is less well known, the defect size, at which the repair decisions are taken, is considerably larger before there can be a confidence in the inspection information. Repairs carried out an earlier stage when defects are smaller are less costly when larger defects have to be removed. The possibility of the repair is also influenced by the ability to carry out an inspection of the repair to the appropriate standard. Qualification can establish whether such inspection is possible and hence pave the way to repair rather than replacement.

In Paks NPP one of the proof for extending it in service inspection period safely by adopting ASME code is to qualify the safety relevant inspections taking into consideration eight years inspection interval instead of four years that is recently in force. The economic benefit achieved using longer ISI intervals will significantly exceed the costs invested into inspection qualification.

2.3 *Risk-informed ISI*

Traditionally, ISI codes, including ASME Section XI, specify the locations, frequency and methods of NDE based primarily on the type and safety significance of the component. However, it has been recognized that many examinations have been carried out on component locations where no defect detected, which means that many resources have been spent on sites of low or negligible risk for plant safety. Risk is defined as the product of the measure of the consequence resulting from a failure and the probability of that failure occurring within a given time. In this context the goal of risk-informed ISI (RI-ISI) can be defined as addressing and in ideal case, reducing the risk posed by failure of pressurized components. To reduce the risk, any ISI must first be directed at the locations of highest risk and must then be capable of detecting and sizing possible flaws. The latter one is the area of inspection qualification.

The failure potential is directly connected with specific damage mechanisms and the likelihood that particular damage mechanisms could be active in particular locations. Probabilistic Risk Assessment (PRA) methods are involved in determining the risk impact of failures in piping systems. Degradation mechanisms may be erosion corrosion, local corrosion, stress corrosion cracking, thermal fatigue, vibration fatigue, etc. The consequence assessments are made by evaluating the contribution to cumulative core damage frequency. If the probability of failure occurrence and consequence of failure are known, the risk can be calculated or ranked (from the highest to the lowest). There are different ways of the graphical representation of risk categories, for instance risk plots, risk matrix or risk diagrams. A usual risk matrix is shown in Figure 2, see ENIQ (2005).

			Conditional Consequence				
			Very Low	Low	Medium	High	Very High
			$<10^{-6}$	$10^{-6}-10^{-5}$	$10^{-5}-10^{-4}$	$10^{-4}-10^{-3}$	$>10^{-3}$
Failure Probability	Very High	$>10^{-4}$					
	High	$10^{-5}-10^{-4}$					
	Medium	$10^{-6}-10^{-5}$					
	Low	$10^{-7}-10^{-6}$					
	Very Low	$<10^{-7}$					

Figure 2. Risk matrix (values in the table are illustrative).

The first step of developing an RI-ISI program is the identification of risk outliers (areas having much higher risk than the overall mean risk level for all areas). Then, the risk associated with these areas is compared with the risk associated with the sites forming the scope of the current (deterministic) ISI program. The total risk being addressed by the selected areas should address at least as much but preferably more than the total risk addressed by the current ISI program. When moving to an RI-ISI program, at least risk neutrality or, better, risk reduction should be achieved. Since ISI at NPPs leads to radiation exposure to the NDE personnel, the different RI-ISI options with the same risk reduction can mean different total radiation exposure. When faced with such choice, the RI-ISI program option that gives as little radiation exposure as possible should be chosen.

Of course, RI-ISI is reducing costs by effectively concentrating limited inspection resources on safety of systems and locations. Nowadays, many countries in the world moving to or having already been introduced RI-ISI. In the USA, e.g. 80% of the nuclear units are committed to implement RI-ISI. Experience shows that RI-ISI for piping can save plants up to USD 300 000 per year in unnecessary inspection costs, while reducing by up to 75% the number of inspections that must be performed in radiological area. This, in turn, can reduce worker dose rates up to 0.6 Sv over a 10-year inspection cycle (not only NDE but preparatory works like scaffolding, removing returning isolations, etc. collect dose), see Balkey & Closky (2000).

VVER plants have started to launch pilot studies on RI-ISI; Finland and the Czech Republic are eminent in this field. Paks NPP has decided to establish a pilot project on a piping system, the project of preparatory activities is in progress. About quantifiable economic benefit in association with RI-ISI can be reported after accomplishing and evaluating the pilot project.

References

Balkey, K.R. & Closky, N.B. (2000) Implementation of risk-informed in-service inspection. *Nuclear News* No. 5, 35-38

Engl, G. & Trampus, P. (2002) Criteria and Recommendations for ISI Effectiveness Improvement. In: *Proc. Joint EC-IAEA Techn. Meeting on Improvements on In-service Inspection Effectiveness, Petten.*

ENIQ (2005) European Framework Document for Risk-informed In-service Inspection. EUR 21581 EN, Luxembourg

ENIQ (2007) European Methodology for Qualification of Non-Destructive Testing (third issue). EUR 17299 EN, Luxembourg

IAEA (1998) Methodology for Qualification of In-Service Inspection Systems for WWER Nuclear Power Plants. IAEA-EBP-WWER-11, IAEA, Vienna

PISC (1992): Summary of the Three Phases of the PISC Programme. Report No. 17

Somogyi, G., Klausz, G., Szabó, D. & Trampus P. (2006) Qualification of NDT Systems in Hungary. In: *Proc. 9th Eur. Conf. on NDT, Berlin, DGZfP, BB 103-CD, We 1.4.2.*

Section 13
Applied mechanics

13.1 Analytical Strength Analysis of Strands Applied for High Pressure Hoses

István Barkóczi[1], László Sárközi[2]
[1]*Fux Ltd. Miskolc, Hungary, general manager, research engineer,* [2]*University of Miskolc, Regional Adult Education Centre, director*

Abstract

The intention of this paper is to show some analytical and numerical techniques and typical results have been applied and gained for the so called SZ and ZZ type 1+6+12 wire structures hose strands to get in their analysis for axial tension, twisting and bending. Analytically the well established Costello – Velinsky theory (see Costello et al. 1974, 1979, 1983, Velinsky, 1985) was applied while the FE modelling and calculation was proceeded by the Marc commercial software. The results of the investigation include not only the geometrical data like extension, rotation, curvature etc. but also the typical stress coordinates and the effective Young modulus and Poisson ratio of the strands under investigation. Both, so called free end loading and fixed end loading tasks have been investigated. The results have been compared with each other and also with experimental measurements.

Keywords: *strand wire, reinforced elastomer composites*

1 Introduction

The high pressure hoses used in the oil industry are typical examples of reinforced elastomer composites where the main strength carrying capacity is assured by special strands made from strong wires. There are two types of standardized flexible pipes particularly used in offshore industry: unbonded and bonded flexible pipes. The big plant of Phoenix Rubber Industrial Ltd. in Hungary, Szeged is the world-wide market leader of these types of products for which the FUX Ltd. Miskolc is the main supplier of the strands. Figure 1 shows two examples of hoses with two and four layer of strands. Based on the very rigorous and high level requirements of the oil industry the constructional and technological parameters have to be considered and designed in a sophisticated manner to include these constituents.

Figure 1. Axonometric views of two and four layers high pressure flexible hoses

For the improvement of the hoses and internally the reinforcements the research engineers have succeeded in lifting up the minimum tensile strength of the typical 19 wires strand from 13540N up to 17100N. However, this is currently a limitation as further increase would reduce the toughness, which would not be acceptable. The possibility for further improvement could be the structural design and manufacturing of the strands. An intensive EU supported project is in progress /GVOP-3.1.1.-2004-05-0171/3.0 in Hungary/, in which the Bay Zoltán Institute of Logistics and Production Engineering, Miskolc, the Flexib Ltd. Budapest, the University of Miskolc, Faculty of Mechanical Engineering and the two above mentioned companies are participating.

It should be emphasised, that the theoretical results, for example the axial and torsional stiffness properties of the strands have got importance from the point of further analytical investigation of the hoses as well, that is a more accurate composite modelling of the elastomer - reinforcements will be possible for an improved design of the hoses. (Sárközi et al. 1995, 1996)

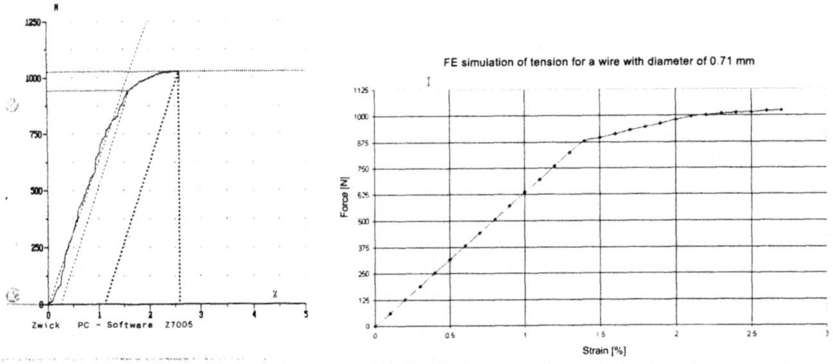

Figure 2. Test and FE diagram of a coppered wire sample with 0.71 mm diameter

In this paper three different form strands are shown from which the first is a simple version with 1+6 wires, first subversion: coppered (1x0.71+6x0.71), and second subversion: galvanized zinc-plated strand with a nominal diameter of 2.1mm (1x0.76+6x0.71), while the second is a "classical" SZ type 1x0.71+6x0.71+12x0.71 coppered structure. S means right and Z means left direction pitch of the curbs. Just for the above mentioned improvement of the different type of hoses this SZ type reinforcement structure was intended to be substituted by a new ZZ type galvanized zinc-plated strand with nominal diameter of 0.36mm (1x0.76+6x0.71+12x0.71) which is the third structure now under investigation. The simplified circular cross sections of the structures can be seen in Figure 3.

3 Analytical strength analysis based on the Costello-Velinsky theory

Based on the scientific literature, in the elastic range the Costello – Velinsky theory (Costello et al. 1974, 1979, 1983, Velinsky, 1985) seems to be an excellent analytical tool for modelling the mechanical behaviour of the strands under different kinds of loading. In the case of axial tension plus torsion the theory assumes

- the linear elastic behaviour,
- the existence of the Poisson effect, that is the diameters of the wires are changing under loading,

- though the helical angles, that is the pitches of the curbe wires are large, their changes and also the strains are small, so geometrically linearised equations can be applied.

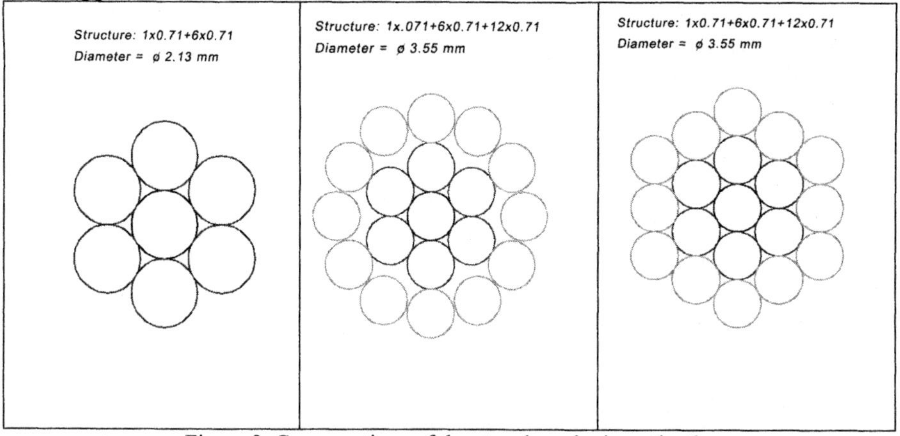

Figure 3. Cross sections of the strands under investigation

The contact conditions are not a priori included in the theory however, after the calculation of the structure, in the knowledge of the new geometry caused by the loading, the contact Hertz stresses can determined. The simple basic equation system of given structure seems like this

$$\frac{F}{AE} = C_1\varepsilon + C_2\beta \tag{1}$$

and $$\frac{M_t}{ER^3} = C_3\varepsilon + C_4\beta$$

where F and M_t the axial loading and torsional moment at a given cross section ε and β strain and rotation also in axial direction

The total metal cross section of the strand is

$$A = \Sigma \pi R_i^2$$

R is the nominal radius of the strand cross section, R_i the radius of the different wires C_1, C_2, C_3 *és* C_4 are parameters which should be determined by a given technique (Costello et al. 1974, 1979, 1983, Velinsky, 1985).
Without going into the details, in frame of the project special algorithms have been developed based on the literature by which the given strands can be analysed for tension and for other types loading. Looking at (1.) two basic type of problems can be constructed and solved:

- if the axial loading F is given and the β is prescribed to be zero then the axial strain ε and the reactional torsional moment M_t should be determined. This is the *fixed end loading* task, and
- if the axial loading F is given and the M_t torsional loading is zero, i.e. the strand can be freely rotated, then the axial strain ε and the β should be determined. This is the *free end loading* task.

Of course, in each case all the geometrical data, like axial strains per wire constituents, normal and binormal direction curvatures, twist per unit length,

proportion of the total loading between of the different curbes, stress coordinates per wires etc. are determined plus the so called effective elastic modulus of the given strand can be determined which can be also compared with practical measurements. A very important fact for the analysts: the computer time on an everyday PC is not more than one second, so it is easy to make parameter investigations.
In this sense number of investigations has been performed, from which some examples will be shown.

3.1 *Analysis of 1+6 strands with different core wire diameter*

For the two subversions of the seven wire strands were investigated under an axial loading of 4000N both for fixed and free end loading cases. The pitches of the curbes were in both cases 36 mm. The most characteristic results are summarised in Tablet 1. Studying the data of the tablet it is clear, that in case of given axial load the strain is smaller if the task is "fixed end loading". It means in this case, that the axial stiffness of the structure is approximately 10-20% higher if the end of the strand sample is not allowed to rotate. Similarly, the effect of the increase of the diameter of the middle wire makes such kind of changes which effect on the Hertz contact stresses as well. The calculations signed a maximum normal stress in the core wire which was 1955 MPa in the coppered strand but decreased to 1812 MPa for the improved structure. The changes for the effective elasticity modulus can also be seen in the Table.

Table 1. Some results for the 1+6 structures

		F [N]	M_t [Nmm]	ε	β	E_{eff} [Mpa]	σ_c [MPa]
Galva-nized zinc-plated	fixed end loading	$4*10^3$	248,6	0,0097	0	$1,8621*10^5$	-906,2
	free end loading	$4*10^3$	0	0,0115	-0,0275	-	
coppered	fixed end loading	$4*10^3$	210,9	0,0104	0	$1,867*10^5$	-809,4
	free end loading	$4*10^3$	0	0,0121	-0,0265	-	

Some parametric investigations have also been carried out for changes of the pitch of the galvanized strand from 30 mm up to 44 mm. Figure 4 informs about the changes of the strains and the rotations (if there is) and Figure 5 about the Hertz stresses.

3.2 *The 1+6+12 structure strands manufactured by the FUX Ltd.*

Studying the linear diagrams of Figure 6, number of important facts can be seen to be concerned of their behaviour and which are important from the point of the final characteristics of the hoses where they will be built in. By comparison for example the fixed end loading diagrams for an axial loading of 10000 N, the reaction torsion moment for the classical SZ strand is -1284 Nmm, while for the new ZZ strand construction it is 1984 Nmm in the opposite direction. Qualitatively, the different behaviour for the fixed and the free end loading cases are not surprising but its measure for the ZZ construction does. It can be seen for example that for a 1% strain

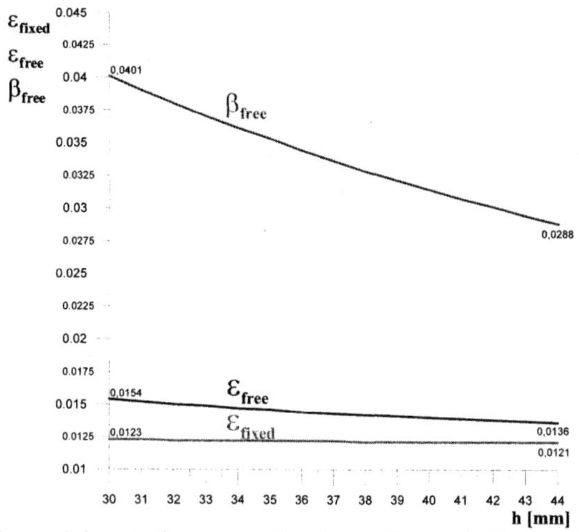

Figure 4. The strains and the rotation versus the pitch of the curbe for a galvanized 1+6 wires strand.

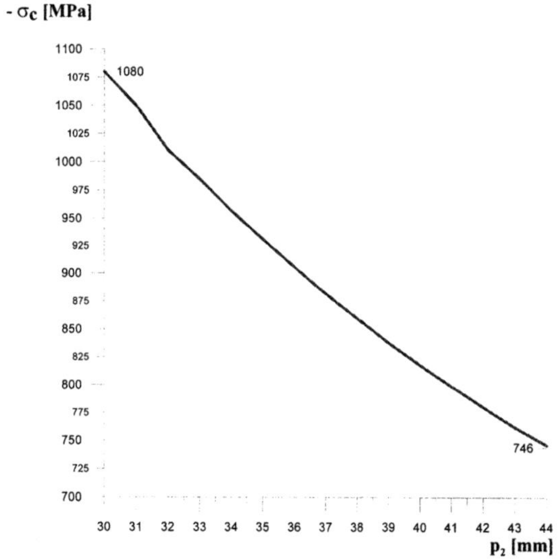

Figure 5. The Hertz stresses versus the pitch of the curbe for a galvanized 1+6 wires strand

twice as big axial loading is needed for the fixed end loading case then for that of the free end loading.

4 Finite element analysis of the 19 wires structures

Due to the number of well known preferencies the SZ and the ZZ structures have been analysed by the Finite Element Method as well. The task is outstandingly challenging because practically a boundary value task of so called many bodies frictional contact problem have to be solved taken not only the material but also the geometrical non-linearities into consideration. The work is huge for which the details will be published later in more scientific papers. However, the feature of the finite element mesh, using eight noded 3D elements are shown in Figure 7. for the SZ and in Figure 9 for ZZ constructions. These pictures show at the same time the special feature of the typical point contact for the SZ and the typical line contact for the ZZ strands. In the calculation, the frictional ratio was selected to be 0.115, while the independently measured fiber test was applied both in the linear and non-linear ranges of the loading history.

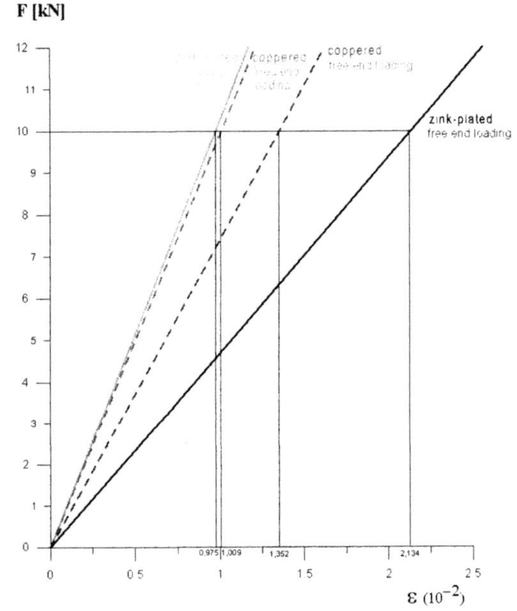

Figure 6. The axial force versus strain diagrams for the 1+6+12 strands

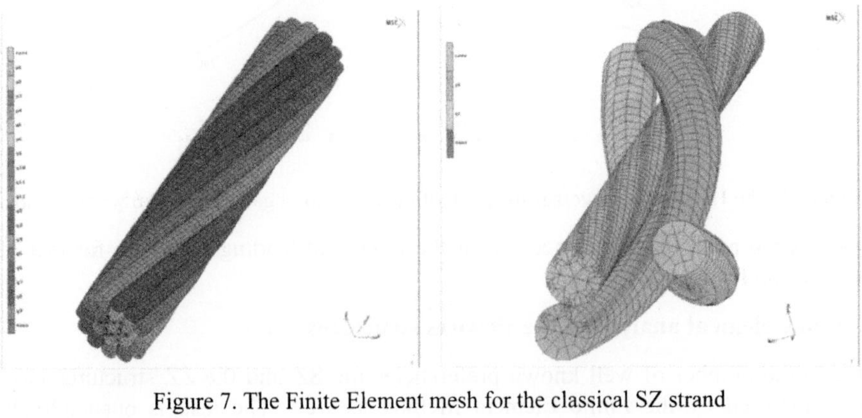

Figure 7. The Finite Element mesh for the classical SZ strand

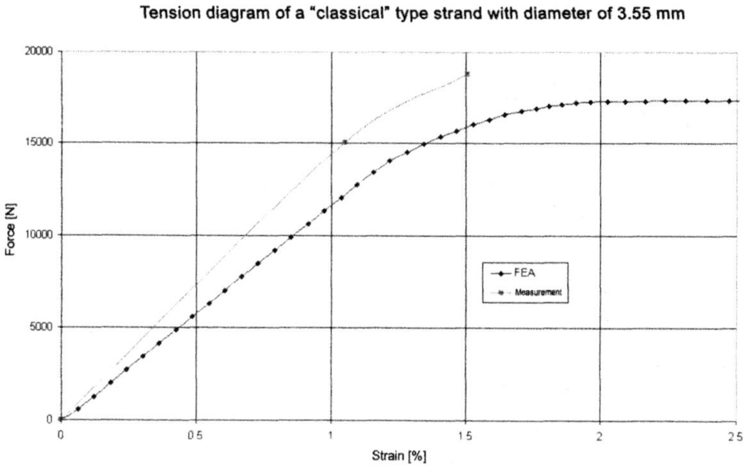

Figure 8. The tension diagram for the classical SZ strand

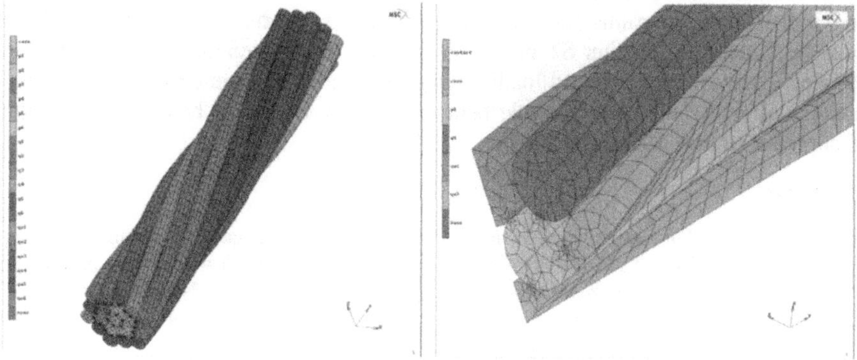

Figure 9. The Finite Element mesh for the compact ZZ strand

5 Conclusion

As a result of the analytical, the finite element and practical work it can be concluded, that the Costello-Velinsky theory and procedure for the relatively small diameter elastomer reinforcement strands can be applied with the same level of effectiveness as for general strands and ropes. The technique is concerned to the elastic range.

Up to 1-1.5 % strain the behaviour of the strands is linearly elastic.
For purposes of improvement of the strands both fixed end loading and free end loading tasks should be composed and solved. The comparison of the results can give a valuable information for the expected behaviour of the reinforcements.

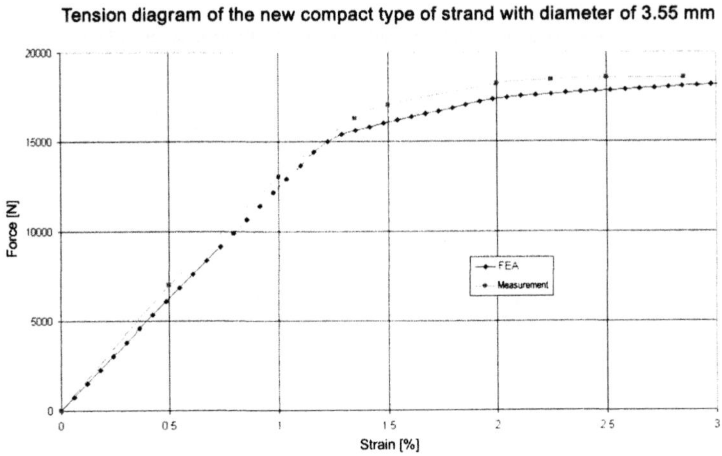

Figure 10. The tension diagram for the compact ZZ strand

The measured axial loading – strain diagrams (Figure 8 and 10) and the calculated fixed end loading diagrams (Figure 6) do not explain the different behaviour of the hoses dependent on whether SZ or ZZ reinforcements are included. At any rate, the difference between the free end loading diagrams (see also Figure 6) gives probably an explanation for the more flexible behaviour of the improved hoses including the ZZ compact strands (Barkóczi 2007).

References

Barkóczi, I. (2007) *Wire and stranded wire*, FUX Zrt. Miskolc, Hungary. (in Hungarian).

Costello, G.A. & Philips, J.W. (1974) A more exact theory for twisted wire cables. *Journal of the Engineering Mechanics Division*, ASCE, 100. (EM5, Proc.Paper 10856), 1096-1099,

Costello, G.A. & Miller, R.E. (1979) Lay effect of wire rope. *Journal of the Engineering Mechanics Division*, ASCE,105, (EM4, paper 14753) 597-608,

Costello, G.A. (1983) Stresses in multilayered cables. *Journal of Engineering Resources Technology, Trans. the ASME*, **105** 337-340,

Sárközi, L., Nándori F., Szabó T. & Németh T. (1995) Investigation of Effects of Constructional Parameters on the *Mechanical Behaviour of Truck Tires. IRC'95 Conference.* Kobe, Japan, Proceedings pp. 695-698. 10. 23-27.

Sárközi, L., Nándori F., Szabó T. & Kriston S. (1996) A Special Purpose-Oriented Finite Element Package for Strength Analysis of Agricultural Tyres. *Kautschuk-Herbst-Kolloqium '96.* Hannover. Proceedings 10.24-26. pp. 101-108.

Sárközi, L., Nándori F. & Szabó T. (1996) FE Analysis of High Pressure Rubber Hoses for Tension and Bending. Proceeding of the *International Scientific Conference on Numerical Methods in Continuum mechanics*. High Tatras, Stara Lesna. Slovakia. 09.26-28. pp. 226-270.

Velinsky, S.A. (1985) Analysis of fiber core wire rope. *Journal of Energy Resourses Technology, Trans. ASME*, **107** No. 3, 388-393.

13.2 Optimization of Inelastic Cylindrical Shells with Stiffeners

Jaan Lellep, Sander Hannus, Annika Paltsepp
Institute of Mathematics, Tartu University
2 Liivi str., 50409 Tartu, Estonia

Abstract

A method of optimal design of inelastic cylindrical shells with absolutely rigid hoop stiffeners is suggested. The shells under consideration are subjected to internal pressure loading and axial dead load. Taking geometrical non-linearity of the structure into account optimal locations of hoop stiffeners are determined so that the cost function attains its minimum value. A particular problem of minimization of the mean deflection of the shell with cracks at the cross sections where stiffeners are located is treated in a greater detail.

Keywords: *cylindrical shell, plasticity, optimization, crack.*

1 Introduction

Problems of optimal design of supports to elastic structures are first formulated by Mroz & Rozvany (1975). Optimality conditions for elastic supports were derived for beam and frame structures by Bojczuk & Mroz (1998) who developed the method for simultaneous optimization of topology, configuration and cross-sectional dimensions of structures. The paper Bojczuk & Mroz (1998) is an extension of earlier works by Garstecki & Mroz (1987), Mroz & Lekszycki (1981), Lepik (1978), also Szelag & Mroz (1978) devoted to optimal design of intermediate supports to elastic structures. Olhoff & Åkesson (1991) studied the problem of minimal stiffness of optimally located supports providing maximal eigenfrequencies of beams and problems of maximal buckling loads in columns. Layout optimization of structures was studied by Kaliszky & Logo (2006, 2003).

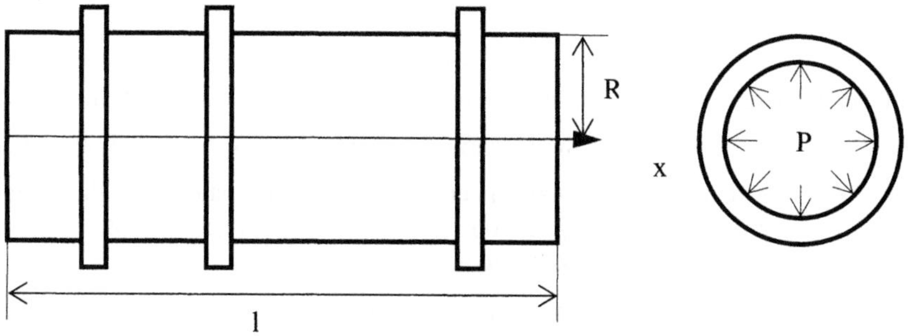

Figure 1. Geometry of the shell.

Optimal design of ideal plastic cylindrical shells investigated by Cinquini & Kouam (1983). Cinquini and Kouam considered cylindrical shells made of Tresca material and used the geometrically linear concept of the structure. Geometrically non-linear pure plastic cylindrical shells were studied by Lellep (1984, 1985) and Lellep & Hannus (1992). Making use of the variational methods of the control theory an optimization procedure was developed in Lellep (1985) for shells which obey piece wise linear yield conditions. In Lellep & Hannus (1992) the method was extended to non-linear yield surfaces. In earlier papers shells without defects were considered. However, it is well known in the fracture mechanics Broek (1988), Broberg (1999) that due to repeated or extremal loading the material of structures is weakened by defects. In the present paper the influence of eventual cracks at supports is taken into account. The cracks are assumed to be part-through surface cracks which remain stationary. This concept has been used by Kumar and Petroski (1985), Petroski (1981), Yu and Chen (1999, 2000) to study the motion rigid plastic beams subjected to impact loading.

2 Formulation of the problem

Let us consider a circular cylindrical shell or radius R and length l subjected to internal pressure loading of intensity P and axial dead load N. The right hand end of the shell is simply supported whereas the left end is clamped.

Let the origin of coordinates be located at the left end of the tube. It is assumed that at $x = S_j$ where $j = 1,...,n$ absolutely rigid hoop stiffeners are located. Since the width of stiffeners is small enough it is reasonable to expect that the transverse deflection $W(S_j) = 0 \quad j = 1,...,n$.

We are looking for the positions of stiffeners for which the cost function (r is a positive integer)

$$J = \int_0^l W^r dx \tag{1}$$

attains its minimal value.

The material of the tube is assumed to be a perfect plastic material obeying the yield condition suggested by Lance and Robinson (1973). This condition can be considered as a piece wise linear approximation of yield conditions by Hill and Tsai-Wu (see Jones, 1999) used for the calculation of the failure of fiber reinforced composites.

It is well known in fracture mechanics that sharp corners and extreme loading entail cracks (see Broberg, 1999). We assume herein that circular cracks of length c_j are located at $x = S_j$ ($j = 1,...,n$). In the foregoing, we assume that $S_0 = 0$. Following the ideas of Yu and Chen (1999, 2000) the cracks are treated as stationary flaws, no attention will be paid to crack propagation. However, the changes of cross section dimensions caused by cracks are taken into account.

The geometrical non-linearity is taken into account, as well, when minimizing the cost function (1) with appropriate equilibrium equations and the associated flow law. It is thus reasonable to assume that the pressure loading P is high enough to cause plastic deformations in each part of the shell, e.g.

$$P - P_j \geq 0 \tag{2}$$

for $j = 0,...,n$. In Eq. (2) $P_j = P_j(S_j, S_{j+1})$ stands for the initial yield load for the section of the shell located between planes $x = S_j$ and $x = S_{j+1}$.

3 Governing equations

Equilibrium equations of a shell element can be presented as (see Lellep, 1984, 1985)

$$\frac{d^2M}{dx^2} - N_1\frac{d^2W}{dx^2} + \frac{N_2}{R} - P = 0 \tag{3}$$

where N_1, N_2 stand for membrane forces and M is the bending moment. The strain components corresponding to Eq. (3) are

$$\varepsilon_1 = \frac{dU}{dx} + \frac{1}{2}\left(\frac{dW}{dx}\right)^2, \qquad \varepsilon_2 = \frac{W}{R} \tag{4}$$

whereas the curvatures are given as

$$\kappa_1 = \frac{d^2W}{dx^2}, \qquad \kappa_2 = 0. \tag{5}$$

The tubes under consideration are made of a fiber reinforced inelastic material. Let the ratio of yield stresses in the direction of fibers and in the transverse direction, respectively, be α. In the following we shall consider the cases of circumferential and axial orientation of fibers in the matrix material. In this case the stress profile lies on the face

$$N_2 = kN_0 \tag{6}$$

of corresponding yield surface. On the plane (Eq. 6)

$$|M| \le \frac{\alpha}{k}M_0. \tag{7}$$

In Eqs. (6, 7) $k=\alpha$ in the case of circumferential orientation of fibers and $k=1$, provided fibers embedded in the matrix have the axial direction. Here N_0 and M_0 denote limit force and moment, respectively.

According to the associated flow law and the stress regime on (Eq. 6) has $\varepsilon_1 = \kappa_1 = \kappa_2 = 0$, $\varepsilon_2 \ge 0$. It means that if the stress state corresponds to the plane Eq. 6 and 7 are satisfied as a strict inequality then according to Eq. 5 the transverse deflection is a linear function of the coordinate x. However, if the stress state reaches the ridge where $M = \pm\frac{\alpha}{k}M_0$, then the curvature κ_1 remains unspecified from the gradientality law.

It is reasonable to use following non-dimensional quantities

$$\xi = \frac{x}{l}, \qquad s_j = \frac{S_j}{l}, \qquad m = \frac{M}{M_0}, \qquad p = \frac{RP}{N_0},$$

$$p_j = \frac{RP_j}{N_0}, \qquad n_{1,2} = \frac{N_{1,2}}{N_0}, \qquad w = \frac{N_0}{M_0}W, \qquad u = \frac{N_0^2 l}{M_0^2}U,$$

$$\omega = \frac{N_0 l^2}{M_0 R}, \qquad \alpha_j = \frac{a_j}{l}, \qquad \beta_j = \frac{b_j}{l}. \tag{8}$$

Here a_j and b_j stand for coordinates of internal points of the interval $[S_j, S_{j+1}]$ so that $S_j < a_j < b_j < S_{j+1}$ whereas bending moment M has its minimum value at each $x \in [a_j, b_j]$.

Making use of Eqs. 8 and 6 one can present equilibrium equations as

$$m'' - n_1 w'' + \omega(k - p) = 0 \tag{9}$$

where primes denote the differentiation with respect to ξ.

4 Deflected shape of the shell

It is reasonable to assume that the shell operates in the post-yield range. Let us consider a section of the shell lying between stiffeners at $\xi = s_j$ and $\xi = s_{j+1}$ ($j=0,\dots,n$) in a greater detail. Evidently, at $\xi = s_j$ ($j=0,\dots,n$) the stress state calls forth stationary hinges so that for $j=0,\dots,n$

$$m(s_j) = \frac{\alpha}{k}\left(1 - 2\delta_j + \delta_j^2\right) \tag{10}$$

where $\delta_j = c_j/h$.

The stress state between hinges is expected to be such that $|m| < \alpha/k$ for $\xi \in (s_j, \alpha_j)$ and $\xi \in (\beta_j, s_{j+1})$ whereas $m = -\alpha/k$ for $\xi \in (\alpha_j, \beta_j)$. Here α_j and β_j are such that $s_j < \alpha_j < \beta_j < s_{j+1}$.

According to these expectations it follows from (3.7) that

$$-n_1 w'' + \omega(k - p) = 0 \tag{11}$$

for $\xi \in (\alpha_j, \beta_j)$ and

$$m'' + \omega(k - p) = 0 \tag{12}$$

for $\xi \notin (\alpha_j, \beta_j)$ where $j=0,\dots,n$.

In addition to Eq. 12 one has $\kappa_1 = 0$ for $\xi \notin (\alpha_j, \beta_j)$. In these regions $w'' = 0$ and thus

$$w = A_j \xi + B_j \tag{13}$$

for $\xi \in (s_j, \alpha_j)$ and

$$w = C_j \xi + D_j \tag{14}$$

for $\xi \in (\beta_j, s_{j+1})$, where $j=0,\dots,n$.

In the central zone for $\xi \in (\alpha_j, \beta_j)$ it follows from Eq. 11 that

$$w = \frac{\omega}{2n_1}(k - p)\xi^2 + E_j \xi + F_j. \tag{15}$$

Since the deflection w vanishes at supports one easily obtains

$$B_j = -A_j s_j, \qquad D_j = -C_j s_{j+1}. \tag{16}$$

Thus according to Eqs. 13 – 16

$$w=\begin{cases} A_j(\xi-s_j), & \xi\in(s_j,\alpha_j),\\ \dfrac{\omega}{2n_1}(k-p)\xi^2+E_j\xi+F_j, & \xi\in(\alpha_j,\beta_j),\\ C_j(\xi-s_{j+1}), & \xi\in(\beta_j,s_{j+1})\end{cases} \tag{17}$$

for each $j=0,\dots,n$.

Satisfying requirements of continuity of w and w' at $\xi=\alpha_j$ and $\xi=\beta_j$ yields

$$w=\frac{-\omega(k-p)}{2n_1}\begin{cases} \dfrac{\beta_j-\alpha_j}{s_j-s_{j+1}}(\alpha_j+\beta_j-2s_{j+1})(\xi-s_j), & \xi\in[s_j,\alpha_j],\\ \dfrac{\beta_j-\alpha_j}{s_j-s_{j+1}}(\alpha_j+\beta_j-2s_{j+1})(\xi-s_j)-(\xi-\alpha_j)^2, & \xi\in[\alpha_j,\beta_j],\\ \dfrac{\beta_j-\alpha_j}{s_j-s_{j+1}}(\alpha_j+\beta_j-2s_j)(\xi-s_{j+1}), & \xi\in[\beta_j,s_{j+1}]\end{cases} \tag{18}$$

The relations (Eq. 18) hold good for each $j=0,\dots,n$ in order to determine unknown parameters α_j and β_j one has to integrate equations (12) in the regions $\lfloor s_j,\alpha_j\rfloor$ and $\lfloor\beta_j,s_{j+1}\rfloor$.

Direct integration yields

$$m=\frac{-\omega}{2}(k-p)\xi^2+C_{1j}\xi+C_{2j} \tag{19}$$

for $\xi\in\lfloor s_j,\alpha_j\rfloor$ and

$$m=\frac{-\omega}{2}(k-p)\xi^2+C_{3j}\xi+C_{4j} \tag{20}$$

for $\xi\in\lfloor\beta_j,s_{j+1}\rfloor$.

For determination of arbitrary constants C_{1j}, C_{2j}, C_{3j}, C_{4j} one has following requirements. Since in the central part (α_j,β_j) the inequality (3.5) undergoes to equality

$$m(\alpha_j)=-\frac{\alpha}{k},\qquad m(\beta_j)=-\frac{\alpha}{k} \tag{21}$$

and $$m'(\alpha_j)=0,\qquad m'(\beta_j)=0 \tag{22}$$

for each $j=0,\dots,n$.

Making use of Eqs. 19 – 22, one easily obtains that

$$\begin{aligned} C_{1j}&=\omega(k-p)\alpha_j, & C_{2j}&=-\frac{\omega}{2}(k-p)\alpha_j^2-\frac{\alpha}{k},\\ C_{3j}&=\omega(k-p)\beta_j, & C_{4j}&=-\frac{\omega}{2}(k-p)\beta_j^2-\frac{\alpha}{k}.\end{aligned} \tag{23}$$

Substituting Eq. 23 in 19, Eq. 20 yields

$$m = \begin{cases} \frac{\omega}{2}(p-k)(\xi-\alpha_j)^2 - \frac{\alpha}{k}, & \xi \in [s_j, \alpha_j], \\ -\frac{\alpha}{k}, & \xi \in [\alpha_j, \beta_j], \\ \frac{\omega}{2}(p-k)(\xi-\beta_j)^2 - \frac{\alpha}{k}, & \xi \in [\beta_j, s_{j+1}] \end{cases} \tag{24}$$

for $j = 0,...,n$. As it was assumed above at $\xi = s_j$ ($j = 0,...,n$) are located stationary hinge circles. Thus Eq. 11 must be met for each $j = 0,...,n$. However, for $\xi = s_{n+1} = 1$ one has $m(s_{n+1}) = 0$. Making use of Eq. 24 and satisfying these requirements leads to results

$$\alpha_j = s_j + \sqrt{\frac{4\alpha(1-\delta_j+\delta_j^2)}{k\omega(p-k)}} \tag{25}$$

for $j = 0,...,n$ and

$$\beta_j = \begin{cases} s_{j+1} - \sqrt{\frac{4\alpha(1-\delta_j+\delta_j^2)}{k\omega(p-k)}}, & j = 0,...,n-1, \\ 1 - \sqrt{\frac{2\alpha}{k\omega(p-k)}}, & j = n. \end{cases} \tag{26}$$

It is worthwhile to mention that Eqs. 25 and 26 hold good under the condition that the load intensity meets inequalities Eq. 2. If in Eq. 2 an equality takes place then the loading corresponds to the onset of plastic deformations. In this case $\alpha_j = \beta_j$. These equalities with Eqs. 25, 26 admit to define limit loads as

$$p_j = k + \frac{16\alpha(1-\delta_j+\delta_j^2)}{k\omega(s_{j+1}-s_j)^2} \tag{27}$$

for $j = 0,...,n-1$ and

$$p_n = k + \frac{2\alpha}{k\omega(1-s_n)^2}\left[1+\sqrt{2(1-\delta_n+\delta_n^2)}\right]^2. \tag{28}$$

If $p < p_j$ then the section of the shell between $\xi = s_j$ and $\xi = s_{j+1}$ remains rigid and $w \equiv 0$. Evidently, the minimization of the cost function Eq. 1 is meaningless if the bulk of shell is not deformed.

5 Optimal location of stiffeners

Let us consider the problem of minimization of the cost function (Eq. 1) in the case when $r = 1$. Substituting Eq. 18 in Eq. 1 gives

$$J = \frac{\omega}{2n_1}(p-k)\left\{\sum_{j=0}^{n}\left[\frac{\beta_j-\alpha_j}{s_j-s_{j+1}}\right]\left\{\frac{1}{2}(\alpha_j-s_j)^2(\alpha_j+\beta_j-2s_{j+1})+\right.\right.$$
$$\left.+\frac{1}{2}(\alpha_j+\beta_j-2s_{j+1})\left[(\beta_j-s_j)^2-(\alpha_j-s_j)^2\right]-\frac{1}{3}(s_j-s_{j+1})(\beta_j-\alpha_j)^2-\right.$$

$$-\frac{1}{2}(\alpha_j+\beta_j-2s_j)(\beta_j-s_{j+1})^2\Big\}\Big\}\cdot\frac{lM_0}{N_0}. \tag{29}$$

Here the parameters α_j, β_j are given by Eqs. 25 and 26.

Taking Eqs. 27 and 28 into account the constraints Eq. 2 can be cast into the form

$$p-k-\frac{16\alpha(1-\delta_j+\delta_j^2)}{k\omega(s_{j+1}-s_j)^2}\geq 0; \qquad j=0,...,n-1 \tag{30}$$

and

$$p-k-\frac{2\alpha}{k\omega(1-s_n)^2}\left[1+\sqrt{2(1-\delta_n+\delta_n^2)}\right]^2\geq 0\cdot \tag{31}$$

The non-linear programming problem, which consists in the minimization of the cost function accounting for inequality constraints Eqs. 30, 31 can be solved with the aid of multipliers of Lagrange.

6 Discussion and conclusions

The results of calculations are presented in Fig. 2 – Fig. 4. The curves presented in Fig. 2 – Fig. 4 correspond to following values of geometrical and loading parameters: $\omega=8$; $p=11$; $n_1=0{,}1$; $\alpha=1{,}2$.

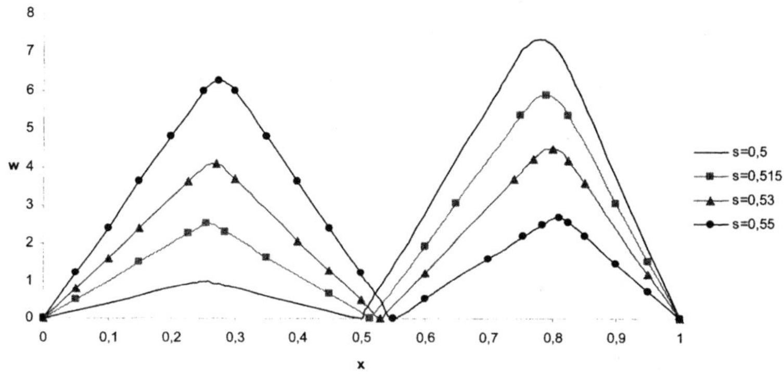

Figure 2. Deflections of the shell with a stiffener.

In Fig. 2, 3 deflected shapes of shells with a single stiffener are shown. Fig. 2 shows the influence of the position of the stiffener on the distribution of transverse deflection. Different curves in Fig. 2 correspond to different positions of the rigid hoop stiffener. It can be seen from Fig. 2 that in the case of the optimal position of the stiffener the areas of left and right hand configurations are the same (s=0,51). The peaks of deflections are approximately the same.

Fig. 2 corresponds to the case where no cracking takes place.

In Fig. 3 the deflected shapes of the shell are presented for different values of the crack length.

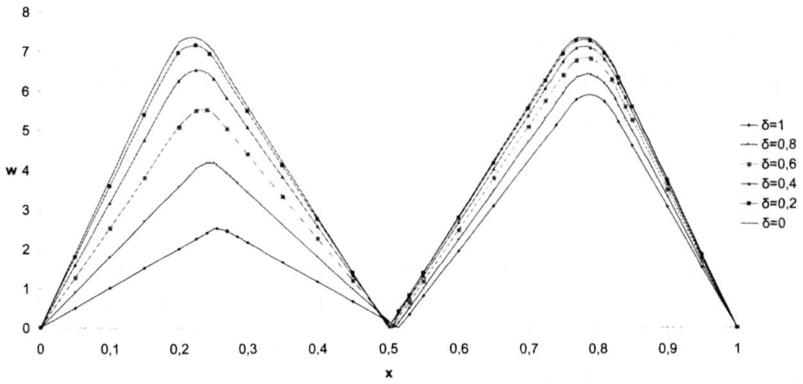

Figure 3. Deflections of the shell with cracks.

It can be seen from Fig. 2, 3 that when the stiffener moves towards the simply supported end maximal deflection of the left part of the shell increases and that of the right hand part of the shell decreases, as might be expected.

In Fig. 4 distributions of the bending moment m_1 are presented for different positions of the hoop stiffener.

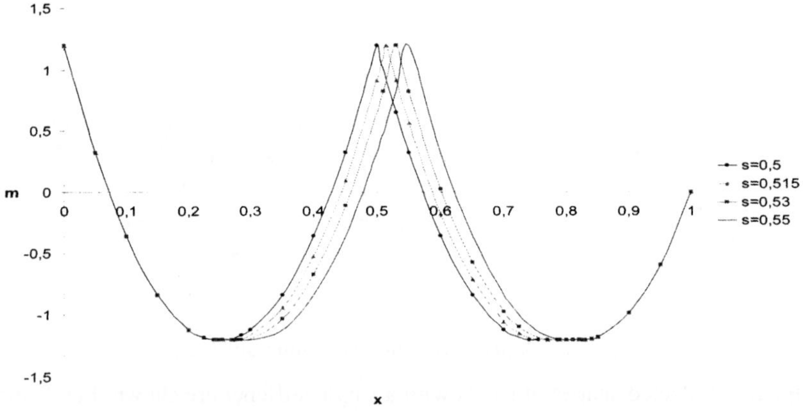

Figure 4. Bending moments of a shell without any crack.

In Fig. 4 bending moment distributions are presented. It can be seen in Fig. 4 that a change of the position of the stiffener influences the bending moment distribution only in the near region of the stiffener.

Acknowledgement

The partial support of the Estonian Science Foundation through grant № ETF7461 is acknowledged.

References

Åkersson, B. & Olhoff, N. (1988) Minimum stiffness of optimally located supports for maximum value of beam eigenfrequencies. *J. Sound Vibr.* **120** 457-463.

Bojczuk, D. & Mroz, Z. (1998) On optimal design of supports in beam and frame structures. *Struct. Optim.* **16** No. 1, 47-57.

Broek, D. (1988) *The Practical Use of Fracture Mechanics*, Dordrecht: Kluwer.

Broberg, K. B. (1999) *Cracks and Fracture.* New York: Academic Press.

Chen, F. L. & Yu, T. X. (1999) Dynamic behaviour of a clamped plastic beam with cracks at supporting ends under impact. Trans. ASME. *J. Pressure Vessel Technol.* **121** 406-412

Cinquini C. & Kouam, M. (1983) Optimal plastic design of stiffened shells. *Int. J. Solids Struct.*, **19** No. 9, 773-783.

Daniel, I. M. & Ishai, O. (1994) *Engineering Mechanics of Composite Materials.* Oxford: Oxford Univ. Press.

Garstecki, A. & Mroz, Z. (1987) Optimal design of supports of elastic structures subjected to loads and initial distortions. *J. Struct. Mech.* **15** 47-68.

Jones, R. (1999) *Mechanics of Composite Materials.* Philadelphia: Taylor and Francis.

Kaliszky, S. & Logo, J. (2006) Optimal design of elasto-plastic structures subjected to normal and extremal loads. *Comp. Structures* **84** 1770-1779.

Kaliszky, S. & Logo, J. (2003) Layout optimization of rigid-plastic structures under high intensity, short-time dynamic pressure. *Mechanics Based Design Struct. Mach.* **31** No. 2, 131-150.

Kumar, S. & Petroski, H. J. (1985) Plastic response to impact of a simply supported beam with a stable crack. *Int. J. Impact Eng.* **3** No. 1, 27-40.

Lance, R. H. & Robinson, D. N. (1973) Plastic analysis of filled reinforced circular cylindrical shells. *Int. J. Mech. Sci.* **15** No. 1, 65-79.

Lellep, J. (1984) Optimal location of additional supports for plastic cylindrical shells subjected to impulsive loading. *Int. J. Non-Linear Mech.* **19** No. 4, 323-330.

Lellep, J. (1985) Optimal location of additional supports for a geometrically non-linear plastic cylindrical shell. *Prikl. Mekhanika* **1** 60-66.

Lellep, J. (1985) Parametrical optimization of plastic cylindrical shells in the post-yield range. *Int. J. Eng. Sci.* **23** No. 12, 1289-1303.

Lellep, J. & Hannus, S. (1992) Optimal locations of rigid stiffeners for a geometrically non-linear plastic cylindrical shell. *Tartu Ülik. Toimetised* **939** 70-78.

Lepik, Ü. (1978) Optimal design of beams with minimum compliance. *Int. J. Non-Linear Mech.* **13** 33-42.

Mroz, Z. & Lekszycki, T. (1981) Optimal support reaction in elastic frame structures. *Comput. Struct.* **14** 179-185.

Mroz, Z. & Rozvany, G. I. N. (1975) Optimal design of structures with variable support positions. *J. Optim. Theory. Applic.* **15** 85-101.

Olhoff, N. & Åkesson, B. (1991) Minimum stiffeness of optimally located supports for minimum value of column buckling loads. *J. Struct. Optim.* **3** 163-175.

Petroski, H. J. (1981) Simple static and dynamic models for the cracked elastic beam. *Int. J. Fracture* **17** R71-R76

Szelag, D. & Mroz, Z. (1978) Optimal design of elastic beams with unspecified support positions. *ZAMM* **58** 501-510.

Yu, T. X. & Chen, F. L. (2000) Failure of plastic structures under intense dynamic loading: modes, criteria and thresholds. *Int. J. Mech. Sci.* **42** 1537-1554.

13.4 The Result of Developing and Building of a Many Sheaves Cable System Analyzer Machine and a Sheave-wear Machine

András Malik, János Németh
University of Miskolc, H-3515 Miskolc, Hungary,
malikandras@vipmail.hu

Abstract

Sometimes unexpected overloads start up in many sheaves cable systems by synchrodrive. These overloads cannot be seen before and cause the breakage of the rope(s). There are many parts to be analyzed, but the most important is the wear of the sheaves. The influences of the different diameter of the sheaves and the sheaves wearing by variable loads are not fully explored. Design of a many sheaves examiner machine and a sheave-wear-machine was necessary to reach further results.

Keywords: *many sheaves cable system, wire rope, wearing of sheaves, breakage by overload*

1 Introduction

Sometimes unexpected overloads start up in many sheaves cable systems by synchrodrive. The wearing of the sheaves causes overloads. Usually the wearing of the first and the last sheaves is bigger than the middle ones because of the force transfer (Czitary 1962). This is caused because the diameters of the side sheaves are less. A many sheaves examiner machine and a sheave-wear-machine were developed. With these machines the influence of sheaves with different diameters and the process of the sheave wearing by variable loads can be analyzed. The first aim was by either developing that the machines had to be near to realistic conditions or that parameters can be set up later.

2 The 7 sheaves examiner machine

A 7 sheaves examiner machine is needed to analyze the overloads. The influence of the variable diameters can be registered with this machine. This machine is a prototype.

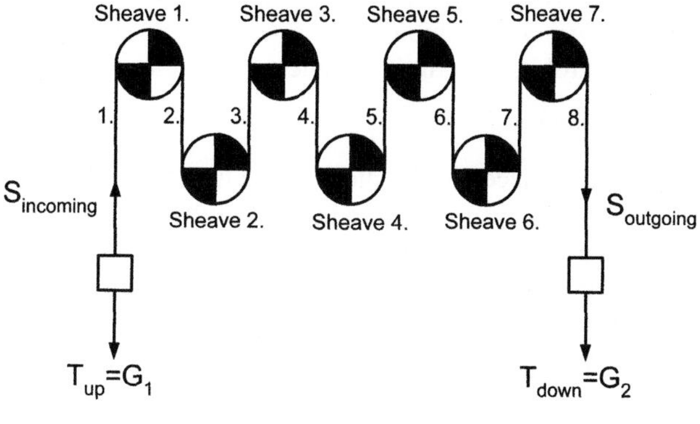

Figure 1. The kinematic draft of the 7 sheaves machine.

2.1 *Aims*

There are a lot of aims to be accomplished such as using variable type of springs, with other and other loads, using sheaves with variable diameters, by variable speeds.

2.2 *The produced machine*

The machine is fully complete the tasks and aims. First a number of sheaves had to be defined. The effect of normal conditions has also to appear. Seven sheaves were the acceptable and sure choice why the effect cannot be seen by 3 sheaves or less. The kinematic draft of the machine is shown in Figure 1. A spring is used instead of wire rope to achieve the simply and correct measuring. The distance of spring threads can be registered with a camera more easily than bearing pressure by wire rope with force register. The linear characteristics are almost the same but the main difference between the rope and the spring is in the coefficients of elasticity and cross contractions. The coefficients of elasticity can be recast and cross contractions

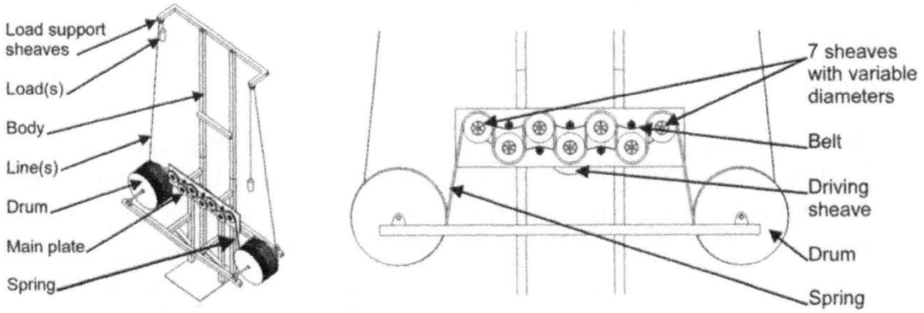

Figure 2. The developed 7 sheaves machine

are diagnosable with using several types of springs. The result of this developing can be seen in Figure 2.

2.2 *Running*

The springs superpose on the sheaves which are lubricated. The ends of the spring are attached to the drums. The drums have a less diameter parts too because the lines are fastened to them. Variable loads can be placed at the other end of the lines. The upper body of the 7 sheaves machine is enough tall to ensure right way for the loads and fixed to the wall. An electric motor drives the "driving sheave" and a frequency changer is used to control it. Two limit switches are used for reversing.

2.3 *Procedure of measuring*

At first, the spring has to placed to the sheaves and than the ends have to fixed to the drums. After that the loads has to fixed to the end of the lines. The following step is the start. The system and the spring need time because of the transient process. The machine and the spring need 2 or 3 back-and-forth running. The photo has been taken before the reversing. First, all of the threads were analyzed but that was too much information. When only the horizontal branches were measured, than the result was the same, so the threads need to be measured only in the 8 branches.

2.4 *Experiences*

The loads waving after the placing and at start because of the transient process that could lead to the breakage of the line. It is necessary to wait a little bit before the start. Checking of the fixing of the springs is very important. The false fixing can cause the falling off of the spring from the sheaves. The consequent of spring falling could be the damage of spring or the loads could fall to the drums. The machine is running and several experiences and measuring can be made with this examiner.

3 Sheave-wear machine

The analysis of sheaves wearing can realizable with a sheave-wear machine. A wire rope has to wear the sheaves as in realistic situations. This machine will be a prototype too.

3.1 *The first aims (Keresztesi & Otrosinka 2007)*

The target was to build a machine working as the kinematic draft in Figure 3. The most important point was to achieve that the parameters can be set one by one. These parameters, for example variable loads, the diameter, the material, the groove

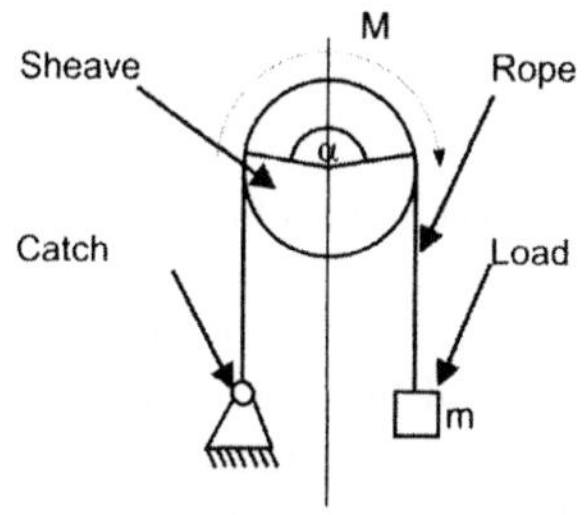

Figure 3. The kinematic draft of the sheave-wear machine

and the lubrication of the sheave, the material, the structure, the diameter and the lubrication of the wire rope, controllability of torque or revolution number, the change of the direction and the sheave are considered. The diameter, the material and the groove of the sheave is defined by the production. Modifying of sheave parameters is possible with new production. The new production is the best solution

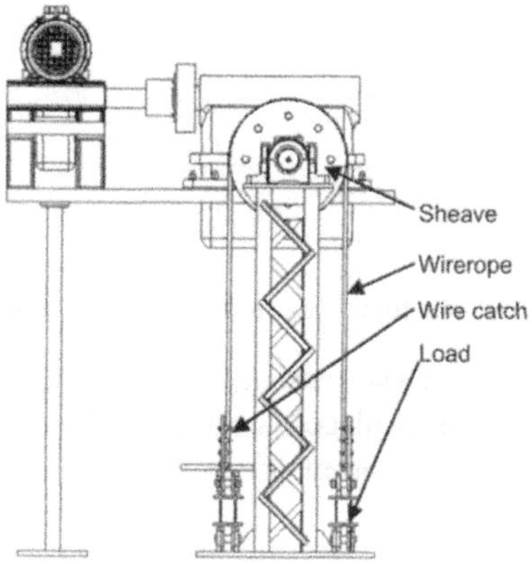

Figure 4. The produced sheave-wear machine

again by wire ropes. The revolution number can be controlled easily with frequency changer. The 3 main aims are as follows: variable sheaves can be wearing with variable wire ropes by variable revolution number.

3.2 *The built machine*

The machine satisfies almost every function. All of the main parts are separable in Figure 4 as it is seen on kinematic draft in Figure 3. Variable sheaves and variable wire ropes can be used because of the special solutions. For example, the variable sheaves could be used because it is fixed with 8 screws on a centre ring. The usability of the sheaves is depending only from these parameters. There is a very important mark. The radius of the groove on the sheave and the diameter of the rope have to be checked before every start because the wrong dimensions can cause the

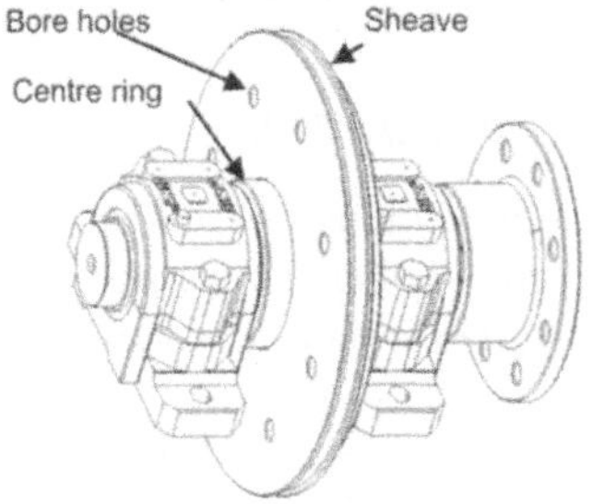

Figure 5. The fixing of the sheave

damage of the machine. The fixing of the sheave can be seen in Figure 5. Hook-eyes and pins are used for fixing the end of the ropes. The end of the rope is folded back and fixed with an iron ring by a hook-eye process. There is an iron plate at the inner side of the hook to accept the loads better. The loads are given to the system by a balance arm. The fixing of the rope is at the 1/10 part of the arm, so if a mass is placed to the end of the arm, 10 times bigger force starts up in the rope (Figure 6). Beside the bearings L-bars and double tee-sections are used for fully support of the

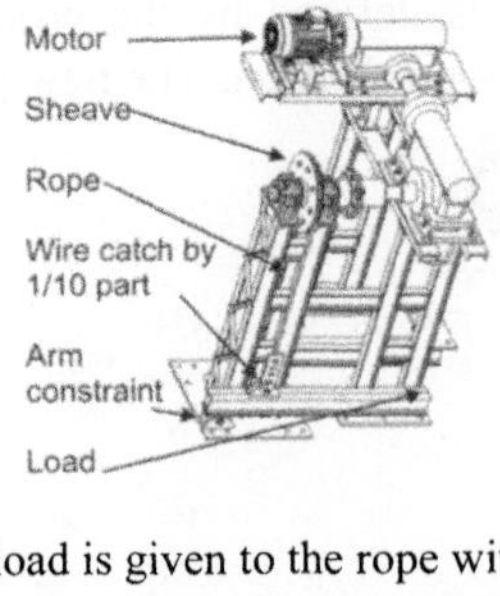

Figure 6. The load is given to the rope with arm mechanism

sheaves. The appliance of these was needful because the load capacity has to be enough big. (If 300kg mass is placed to the end of the arm, then 3t starts up in the rope.) An asynchronous induction motor drives the machine and it is controlled by a frequency converter. Two worm drives help the system for reduction of the revolution speed beside the frequency converter.

3.3 *Errors*

The wearing machine is a prototype, what is running and working, but it may contain mistakes from aspect of the measuring. These errors can be detected after the assembling by empirical ways. The first negative experience was that the sheave does not take out the abrasion material from the contact zone. So this material stays between the sheave and the rope. Later, it can get into the strands and wires of the rope. After that the attributes of the wire have changed. Too much abrasion material can welded to the sheave with cold-welding and can change the diameter of the sheave. The second experience is about the arm. The arm starts to weaving after placing the loads and that is clearly visible. This waving is subsiding slowly and causes an ellipsoid shape sheave. The third experience is about the too much measuring information. The radius of the sheave is measured for 1860 times per rev and contains a lot of mistakes. Only 8 measures per rev can be enough to draw a trend line. The next experience is that the variable diameter of the ropes can cause centering failures. The last experience is a temperature problem. The heat what is coming from the abrasion process is not leaving with the moving rope because the rope is not moving. This heat causes no problem at realistic situation. This heat can change the quality of the material of the sheave and the rope. The list of experiences is as follows: the abrasion material is staying, the load of the rope is waving, too much measured information, catching causes problems, heating of the rope.

3.4 *Theory of wear analyses (Szűcs 2007)*

The knowledge of wear analyses is necessary to approach the realistic conditions as close as possible. The machine can be upgraded with this information. There are two bodies by wearing analysis: the first is the test piece and the second is the check piece. The rope is the check piece and the sheave is the test piece at this machine. The sheave has abrasion resistance but the rope has abrasive power. The produced machine is rope-wearing machine with these aspect and it is absolutely opposite of the sheave wear machine. One of the best solutions is that the rope has to move too. One revolution of the sheave has chosen as a basic cycle. The ratio of the sheave way and the rope has to be close to the real ratio. The rope motion with a perimeter of a sheave belongs to one revolution by realistic condition. So ratio of the rope way and the sheave way is ~1. This ratio was 0 by the first machine because the rope was caught. This ratio and the realistic ratio can be seen in Figure 7. An upgrade is

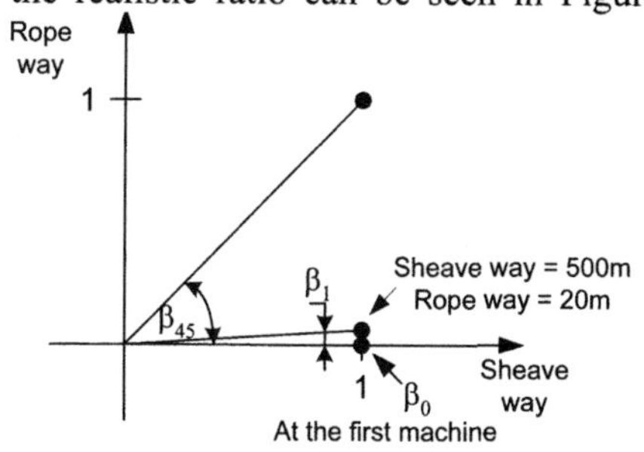

Figure 7. The ratio of the sheave way and the rope way

needed because the ratio has to be between 0 and 1. The ratio does not have to be too close to 0 because the mentioned problems will appear again and too close to 1 neither because the wearing time will be too long (the wearing processes are very slow by realistic speeds and loads by this machine.) Finding the optimum is needed for example: 20 m rope way belongs to 500 m sheave way. The β angle shows the angle of the ratio of rope way and sheave way. It can be seen in the Figure 6 that the angle of ratio was 0 by the first machine. The realistic conditions can be closed when this angle is close to 45°. A function or a curve can be drawn with a continuous closing. That could be used for the calculation of real wearing by real loads. The machine needs to redesign.

3.5 *The new aims*

The first steps are the defining of the new aims and rethinking of the old ones. The main target is that the rope has to move but the load can be given to the rope as easily as it was at the first machine. The rope can move forward and back (Figure 8).

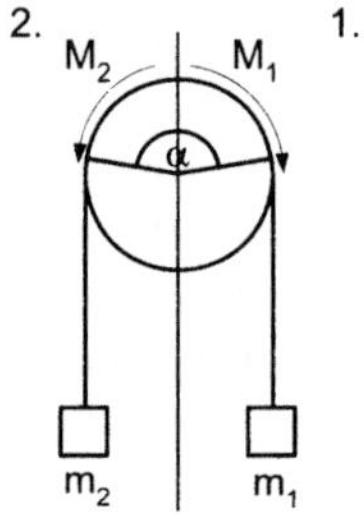

Figure 8. The modified kinematic draft

If the rope is moving then it will take the heat away, so the problem of overheating will not appear again. The next advantage is that the rope will take away the abrasion material too, what might be cleaned with a brush from the wire rope. Using one or two hydraulic cylinder makes the setup of the load(s) more correct and the load(s) will be closer to constant load(s). The catching of the ropes will not be more difficult as was before because it can be fixed to the head of the piston easily. The number of measuring has to be decreased to 8 per 1 revolution but a control measuring is needed with micrometer after every 1000 revolutions.

3.6 *The new drawings and selecting the best one*

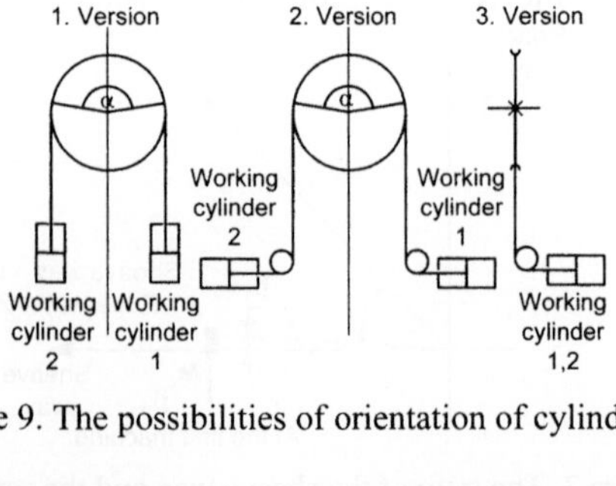

Figure 9. The possibilities of orientation of cylinders

The new drawing can separate to two parts. These are the orientation of the hydraulic cylinder(s) and the catch and the spacing of the rope. The cylinder can be placed horizontally or vertically (ideal orientation). The possibilities can be seen on Figure 9. The first version is space saving, and the influence of the gravity causes no problems, so the vertical placing of the cylinders are the best. The heightening of the machine does not need any special engineering solution. The versions of catching and spacing can be seen on Figure 10. The catching can be on the head of piston or on drums. The cylinders or also the drums with hydraulic motor can drive the rope or give the load to it. The load can be given to the rope by vertical cylinders with catching or with sheaves at the end of the piston. The system with sheaves is more advantageous because continuous measuring is possible. But the rope way is enough little so the first version is optional too. The machine can be stopped or turned when the piston gets its own end position. The rope can be spaced easily after relaxation. The rope might be damaged at the catching points, so these parts of the rope can not

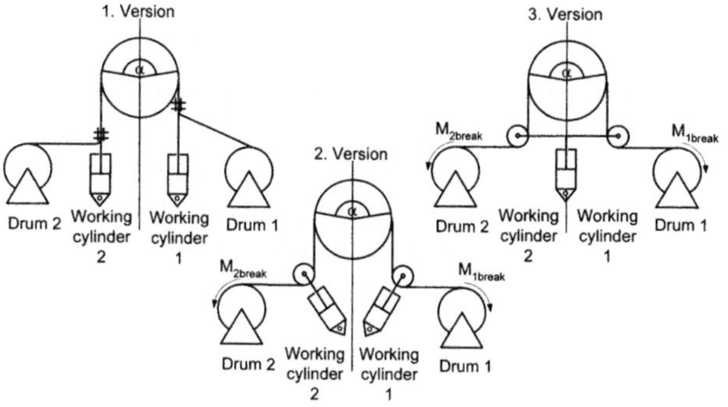

Figure 10. The versions of catching

wear the sheave. The second and the third version have a big disadvantage what is the hydraulic drives. These drives are very expensive. This is too much cost what can be avoiding by the first version. The best was from aspect of the catching and spacing of the rope and the orientation of the cylinders the first version on Figure 10.

3.7 *The designed and produced new machine*

The target is that the stroke has to be as long as the half perimeter of the sheave. The angle of wrap is 180 °, so it is necessary that the rope way can be as long as the half perimeter of the sheave. The rope wears the sheave under this angle. When the rope has moved with the distance of the half perimeter than a reversing or a spacing (stepping) can do. The rope has to move 0.75m. If the ratio of the rope way and the sheave way is defined than the speeds and the time of reversing or stepping can be calculated. These parameters can be set later too. The distance between the bottom of the sheave and the base plate is 1,6m. The min motion is 0,75m, so lifting of the machine is needed. 0,5m lifting is recommended for sure function (Figure 11). The stability is the main aim by the catching on the head of the piston. The rope may injure because the caught part of the rope does not wear the sheave but the catch has to be stable with injured rope too. A hydraulic support is needed for the hydraulic cylinders what is functional on max 160 bar. Each of the cylinders gives 3t load to the rope. Using of limit switches makes the control simple and can makes the reversing running easily. The machine support needs strengthening because of the

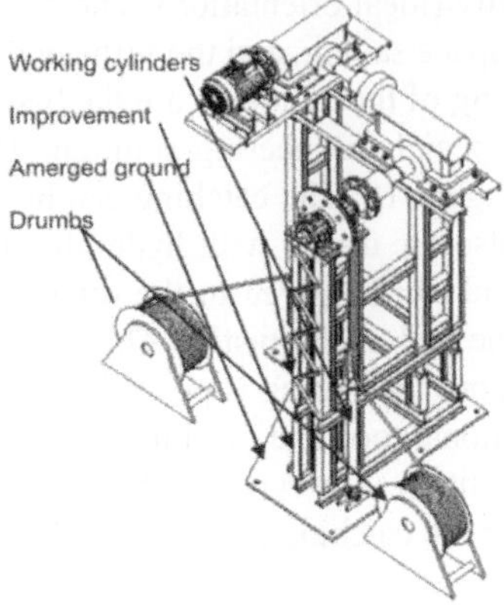

Figure 11. The modified wear machine

lifting. There is a main plate because of the main oscillation but a good frame structure would be better. Bigger unexpected oscillation can be appeared without these modifications. The stepping of the ropes is more simply with drums. The length of the rope on the drums is enough big to make all of the measurement with other and other parameters. The produced machine has all of the parts and criteria what were at the kinematical draft in Figure 8. Test running and test measuring will

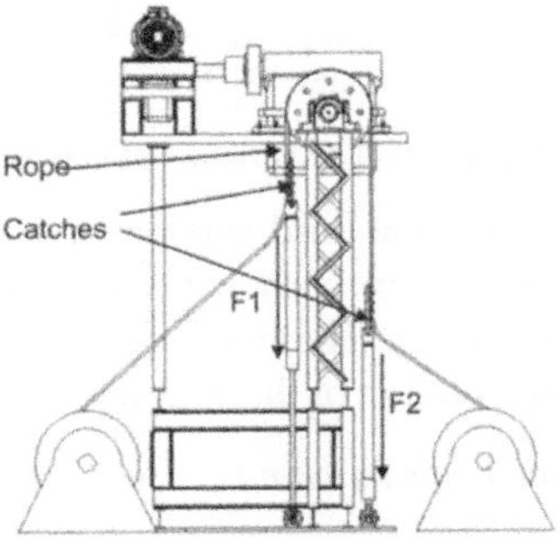

Figure 12. The final produced machine

follow the upgrade. A lot of interesting results could be collect with variation of the parameters after that checking process.

Acknowledegement

This work received support from the CASAR Drahtseilwerk Saar Gmbh, Germany.

References

Czitary, E. (1962) *Seilschwebebahnen*, Wien: Springer-Verlag

Keresztesi, J. & Otrosinka, T. (2007) *Personal consultations about many sheaves systems*, Miskolc.

Szűcs, J. (2007) *Personal consultations about theory of wear analyses*, Miskolc.

13.4 Safety Factors in Design of Steel Members for Accidental Fire Situation

Mariusz Maślak[1], Tomasz Domański[2]
Cracow University of Technology, PL 31-155 Cracow, Warszawska 24, Poland
e-mail: [1]mmaslak@pk.edu.pl, [2]doman@usk.pk.edu.pl

Abstract

Standard EN 1993-1-2 gives only a simplified methodology for evaluation of real safety level of steel members during fire. The authors of the article suggest in this field a new approach, based on the probabilistic concept, which seems to be more adequate. The purpose of our analysis is estimation of safety factors for fire conditions if values of member failure probability are accepted. Two different ways of definition of member failure are considered independently. The first one is connected with the critical temperature of steel; whereas, the second directly with the fire moment (in time units) when member carrying capacity, reduced in high temperature, becomes insufficient. Design methodology presented by the authors is user friendly and can be applied for calibration of values of partial safety factors in accidental fire situation for particular reliability classes suggested by EN 1990.

Keywords: *fire, safety factors, failure probability*

1 Introductory remarks

In accordance with the basic safety condition taken from EN 1991-1-2, which is also confirmed in a particular standard EN 1993-1-2 (referred to the analysis of the behaviour of steel elements during fire), it is accepted that the structural member in fire situation is capable to carry all applied loads if:

$$E_{fi,d} \leq R_{fi,d,t} \tag{1}$$

where $E_{fi,d}$ is the design value of action effect, calculated for unfavourable load combination; whereas, $R_{fi,d,t}$ means the design value of member resistance, related to the t_{fi} fire moment. Quantity $E_{fi,d}$ can be determined as a sum of component effects; however, accidental load combination rule should be taken into consideration, which gives:

$$E_{fi,d} = G_k + (\psi_{1,1} \text{ or } \psi_{1,2})Q_{k,1} + \sum_{i>1} \psi_{2,i} Q_{k,i} \tag{2}$$

In such a formula G_k is the characteristic value of permanent load; while, $Q_{k,i}$ - the characteristic value of the *i*-th variable load ($i = 1,...,n$). Finally, factors $\psi_{1,i}$ and $\psi_{2,i}$ are suitable coefficients of load combination, for *frequent* and for *quasi-permanent* values of variable actions, respectively. Furthermore, let us underline that index $i = 1$ means that the considered variable action is treated as a leading among the others.

It is usually accepted that value $E_{fi,d}$ remains constant during the whole fire time, provided that the analysed steel member has the possibility of unlimited thermal deformation. This is only an approximation of the reality because no load changes generated by evacuation of occupants or furnishings combustion are taken into account. On the other hand, always, if unlimited thermal deformations are restrained in any way, thermally induced additional internal forces and moments must occur, which is the source of changeability of the conclusive value of a reliable action effect in fire.

Design value of member resistance $R_{fi,d,t}$ is obtained based on its characteristic value $R_{fi,k,t}$:

$$R_{fi,d,t} = \frac{R_{fi,k,t}}{\gamma_{M,fi}} \tag{3}$$

According to EN 1991-1-2 there is a suggestion to accept $\gamma_{M,fi} = 1{,}0$. In simple load cases the value $R_{fi,k,t}$ is directly proportional to the steel yield point $f_{y,k}$. The proportionality factor is usually member section area A (in the case of tension or compression) or bending modulus: plastic W_{pl} - if the section is classified in class 1 or class 2, or elastic W_{el} - when the section belongs to class 3 (in the case of bending). When steel temperature Θ_a grows, the value of $f_{y,k}$ decreases as follows:

$$f_{y,k,\Theta} = k_{y,\Theta} f_{y,k,20} \tag{4}$$

The steel yield point $f_{y,k,20}$ defined for the room temperature $\Theta_a = 20\ ^{\circ}\mathrm{C}$ is then adopted as a reference value. Values of the reduction coefficients $k_{y,\Theta}$ for particular steel temperatures Θ_a are given in EN 1993-1-2. In the considered simple load case we have:

$$R_{fi,k,t,\Theta} = R_{fi,k,t}(\Theta_a) = k_{y,\Theta} R_{fi,k,t,20} \tag{5}$$

Acceptance of constant value of partial safety factor $\gamma_{M,fi}$ for the whole of fire duration (particularly if $\gamma_{M,fi} = 1{,}0$) allows to use the formula:

$$f_{d,\Theta} = \frac{f_{yk,\Theta}}{\gamma_{M,fi}} = \frac{k_{y,\Theta} f_{yk,20}}{\gamma_{M,fi}} = k_{y,\Theta} f_{d,20} \tag{6}$$

which means that:

$$R_{fi,d,t,\Theta} = R_{fi,d,t}(\Theta_a) = k_{y,\Theta} R_{fi,d,20} \tag{7}$$

2 Safety evaluation by means of semi-probabilistic approach

For more detailed analysis the probability-based approach is necessary. In general, resistance $R_{fi,\Theta}$ is the random variable. It is usually described by means of log-normal probability distribution $\mathrm{N}(\breve{R}_{fi,\Theta}, \upsilon_R)$, where $\breve{R}_{fi,\Theta}$ is its median value; whereas, υ_R is the log-normal coefficient of variation. Let us notice that values of the reduction coefficient $k_{y,\Theta}$ are strictly fixed for particular temperatures Θ_a. For this reason in further analysis they are treated as the deterministic (not-random) parameters. However, the more precise study, when such coefficients are considered as the random variables, is also possible in this field (Holicky 2005).

Consequently, if the proportionality between $R_{fi,20}$ and $f_{y,20}$ occurs, the following relation is true:

$$\breve{R}_{fi,\Theta} = k_{y,\Theta} \breve{R}_{fi,20} \tag{8}$$

It has been assumed that log – normal coefficient of variation υ_R does not depend on steel temperature Θ_a which means that:

$$\upsilon_{R,\Theta} = \upsilon_{R,20} = const \tag{9}$$

This hypothesis (called H_0) has been verified by the second of the authors (Domański 1987) who has used the statistic methods. Constant value of the standard deviation of the steel resistance σ_R during the whole of fire time has been tested as an alternative hypothesis H_1. In such a case the log-normal coefficient of variation υ_R would be increasing when steel temperature Θ_a grows. Laboratory tests of steel resistance have been performed for selected temperatures Θ_a (20, 300, 400 and 500 $^\circ C$). Specimens taken from various steel members (bars, hot-rolled sections), made from different steel grades (S235JR, S355JR), have been prepared. The empirical variances obtained from these experiments have been relatively large, because of many factors taken into consideration. The analysis of variance has been applied to separate gross laboratory errors. Finally, the *Bartlett* statistic test has been used to verify Eq. 9 and, as a result, the statement that there is no reason to question hypothesis H_0, has found the confirmation.

Design value of the resistance, calculated in relation to room temperature $\Theta_a = 20\ ^\circ C$, can be obtained independently from the following formula:

$$R_{fi,d,20} = \breve{R}_{fi,20}\, exp\left(-\beta_{R,20,req}\upsilon_R\right) \tag{10}$$

However, if steel temperatures are higher, in other words when $\Theta_a > 20\ ^\circ C$, then we have:

$$R_{fi,d,\Theta} = \breve{R}_{fi,\Theta}\, \exp\left(-\beta_{R,\Theta,req}\upsilon_R\right) = k_{y,\Theta} \breve{R}_{fi,20}\, \exp\left(-\beta_{R,\Theta,req}\upsilon_R\right) \tag{11}$$

Thus:

$$\frac{R_{fi,d,\Theta}}{R_{fi,d,20}} = \frac{k_{y,\Theta}\, exp\left(-\beta_{R,\Theta,req}\upsilon_R\right)}{exp\left(-\beta_{R,20,req}\upsilon_R\right)} = k_{y,\Theta}\, exp\left[\left(\beta_{R,20,req} - \beta_{R,\Theta,req}\right)\upsilon_R\right] = \zeta k_{y,\Theta} \tag{12}$$

Parameters $\beta_{R,20,req}$ and $\beta_{R,\Theta,req}$, specified in Eq. 10 and Eq. 11, are the required values of partial safety indices, determined for the room temperature and for fire temperatures, respectively. They are defined in relation to the maximum, acceptable by the user of the structure, value of probability of failure $p_{f,ult}$. Let us notice that in such an approach the failure is identified with the time moment under fire conditions when the structural element completely loses the possibility to carry all the imposed loads, together with the actions which are thermally generated. Nevertheless, in general it may be determined also in different ways, for example basing on the acceptance of ultimate value of member deformation. There is the following relation if random resistance $R_{fi.\Theta}$ is described by means of normal or log-normal probability distribution:

$$p_{f,ult} = \Phi\left(-\beta_{R,req}\right) \rightarrow \beta_{R,req} = -inv\Phi\left(p_{f,ult}\right) \tag{13}$$

Symbol $\Phi(\ )$ means in such a formula the cumulative distribution function of standardized normal probability distribution. It is so called the *Laplace* function, easy to find in ordinary statistical tables. Furthermore, the notation $inv\Phi$ is understood as an inverse function of Φ. To sum up, the fact that the commonly accepted equation $\beta_{fi,20,req} = \beta_{fi.\Theta.req} = \beta_{fi,req} = const$ is true only if the assumption that partial safety factor $\gamma_{M.fi}$ is constant during the whole of fire time is fulfilled, has to be underlined. Only in this case in Eq. 12 $\zeta = 1{,}0$, which means that Eq. 7 and Eq. 8 are equivalent to each other. Generally, for ordinary safety requirements, it is usually specified $\beta_{R,req} = 3$, which simultaneously means that $p_{f,ult} = 0{,}00135$ (according to Eq. 13).

Design resistance $R_{fi,d,\Theta}$ of structural member decreases with growing steel temperature Θ_a according to Eq. 7, analogously to the reduction of steel yield point. Adequate action effect $E_{fi,d,\Theta}$ remains constant, provided that the assumptions made in the previous part of this article are fulfilled. However, let us notice that this value can also significantly increase if additional thermally induced internal forces and moments, disadvantageous from the safety point of view, occur in structural member with growing intensity. As a result of both kinds of variation the safety margin under fire conditions decreased. Finally, for $t_{fi} = t_{fi.d}$ (this fire moment is called the member fire resistance) its value totally vanishes with given probability of failure $p_f = p_{f.ult}$. Then also occurs $E_{fi.d.\Theta} = R_{fi.d.\Theta}$.

Probability p_f (particularly its limit value $p_{f.ult}$) should be understood as a conditional probability, with the condition that fire has taken place. Precise information about the type of considered probability is absolutely necessary, because at least two kinds of such probabilities can be distinguished (Maślak & Domański 2007):

- probability of failure p_f, caused by fire if it is known that fire ignition has occurred and; moreover, this fire has reached the flashover point (it may be described as a fully developed fire),
- probability of failure p_{ff}, caused by fire which can take place; however, the designer has no information about its ignition and flashover.

Relation between p_f and p_{ff} is given by *T. T. Lie* (Lie (1972)):

$$p_{ff} = p_t p_f \tag{14}$$

where p_t means probability of fire occurrence (not only of fire ignition but also reaching the flashover point). Not only quantitative but also qualitative distinction between probabilities p_f and p_{ff} seems to be very significant. Even if conditional probability p_f is large, probability p_{ff} is usually quite small and does not seem to be apprehensive, because in reality the value of probability p_t is also slight (Maślak 2005). Similar conclusions can be derived also from other investigations (for example Holicky & Schleich (2001)). However, quantity p_{ff} can also be considered as a conditional probability. Both values p_f and p_{ff} allow the designer to evaluate the real safety level, but with the assumption that she/he knows that failure will occur resolutely as a result of fire action. Meanwhile, the construction can be destroyed also in a situation when fire has not appeared at all. If the probability of such an event is described as p_{f0}, then, finally, the probability of construction collapse p_{fff} can be calculated as:

$$p_{fff} = (1 - p_t) p_{f0} + p_t p_f \tag{15}$$

Eq. (15) follows directly from the scheme of *Bernoulli* sampling with two samples.

3 A new probabilistic concept to assess the member fire resistance

The basic advantage of described above methodology of steel member safety evaluation in accidental fire situation is its simplicity. Semi-probabilistic approach, which is a base of the limit states concept, commonly used in structural design, is there applied. However, it is well known that solutions obtained in this way can give only the approximate estimations of real safety level. The fact that even though such assessments are always safe, they lead to uneconomical design, is necessary to underline. In the authors' opinion, it is possible to obtain more reliable safety evaluations, without excessive complication of the calculations. For this purpose the application of the classical probability approach is proposed. The failure is now recognized as an up-crossing of the level of random member resistance $R_{fi,\Theta}$, reduced in given steel temperature Θ_a, by random value of the action effect E_{fi}. In consequence we have two fully separate random variables, E_{fi} and $R_{fi,\Theta}$. For this reason, considering the density function of two-dimensional normal probability distribution $f(E_{fi}, R_{fi,\Theta})$ is necessary to precisely evaluate the real safety level in given steel temperature. Let us notice that in codified formats, in general only simplified approach is applied, in which one-dimensional marginal density distributions, $f(E_{fi})$ and $f(R_{fi,\Theta})$, are taken into account. However, if the new random variable:

$$\gamma_{\Theta} = \frac{R_{fi,\Theta}}{E_{fi}} = \frac{k_{y\Theta} R_{fi,20}}{E_{fi}} \tag{16}$$

is defined, then taking into consideration only the density function of one-dimensional probability distribution $f(\gamma_\Theta)$ is sufficient. The event that $\gamma_\Theta \geq 1$ is in this case interpreted as a member survival; whereas, the value $\gamma_\Theta < 1$ means its failure. Value of probability p_f can be calculated by means of the concept of global safety index β_Θ, where:

$$\beta_\Theta = \frac{ln\breve{\gamma}}{\upsilon_\Theta} \tag{17}$$

$$\breve{\gamma} = \frac{k_{y\Theta}\breve{R}_{fi,20}}{\breve{E}_{fi}} \quad \text{and} \quad \upsilon_\Theta = \sqrt{\upsilon_R^2 + \upsilon_E^2} \tag{18}$$

There is an important qualitative difference between the global safety index β_Θ, defined in Eq. (17), and partial safety indices: β_R - connected with the member resistance (applied in Eq. (10) and Eq. (11)), and β_E - defined for the action effect, respectively.

The modal value of member resistance $\breve{R}_{fi,20}$, calculated with reference to room temperature, is in considered simple load case also proportional to adequate modal value of random steel strength $\breve{f}_{y,20}$. Parameters of distribution of random action effects (modal value and log-normal coefficient of variation) can be estimated as follows:

$$\breve{E}_{fi} \cong \overline{G} + \sum_i \overline{Q_i} \quad \text{and} \quad \upsilon_E \cong \sqrt{\upsilon_G^2 + \sum_i \upsilon_{Qi}^2} \tag{19}$$

where G means permanent load; whereas Q_i - i-th variable load. Finally:

$$p_f = \Phi(-\beta_\Theta) \tag{20}$$

It should be noticed that both action effect E_{fi} and member resistance $R_{fi,\Theta}$ depend on the steel temperature Θ_a. That is the reason that these variables are correlated in statistical sense and significantly complicated analysis is necessary to precisely describe the shape of function $p_f = p_f(\Theta_a)$. However, the authors propose the application of a simplified approach in which evaluations of probability p_f in relation to fixed values of steel temperature, are obtained. The fact that in such design methodology the steel temperature Θ_a cannot be taken into account as a random variable, because it is now only the design parameter (which is not random in formal sense), must be underlined. Dependence $p_f = p_f(\Theta_a)$ can be determined indirectly by means of multiple recurrent calculations, made for succeeding Θ_a values. On the other hand, values of ultimate failure probability $p_{f,ult}$, acceptable by the user of the structure, can be determined arbitrarily. It is necessary to pay attention that these values are explicitly connected with adequate required values of $\beta_{\Theta,req}$ index. However, according to EN 1990, when ordinary safety requirements are taken into consideration (so called reliability class *RC2*), value $\beta_{\Theta,req} = 3{,}8$ should be accepted. It is an equivalent of the ultimate failure probability value equal to $p_{f,ult} = 7{,}235 \cdot 10^{-5}$. For another reliability classes, different ultimate parameters are defined, particularly:

- for class *RC1* – reduced safety requirements:
  $\beta_{\Theta,req} = 3{,}3$, then $p_{f,ult} = 48{,}342 \cdot 10^{-5}$
- for class *RC3* – special safety requirements:
  $\beta_{\Theta,req} = 4{,}3$, then $p_{f,ult} = 0{,}854 \cdot 10^{-5}$.

Slightly different values of $\beta_{\Theta,req}$ are recommended in the JCSS document. For example, if we consider the structural member, for which the consequence of failure can be determined as average and, moreover, the cost of ensuring the required safety level is contained not far from the typical cost (characteristic for ordinary conditions), then it is suggested to apply $\beta_{\Theta,req} = 4{,}2$, which is an equivalent of the probability $p_{f,ult} \approx 10^{-5}$.

Consequently, the global safety condition can be described as follows:

$$p_f \leq p_{f,ult} \tag{21}$$

or in an equivalent way:

$$\beta_\Theta \geq \beta_{\Theta,req} \tag{22}$$

Temperature Θ_a for which $p_f = p_{f,ult}$ occurs (then also $\beta_\Theta = \beta_{\Theta,req}$) is named the critical temperature of the member, $\Theta_{a,cr}$. Let us notice that checking the global safety condition by means of Eq. 21 or Eq. 22 resolves itself into the verification of the following inequality:

$$\Theta_a < \Theta_{a,cr} \tag{23}$$

However, the steel temperature Θ_a is not the only parameter that allows the designer to verify the safety condition of the steel member in fire. In many cases determination of the time period t_{fi}, which can be used by the user of a structure under fully developed fire conditions to safely evacuate from the fire compartment (with the acceptable fixed value of ultimate probability of failure $p_{f,ult}$), seems to be more useful. Particular steel temperature value $\Theta_{a,cr}$ may be in this way obviously linked with the fire moment $t_{fi,d} = t_{fi}(\Theta_{a,cr})$, in which member failure occurs. The time period, calculated from fire flashover t_0 to $t_{fi,d}$ moment, is commonly called the member fire resistance. In classical engineering approach the intensity of member fire exposure is described as a function of the fire time by means of the assumption of a model temperature-time curve $\Theta_a - t_{fi}$. Every fire moment, which has been chosen by the designer, may be explicitly connected with adequate steel temperature; therefore, such a relation can be interpreted as a mapping in a mathematical sense. Consequently, Eq. 21, Eq. 22 and also Eq. 23, can be described otherwise, in time units:

$$t_{fi,d} \geq t_{fi,req} \tag{24}$$

where the required value of member fire resistance is $t_{fi,req}$, given for buildings from the national annexes.

4 Conclusions

The approach presented in this paper allows the designer to assess the real safety level under fully developed fire conditions in a way which seems to be more objective and complete in comparison with the classical solutions, applied in national codes. Moreover, design methodology proposed in the article remains relatively simple and not too much time consuming for structural designers. Partial safety factors, which are commonly used in current standard recommendations, have been replaced by the maximum, possible to accept values of ultimate probability of failure $p_{f,ult}$. Determination of such values gives us the opportunity to make the adequate safety analysis particularly if different levels of reliability requirements have to be taken into account. It is consistent with formal suggestions given in EN 1990. For these reasons the solutions described above can be considered as a base helpful in the process of calibration and verification of typical parameters applied in the reliable fire safety analysis.

References

Domański T. (1987) *Probabilistic methodologies of fire resistance assessment of steel structural members* (in Polish), Doctor's Thesis, Cracow University of Technology.

Holicky M. (2005) Calibration of the combination factors for fire design situation. In: *Proc. Int. Conf. on Structural Safety and Reliability (ICOSSAR), Rome, Italy.*

Holicky M. & Schleich J.–B. (2001) Modelling of a Structure under Permanent and Fire Design Situation. In: *Proc. of the IABSE Int. Conf. "Safety, Risk, Reliability – Trends in Engineering", Malta*, pp. 1001-1006.

Lie T. T. (1972) Optimum fire resistance of Structures. *Journal of the Structural Division*, **98**, No. ST1, pp. 215-232.

Maślak M. (2005) Failure Probability of Building Load-bearing Structure under Fire Conditions (in Polish). *Czasopismo Techniczne*, 12-B/2005, pp.67-80.

Maślak M. & Domański T. (2007) On the problem of safety evaluation in design of steel members for accidental fire situation. In: *Proc. of XI Int. Sci. Conf. "Current Issues of Civil and Environmental Engineering, Lviv – Koszyce – Rzeszów"*, Lviv, Ukraine.

EN 1990: Eurocode – Basis of Structural Design.

EN 1991-1-2: Eurocode 1: Actions on Structures – Part 1-2: General Actions – Actions on Structures Exposed to Fire.

EN 1993-1-2: Eurocode 3: Design of Steel Structures – Part 1-2: General Rules – Structural Fire Design.

JCSS (2001) Probabilistic Model Code, Joint Committee on Structural Safety.

13.5 Analysis of Random Limit Load and Reliability of Hyperstatic Trusses

Jan Rządkowski, Krzysztof Mierzwa
University of Wroclaw, Poland, jan.rzadkowski@pwr.wroc.pl ,
krzysztof.mierzwa@pwr.wroc.pl

Abstract

Limit load analysis of hyperstatic trusses is faced with real quantitative difficulties even for low statically indeterminate trusses. These quantitative problems result from a large number of plastic collapse mechanisms that can occur in the hyperstatic trusses. The presented method based on linear programming provides facilities for algorithmic determination of plastic collapse mechanisms of a truss. In combination with Monte Carlo method it provides more accurate assessment of reliability of hyperstatic elastic – plastic structures. An example of reliability assessment of a hyperstatic truss is also given.

Keywords: *reliability, limit load, hyperstatic truss*

1 Theoretical basis of plastic collapse mechanisms: analysis of a hyperstatic truss

From the point of view of statics, the limit load of elastic – plastic structures is connected with a moment of the change of a static structure into a kinematical chain (Borkowski 1985). In the case of metal trusses the plastic collapse mechanism (PCM) occurs in the moment of reach the load capacity of the last truss member consisting a set ***A***:

$$\pm F = \pm R \tag{1}$$

The set ***A*** is a such minimal set of truss members in which replacement of limit load of a last member by the pair of forces $F = N$ causes that the stiffness matrix [*K*] of constitutive equation (Borkowski 1985, Bródka et al. 1985, Charnes et al. 1959, Čyras 1974)

$$\{b\} = [K]\{\delta\} \tag{2}$$

become singular, so the value of matrix determinant

$$\det[K] = 0 \tag{3}$$

where: $\{b\}$ – external loads vector,
$\{\delta\}$ – displacements of nodes vector.

The system of constitutive equations forming the matrix [K] can be analysed on the numerical way by using methods based on step – by – step (s-b-s) procedure.

The second method of searching of possible plastic collapse mechanismes is an analysis of a system of equations connecting the vector of truss bar strains $\{\Delta\}$ with displacements of nodes vector $\{\delta\}$ by using consistency matrix [*C*]. In the small deformation range the system of equations can be expressed in the form:

$$\{\Delta\} = [C]\{\delta\} \tag{4}$$

For the geometrically invariant truss the displacements of truss nodes $\delta \neq 0$ are possible only for $\Delta \neq 0$ for some truss nodes at least. From the algebraic point of view it means, that the system of equations (4) for the null vector of bar strains $\{0\}$ has only the zero solution:

$$\{\delta\} = 0 \tag{5}$$

The system of equations (4) for the null vector of bar strains $\{0\}$ can be expressed in the form (Bródka et al. 1985, Karczewski & Konig 1979)

$$\begin{aligned} &C_{1,1}\delta_1 + C_{1,2}\delta_2 + \ldots + C_{1,k}\delta_k = 0 \\ &C_{2,1}\delta_1 + C_{2,2}\delta_2 + \ldots + C_{2,k}\delta_k = 0 \\ &\ldots\ldots\ldots\ldots\ldots\ldots\ldots\ldots\ldots\ldots\ldots \\ &\ldots\ldots\ldots\ldots\ldots\ldots\ldots\ldots\ldots\ldots\ldots \\ &C_{n,1}\delta_1 + C_{n,2}\delta_2 + \ldots + C_{n,k}\delta_k = 0 \end{aligned} \tag{6}$$

where C_{ij} are elements of consistency matrix [C] for $i = 1,2,...,n$; $j = 1,2,...,k$. and k is the number of possible displacements of all truss nodes:

$$k = 3m - p \tag{7}$$

where m is the number of truss nodes and p is the number of support constrains. From Sylwester theorem (Bródka et al. 1985, Karczewski & Konig 1979, Karczewski 1980) is known, that the system of equations (6) has the zero solution only in the case, when the matrix grade q (the greatest number of linear independent columns) is equal to the number of equations k:

$$q[C] = k\,. \tag{8}$$

The truss is geometrically invariant when the condition (8) is satisfied, and whereas inequality occurs

$$q[C] < k \tag{9}$$

then for such a situation there are possible plastic collapse mechanisms (PCMs) of the truss in the number equal all possible non zero solutions of the system of equations (6) (Bródka et al. 1985, Karczewski & Konig 1979).

The number l of PCMs is different than the number z of possible sets $\boldsymbol{A}$ because of possible combinations of $+R = R_c$ and $-R = R_T$ limit loads of truss members consisting sets $\boldsymbol{A}_{1,\ldots,z}$

2 Use of linear programming for the determination of plastic collapse mechanisms

For the small deformation of the truss, having not significant influence for the value of internal forces in truss members, it is possible to make an assumption that the truss geometry for each PCM is known. Usually, it is accepted that limit state geometry of truss is the same as for not loaded structure. Then there are known direction cosines of truss members being coefficients of internal forces.

The equilibrium equations in truss nodes can be expressed in a matrix form (Bródka et al. 1985, Karczewski 1980)

$$[A]\{s\} = \{b\}, \tag{10}$$

In limit states determined by sets $A_{1,...,z}$, where limit loads of truss members are replaced by the pair of forces (1), the equation (10) can be expressed as below:

$$[B]\{s_B\} + [N]\{s_N\} = \{b\} \tag{11}$$

where: $[A] = (a_1, a_2, ..., a_n)$ – matrix of expected coefficients of internal forces, where $a_j = (a_{1j}, a_{2j}, ..., a_{mj})$ for $m = 1,...,n$;
$[B]$ – matrix of expected coefficients of internal forces S_i in truss members not consisting analyzed set A (base matrix of LP problem);
$[N]$ – matrix of expected coefficients of internal forces F_i in truss members consisting analyzed set A;
$\{b\}$ – $(b_1, b_2, ..., b_n)$ – external force vector;
$\{s\}$ – $(S_1, S_2, ..., S_n)$ – internal force vector S_i;
$\{s_N\}$ – vector of internal forces $F_{1,2,...,k}$ in truss members consisting analysed set A;
$\{s_B\}$ – internal force vector of the static determinable part of the structure.

The matrix $[B]$ is m-by-n non-singular matrix; then there is the inverse matrix $[B]^{-1}$, and there is possible to express $\{s_B\}$ depending on $\{s_N\}$ (Kowal & Rzadkowski 1985, Rzadkowski 1994)

$$\{s_B\} = [B]^{-1}\{b\} - [B]^{-1}\{s_N\} \tag{12}$$

The limit load of the truss is in linear programming (LP) goal function given by following equation:

$$\min_l (N_l) = \min_l s_0 = [c]\{s\} \tag{13}$$

The most interesting group of possible solutions of equation (13) are so called base solutions, which fulfil the limit loads characterized each one of l PCMs. Because of the matrix $[B]^{-1}$ is identity one, the final result of conducted LP procedure is (Čyras 1974, Tanaka 1962, Telega 1971)

$$\min s_1 = \min\{s_B\} = \min\{b\} = \min N_l = N_R \tag{14}$$

Introducing the interchanging modifications of the goal function in a form:

$$\min\{s_B\} = \min s_2, \ldots, \min s_{l-1} \tag{15}$$

we can get values of limit loads characterized all PCMs.

Limit load analysis of hyperstatic trusses is faced with real quantitative difficulties even for low statically indeterminate trusses. These quantitative problems result from a large number l of plastic collapse mechanisms (PCM) that can occur in a hyperstatic truss even for a small degree of static indetermination.

Listed methods have good and bad points. "Classical" s-b-s method allows for determination only one PCM of minimal capacity min $N_l = N_R$. In this case simple assessment of parameters of random N_R value is possible only for PCM consisting of a one element.

S-b-s procedure is commonly used for determination of random parameters of N_R value in conjunction with Monte Carlo (MC) method. The main inconvenience of different modifications of MC method, is a long time of numerical calculations. There is also possible overlooking some PCM-s.

Linear programming was used for limit analysis of stiff – plastic structures. LP methods can be used for determination N_R only for elastic – plastic trusses non sensitive for the influence of displacements on stability of a structure. Also proper assessment of parameters of random N_R is possible only for one element PCM.

3 Assessment of a truss reliability

The most simple heuristic model of statically determinate steel (i.e. elastic – plastic) truss is a chain consisted of $i = 1,\ldots, n$ links of different random capacity N_i as well as of different effort σ_i of links. According to the rule of the weakest link the load capacity of the chain N_R equals the least resistance of its elements. The probability of ω of the truss failure (failure frequency) is calculated as follows:

$$\omega = 1 - \prod_{i=1}^{n}(1-\omega_i) \approx \sum_{i=1}^{n}\omega_i \tag{16}$$

where $\omega_i = Pr\,(\sigma_i < N_i)$ is the probability of failure of the i-th link.

For redundant trusses, the probability of failure can be estimated as for r-out-of-n heuristic model. This model represents a system of n elements, r of which must be operable for the system to succeed. For a redundant truss consisted of n elements, r elements in elastic range of its work and formed stationary system is required to function the truss. The probability of the redundant truss failure is:

$$\omega = 1 - \sum_{k=r}^{n}\binom{n}{k}\overline{R}^{k}(1-\overline{R})^{n-k} \tag{17}$$

where $\overline{R}$ is the reliability of truss bars under axial load. Then the minimal size set of $m = n - r$ truss elements working in plastic range and ensured geometrical variability of the truss is plastic collapse mechanism (PCM).

The above-mentioned equation is valid for trusses made from elements of the same reliability.
For the analyzed truss, when elements are of different reliability values, the equation (17) takes the form:

$$\omega = 1 - \prod_{i=1}^{10}(1-\omega_i) - \sum_{k=5}^{10}\left(\frac{\omega_k}{1-\omega_k}\prod_{i=1}^{10}(1-\omega_i)\right) \tag{18}$$

The third way of assessment of probability failure ω is using Monte Carlo method. Different modifications of MC method are given in [14]. In the numerical example below, the LP method and MC method is used.

4 Numerical example

For the truss redundant outer and inner showed in Figure 1, build up from elements of limit loads given in Table 1, PCMs and its limit loads have been determined by using LP method.

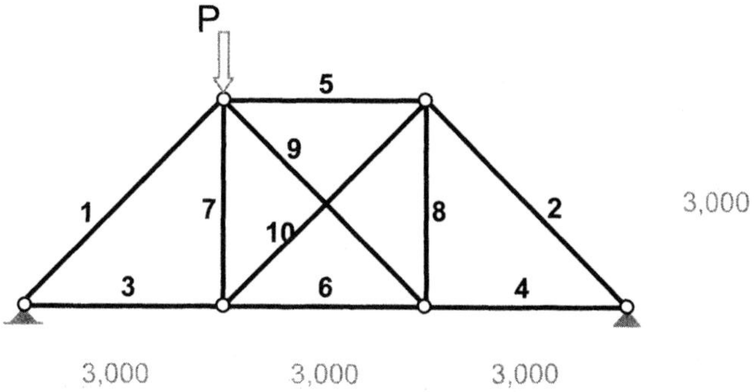

Figure 1 Geometry of analyzed elastic – plastic truss

Table 1. Limit loads of truss elements

Limit load of element [kN]	Number of the truss element									
	1	2	3	4	5	6	7	8	9	10
compression	-70.00	-70.00	-7.00	-70.00	-21.00	-70.00	-70.00	-70.00	-70.00	-70.00
tension	70.00	70.00	7.00	70.00	21.00	70.00	70.00	70.00	70.00	70.00

Results of conducted linear programming limit load truss analysis are presented in Table 2.

Results of probability density function (PDF) of the limit load $N_R = min\ N_i$ of analyzed truss obtained by using MC procedure are given in Figure 2. The PCM of the truss consists of two elements and is not the base solution for LP analysis of the truss limit load.

On the base of conducted limit load analysis was estimated probability of failure ω and reliability $R = 1 - \omega$ of the truss. Results of estimation ω, conducted as for heuristic models and by using MC method, are presented in Table 3. Results were obtained with the assumption that standard deviation of random limit load truss capacity is 10%.

Table 2. Results of analysis

PCM No.	Limit load N_i of the truss [kN]	Capacities of truss elements forming PCM [kN]	
1	46,980	$N_1 = -70,00$	-
2	74,231	$N_3 = 7,00$	-
3	105,888	$N_5 = -21,00$	$N_9 = -70,00$
4	129,205	$N_5 = -21,00$	$N_{10} = -70,00$
5	136,618	$N_5 = -21,00$	$N_8 = 70,00$
6	148,620	$N_2 = -70,00$	-
7	252,205	$N_5 = -21,00$	$N_6 = 70,00$
8	272,989	$N_5 = -21,00$	$N_7 = 70,00$
9	296,764	$N_9 = -70,00$	$N_{10} = 70,00$
10	357,764	$N_{10} = 70,00$	$N_8 = 70,00$
11	358,434	$N_5 = -70,00$	$N_6 = -70,00$
12	378,378	$N_4 = 70,00$	-
13	419,464	$N_7 = -70,00$	$N_8 = 70,00$

5 Discussion of results and conclusions

In many cases the most probable plastic collapse mechanism and minimal limit load of the truss, derived from equilibrium equations in truss nodes by using linear programming, can be the same as obtained by step–by–step method. Conducted limit analysis by s – b – s method showed, that the limit state of the truss is beginning at the moment when for element No. 5 occurs $F_5 = N_5$. The truss is changing into kinematical chain when for element No. 1 occurs $F_1 = N_1$. The so called base solution of LP problem is different than solution obtained using s – b – s method.

Table 3. Probability of failure ω (failure frequency) and reliability R of the truss

random parameters		assessment of reliability and probability of failure heuristic model		
median of force [kN]	reliability failure frequency	Monte Carlo	mixed (18)	chain consisted (16)
20	ω	0,06023	0,06083	0,06241
	R	0,93977	0,93917	0,93759
27	ω	0,12977	0,13129	0,14373
	R	0,87023	0,86871	0,85627
34	ω	0,22821	0,23177	0,29029
	R	0,77179	0,76823	0,70971

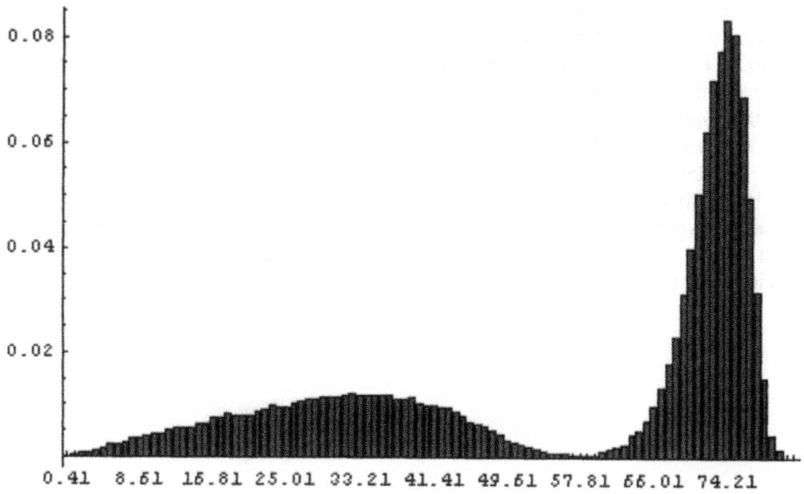

Figure 2 An example of probability density function of the limit load N of analyzed truss for two elements PCM obtained by using MC method.

Such a situation significantly influences the accurateness of probability of failure ω and reliability R of the truss assessment. The failure frequency ω for heuristic models was calculated as for the system of a "hut curve" probability density function (PDF), not for PDF given in Figure 2. It means, that for ω and R truss assessment is proper only s – b – s and Monte Carlo method. The limit load by LP analysis may be used for selecting a set of random load capacity truss elements for MC calculations.

References

Borkowski, A. (1985) *Static analysis of rod structures in elastic and plastic state* (in Polish) PAN IPPT, PWN, Warszawa – Poznań

Bródka, J., Czechowski, A., Grudka, A., Karczewski, J., Kordjak, J., Kowal, Z., Kwaśniewski, K. & Łubiński, M. (1985) *Przekrycia strukturalne* (Structural Roofs) (in Polish) Arkady, Warszawa

Charnes, A., Lemke, E. & Zienkiewicz, O.C. (1959) Virtual work, linear programming and plastic limit analysis, *Proc. Royal Soc.*, London **25** 1A, 110 -119.

Čyras, A.A. (1974) *Metods of linear programming in the investigation of elasto-plastic systems* (in Russian) Strojizdat, Leningrad

Karczewski, J. & Konig, J.A. (1979) Geometric analysis of trusses (in Polish) *Inżynieria i Budownictwo* 1979 No.1.

Karczewski, J. (1980) The limit load of space trusses, *Archiwum Inżynierii Lądowej* 1980 No.1

Kowal, Z. & Rządkowski, J. (1985) Estimation of the ultimate loading capacity of trusses (in Polish) In: *II Konferencja „ Problemy Losowe w Mechanice Budowli"*., Gdańsk, październik 1985, 121 – 127

Maier, G. (1975) Metods of mathematical programming in the analysis of elasto-plastic structures (in Polish) *Arch. Inż. Ląd.* **21** 387 – 411.

Rządkowski, J. (1987) Analysis of plastic mechanisms of limit states of hyperstatic trusses (in Polish) In: *XXXIII Konferencja Naukowa KILiW PAN i KN PZITB*, Gliwice-Krynica 1987, 153 -158.

Rządkowski, J. (1994) Quantitative problems of ultimate capacity analysis of multi – redundant bar structures by linear programming., In: *Proc. of the Third Interuniversity Research Conference.* Technical University of Wrocław, Poland; Eindhoven University of Technology, The Netherlands, Wrocław, Poland, 1994, pp. 225-230.

Rządkowski, J. (1980) Random ultimate bearing capacity of elasto-plastic space truss., In: *Third International Conference on Space Structures., Space Structures Research Center University of Surrey, Guildford* 1980, pp. 561-566.

Tanaka, H. (1962) Automatic analysis and design of plastic frames., *Report of the Inst. of Ind. Sc.*, Tokyo **3**, No. 12

Telega, J.J. (1971) Using the linear programming for determination of ultimate loading capacity of structures (in Polish) *Mech. Teor. i Stos.* **1**, No.9, 9-52.

Author index

Printed and bound by CPI Group (UK) Ltd, Croydon, CR0 4YY

08/05/2025

01864856-0001